Commemorating the Launch of the Section "Marine Toxins"

Commemorating the Launch of the Section "Marine Toxins"

Guest Editors

Andrew Turner
Panagiota Katikou

MDPI

Basel • Beijing • Wuhan • Barcelona • Belgrade • Novi Sad • Cluj • Manchester

Guest Editors

Andrew Turner
Health and Welfare Division
CEFAS
Weymouth
UK

Panagiota Katikou
Department of Hygiene and Technology of Food of Animal Origin and Toxicology
Hellenic Agricultural Organization—DIMITRA
Thermi
Greece

Editorial Office

MDPI AG
Grosspeteranlage 5
4052 Basel, Switzerland

This is a reprint of the Special Issue, published open access by the journal *Marine Drugs* (ISSN 1660-3397), freely accessible at: https://www.mdpi.com/journal/marinedrugs/special_issues/NBO217EWL2.

For citation purposes, cite each article independently as indicated on the article page online and as indicated below:

Lastname, A.A.; Lastname, B.B. Article Title. *Journal Name* **Year**, *Volume Number*, Page Range.

ISBN 978-3-7258-6988-6 (Hbk)
ISBN 978-3-7258-6989-3 (PDF)
https://doi.org/10.3390/books978-3-7258-6989-3

Cover image courtesy of Panagiota Katikou

Contents

marine drugs

Editorial

Commemorating the Launch of the Section "Marine Toxins"

Andrew Turner [1,*] and Panagiota Katikou [2,*]

1 Toxins and Contaminants Domain, Health and Welfare Division, CEFAS, Barrack Road, Weymouth, Dorset DT4 8UB, UK
2 Directorate of Research, Innovation and Education, Ministry of Rural Development and Food, Hapsa & Karatasou 1, 54626 Thessaloniki, Greece
* Correspondence: andrew.turner@cefas.gov.uk (A.T.); pkatikou@otenet.gr (P.K.)

Marine toxins are natural compounds produced by a variety of marine organisms, including microalgae, bacteria, and macroalgae. These compounds have wide-ranging structural chemistries and biological activities. Many of the biological effects negatively impact health, including toxicity to humans and animals, as well as whole marine ecosystems. Toxins identified as hazards in the marine environment range from small polar compounds to high-molecular-weight lipophilic compounds. Toxicological effects are also wide-ranging, with some of these toxin groups resulting in human fatalities in the absence of proper risk management. Conversely, some of these compounds have shown promise for applications in medicine, biotechnology, and agriculture, highlighting the potential benefits that originate from aquatic toxins. This duality underlines the importance of marine toxins as both indicators of ecosystem stress and sources of bioactive compounds of practical interest.

The realities of the impacts of these toxins have resulted in the development of hazard monitoring and risk management strategies, together with well-developed regulatory programmes, across the world. Many of these toxins are known to bioaccumulate in the flesh of marine organisms that are consumed by humans as food products, notably bivalve molluscs, crustaceans, and fin fish. Over the past 20 years, many technological developments have taken place, empowering industry and regulatory authorities to develop monitoring approaches for assuring the safety of seafood products. Most notably, there have been significant developments in the move away from the use of live animal bioassays for food safety testing toward using instrumental chemical detection methods for official control testing of bivalves. Yet, in addition to well-recognised regulated toxins, other families of marine toxins that demonstrate toxicity effects in mammals and other organisms have emerged. Researchers and regulators around the world are working hard to develop approaches to monitoring such toxins, including approaches which are beneficial in their risk management. In addition, there is much work focused on the determination of the exact sources of such toxins, which include microalgae of wide-ranging ecologies, as well as multiple species of marine bacteria. In many cases, the full production mechanisms of toxin production are not yet fully understood; nor are the impacts of environmental changes on the prevalence and toxicity of toxin producers. The mammalian toxicity mechanisms for some emerging toxin hazards also require further elucidation.

The section "Marine Toxins" was instigated in June 2022 to provide a comprehensive publication platform for scientists working within multi-disciplinary fields linked to marine-derived toxins. As of January 2026, there have been 14 Special Issues and a total of 121 published manuscripts, highlighting the relevance of this section to aquatic science throughout the world. It provides a platform for researchers to share their latest findings

 https://doi.org/10.3390/md24020073

on the isolation, analysis, characterization, biosynthesis, and biological activities of these bioactive molecules. It also includes work conducted by researchers focusing on the potential benefits to be gained from the incorporation of marine toxins into drugs of therapeutic benefit. In this way, this section explicitly connects fundamental toxin research with applied, regulatory, and public health perspectives.

The call for submissions for this Special Issue, "Commemorating the Launch of the section Marine Toxins", opened in April 2023 and closed in March 2025. The final publication highlights and celebrates the importance of monitoring and studying these toxins, with a focus on new detection approaches, emerging hazards in new geographical regions, toxicology and the physiological effects of toxins in marine organisms and mammals, method refinements to improve toxin detection, the characterization of microalgal toxin producers, and the impact of environmental parameters on toxin-producing organisms. Together, these articles demonstrate the ongoing need to develop our shared understanding of the toxicological impacts of marine toxins on human, animal, and ecosystem health, as well as the importance of determining toxin production mechanisms and, ultimately, suitable risk monitoring approaches to mitigating these important hazards within our marine environments. Collectively, the contributions illustrate the breadth of marine toxin research, spanning molecular mechanisms, analytical innovation, ecological drivers, and implications for human and animal health.

The following section provides a brief overview of each of the eleven contributions to this Special Issue, together with links to the full published articles. These cover many of the most important marine toxin impacts, including domoic acid, saxitoxins, tetrodotoxins, ciguatoxins, and azaspiracids, as well as one paper on the toxicity of jellyfish venom.

Contribution 1:

Nutrient Deficiency Impact on the Cellular and Metabolic Responses of Saxitoxin-Producing *Alexandrium minutum*: A Transcriptomic Perspective. *Mar. Drugs* **2023**, *21*(9), 497; https://doi.org/10.3390/md21090497.

The article by Akbar et al. describes work conducted to understand the impacts of nutrient-deficient environments on the saxitoxin-producing dinoflagellate *Alexandrium minutum*. Whilst a great deal of information has been reported in the literature on toxin production by this species, the authors identified a need to assess the impacts of nitrogen and phosphorous depletion on the organism. Transcriptomic responses were assessed by the study team, resulting in the identification of specific differentially expressed genes for both nitrogen and phosphorous deficiency, with gene set enrichment analysis revealing enriched gene ontology terms associated with nutrient deficiency. This work was conducted using laboratory cultures, and it highlighted cellular and metabolic changes in the dinoflagellate, developing our understanding of the impacts of environmental changes on the ecophysiology of this toxin-producing organism.

Contribution 2:

Monthly Variation of Tetrodotoxin Levels in Pufferfish (*Lagocephalus sceleratus*) Caught from Antalya Bay, Mediterranean Sea. *Mar. Drugs*, **2023**, *21*(10), 527; https://doi.org/10.3390/md21100527.

Tetrodotoxins (TTXs) have long been associated with risks to human health following the consumption of pufferfish species, most notably in Japan following consumption of the pufferfish delicacy known as *Fugu*. Pufferfish are not limited to Asia, however, with invasive pufferfish species well recognized as spreading to other continents, notably into the Mediterranean Sea, where they impact both native species and human health. Consequently, there is a need to assess TTX concentrations in non-native pufferfish within European waters to assess likely impacts on human health for pufferfish consumers in the region. In this study, the authors conducted an assessment of TTX levels using high-resolution mass spectrometry, through the analysis of 110 pufferfish sampled monthly over a twelve-month period. Fish were dissected and different organs analysed, with quantitative analysis confirming the highest toxin concentrations in the gonads, followed by the liver and edible tissue, respectively. Seasonal variability was also assessed, along with differences in toxin levels between male and female fish. The results confirmed significant health risks for pufferfish consumers in the region.

Contribution 3:

Physiological Effects of Oxidative Stress Caused by Saxitoxin in the Nematode *Caenorhabditis elegans*. *Mar. Drugs* **2023**, *21*(10), 544; https://doi.org/10.3390/md21100544.

A high proportion of research work conducted on the potent marine neurotoxin saxitoxin (STX) relates to its impacts on mammalian health, most notably the health of human seafood consumers. Impacts on other animals such as birds and sea mammals are also well reported throughout the world. This study focusses instead on evaluating the toxicity of STX to nematodes, specifically *Caenorhabditis elegans*, a species of soil-dwelling roundworm. Worms were exposed to STX over a range of concentrations for 24 h and paralysis symptoms were observed but with the majority recovering. The team assessed multiple aspects of the nematode response, including analyses of lifespan, brood size, growth ability, reactive oxygen species, and adenosine triphosphate levels, together with the overexpression of green fluorescent protein. Declines were observed in the exposed worms, which, together with other exposure experiments, highlighted evidence of significant oxidative stress on *C. elegans* following STX exposure, indicating its potential use in ongoing assessment of invertebrate health following neurotoxin exposure.

Contribution 4:

Comparative Analysis of Tentacle Extract and Nematocyst Venom: Toxicity, Mechanism, and Potential Intervention in the Giant Jellyfish *Nemopilema nomurai*. *Mar. Drugs* **2024**, *22*(8), 362; https://doi.org/10.3390/md22080362.

This study reports the findings of Geng et al., who assessed the toxicity and mechanism of toxicity in tentacle extract (TE) and nematocyst venom extract (NV) from the giant jellyfish *Nemopilema nomurai*. This scyphozoan is a major species of concern in China, Korea, and Japan and is widely associated with stings in human bathers, although the full toxicity effects and biological actions of the venoms have yet to be fully elucidated. In this study, the authors recognized the need to assess and compare the effects of venoms from TE and NV, investigating venom composition and toxicity. Multiple assessments were conducted, which showed the impacts of these important venoms, with the authors furthermore assessing the methodological impacts of extraction methods, together with the potential promising application of an antidote.

Contribution 5:

Morphological, Toxicological, and Biochemical Characterization of Two Species of *Gambierdiscus* from Bahía de La Paz, Gulf of California. *Mar. Drugs* **2024**, *22*(9), 422; https://doi.org/10.3390/md22090422.

Ciguatera Fish Poisoning (CFP) is known to be the most prevalent marine toxin-triggered seafood poisoning syndrome in the world, with reports indicating that more than 50,000 people are affected annually, with the majority of incidents reported in tropical and sub-tropical regions of the Pacific, Indian, and Caribbean Oceans. The ciguatoxins and sometimes maitotoxins that cause these human illnesses are produced by a range of benthic dinoflagellates of the genera *Gambierdiscus* and *Fukuyoa*. Multiple *Gambierdiscus* species have been previously identified as ciguatoxin (CTX) producers, with this list expanding as more investigations are conducted. In this study, the authors investigated new isolates of two species (identified as *G.* cf. *caribaeus* and *G.* cf. *carpenteri*) from the southern Gulf of California. The assessments included morphological characterization using light microscopy and scanning electron microscopy, pigment and amino acid analysis and toxicity through a receptor binding assay, and the application of several biological assays. CTX-like toxicity was confirmed in all strains, showing potential risks to human fish consumers in the region from ciguatera poisoning.

Contribution 6:

Marine Algal Toxins and Public Health: Insights from Shellfish and Fish, the Main Biological Vectors. *Mar. Drugs* **2024**, *22*(11), 510; https://doi.org/10.3390/md22110510.

The impacts of marine toxins have long been recognised on the health of the human seafood consumer, with reports of poisonings stretching back several hundred years in some instances. Similarly, there are multiple reports linking the presence of certain pelagic and benthic microalgal species to both toxin production in marine waters and the subsequent bioaccumulation of toxins in wide-ranging marine organisms, most notably, invertebrates such as bivalve molluscs. In this paper, the authors present a review of the current knowledge regarding public health impacts of marine biotoxins, including reports of exposure routes from both shellfish and fish, which are shown to be the most common biological vectors for human poisonings worldwide. Changes in HAB/toxin occurrence are also described, as well as the requirements for a great understanding of the toxicological impacts of major seafood toxin poisoning syndromes, specifically, DSP, PSP, ASP, NSP, and CFP.

Contribution 7:

Optimization of the Extraction Protocol for Pacific Ciguatoxins from Marine Products Prior to Analysis Using the Neuroblastoma Cell-Based Assay. *Mar. Drugs* **2025**, *23*(1), 42; https://doi.org/10.3390/md23010042.

Ciguatera Poisoning (CP) is widely recognised as having a huge impact on human health globally, most notably in developing countries and regions. The impacts on health have been reported most commonly in countries on the Pacific Ocean, where ciguatoxins (CTXs) produced by a range of benthic dinoflagellates are responsible for bioaccumulation in carnivorous fish species, which are sometimes the only source of protein for local populations. Risk management in these fish is notoriously complex, given the high toxicity originating from even very low toxin concentrations, as well as the highly variable chemistries of wide-ranging CTX analogues, with variations reported between different regions and different fish species. Whilst multiple analytical methods have been published for Pacific CTX (P-CTX) determination, the best setup for many of them remains unknown, which is understandable, given the absence of certified reference standards for the key toxin analogues. Consequently, the authors of this study sought to better understand the performance of these techniques through the evaluation of six extraction and purification protocols, in order to assess method recovery following analysis of fish tissues spiked with P-CTX standard. Better method performance was reported, together with recommendations for future improvements in both cell toxicity and mass spectrometric analysis.

Contribution 8:

The Toxic Effects of Environmental Domoic Acid Exposure on Humans and Marine Wildlife. *Mar. Drugs* **2025**, *23*(2), 61; https://doi.org/10.3390/md23020061.

One of the most widely cited examples of marine toxin-related human poisonings from seafood consumption was in Canada in the 1980s, where sickness resulted from consumption of molluscs containing domoic acid (DA). This toxin is produced naturally by some species of the marine diatom *Pseudo-nitzchia* and some red macroalgal species and results in the poisoning syndrome termed Amnesic Shellfish Poisoning (ASP). This toxin not only is a significant and regulated threat to human health around the world but has been linked to multiple incidents of animal health impacts in a large variety of organisms such as sea mammals and seabirds. The authors of this paper sought to review these impacts with a focus on both humans and animals, providing an overview of DA production, toxicity including clinical signs and medical treatment approaches, toxicokinetic properties, susceptibility to DA toxicosis and detection methods used for risk management purposes. Overall, the review provides a timely reminder of all the important impacts that this low-molecular-weight toxin has on both human and animal health throughout the world.

Contribution 9:

First Confirmed Occurrence of Ciguatera Poisoning in the UK from Imported Pinjalo Snapper (*Pinjalo pinjalo*). *Mar. Drugs* **2025**, *23*(2), 67; https://doi.org/10.3390/md23020067.

Ciguatera Fish Poisoning (CFP) is traditionally associated with outbreaks in tropical and sub-tropical regions of the world, with poisonings associated with human consumption of carnivorous fish species which have accumulated ciguatoxins produced by various benthic dinoflagellate species. However, over the last decade, several reports have been published describing human poisonings in fish consumers from temperate latitudes, where the causative products were fish imported into these regions from Pacific or Indian ocean territories. This study presents evidence for the first recorded outbreak of CFP in the UK, which occurred following the import of fish steaks labelled as Red Snapper (*Lutjanus bohar*) from the Indian Ocean. Clinical symptoms were recorded in three affected people, with the leftover fish tissues subsequently assessed for CTX presence using both cytotoxicity and mass spectrometry tools. The authors used a neuroblastoma assay to confirm sodium channel activity consistent with CFP-like activity, with mass spectrometry evidencing the presence of tri- and di-hydroxylated P-CTX 3C congeners. Together the evidence highlighted significant risks from consumption of imported fish fillets in geographical regions where CFP incidents are unexpected.

Contribution 10:

Comparative Analysis of Intestinal Microbiota Between Tetrodotoxin-Containing and Tetrodotoxin-Free *Takifugu rubripes*. *Mar. Drugs* **2025**, *23*(4), 140; https://doi.org/10.3390/md23040140.

The well-known presence of TTXs in pufferfish has long been reported as potentially relating to symbiotic bacteria held within the intestine of the host pufferfish, even though the specific bacterial species and the processes involved in this production and accumulation are yet to be elucidated. Consequently, in this study, the authors report on an experiment where the intestinal microbiota was analysed in both toxic and non-toxic pufferfish of the species *Takifugu rubrupes*. The results revealed that significant differences existed between the toxic and non-toxic fish, with key genera identified. As a consequence, the authors proposed that future toxin-screening approaches might include the use of symbiotic bacteria screening for biomarker assessment for food safety in pufferfish.

Contribution 11:

High-Affinity Aptamers and Their Specificity for Azaspiracid-2 Using Capture-SELEX. *Mar. Drugs* **2025**, *23*(5), 183; https://doi.org/10.3390/md23050183.

Alongside the vast development in recent decades of state-of-the-art laboratory tools for the detection and quantitation of regulated and non-regulated marine toxins in seafood products, there is the need for Food Business Operators to conduct End-Product Testing (EPT) of shellfish batches for marine toxin detection, as part of their hazard monitoring

https://doi.org/10.3390/md24020073

processes. Multiple commercial EPT assays are available around the world for the majority of regulated toxin families, but to date, no such tools have been commercialised for the azaspiracids: toxins that are incorporated into the lipophilic toxin family. This study reports on the development of approaches focussed on finding an aptamer-based solution for AZA detection, specifically by screening aptamers specific to the AZA analogue, AZA-2, in tandem with an in silico determination of aptamer binding sites and stability of aptamer–toxin complexes. As a consequence, two aptamers were obtained, dissociation constants defined, and the application to a new detection method for AZA detection discussed.

Conflicts of Interest: The authors declare no conflicts of interest.

List of Contributions:

1. Akbar, M.A.; Mohd Yusof, N.Y.; Usup, G.; Ahmad, A.; Baharum, S.N.; Bunawan, H. Nutrient Deficiency Impact on the Cellular and Metabolic Responses of Saxitoxin Producing *Alexandrium minutum*: A Transcriptomic Perspective. *Mar. Drugs* **2023**, *21*, 497. https://doi.org/10.3390/md21090497.
2. Kosker, A.R.; Karakus, M.; Katikou, P.; Dal, İ.; Durmus, M.; Ucar, Y.; Ayas, D.; Özogul, F. Monthly Variation of Tetrodotoxin Levels in Pufferfish (*Lagocephalus sceleratus*) Caught from Antalya Bay, Mediterranean Sea. *Mar. Drugs* **2023**, *21*, 527. https://doi.org/10.3390/md21100527.
3. Wu, H.; Prithiviraj, B.; Tan, Z. Physiological Effects of Oxidative Stress Caused by Saxitoxin in the Nematode *Caenorhabditis elegans*. *Mar. Drugs* **2023**, *21*, 544. https://doi.org/10.3390/md21100544.
4. Geng, X.-Y.; Wang, M.-K.; Hou, X.-C.; Wang, Z.-F.; Wang, Y.; Zhang, D.-Y.; Danso, B.; Wei, D.-B.; Shou, Z.-Y.; Xiao, L.; et al. Comparative Analysis of Tentacle Extract and Nematocyst Venom: Toxicity, Mechanism, and Potential Intervention in the Giant Jellyfish *Nemopilema nomurai*. *Mar. Drugs* **2024**, *22*, 362. https://doi.org/10.3390/md22080362.
5. Fernández-Herrera, L.J.; Núñez-Vázquez, E.J.; Hernández-Sandoval, F.E.; Ceseña-Ojeda, D.O.; García-Davis, S.; Teles, A.; Virgen-Félix, M.; Tovar-Ramírez, D. Morphological, Toxicological, and Biochemical Characterization of Two Species of *Gambierdiscus* from Bahía de La Paz, Gulf of California. *Mar. Drugs* **2024**, *22*, 422. https://doi.org/10.3390/md22090422.
6. Yuan, K.-K.; Li, H.-Y.; Yang, W.-D. Marine Algal Toxins and Public Health: Insights from Shellfish and Fish, the Main Biological Vectors. *Mar. Drugs* **2024**, *22*, 510. https://doi.org/10.3390/md22110510.
7. Yon, T.; Cruchet, P.; Viallon, J.; Murray, J.S.; Passfield, E.; Chinain, M.; Darius, H.T.; Roué, M. Optimization of the Extraction Protocol for Pacific Ciguatoxins from Marine Products Prior to Analysis Using the Neuroblastoma Cell-Based Assay. *Mar. Drugs* **2025**, *23*, 42. https://doi.org/10.3390/md23010042.
8. Krasner, A.E.; Martinez, M.E.; Field, C.L.; Fire, S.E. The Toxic Effects of Environmental Domoic Acid Exposure on Humans and Marine Wildlife. *Mar. Drugs* **2025**, *23*, 61. https://doi.org/10.3390/md23020061.
9. Turner, A.D.; Maskrey, B.H.; Stone, D.; Mudge, E.M.; Robertson, A. First Confirmed Occurrence of Ciguatera Poisoning in the UK from Imported Pinjalo Snapper (*Pinjalo pinjalo*). *Mar. Drugs* **2025**, *23*, 67. https://doi.org/10.3390/md23020067.
10. Zhang, H.; Hou, J.; Zhu, Y.; Wu, B.; Ren, J.; Sun, Z.; Liu, X. Comparative Analysis of Intestinal Microbiota Between Tetrodotoxin-Containing and Tetrodotoxin-Free *Takifugu rubripes*. *Mar. Drugs* **2025**, *23*, 140. https://doi.org/10.3390/md23040140.
11. Yang, J.; Li, X.; Sun, W.; Cui, Y.; Chen, H.; Yang, Y.; Sun, M.; Wang, L. High Affinity Aptamers and Their Specificity for Azaspiracid-2 Using Capture-SELEX. *Mar. Drugs* **2025**, *23*, 183. https://doi.org/10.3390/md23050183.

https://doi.org/10.3390/md24020073

marine drugs

Article

First Confirmed Occurrence of Ciguatera Poisoning in the UK from Imported Pinjalo Snapper (*Pinjalo pinjalo*)

Andrew D. Turner [1,*], Benjamin H. Maskrey [1], David Stone [1], Elizabeth M. Mudge [2] and Alison Robertson [3,4]

1 Centre for Environment Fisheries and Aquaculture Science (Cefas), Barrack Road, Weymouth DT4 8UB, UK; ben.maskrey@cefas.gov.uk (B.H.M.); david.stone@cefas.gov.uk (D.S.)
2 Biotoxin Metrology, National Research Council Canada, Halifax, NS B3H 3Z1, Canada; elizabeth.mudge@nrc-cnrc.gc.ca
3 Dauphin Island Sea Lab, 101 Bienville Blvd, Dauphin Island, AL 36528, USA; arobertson@disl.org
4 Department of Marine Sciences, University of South Alabama, University Avenue N., Mobile, AL 36688, USA
* Correspondence: andrew.turner@cefas.gov.uk

Abstract: Three people in England consumed fish steaks labeled as Red Snapper (*Lutjanus bohar*) originating from the Indian Ocean. Within 12 h, all three experienced sickness including nausea, vomiting, diarrhea, as well as myalgia and paresthesia. Three steaks from a single package of fish obtained from a grocery store were consumed, leaving one uneaten, which was submitted for analysis. Cytotoxicity testing via the mouse neuroblastoma assay confirmed the presence of sodium channel specific activity consistent with a ciguatoxin standard, and the levels detected were above established guidance limits for safe consumption. Chemical detection using liquid chromatography coupled with high-resolution mass spectrometry of both intact toxins and periodate oxidation products was used to confirm the presence of chromatographic peaks consistent with tri- and dihydroxylated Pacific ciguatoxin 3C congeners. Taking the shared medical symptoms of patients, the recent dietary history, and the known potential for ciguatera poisoning to occur in snapper species, the subsequent evidence for CTX-like activity and CTXs in the same fish sample provides very strong evidence that the fish steaks consumed were similarly contaminated with CTXs. Furthermore, given the levels reported, such toxicity would be expected to cause intoxication in humans. Fish species identification based on DNA barcoding confirmed that the fish products were mislabeled, with the tissues instead being the *Pinjalo* snapper, *Pinjalo pinjalo*. This is the first confirmed ciguatera poisoning incident in both the UK and from the *Pinjalo* snapper and highlights the need for monitoring of these emerging toxins in reef fish imports to prevent future human intoxication.

Keywords: ciguatera; ciguatoxins; fish toxins; snapper; *Pinjalo pinjalo*; Europe; United Kingdom

1. Introduction

It is estimated that 10,000 to 50,000 people are affected by ciguatera poisoning each year, although it is believed that only 2–10% of cases are reported [1–4]. Ciguatoxin-group toxins (CTXs) are a diverse family of toxins with up to ~40 congeners of ladder-like polyether compounds containing 13–14 ether rings reported with varying levels of conformation. CTXs have been reported in reef-associated organisms in the Pacific, Caribbean and Indian Oceans and are traditionally classified as (P), (C) or (I) CTX-group toxins, respectively [5] (Figure 1). They occur through the enzymatic oxidation and biotransformation of precursor molecules produced by the benthic dinoflagellates *Gambierdiscus* spp. and *Fukuyoa* spp., which typically thrive on biotic and artificial substrates in shallow marine environments

(<30 m) [6–13]. Ciguatera is often associated with the consumption of large predatory reef fish [5,9,14–19], although invertebrate vectors have also been reported [20,21]. The toxins have been detected in a variety of fish genera and species spanning multiple trophic levels across the world [5,14,16,17]. Though stability studies are currently lacking, CTXs are thought to be heat- and low-temperature-stable based on a few reports, so it has been assumed that they cannot be destroyed through routine cooking or freezing [22,23].

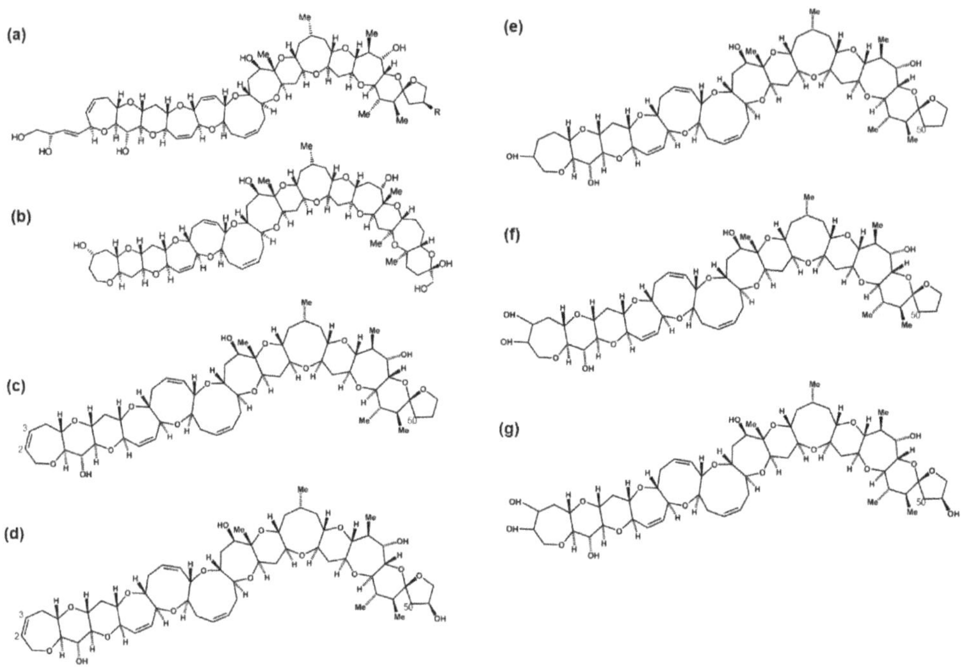

Figure 1. Example chemical structures of Pacific and Caribbean ciguatoxin groups; (**a**) Pacific ciguatoxin 1 (CTX-1B) (R=OH) and Pacific ciguatoxin 3 (54-deoxy-CTX-1B) (R=H) (**b**) Caribbean ciguatoxin-1 (C-CTX-1) (**c**) CTX-3C (**d**) 51-hydroxy-CTX-3C (**e**) 2-hydroxy-CTX-3C (**f**) 2,3-dihydroxy-CTX-3C (**g**) 2,3,51-trihydroxy-CTX-3C.

Ingestion of the toxins cause the syndrome ciguatera, which is characterized by a complex range of clinical signs, including gastrointestinal, neurological and cardiovascular effects, and death can occur from cardiac or respiratory failure in rare cases [24–26]. The voltage-gated sodium channel has been identified as the primary molecular target for the toxins, with toxin binding causing prolonged opening of the channel and influx of sodium to the cell. The situation is further complicated by the variability in symptoms experienced depending on the region, consumption patterns, and culturally specific behaviors of consumers [4,27–36]. Differences in clinical presentations, severity, and duration of illness have been recorded, even among consumers of the same fish [5,37]. The toxins are extremely potent, with LD_{50} in mice (I.P.) equivalent to as low as 0.9 μg/kg depending on the congener [38]. Whilst no regulatory limits or methods of analysis are set currently within Europe or the United States, European Commission (EC) regulations do require that checks are made to ensure that fishery products do not contain biotoxins such as CTX [39]. The USA FDA has proposed guidance levels of <0.1 μg/kg for C-CTX-1 and <0.01 μg/kg for CTX-1B, but these are not regulations, and the guidance only suggests the avoidance of highly endemic regions and/or high-risk species.

In recent years, there have been increasing numbers of reports describing evidence for the presence of CTXs and/or the causative dinoflagellates within European waters [4,32,40–47]. *Gambierdiscus* spp. have been detected as far east as Crete in 2003 [41] and as far west as the Canary Islands and Madeira in 2008 [42,48,49]. In the Atlantic, human intoxications have been linked with consumption of Amberjack (*Seriola rivoliana*) and *S. dumerili* and *S. fasciata*, in the Canary Islands and Madeira, respectively [40,50,51]. CTXs were also reported in *S. fasciata* from the Selvegen Islands, Portugal, and implicated in poisonings in the Canary Islands [43]. Other fish species implicated in the region include Island grouper (*Mycteroperca fusca*), Dusky grouper (*Epinephelus marginatus*), Barred hogfish (*Bodianus scrofa*) and grey triggerfish (*Balistes capriscus*) [47]. More recently, LC-HRMS has been used to confirm the presence of C-CTX analogues including hydroxy-, didehydro- and methoxy-metabolites [46]. Overall, the majority of reports of ciguatera occurring in Europe predominantly relate to travel to endemic countries bordering known hot spots of the syndrome, where the causative algae are known to occur [52,53] or from imported fish (*L. bohar*) from Indo-Pacific waters such as have been reported in Germany and France [18,54–57].

In June 2017, a suspected ciguatera incident occurred in England associated with the consumption of fish marketed as red snapper (*L. bohar*) steaks. Three people were affected from the same family, all of whom consumed separate steaks from the same product package (Figure 2). Investigations revealed that the fish were harvested and processed in India and imported to the UK. The Food Standards Agency of the UK were made aware of the incident (reference number N_3861), and a product withdrawal was put in place for the brand associated with the intoxication. Two of the three family members were admitted to hospital with symptoms of diarrhea and vomiting, followed by paresthesia and shortness of breath. The reporting doctor provided a presumptive diagnosis in the absence of analytical confirmation that ciguatera was the likely cause of illness.

Figure 2. Images of implicated fish steak and packaging.

To date, few reports of ciguatera originate from India or from fish harvested from the Indian Ocean, with a recent review highlighting the lack of CTX or CTX-producing microalgae-related data from the region [18]. Human poisonings have occurred mainly in and around coastal island regions and reef habitat of the SW Indian Ocean, incorporating Comoros, Mayotte, La Reunion, Mauritius, Rodrigues and Seychelles, with a few others in the north, specifically southern India [18] and Malaysia [58]. From the reported incidents, however, some include fatal intoxications following consumption of shark tissues from Madagascar [4,59–63], as well as poisonings from reef fish in Reunion Island [64] and in mainland India [65,66]. Reef fish from the genus Lutjanus (snapper) were assessed for CTX presence by [33,34], who demonstrated CTX-like in vivo biological and sodium-channel activity, and CTXs by LC-MS, from two separate snapper species.

Following the reporting of the event in the U.K., one uneaten fish steak from the implicated batch was sent to the Centre for the Environment, Fisheries and Aquaculture science (Cefas) for assessment. The fish steak was subsequently analyzed for the presence of CTXs using multiple methods. This paper reports the subsequent investigations conducted to establish the cause of the poisonings following the consumption of the imported fish steaks in the U.K. The protocol adopted followed optimized strategies to those previously reported by the United States Food and Drug Administration (US FDA) for suspected ciguatera incidents, specifically involving fish species identification through DNA barcoding, semi-quantitation of CTX activity through use of an in vitro mouse neuroblastoma cell proliferation assay (N2a-MTT), and molecular confirmation of toxin presence with liquid chromatography and mass spectrometry [5,10,17,29,45,67–72].

2. Results

2.1. Fish Speciation

Three steaks from a packet of four were consumed, leaving one uneaten in frozen storage, which was subsequently submitted for DNA barcoding. Fish tissue samples were processed for DNA extraction and barcoding for species identification using partial cytochrome oxidase I gene sequence analysis [73–76]. The full 536 nucleotide sequence generated from the amplified product shared 100% nucleotide identity with nine partial COI sequences submitted to GenBank for Pinjalo snapper (*Pinjalo pinjalo*) including accessions MT076707, KX060042, OQ385992 and KP856805. The 536 nucleotide sequence shared <86.2% nucleotide identity to the partial COI sequences available for *Lutjanus bohar* on GenBank.

2.2. N2a-MTT Cytotoxicity Assay

The fish extracts showed positive CTX-like activity by the N2a-MTT assay. A dose-dependent loss in cell viability of N2a cells was observed following treatment with the snapper extract in the presence of ouabain and veratrine, indicating sodium channel-specific activity consistent with a CTX-3C standard (Figure 3). Cell viability was not affected in the non-sensitized cells, indicating that no other toxin class capable of affecting neuroblastoma cells was likely present in the snapper sample extracts. Data were highly reproducible across replicate assays performed on the same date (<5% CV) and across different days with different cell passages (<12% CV). Quantification based on the IC_{50} comparison between the sample and CTX-3C standard curves revealed a concentration of 1.81 μg CTX-3C eq./kg tissue, though the toxicity equivalence between CTX-3C and the causative toxins in our samples could not be determined and was outside the scope of this study since purified analytes would be required and were not available.

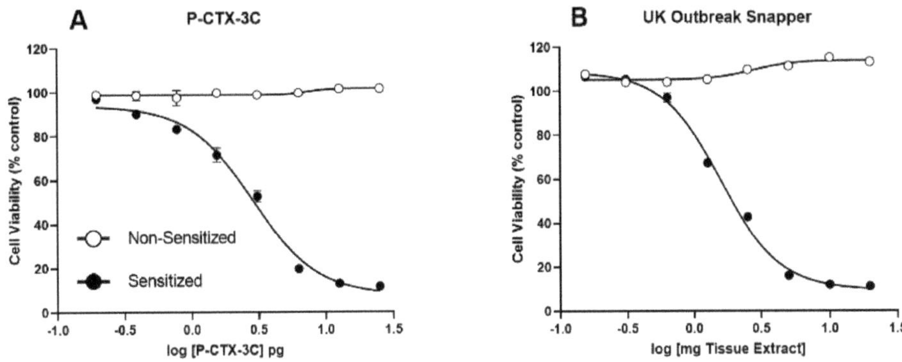

Figure 3. Representative dose-response curves from the N2a-MTT assays performed during this study. N2a cells sensitized with ouabain (O) and veratrine (V) in addition to standard or sample extract dilutions (solid black circles) and non-sensitized cells that received no additional dosing with OV (open circles). The response observed for (**A**) commercially available CTX-3C standard (Wako chemicals); and (**B**) extracts of the UK Snapper (*P. pinjalo*) are shown. Triplicate values and the standard error of the mean are plotted in all cases; however, in some instances, the error bars are obscured by the symbols.

2.3. *Liquid Chromatography-High Resolution Mass Spectrometry (LC-HRMS)*

2.3.1. Full Scan

Full-scan analysis of the SPE-cleaned snapper tissue extract by UHPLC-HRMS revealed the presence of distinct chromatographic peaks eluting at 6.84, 7.59 and 8.61 min consistent with the masses of tri-, di- and mono-hydroxylated CTX-3C, respectively (Figure 4 and Table 1). Furthermore, the Na^+ adduct for 2,3,51-trihydroxy-CTX-3C was also obtained (m/z 1095.5499) with an identical chromatographic retention time to the $[M + H]^+$ ion at 6.84 min. The presence of a small peak at 7.59 min at m/z consistent with the mono- and di- hydroxy CTX-3C is believed to be due to loss of H_2O within the electrospray source. Interrogation of the data for the presence of ions consistent with the exact masses of CTX-3C and other CTX-3C group congeners identified potential low levels peaks with ions at m/z 1059.5887 (5.7 min), 1055.5574 (5.25 min) and 1041.5781 (5.83 min) consistent with masses for A-*seco*-51-hydroxy-CTX-3C, 51-hydroxy-2-oxo-CTX-3C and 2-hydroxy-CTX-3C, respectively. No fragmentation of parent ions was performed. No peaks were detected with exact masses consistent to CTX-4A group and the C-CTX group analogues. No NH_4^+ adducts were observed for any of the hydroxylated CTX-3C congeners. These peaks were cross referenced against the bulk seized material and previously reported hydroxylated CTX-3C analogues in contaminated tissues from poisonings associated with *L. bohar* from Vietnam [57] (Supplementary Information). No C/I-CTXs were detected.

Table 1. Summary of HRMS identification of CTX3C congeners.

CTX3C Congener	Formula	Ion	Theoretical m/z	Observed m/z	Mass Error (ppm)	Retention Time (min)
2,3,51-trihydroxy-CTX-3C	$C_{57}H_{84}O_{19}$	$[M + H]^+$	1073.5680	1073.5663	−1.58	6.84
2,3,51-trihydroxy-CTX-3C	$C_{57}H_{84}O_{19}$	$[M + Na]^+$	1095.5499	1095.5507	−0.73	6.84
2,3-dihydroxy-CTX-3C	$C_{57}H_{84}O_{18}$	$[M + H]^+$	1057.5730	1057.5747	1.61	7.59
51-hydroxy-CTX-3C	$C_{57}H_{82}O_{17}$	$[M + H]^+$	1039.5625	1039.5602	−2.21	8.61

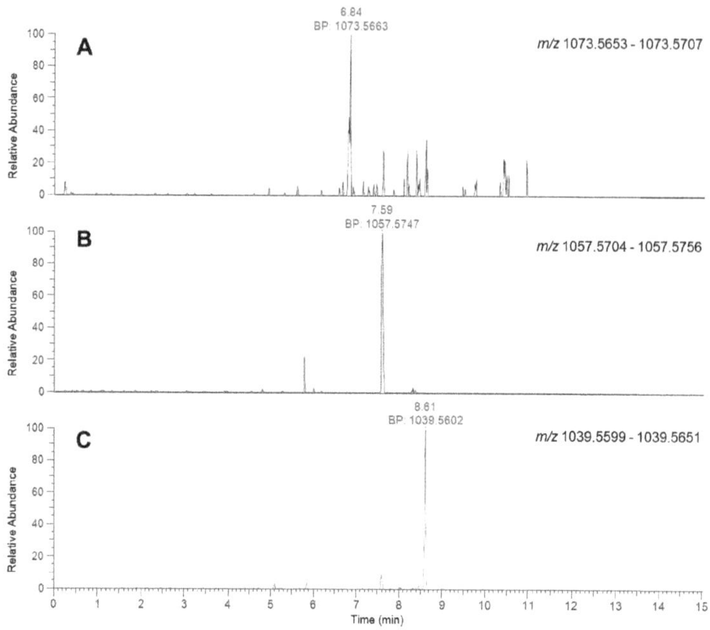

Figure 4. UHPLC-HRMS analysis of hydroxylated CTX-3C analogues in SPE-cleaned snapper tissue extracts, showing base peak chromatograms of (**A**) $[M + H]^+$ of 2,3,51-trihydroxy-CTX-3C, (**B**) $[M + H]^+$ of 2,3-dihydroxy-CTX-3C, (**C**) $[M + H]^+$ of 51-hydroxy-CTX-3C.

Analysis of a sub-sample of the fish tissue was extracted and analyzed for paralytic shellfish poisoning (PSP) toxins and tetrodotoxin (TTX), with results showing no detectable levels of these toxins.

2.3.2. Reaction with Periodate

Periodate was used here for the oxidative cleavage of 1,2-diols as demonstrated previously by the oxidation of C-CTXs [77]. Periodate oxidations were applied to 20 µL aliquots of the SPE-cleaned fish extracts, looking for the oxidation products of the hydroxylated CTX-3C analogues. Given the expected carbon bond cleavage between the 2,3 positions on the di- and tri-hydroxylated CTX-3C molecules, reaction products were anticipated to result in the loss of 2H from the original congeners following the formation of two carbonyl moieties (Supplementary Materials, Figure S2). UHPLC-HRMS confirmed the presence of chromatographic peaks eluting at 6.71 min and 6.75 min for the reaction products of 2,3,51-trihydroxy-CTX-3C (m/z 1071.5680) and 2,3-dihydroxy-CTX-3C (m/z 1055.5730), respectively, with a mass error of <2 ppm (Figure 5).

2.4. Medical Observations

Three female patients were assessed in hospital, with two being held under observation for two days (patients 1 and 2). Patient 3 was observed on day 1, before discharge, and a follow-up hospital visit on day 2. The ages of the three patients were 20, 45 and 11, respectively. The two older females presented similar complaints, including vomiting, diarrhea, myalgia (muscle pain), shortness of breath and paresthesia (which included skin pain, itching, numbness, as well as pins and needles). The third younger patient also experienced bone pain in the right shin, some reduced power in the legs, and diarrhea. Intravenous fluids were administered to patients 1 and 2 but no further treatment was provided. Blood analyses for full blood count, kidney function, i.e., urea and electrolytes, and liver function tests were all considered in the normal range. No neurological anomalies were noted in patient 3. Pain relief medications (codeine, paracetamol, ibuprofen) were provided for patients 1 and 2 following hospital discharge and patient 2 was also provided with antihistamines. Table 2 summarizes the medical reports from each of the three patients. Before any testing commenced, these cases were determined by clinicians to be associated with ciguatera poisoning due to the recent consumption of tropical fish and symptoms deemed consistent with ciguatera.

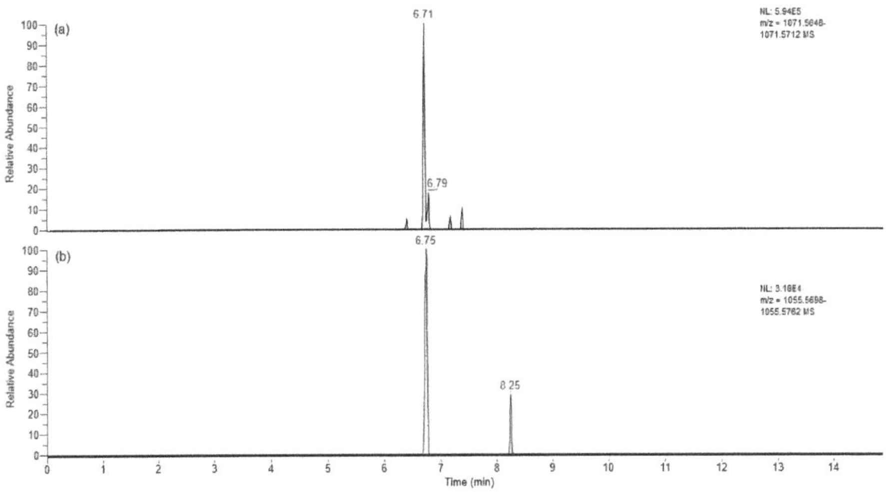

Figure 5. UHPLC-HRMS analysis of periodate oxidation products of (**a**) $[M + H]^+$ of 2,3,51-trihydroxy-CTX-3C, (**b**) 2,3-dihydroxy-CTX-3C.

Table 2. Summary of medical reports from three patients in suspected ciguatera poisoning cases.

	Patient 1	Patient 2	Patient 3
Admission date	22 June 2017	22 June 2017	23 June 2017
Discharge date	24 June 2017	24 June 2017	Not recorded
Sex	Female	Female	Female
Year of birth	1996	1972	2003
Presenting complaint	Vomiting, myalgia	Nausea, vomiting, diarrhea, myalgia	Pain in the right shin
Allergies	None known	None known	None known
Co-morbidities	Asthma	Hypertension	None known
post-discharge	Food poisoning: likely/potential ciguatera poisoning	Food poisoning: likely/potential ciguatera poisoning	Ciguatera poisoning: non-specific pains
Additional analyses conducted	Full blood count (FBC), urea and electrolytes (U&E), liver function test (LFT) = normal	FBC, U&E, LFT: No apparent disorder	U&E: Urea 2.3 mmol/L, Creatinine 50 µmol/L, Sodium 142 mEq/L, Potassium 3.4 mmol/L
	Creatinine kinase = normal	Electrocardiogram: No apparent distress	
	C-Reactive Protein 8.7 mg/L = mild elevation		
Procedures	Ward based care	Ward based care	Blood taken for analysis.
	Intravenous fluid infusion	Intravenous fluid infusion	
Post-discharge medications	Paracetamol, codeine	Chlorphenamine, ibuprofen	None
Follow up	No	No	No

3. Discussion

3.1. Location

Three females experienced sickness, including nausea, vomiting, diarrhea, myalgia, and paresthesia, following the consumption of snapper steaks. The fish had been imported to the UK through a Midlands-based food distributor. The original source of the fish was the Indian Ocean, FAO area 51, with the fish processed in a food production and packaging facility in Alappuzha, Kerala, southern India prior to export. No information was made available concerning the origin of the fish by the food processors upon repeated request. FAO major fishing area 51 is situated in the western Indian Ocean, extending from the Suez Canal and the Iranian Gulf in the north (latitude 30° North) down to the most southerly extent of latitude 45° South. The zone is split into eight sub-areas, incorporating the Red Sea, the Gulf, W. Arabian Sea, E. Arabian Sea and Laccadives, Somalia, Kenya and Tanzania, Madagascar and Mozambique Channel, Oceanic and Mozambique (Figure 6). Consequently, without any further information on the location of harvest/origin of the snapper, it is clear that the implicated samples may have originated from anywhere within a large geographical range [78]. The origin within this zone is therefore largely unknown, but expected to relate to reef-associated areas in warmer waters.

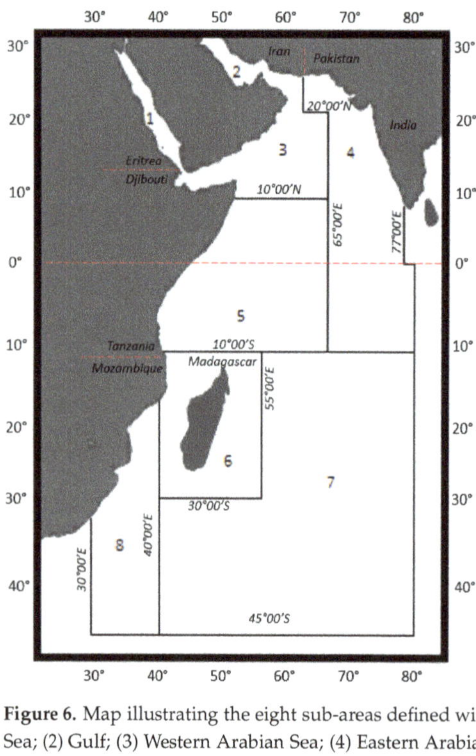

Figure 6. Map illustrating the eight sub-areas defined within FAO Major Fishing Area 51: (1) Red Sea; (2) Gulf; (3) Western Arabian Sea; (4) Eastern Arabian Sea; (5) Somalia, Kenya and Tanzania; (6) Madagascar and Mozambique Channel; (7) Oceanic; (8) Mozambique [78].

Ciguatera is frequently associated with poisoning incidents in coastal areas and islands of the Pacific and Caribbean Oceans with relatively few poisonings recorded in synonymous regions of the Indian Ocean [32,79], although earliest reports from this region date back to the 17th century [80]. Previous severe outbreaks in Madagascar (sub-area six, Figure 5) have been associated with the consumption of sharks, resulting previously in high fatality rates of more than 20%, with the apparent differences in clinical symptoms between those cases and classic ciguatera. This has raised the possibility of other toxins, carchatoxins, causing intoxication following consumption of sharks [59–62], although more recently, I-CTXs have been identified using high-resolution mass spectrometry in a bull shark following another fatal food poisoning in Madagascar [63]. Reef fish CTXs were isolated previously from samples of two red snappers/bass (*Lutjanus bohar*) and two Red Emperors (*Lutjanus sebae*) taken from the Soudan fishing banks approximately 100 miles to the north east of Mauritius, located in the north-west of sub-area seven of FAO zone 51 (Figure 5) [33,34]. Ciguatera has also been reported west of Mauritius in Reunion Island, where 375 notified cases of poisonings were reported between 1986 and 1994, even though the sale of reef fish is banned [64]. It is believed that the majority of the fish captured and eaten in Reunion originated from offshore coral banks north of Mauritius, such as the Soudan, Nazareth and Saya de Malha [64]. Most recently, several reports have originated from mainland India, bordering sub-zone 4, describing human intoxications in SW India following consumption of *L. bohar* [81,82], with poisonings in Europe also occurring following consumption of snappers originating from processing establishments in India [57,65,81,83–85]. With ciguatera not being reported in India prior to 2016, there are implications for climate-change-related shifts in benthic dinoflagellate communities and a consequent need to monitor coral reef fish species for CTXs [65].

3.2. Fish Species

Following DNA barcoding, the fish species was confirmed as being the *Pinjalo* snapper (*Pinjalo pinjalo*) in contrast to the marketed species of Red Snapper (*L. bohar*). Such mislabeling seems not to be uncommon, given similar issues in Germany where snappers sourced from FAO zone 71 (western central Pacific Ocean) labeled as *L. malbaricus* were tested and found to be *L. bohar* [57], and all of which are red-colored snapper species. The *P. pinjalo* from this study are reef-dwelling carnivores, which are typically trawled to a depth of 100 m but can be found in schools in the open ocean in a depth range of 15–100 m [86,87]. Diet has been reported as primarily benthic and planktonic invertebrates as well as small fishes, though ontogenetic dietary shifts over the lifespan of the fish may also depend on habitat and prey availability. *Pinjalo* snapper have been reported to reach a maximum length of 80 cm [88,89] but are listed as a minor commercial species and harmless in terms of human threat on FishBase.org and related databases. However, many other snappers from the family Lutjanidae have been associated with ciguatera and the noted habitat and feeding patterns fit with the detection of CTXs and ciguatoxicity in this species. Previous reports of ciguatera in the west Indian Ocean include over 30 different fish species, the majority of which were large predators such as groupers [64] and other snappers [33]. To our knowledge, however, this study reports the first confirmed occurrence of ciguatera in *P. pinjalo*.

3.3. Testing Methods

For reasons relating to both the absence of commercially available reference standards for an appropriate range of CTX analogues and the complexities associated with developing fit-for-purpose detection methods for ciguatera, there are no formal fully validated methods available for testing implicated fish samples [43,44,90,91]. In addition, complications relate to the low concentrations of toxins present in fish tissues and to the fact that P-, C- and potentially I-CTX variants are structurally different so that methods developed to detect one may not be applicable to another [43,44]. Testing for CTXs traditionally involved the use of live animal (mouse) bioassays, which give a measure of total toxicity [14,92–94], but with limited sensitivity, specificity, and ethical concerns. Biomolecular methods such as those utilizing antibodies, the measurement of cytotoxicity through in vitro assays in clonal mammalian neuronal cell lines, radioligand binding receptor assays utilizing tritiated brevetoxin, and the detection of the CTXs themselves through application of chemical detection methods are other approaches which have received wide usage in more recent years [3,10,17,22,32,38,67–69,95–102]. Out of the biomolecular approaches, the sodium channel dependent cell viability assay utilizing murine neuroblastoma cells (N2a) in a MTT assay (N2a-MTT or N2a-CBA) is currently the most widely applied, given its sensitivity and applicability to high throughput screening for CTX-like activity whilst also providing an estimate of composite toxicity [10,32,103–105]. Whilst a range of chemical detection methods have been reported, the use of liquid chromatography with tandem mass spectrometry (LC-MS/MS) and high-resolution mass spectrometry (LC-HRMS) are the most applicable to detection of low concentrations of toxins [13,23,33,43–47,63,70,85,99,102,106–108]. Therefore, for this assessment, two approaches were taken. Fish tissue samples were subjected to both N2a-MTT cytotoxicity assessment for the presence of sodium channel activating toxins as well as LC-HRMS analysis for the confirmation of CTXs. This tiered approach is the current accepted method for regulatory analysis of CTX in fish in the United States, where ciguatera poisonings occur with frequency in Florida, Puerto Rico, and the US Virgin Islands [102,109,110].

3.4. Fish Toxicity

Assessment of replicate extracts using the N2a-MTT cell proliferation assay confirmed the presence of sodium channel specific activity in sub-samples taken from the *P. pinjalo* steak implicated in the UK ciguatera poisonings. At the levels detected, human illness would be expected following the consumption of this fish; however, since this fish was sourced from the Indian Ocean, caution should be applied, and the quantification can only be used as a qualitative metric. It should also be noted that the toxicity equivalence between CTX-3C and CTX-1B is also yet to be verified, but having the CTX-3C standard was the only choice available. Since no quantified Caribbean or Indian Ocean CTX standards were available commercially at the time of this study, a C-CTX fish reference material that was prepared at the Dauphin Island Sea Lab (see methods for preparation and testing) was used for qualitative comparison. Ciguatoxicity observed by N2a-MTT assay in a fish matrix containing C-CTX-1 and -2 was consistent with both CTX-3C and the implicated fish sample in this study. It should be noted, however, that N2a-MTT data alone cannot confirm the structural identity or final concentration of any or all CTX congeners that may be present in the implicated sample. However, this confirms that sodium channel specific activity using a standardized method is consistent between a commercially available CTX-3C standard, a reference material confirmed to contain C-CTX-1 and -2 (by LC-MS/MS and LC-HRMS), and the implicated UK sample.

3.5. UHPLC-HRMS Analysis

Without quantified and certified standards, the assessment of CTXs in the study was conducted based on known molecular masses, mass spectrometric characteristics and cross verification against other poisoning tissue extracts. Chemical detection using UHPLC-HRMS was used to confirm the presence of chromatographic peaks indicative of Pacific, Caribbean and/or Indian CTXs using evidence from prior studies [33,44,57,63,85,102]. The absence of PSP toxins, or TTX confirmed that the toxicity and illness did not originate from these hydrophilic shellfish biotoxins.

High-resolution MS has previously provided accurate mass ion chromatograms to evidence the presence of CTXs from multiple global regions [45,57,63,77,111,112]. In our mass spectrometer, full-scan UHPLC-HRMS analysis showed the presence of chromatographic peaks with accurate masses with less than 2 ppm mass error, indicative of tri- di- and mono-hydroxylated CTX-3C congeners. Additional evidence was obtained for the presence of 2,3,51-trihydroxy-CTX-3C and 2,3-dihydroxy-CTX-3C through oxidative cleavage using periodate, with UHPLC-HRMS showing expected oxidation products of both compounds, as described previously for confirmation of C-CTX-1, 2, 3, and 4 in fish tissues [77]. Consequently, the 51-hydroxy-CTX-3C could not be confirmed using both approaches. Confirmation of 2,3-51-tryhydroxy-CTX-3C and 2,3-dihydroxy-CTX-3C therefore concurs with the findings of [57] who reported these analogues, in addition to 51-hydroxy-CTX-3C, M-*seco*-CTX-3C, 2-hydroxy-CTX-3C and CTX-3C in samples of red snapper (*L. bohar*) imported into Germany from the western Pacific Ocean and causing a public health concern in 2017 [56]. Such samples were also associated with incidents of human CP, with N2a cytotoxicity results confirming toxicity [55–57]. Other groups reporting similar results also included examples of the same hydroxylated CTX-3C congeners in a range of fish species, including *L. bohar* from the tropical western Pacific region [25,70,71,85].

Overall, this study provides further evidence for the presence of hydroxylated CTX-3C congeners in snapper fillets imported from the Indian Ocean/western Pacific region, evidencing multiple structural toxin types from the region including both P- and I/C-CTXs. The I-CTXs initially reported by [33,34] are proposed to be different congeners from C-CTXs based on subtle changes in clinical effects following I.P injection, differing product ion ratios in the mass spectrometer, and earlier elution characteristics for I-CTX following preparative-scale chromatographic purification. It is not known, however, whether these subtle differences result from matrix-related effects or do indeed indicate the I-CTXs to be structurally distinct from the previously reported C-CTX congeners. Our data in combination with prior studies of CTXs conducted in fish collected from the Indian Ocean have reported P-CTX congeners as the detectable CTXs responsible for Europe-wide human health impacts as a consequence of Indian Ocean fish imports. Going forward, purification and isolation of these toxins will be needed to fully elucidate the chemical structures of all analogues potentially present in these tissues, which will be pursued in future studies.

3.6. Symptoms and Overall Assessment

Symptoms experienced by human victims of ciguatera are numerous, varied, and complex. Following activation of the voltage-gated sodium channels in cell membranes, increasing sodium ion permeability, the consequent depolarization of the nerve cells results in a poisoning response. Ciguatoxins cause a range of gastrointestinal, cardiovascular and neurological symptoms, generally with a low mortality rate [113], but which can last between several days and several months [10]. As such, many of the ciguatera symptoms differ markedly in comparison to other marine-borne natural toxin poisonings such as Paralytic, Diarrhetic and Amnesic Shellfish Poisoning [114], although some such as the gastrointestinally related are shared [10]. A further complication results from the variability in clinical features depending on the geographical source of the fish consumed [10,64,115,116], though there is significant individual variability reported between vectors, and between patients. Ciguatera diagnosis is generally performed by combining evidence for consumption of fish species implicated or associated with the potential presence of CTXs, scientific data demonstrating the presence of toxins and/or ciguatoxicity, and the presentation of acute clinical symptoms indicative of the syndrome [117]. A universal case definition has been suggested for identification of ciguatera, which is applicable to the different symptoms experienced throughout the world, depending on the geographical source of the implicated fish [10]. These include clinical criteria, i.e., the patient has consumed marine reef fish previously associated with ciguatera, and reports neurological symptoms which may include any combination of paresthesia, dysesthesia, pruritus, allodynia, myalgia, arthralgia, dizziness and even painful ejaculation [10,24,32,118] with onset generally up to 24 h after fish consumption, gastrointestinal symptoms (nausea, vomiting, diarrhea), usually preceding the neurological symptoms between 1 and 12 h after fish consumption. There is also the possibility of cardiovascular symptoms such as hypotension and bradycardia. In addition to the symptoms, classification requires the laboratory confirmation of CTXs in implicated fish meal remnants. While necessary for case confirmation, the ability to obtain a meal remnant for testing is often challenging and this requirement could result in a large underestimation of cases.

In this study, all three patients exhibited gastrointestinal symptoms with vomiting and/or diarrhea within six hours of consuming the fish. Patients 1 and 2 both reported neurological symptoms such as myalgia and paresthesia including skin pain, intense pruritus, numbness, and pins and needles. Patient 3, in addition to gastrointestinal symptoms, experienced pain in the shin bone of one leg as well as reduced strength. Neuropsychological or cardiac symptoms were not reported by any of the patients in this study. As such, given the occurrence of both gastrointestinal and neurological symptoms, all three patients met the clinical criteria for ciguatera diagnosis. Taking the shared medical symptoms, and the known food consumption patterns, together with the known potential for ciguatera poisoning to occur in piscivorous, reef-associated Lutjanids, the subsequent evidence for CTX-like activity and confirmed CTXs in the same fish samples by UHPLC-HRMS provides the evidence for this fish to be the cause of human illness in this event. With two or more cases epidemiologically related, this provides the evidence needed (based on [10]) for confirmation of the first ciguatera poisonings in the UK. Unfortunately, no long-term follow-up with the three patients was carried out by the medical practitioners involved, so post-assessment of disease chronology was not possible.

3.7. Legislation and Health Protection

In terms of the protection of human consumers of fish, EU regulations [39,119] state that checks must take place to ensure that fishery products containing biotoxins such as ciguatera or other toxins dangerous to human health are not available in the market. The same legislation also states that poisonous fish from the specific families *Tetraodontidae* (pufferfish), *Molidae* (molas or ocean sunfish), *Diodontidae* (Porcupine fish/globefish/blowfish), also specifically mentioning the genus *Canthigasteridae* (member of the *Tetraodontidae* family) are not to be available in the market, although these are mentioned due to the specific link between these fish and the likely presence of pufferfish poisoning toxins, tetrodotoxin. Currently, no routine CTX testing is undertaken within the UK or US on imported fish products. In parts of the world most affected by ciguatera, some guidance does exist, relating to the purchase and/or consumption of certain species and/or the prevention of fishing certain species above specific weight limits [120]. In regions belonging to France and the US, neurotoxicity screening and LC-MS/MS confirmation has also been undertaken in a post hoc setting (i.e., after poisonings have occurred), but to our knowledge, there are no routine pre-surveillance mechanisms in place for the prevention of ciguatera in local populations or in imported products [121]. Until reliable, validated, and cost-effective CTX-testing methods can be implemented into routine monitoring programs, the most common approach for mitigating ciguatera is the avoidance of reef fish most commonly associated with illness [24] and more generally, avoidance of reef fish harvested from endemic or emerging high-risk regions [120].

In this study, given the clinical symptoms, the toxicity results and the confirmed fish species implicated in the poisonings, the UK authorities concluded that fish from the same batch should not be released for retail sale. Consequently, the recalled products were not released for further sale and samples provided for further study to enable enhanced levels of future public health protection. In conclusion, we report here for the first time the confirmed intoxication in the UK from ciguatera poisoning as a result of toxic carnivorous reef fish being imported into the country and affecting the local population. We therefore highlight the importance of monitoring for the presence of these toxins, and the need to develop risk management strategies for protecting the human population from highly contaminated ciguateric fish.

4. Materials and Methods

4.1. Fish Speciation

Genomic DNA was extracted from fish samples using an automated system. Briefly, 0.1 g of fish tissue was homogenised in 900 μL of G2 buffer (Qiagen, Manchester, UK) containing 10 μL proteinase K. The sample was incubated at 56 °C for 3 h and the DNA was then extracted from 50 μL of the digest using the DNA tissue mini kit (Qiagen, Manchester, UK) and the EZ-1xl biorobot, as per the manufacturer's instructions. The nucleic acid was eluted in a 60 μL volume. The primers used in the analysis were modifications of those already published [73–75]. A cocktail of 4 primers, Fish COI For1 (AACCAACCACAAAGACATTGGCAC), Fish COI For2 (GACTAATCATAAAGATATCGGCAC), Fish COI Rev1 (TTCAGGGTGACCGAAGAATCAGAA) Fish COI Rev2 (CTCAGGGTGTCCGAARAAYCARAA) was used to amplify a partial COI sequence for confirmation of the species by sequence analysis. Amplifications were performed in a 50 μL reaction volume consisting of 1× GoTaq flexi buffer (Promega, Southampton, UK), 2.5 mM $MgCl_2$ (Merck, Poole, UK), 1 mM dNTP mix, 50 pmol of the forward and reverse primers, 1.25 units of GoTaq DNA Polymerase (Promega, Southampton, UK) and 2.5 μL of the purified DNA template (SOP 2007). The reaction mix was overlaid with mineral oil and after an initial denaturing step (5 min at 95 °C), was subjected to 40 temperature cycles (1 min at 95 °C, 1 min at 60 °C and 1 min at 72 °C) in a Peltier PTC-225 thermal cycler (MJ Research, Braintree, UK) followed by a final extension step of 10 min at 72 °C. The PCR was performed in duplicate. PCR products were resolved on a 2% agarose gel containing ethidium bromide by electrophoresis and visualised under UV light. PCR products generated using the COI primers cocktail were extracted and purified by ethanol precipitation using the freeze and squeeze method and both DNA strands of the amplicon were sequenced using the ABI PRISM BigDye terminator cycle sequencing system (Thermo Fisher Scientific, Hemmel Hempstead, UK) and the primers used in the initial amplification. Sequencing reactions were analyzed on an ABI 3500 xl genetic analyzer (Thermo Fisher Scientific, Hemmel Hempstead, UK). A consensus sequence (with primer derived sequences removed) was determined using Sequencher version 5.4 software (Gene Codes Corporation, Ann Arbor, MI, USA) and the origin of the amplicon sequence identified using the Basic Local Alignment Search Tool (BLAST) facility available at the National Centre for Biotechnology Information (NCBI, Bethesda, MD, USA). Species-level identification was confirmed based on % identity to voucher sequences of *P. pinjalo* and sequences deposited on Genbank (NCBI, Bethesda, MD, USA).

4.2. Chemicals

All chemicals and reagents used for UHPLC-HRMS analysis were LC-MS-grade (Merck, Poole, UK), and all others were HPLC-grade (Merck, Poole, UK) or equivalent. All cell culture reagents, supplements, fetal bovine serum, and media were of the highest cell culture grade available. Mouse neuroblastoma cell lines, Neuro-2a (ATCC® CCL131™), were purchased from the American Type Culture Collection—ATCC (Manassas, VA, USA) and new lines generated through ouabain and veratrine pre-treatment and desensitization to improve detection limits in subsequent assay use as previously described [13]. A CTX-3C standard used for UHPLC-HRMS optimization was obtained from FujiFilm Wako Pure Chemical Corporation (Osaka, Japan).

CTX reference materials for the N2a-MTT assay were prepared at the Dauphin Island Sea Lab from *Scomberomorus cavalla* (King mackerel) and *Sphyraena barracuda* (Great barracuda) collected from long term monitoring sites on the Southeast coast of St. Thomas, US. Virgin Islands in 2014–2016 (permit STT-015-11, Dept. Planning and Natural Resources, Virgin Islands) as previously described [77].

4.3. Fish Samples

The frozen fish steak linked to the ciguatera incident had been kept frozen until cooking and consumption. The package contained four steaks, each around 10 mm in thickness, all cut across the full width of the fish, thereby containing a cross section of edible flesh, together with the outer skin and a segment of backbone. The one remaining fish steak from the purchaser was shipped to Cefas frozen in a temperature controlled cool box. Once received, the tissue was allowed to thaw to room temperature, cut into small chunks with scissors and blended using a high-speed Waring blender. Stringy fragments of skin were removed, and thorough homogenization achieved from the remaining edible flesh. The homogenized tissue sample was divided into sub-samples and re-frozen (−80 °C) prior to subsequent analysis. Red snapper (*L. bohar* from Vietnam) samples were also received from Germany following human poisonings [57,122] and used for cross referencing CTX-3C congener detection (Supplementary Materials).

4.4. Extraction Procedure for Cytotoxicity (N2a-MTT) Assay

Fish flesh samples (2 g) were each extracted with 4 mL acetone twice using a bead mill homogenizer (Beadruptor-24, Omni, Kennesaw, GA, USA) with 2.8 mm ceramic beads. Tissue disruption was conducted during acetone extraction in two cycles of 30 s duration at a speed of 5 m/s. Samples were centrifuged between extractions and supernatants pooled into glass test tubes stored at −80 °C for at least 4 h to precipitate proteins. Clarified samples were then taken to dryness under purified $N_2(g)$, resuspended in 2 mL 90% aqueous methanol and defatted twice with 4 mL hexane before being dried, resuspended in 4 mL $CHCl_2$ and partitioned with equal volumes of water twice. Samples were then dried under $N_2(g)$ and further purified by solid phase extraction (SPE) prior to N2a assay to reduce matrix effects. Agilent (Manchester, UK) Bond Elut SI (1 g/6 mL) cartridges were first conditioned using 6 mL 95% aqueous MeOH, followed by 12 mL MeOH then 12 mL $CHCl_3$. The 1 mL $CHCl_3$ sample was added to the top of cartridge, washed using 12 mL $CHCl_3$ and the toxins were eluted using 12 mL of 10% MeOH: 90% $CHCl_3$. SPE eluants were evaporated to dryness before resuspending in 0.5 mL MeOH.

4.5. N2a-MTT Cell Proliferation Assay

Cleaned fish tissue extracts were examined for ciguatoxicity using the sodium channel-specific MTT assay with mouse neuroblastoma cells (N2a-MTT) to evaluate changes in cell viability following various treatments and controls. Clonal cells were purchased from the supplier (American Type Culture Collection, ATCC, Manassas, VA, USA) and grown to 90% confluency until cell viability was >95% and measured growth rates were stable across 10 passages. Once stable, modified cell lines were prepared via pretreatment with ouabain and veratrine as previously described ([12,13]; Supplementary Materials). All cell lines were authenticated by sequencing to verify species origin and were verified mycoplasma-free before and after use since this type of contamination is difficult to pick up by standard microscopy but can affect cell sensitivity, growth, and downstream assay reproducibility. Assay cell stocks were also limited to maintenance for 2–3 months (i.e., no more than 90 passages) to reduce mutations caused by extended culture. Likewise, antibiotics were not used at any point in the N2a maintenance or cell-based assays to reduce the uncertainty associated with antibiotic induction of xenobiotic gene expression (e.g., [123]), possible mitochondrial dysfunction (e.g., [124]), antibiotic-induced oxidative stress, and other changes to cell growth and biochemistry that could result in anomalous results.

Prior to N2a-MTT assay, replicate flasks of at least two distinct passages of modified N2a cells (DISL-N2a-101) were grown in 175 cm^2 culture flasks until 80–90% confluent. Cells were harvested in trypsin-EDTA (0.025%), deactivated and washed, and then counted and evaluated to meet quality control criteria (cell viability of population > 95%). Cells were then seeded into cell culture treated 96-well polystyrene plates (Celltreat; Pepperell, MA, USA) at 30,000 cells/well. Dose response experiments were conducted after approx. 18 h, once cells reached 50–60% confluency in sample wells. Briefly, ouabain octahydrate (2.5 mM), veratrine hydrochloride (0.25 mM) (OV), and/or sterile phosphate-buffered saline (PBS; pH 7.4) were added to the sample wells containing DISL-N2a-101 cells. The snapper extracts, controls, and reference materials were reconstituted in 5% FBS-RPMI media at a maximum of 20 mg tissue equivalent (TE) dose with a 2.0% MeOH vehicle and serially diluted (20–0.16 mg TE). Standard curves were generated with CTX-3C with a starting dose of 25 pg and 2× serially diluted to provide at least 8 points across the curve. Samples and standards were then applied to the sensitized (dosed with OV) and non-sensitized cells (dosed with PBS) in triplicate across two distinct cell passages. Additional quality assurance and quality controls were incorporated into the workflow to ensure high quality, reproducible data as previously described [13]. Dosed cells were incubated in a humidified environment for 18 h at 37 °C with 5% CO_2. After incubation, thiazolyl blue tetrazolium bromide (MTT) was added to plates and incubated at 37 °C for 25–30 min to produce a formazan product in live cells remaining in the wells. Dimethyl sulfoxide (DMSO) was then added to dissolve the formazan deposits and develop the final colorimetric endpoint. Plates were then read on a 96-well spectrophotometer with Magellan software version 7 (TECAN Sunrise; Morrisville, NC, USA) at a measurement wavelength of 570 nm.

Dose–response curves (8 doses) were constructed following normalization of the data to % OV control as is deemed best practice for MTT based assays and performing a logarithmic transformation of the dose. Full curves from the standards and fish reference materials were compared to the snapper extracts and analyzed by a logistic four parameter nonlinear regression with variable slope (GraphPad Prism, v7, GraphPad Software, Boston, MA, USA). Measurements between replicates with a percent coefficient of variance less than 10% were considered acceptable. The inhibitory concentration at which 50% of the cells remained viable (IC_{50}) were compared with the available commercial standard CTX-3C to provide an estimate in nanograms per gram CTX-3C equivalents. A C-CTX reference fish was also analyzed in parallel to assess similarities and differences between toxin classes; however, a quantified standard was not available for the determination of concentrations in the extracts as C-CTX equivalents.

4.6. Full Scan UHPLC-HRMS

Fish flesh samples (10 g) were extracted following an approach based on [125] with 30 mL MeOH:water (3:2 *v/v*) using an UltraTurrax homogenizer (IKA, Oxford, UK). Tissue disruption was conducted in two cycles of 60 s duration at a speed of 5 m/s. Following homogenization, samples were then boiled for 10 min. Cooled samples were then centrifuged (10 min, 4500× *g*) with the clear supernatant transferred into a clean 50 mL tube and left in the freezer overnight to precipitate insoluble material. The next day, liquid–liquid partitioning was conducted by first adding 20 mL of dichloromethane (DCM), the contents shaken and allowed to separate before transferring the bottom DCM layer into a round-bottom flask. The DCM partitioning was repeated two more times, after which, both times, the DCM layer was added to the same round-bottom flask. The DCM fraction was subsequently rotary evaporated to dryness to ensure any traces of water were removed, before being re-dissolved in 5 mL DCM and transferred to a smaller glass tube. The solution was then re-evaporated using a Genevac miVac Quatro centrifugal concentrator (Fisher Scientific, Loughborough, UK) before re-dissolving in 1.0 mL DCM ready for solid phase extraction (SPE) clean-up.

Normal-phase SPE was conducted using JT Baker Bakerbond 1 g/6 mL Florisil cartridges (VWR, Lutterworth, UK) initially conditioned using 10 mL MeOH/H_2O (95:5 v/v), then 10 mL MeOH, prior to 20 mL DCM. The 1 mL sample was added to the SPE, before rinsing with 10 mL DCM. The cleaned-up sample was then collected in 20 mL MeOH/DCM (1:9 v/v), with the eluant evaporated in the Genevac miVac. The dried residue was subsequently re-dissolved in 1.0 mL MeOH prior to analysis.

UHPLC-HRMS analysis was conducted using a Vanquish UHPLC system coupled to an Orbitrap Exploris 120 high resolution mass spectrometer (Thermo Scientific, Hemel Hempstead, UK) with ciguatoxins separated using a 1.7 μm, 2.1 × 100 mm Waters (Manchester, UK) Acquity UHPLC BEH C18 column in conjunction with a Waters VanGuard BEH C18 1.7 μm 2.1 × 5 mm guard cartridge. The column was held at 40 °C, with samples held in the sample manager at 10 °C. The sample injection volume was 2 μL, and the mobile phase flow rate was 0.40 mL/min, with mobile phases A consisting of 5 mM ammonium formate and 0.1% formic acid in water and B being 100% acetonitrile. The UHPLC gradient was as follows: 58% B starting conditions, holding for 1 min then rising to 100% B at 10 min and holding for 1 further min, before dropping back to 58% B at 11.5 min and equilibrating until the end of the run at 15 min.

The developed HRMS method utilized the following source conditions after optimization using the Wako CTX-3C standard: H-ESI source type, static spray voltage, 4000 V positive ion voltage, sheath gas, aux gas and sweep gas at 50, 10 and 1 arbitrary units, respectively, 325 °C ion transfer tube, 225 °C vaporizer temperature. EASY-IC™ internal mass calibration was used during the analysis, with the LC flow diverted to waste until 0.2 min. Full scan was conducted over m/z 950–1250 using maximum resolution of 120,000 and an RF lens of 70%, with full scan data interrogated using Thermo Scientific FreeStyle analysis software v1.8 (Thermo Scientific, Hemel Hempstead, UK).

In addition, the fish tissue was analyzed for paralytic shellfish poisoning (PSP) toxins and tetrodotoxins, shellfish toxins which also cause toxicity through their action on sodium channels. Standard, validated methodologies were used for these assays [126,127].

4.7. Periodate Oxidation

A 20 μL aliquot of the SPE-cleaned fish extract was subjected to oxidative cleavage of the vicinal diols based on the method of [77]. The reaction was achieved through the addition of 20 μL periodate reagent, prepared freshly from 0.15 M periodic acid, 0.3 M ammonium formate and 0.3 M Na_2HPO_4. Control samples were also prepared including equal volumes fish extract and water, plus equal volumes water and periodate reagent. The mixtures were heated to 45 °C for 35 min, before cooling and analysis by LC-HRMS, as above.

4.8. Medical Observations

Following the hospitalization, anonymized medical records were provided by two out of the three affected people.

5. Conclusions

This study reports the first confirmed occurrence of Ciguatera Poisoning in the UK, following the consumption of fish steaks imported into the country from products harvested in the Indian Ocean. Whilst the fish products were labeled as Red Snapper (*Lutjanus bohar*), DNA barcoding confirmed the species to be the Pinjalo Snapper (*Pinjalo pinjalo*). Three hospitalized patients exhibited gastrointestinal symptoms, with two of these reporting neurological symptoms, all meeting the clinical diagnosis criteria for Ciguatera Poisoning. Toxicity was determined in one of the implicated fish steaks using the N2a-MTT cytotoxicity assay, which revealed quantitative concentration of 1.81 μg CTX-3C equivalents per kg tissue. Whilst no Caribbean or Indian Ocean CTX analogues were detected, high-resolution mass spectrometry confirmed the presence of tri-, di- and mono-hydroxylated CTX-3C congeners to within 2 ppm mass accuracy, with further confirmation obtained through periodate triggered oxidative cleavage and the detection of di- and tri-hydroxy CTX-3C reaction products. Consequently, this study demonstrates the use of multiple detection methods for the confirmation of Ciguatera Poisoning in UK patients following consumption of a mis-labeled fishery product from the Indian Ocean, and the importance of continued surveillance of imports to the UK from high-risk regions.

Supplementary Materials: The following supporting information can be downloaded at: https://www.mdpi.com/article/10.3390/md23020067/s1, Supplementary methods for cross-referencing including sample preparation and LC-HRMS analysis. Figure S1—LC−HRMS extracted ion chromatograms of tri-hydroxyCTX3C ($[M + H]^+$ m/z 1073.5680), di-hydroxyCTX3C ($[M + H]^+$ m/z 1057.5730), 51-hydroxyCTX3C ($[M + H]^+$ m/z 1039.5625) and 2/3-hydroxyCTX3C ($[M + H]^+$ m/z 1041.5781) in (A) *L. bohar* tissue extract from an outbreak in Vietnam, provided by the German Federal Institute for Risk Assessment (Spielmeyer et al., 2022; Loeffler et al. 2022) and (B) *Pinjalo pinjalo* tissue extract from the lot of fish material seized from the UK ciguatera outbreak. Chromatograms are normalized to the most abundant peak (m/z 1057.5730, RT 11.05 min) in each sample. Figure S2—Reaction products following periodate oxidative cleavage of a) di-hydroxylated CTX-3C and b) tri-hydroxylated CTX-3C

Author Contributions: A.D.T., B.H.M. and A.R. designed the study. A.D.T., E.M.M. and B.H.M. prepared samples for LC-MS/MS. A.D.T., E.M.M. and B.H.M. ran UHPLC-HRMS analysis. A.R. prepared samples and performed N2a-MTT analyses and collated incident data. D.S. conducted DNA speciation. A.D.T., A.R., B.H.M. and D.S. performed data interpretation. A.D.T. and A.R. drafted the manuscript. All authors have read and agreed to the published version of the manuscript.

Funding: This research was jointly supported by the Cefas Seedcorn funding (C7366) and the Interreg Atlantic Area Program project Alertox-Net (EAPA-317-2016). Collaborative efforts with AR were supported in part by the U.S. National Science Foundation Partnerships in International Research and Education Program (1743802 NSF PIRE), National Oceanic and Atmospheric Administration ECOHAB (CiguaTOX NA11NOS4780059), and NSF and NIH funded Centres for Oceans and Human Health: Caribbean Centre for Ciguatera Research (NSF 1841811 and NIH: 1P01ES028949-01, A.R.).

Institutional Review Board Statement: Use of de-identified retrospective medical data associated with the ciguatera incident (protocol 20-017, 1550363-1) was reviewed and deemed exempt December 2020 based on exemption 45 CFR 46.104(d)(4) by an Institutional Review Board (IRB) panel, University of South Alabama (AR).

Data Availability Statement: All relevant data is displayed in this manuscript or Supplementary Materials.

Acknowledgments: Samples used in the preparation of reference materials for N2a-MTT were collected in collaboration with Tyler Smith at the University of the Virgin Islands, St. Thomas. Thanks to Lynn Wilking, Katherine Baltzer (Dauphin Island Sea Lab), and Jessica Gwinn (University of South Alabama) who assisted with collections of fish for reference materials in 2014–2019. Astrid Spielmeyer and Chris R. Loeffler from the German Federal Institute for Risk Assessment, National

Reference Laboratory for Monitoring of Marine Biotoxins (Berlin, Germany) provided the sample extracts of CTX3C-group contaminated *Lutjanus bohar*. Christopher O. Miles provided his technical guidance on the periodate cleavage reactions.

Conflicts of Interest: The authors declare no conflicts of interest.

References

1. CDC (Centre for Disease Control and Prevention). Ciguatera Fish Poisoning—Texas, 1998 and South Carolina, 2004. *Morb. Mortal. Wkly. Rep. (MMWR)* **2006**, *55*, 935–937.
2. Fleming, L.E.; Broad, K.; Clement, A.; Dewailly, E.; Elmir, S.; Knap, A.; Pomponi, S.A.; Smith, S.; Solo Gabriele, H.; Walsh, P. Oceans and human health: Emerging public health risks in the marine environment. *Mar. Pollut. Bull.* **2006**, *53*, 545–560. [CrossRef]
3. European Food Safety Authority Panel on Contaminants in the Food Chain (CONTAM). Scientific opinion on marine biotoxins in shellfish—Emerging toxins: Ciguatoxin group. *EFSA J.* **2010**, *8*, 1627.
4. Chinain, M.; Gatti, C.M.I.; Darius, H.T.; Quod, J.-P.; Tester, P.A. Ciguatera poisonings: A global review of occurrences and trends. *Harmful Algae* **2021**, *102*, 101873. [CrossRef]
5. FAO, Food and Agriculture Organization of the United Nations; World Health Organization. *Report of the Expert Meeting on Ciguatera Poisoning: Rome, 19–23 November 2018*; World Health Organization: Geneva, Switzerland, 2020.
6. Yasumoto, T.; Nakajima, I.; Bagnis, R.; Adachi, R. Finding of a dinoflagellate as a likely culprit of ciguatera. *Bull. Jpn. Soc. Sci. Fish.* **1977**, *43*, 1021–1026. [CrossRef]
7. Murata, K.; Legrand, A.M.; Ishibashi, Y.; Yasumoto, T. Structures of ciguatoxin and its congoner. *J. Am. Chem. Soc.* **1989**, *111*, 8929–8931. [CrossRef]
8. Murata, K.; Legrand, A.M.; Ishibashi, Y.; Fukui, M.; Yasumoto, T. Structures and configurations of ciguatoxin from the moray eel *Gymnothorax javanicus* and its likely precursor from the dinoflagellate *Gambierdiscus toxicus*. *J. Am. Chem. Soc.* **1990**, *112*, 4380–4386. [CrossRef]
9. Lewis, R.J.; Holmes, M.J. Origin and transfer of toxins involved in ciguatera. *Comp. Biochem. Physiol. Part C Pharmacol. Toxicol. Endocrinol.* **1993**, *106*, 615–628. [CrossRef] [PubMed]
10. Friedman, M.A.; Fernandez, M.; Backer, L.C.; Dickey, R.W.; Bernstein, J.B.; Schrank, K.; Kibler, S.; Stephan, W.; Gribble, M.O.; Bienfang, P.; et al. An updated review of ciguatera fish poisoning: Clinical, epidemiological, environmental, and public health management. *Mar. Drugs* **2017**, *15*, 72. [CrossRef] [PubMed]
11. Ikehara, T.; Kuniyoshi, K.; Oshiro, N.; Yasumoto, T. Biooxidation of ciguatoxins leads to species-specific toxin profiles. *Toxins* **2017**, *9*, 205. [CrossRef] [PubMed]
12. Mudge, E.M.; Robertson, A.; McCarron, P.; Miles, C.O. Selective and efficient capture and release of vic-diol-containing Pacific and Caribbean ciguatoxins from fish extracts with a boronate affinity polymer. *J. Agric. Food Chem.* **2022**, *70*, 12946–12952. [CrossRef]
13. Mudge, E.M.; Miles, C.O.; Ivanova, L.; Uhlig, S.; James, K.S.; Erdner, D.L.; Fæste, C.K.; McCarron, P.; Robertson, A. Algal ciguatoxin identified as source of ciguatera poisoning in the Caribbean. *Chemosphere* **2023**, *330*, 138659. [CrossRef] [PubMed]
14. Banner, A.H.; Scheuer, P.J.; Sasaki, S.; Helfrich, P.; Alender, C.B. Observations on ciguatera-type toxin in fish. *Ann. N. Y. Acad. Sci.* **1960**, *90*, 770–787. [CrossRef] [PubMed]
15. Banner, A.H.; Helfrich, P.; Piyakarn, T. Retention of ciguatera toxin by red snapper *Lutjanus bohar*. *Copeia* **1966**, *1966*, 297. [CrossRef]
16. Lehane, L.; Lewis, R.J. Ciguatera: Recent advances but the risk remains. *Int. J. Food Microbiol.* **2000**, *61*, 91–125. [CrossRef] [PubMed]
17. Dickey, R.W.; Plakas, S.M. Ciguatera: A public health perspective. *Toxicon* **2010**, *56*, 123–136. [CrossRef]
18. Habibi, N.; Uddin, S.; Dechraoui Bottein, M.-Y.; Faizuddin, M. Ciguatera in the Indian Ocean with special insights on the Arabian Sea and Adjacent Gulf and Seas: A review. *Toxins* **2021**, *13*, 525. [CrossRef]
19. Oshiro, N.; Nagasawa, H.; Nishimura, M.; Kuniyoshi, K.; Kobayashi, N.; Sugita-Konishi, Y.; Ikehara, T.; Tachihara, K.; Yasumoto, T. Analytical studies on ciguateric fish in Okinawa, Japan (II): The Group *Variola albimarginata*. *J. Mar. Sci. Eng.* **2023**, *11*, 242. [CrossRef]
20. Roue, M.; Darius, H.T.; Chinain, M.; Sibat, M.; Amzil, Z. Ability of giant clams to bioaccumulate ciguatoxins from *Gambierdiscus* cells. *Harmful Algae News* **2016**, *55*, 12–13.
21. Darius, H.T.; Roué, M.; Sibat, M.; Viallon, J.; Mahana iti Gatti, C.; Vanderesa, M.W.; Tester, P.A.; Litaker, R.W.; Amzil, Z.; Hess, P.; et al. *Tectus niolticus* (Tegulidae, Gastropod) as a novel vector of ciguatera poisoning: Detection of Pacific ciguatoxins in toxic samples from Nuka Hiva Island (French Polynesia). *Toxins* **2018**, *10*, 2. [CrossRef] [PubMed]
22. Yasumoto, T.; Igarashi, T.; Legrand, A.-M.; Cruchet, P.; Chinain, M.; Fujita, T.; Naoki, H. Structural elucidation of ciguatoxin congeners by fast-atom bombardment tandem mass spectroscopy. *J. Am. Chem. Soc.* **2000**, *122*, 4988–4989. [CrossRef]

23. Abraham, A.; Jester EL, E.; Granade, H.R.; Plakas, S.M.; Dickey, R.W. Caribbean ciguatoxin profile in raw and cooked fish implicated in ciguatera. *Food Chem.* **2012**, *131*, 192–198. [CrossRef]
24. Friedman, M.A.; Fleming, L.E.; Fernandez, M.; Bienfang, P.; Schrank, K.; Dickey, R.; Bottein, M.-Y.; Backer, L.; Ayyar, R.; Weisman, R.; et al. Ciguatera fish poisoning: Treatment, prevention and management. *Mar. Drugs* **2008**, *6*, 456–479. [CrossRef]
25. Oshiro, N.; Tomikawa, T.; Kuniyoshi, K.; Ishikawa, A.; Toyofuku, H.; Kojima, T.; Asakura, H. LC–MS/MS Analysis of Ciguatoxins Revealing the Regional and Species Distinction of Fish in the Tropical Western Pacific. *J. Mar. Sci. Eng.* **2021**, *9*, 299. [CrossRef]
26. Oshiro, N.; Tomikawa, T.; Kuniyoshi, K.; Kimura, K.; Kojima, T.; Yasumoto, T.; Asakura, H. Detection of Ciguatoxins from Fish Introduced into a Wholesale Market in Japan. *J. Food Hyg. Soc. Jpn. (Shokuhin Eiseigaku Zasshi)* **2021**, *62*, 8–13. [CrossRef] [PubMed]
27. De Fouw, J.C.; van Egmond, H.P.; Speijers, G.J.A. *Ciguatera Fish Poisoning: A Review*; RIVM Report 388802 021; RIVM: Bilthoven, The Netherlands, 2001.
28. Pottier, I.; Vernoux, J.P.; Lewis, R. Ciguatera fish poisoning in the Caribbean Islands and Western Atlantic. *Rev. Environ. Contam. Toxicol.* **2001**, *168*, 99–141.
29. Pottier, I.; Hamilton, B.; Jones, A.; Lewis, R.J.; Vernoux, J.P. Identification of slow and fast-acting toxins in a highly ciguatoxic barracuda (*Sphyraena barracuda*) by HPLC/MS and radiolabelled ligand binding. *Toxicon* **2003**, *42*, 663–672. [CrossRef] [PubMed]
30. Wong, C.K.; Hung, P.; Lee, K.L.H.; Kam, K.-M. Study of an outbreak of ciguatera fish poisoning in Hong 272 Kong. *Toxicon* **2005**, *46*, 563–571. [CrossRef] [PubMed]
31. Baumann, F.; Bourrat, M.; Pauillac, S. Prevalence, symptoms and chronicity of ciguatera in New Caledonia: Results from an adult population survey conducted in Noumea during 2005. *Toxicon* **2010**, *56*, 662–667. [CrossRef] [PubMed]
32. Caillaud, A.; de la Iglesia, P.; Darius, H.T.; Pauillac, S.; Aligizaki, K.; Fraga, S.; Chinain, M.; Diogene, J. Update on methodologies available for ciguatoxin determination: Perspectives to confront the onset of ciguatera fish poisoning in Europe. *Mar. Drugs* **2010**, *8*, 1838–1907. [CrossRef] [PubMed]
33. Hamilton, B.; Hurbungs, M.; Vernoux, J.P.; Jones, A.; Lewis, R.J. Isolation and characterisation of Indian Ocean ciguatoxin. *Toxicon* **2002**, *40*, 685–693. [CrossRef] [PubMed]
34. Hamilton, B.; Hurbungs, M.; Jones, A.; Lewis, R.J. Multiple ciguatoxins present in Indian Ocean reef fish. *Toxicon* **2002**, *40*, 1347–1353. [CrossRef] [PubMed]
35. Hamilton, B.; Whittle, N.; Shawc, G.; Eaglesham, G.; Moore, M.R.; Lewis, R.J. Human fatality associated with Pacific ciguatoxin contaminated fish. *Toxicon* **2010**, *56*, 668–673. [CrossRef] [PubMed]
36. Boucaud-Maître, D.; Vernoux, J.P.; Pelczar, S.; Daudens-Vaysse, E.; Aubert, L.; Boa, S.; Ferracci, S.; Garnier, R. Incidence and clinical characteristics of ciguatera fish poisoning in Guadeloupe (French West Indies) 275 between 2013 and 2016: A retrospective cases-series. *Sci. Rep.* **2018**, *8*, 3095. [CrossRef] [PubMed]
37. Tubaro, A.; Sosa, S.; Hungerford, J. Toxicology and diversity of marine toxins. In: *Veterinary Toxicology: Basic and Clinical Principles.* Ed: R.C. Gupta, Elsevier. *Chp* **2012**, *69*, 896–936.
38. Lewis, R.J.; Sellin, M.; Poli, M.A.; Norton, R.S.; MacLeod, J.K.; Sheil, M.M. Purification and characterization of ciguatoxins from moray eel (*Lycodontisjavanicus, Muraenidae*). *Toxicon* **1991**, *29*, 1115–1127. [CrossRef] [PubMed]
39. EC regulation no 625/2017 of the European Parliament and of the Council of 15 March 2017 Amending EC Regulation 854/2004 Laying down Specific Rules for the Organisation of Official Controls on Products of Animal Origin Intended for Human Consumption. Available online: https://eur-lex.europa.eu/eli/reg/2017/625/oj/eng (accessed on 29 December 2024).
40. Perez-Arellano, J.L.; Luzardo, O.P.; Brito, A.P.; Cabrera, M.H.; Zumbado, M.; Carranza, C.; Angel-Moreno, A.; Dickey, R.W.; Boada, L.D. Ciguatera fish poisoning, Canary Islands. *Emerg. Infect. Dis.* **2005**, *11*, 1981–1982. [CrossRef]
41. Aligizaki, K.; Nikolaidis, G. Morphological identification of two tropical dinoflagellates of the genera *Gambierdiscus* and *Sinophysis* in the Mediterranean Sea. *J. Biol. Res. Thessalon.* **2008**, *9*, 75–82.
42. Aligizaki, K.; Nikolaidis, G.; Fraga, S. Is Gambierdiscus expanding to new areas? *Harmful Algae News* **2008**, *36*, 6–7.
43. Estevez, P.; Castro, D.; Pequeno-Valrierra, A.; Leao, J.M.; Vilarino, O.; Diogene, J.; Gago-Martinez, A. An attempt to characterise the ciguatoxin profile in *Seriola fasciata* causing ciguatera fish poisoning in Macaronesia. *Toxins* **2019**, *11*, 221. [CrossRef] [PubMed]
44. Estevez, P.; Castro, D.; Leao, J.M.; Yasumoto, T.; Dickey, R.; Gago-Martinez, A. Implementation of liquid chromatography tandem mass spectrometry for the analysis of ciguatera fish poisoning in contaminated fish samples from Atlantic coasts. *Food Chem.* **2019**, *280*, 8–14. [CrossRef] [PubMed]
45. Estevez, P.; Sibat, M.; Leao-Martins, J.M.; Costa, P.R.; Gago-Martínez, A.; Hess, P. Liquid Chromatography Coupled to High-Resolution Mass Spectrometry for the Confirmation of Caribbean Ciguatoxin-1 as the Main Toxin Responsible for Ciguatera Poisoning Caused by Fish from European Atlantic Coasts. *Toxins* **2020**, *12*, 267. [CrossRef]
46. Estevez, P.; Prieto, J.O.; Burlingame, A.; Gago-Martinez, A. Characterisation of the ciguatoxin profile in fish samples from the Eastern Atlantic Ocean using Liquid Chromatography-High Resolution Mass Spectrometry. *Food Chem.* **2023**, *418*, 135960. [CrossRef] [PubMed]
47. Costa, P.R.; Estevez, P.; Castro, D.; Soliño, L.; Gouveia, N.; Santos, C.; Rodrigues, S.M.; Leao, J.M.; Gago-Martínez, A. New Insights into the Occurrence and Toxin Profile of Ciguatoxins in Selvagens Islands (Madeira, Portugal). *Toxins* **2018**, *10*, 524. [CrossRef]

48. Fraga, S.; Riobo, P.; Diogene, J.; Paz, B.; Franco, J.M. Toxic and potentially toxic benthic dinoflagellates observed in Macaronesia (NE Atlantic Archipelago). In Proceedings of the 11th International Conference on Harmful Algae, Cape Town, South Africa, 15–19 November 2004.
49. Fraga, S.; Rodriguez, F.; Caillaud, A.; Diogene, J.; Raho, N.; Zapata, M. *Gambierdiscus exectricus* sp. nov. (Dinophyceae), a benthic toxic dinoflagellate from the Canary Islands (NE Atlantic Ocean). *Harmful Algae* **2011**, *11*, 10–22. [CrossRef]
50. Boada, L.D.; Zumbado, M.; Luzardo, O.R.; Almeida-Gonzalez, M.; Plakas, S.M.; Granade, H.R.; Abraham, A.; Jester, E.L.E.; Dickey, R.W. Ciguatera fish poisoning on the West Africa Coast: An emerging risk in the Canary Islands (Spain). *Toxicon* **2010**, *56*, 1516–1519. [CrossRef]
51. Otero, P.; Pérez, S.; Alfonso, A.; Vale, C.; Rodríguez, P.; Gouveia, N.N.; Gouveia, N.; Delgado, J.; Vale, P.; Hirama, M.; et al. First toxin profile of ciguateric fish in Madeira Arquipelago (Europe). *Anal. Chem.* **2010**, *82*, 6032–6039. [CrossRef]
52. Sanner, B.M.; Rawert, B.; Henning, B.; Zidek, W. Ciguatera fish poisoning following travel to the tropics. *Z. Fur Gastroenterol.* **1997**, *35*, 327–330.
53. FAO (Food and Agriculture Organization of the United Nations). *Marine Biotoxins*; FAO Food and Nutrition Paper; FAO: Rome, Italy, 2004; Volume 80, pp. 1–287.
54. Mattei, C.; Vetter, I.; Eisenblätter, A.; Krock, B.; Ebbecke, M.; Desel, H.; Zimmermann, K. Ciguatera fish poisoning: A first epidemic in Germany highlights an increasing risk for European countries. *Toxicon* **2014**, *91*, 76–83. [CrossRef] [PubMed]
55. Friedemann, M. Erster Ciguatera-Ausbruch in Deutschland 2012. *Bundesgesundheitsblatt-Gesundheitsforschung-Gesundheitsschutz* **2016**, *59*, 1556–1565. [CrossRef]
56. Friedemann, M. Ciguatera fish poisoning outbreaks from 2012 to 2017 in Germany caused by snappers from India, Indonesia, and Vietnam. *J. Consum. Prot. Food Saf.* **2019**, *14*, 71–80. [CrossRef]
57. Loeffler, C.R.; Spielmeyer, A.; Friedemann, M.; Kapp, K.; Schwank, U.; Kappenstein, O.; Bodi, D. Food safety risk in Germany from mislabeled imported fish: Ciguatera outbreak trace-back, toxin elucidation, and public health implications. *Front. Mar. Sci.* **2022**, *9*, 849857. [CrossRef]
58. Dao, H.V.; Uesugi, A.; Uchida, H.; Watanabe, R.; Matsushima, R.; Lim, Z.F.; Jipanin, S.J.; Pham, K.X.; Phan, M.-T.; Leaw, C.P.; et al. Identification of fish species and toxins implicated in a snapper food poisoning event in Sabah, Malaysia, 2017. *Toxins* **2021**, *13*, 657. [CrossRef]
59. Yasumoto, T. Fish poisoning due to toxins of microalgal origins in the Pacific. *Toxicon* **1998**, *36*, 1515–1518. [CrossRef]
60. Turquet, J.; Quod, J.P.; Pannetier, S.; Ramialiharisoa, A.; Ranaivoson, G.; Hurbungs, M.; Jeannoda, V. *Manuel Méthodologique. Suivi et Prévention des Intoxications par Consommation d'Animaux Marins dans le Sud-Ouest de l'Océan Indien*; Réalisé par ARVAM pour le compte de Commission de l'océan Indien: Quatbon, Mauritius, 2000; 50 p, ISBN 99903-71-01-6.
61. Gatti, C.; Oelher, E.; Legrand, A.M. Severe seafood poisoning in French Polynesia: A retrospective analysis of 129 medical files. *Toxicon* **2008**, *51*, 746–753. [CrossRef] [PubMed]
62. Contamination of Sharks, Especially Tiger and Bull Sharks, by Ciguatoxins: Occurrence, Analytical Methods, Human Cases REPORTED and Ethological Information; ANSES Opinion, Collective Expert Appraisal Report. 2015. Available online: https://www.researchgate.net/publication/296282697 (accessed on 29 December 2024).
63. Diogene, J.; Reverte, L.; Rambla-Alegre, M.; del Rio, V.; de la Islesia, P.; Campas, M.; Palacios, O.; Flores, C.; Caixach, J.; Ralijaona, C.; et al. Identification of ciguatoins in a shark involved in a fatal food poisoning in the Indian Ocean. *Sci. Rep.* **2017**, *7*, 8240.
64. Quod, J.P.; Turqet, J. Ciguatera in Reunion Island (SW Indian Ocean): Epidemiology and clinical patterns. *Toxicon* **1995**, *34*, 779–785. [CrossRef]
65. Rajisha, R.; Kishore, P.; Panda, S.K.; Ravishankar, C.N.; Kumar, K.A. Confirmation of ciguatoxin fish poisoning in red snapper *Lutjanus bohar* (Forsskal, 1775) by mouse bioassay. *Fish. Technol.* **2017**, *54*, 287–290.
66. Karunasagar, I.; Turner, A.D.; Maskrey, B.; Robertson, A.; Shivanagouda Hosagoudar, S.; Rai, P.; Adappa, S.; Hiremath, S.; Kogaluru Shivakumaraswamy, S.; Dechraoui Bottein, M.Y.; et al. Report of a major outbreak of ciguatera fish poisoning in Mangalore, India. In Proceedings of the 18th International Conference on Harmful Algae, Nantes, France, 21–26 October 2018; p. 482.
67. Dickey, R.; Jester, E.; Granade, R.; Mowdy, D.; Moncreiff, C.; Rebarchik, D.; Robl, M.; Musser, S.; Poli, M. Monitoring brevetoxins during a gymnodinium breve red tide: Comparison of sodium channel specific cytotoxicity assay and mouse bioassay for determination of neurotoxic shellfish toxins in shellfish extracts. *Nat. Tox.* **1999**, *7*, 157–165. [CrossRef]
68. Manger, R.L.; Leja, L.S.; Lee, S.Y.; Hungerford, J.M.; Hokama, Y.; Dickey, R.W.; Granade, H.R.; Lewis, R.; Yasumoto, T.; Wekell, M.M. Detection of sodium channel toxins: Directed cytotoxicity assays of purified ciguatoxins, brevetoxins, saxitoxins, and seafood extracts. *J. AOAC Int.* **1995**, *78*, 521–527. [CrossRef] [PubMed]
69. Dechraoui, M.Y.; Tiedeken, J.A.; Persad, R.; Wang, Z.; Granade, H.R.; Dickey, R.W.; Ramsdell, J.S. Use of two detection methods to discriminate ciguatoxins from brevetoxins: Application to great barracuda from Florida Keys. *Toxicon* **2005**, *46*, 261–270. [CrossRef] [PubMed]

70. Yogi, K.; Oshiro, N.; Inafuku, Y.; Hirama, M.; Yasumoto, T. Detailed LC-MS/MS analysis of ciguatoxins revealing distinct regional and species characteristics in fish and causative alga from the Pacific. *Anal. Chem.* **2011**, *83*, 8886–8891. [CrossRef] [PubMed]
71. Yogi, K.; Sakugawa, S.; Oshiro, N.; Ikehara, T.; Sugiyama, K.; Yasumoto, T. Determination of Toxins Involved in Ciguatera Fish Poisoning in the Pacific by LC/MS. *J. AOAC Int.* **2014**, *97*, 398–402. [CrossRef]
72. USFDA. Fish and Fishery Products Hazards and Controls Guidance, Fourth Edition. August 2019. Available online: https://www.fda.gov/food/seafood-guidance-documents-regulatory-information/fish-and-fishery-products-hazards-and-controls-guidance (accessed on 26 January 2025).
73. Bennett, C.T.; Robertson, A.; Patterson, W.F., III. First record of the non-indigenous Indo-Pacific damselfish, *Neopomacentrus cyanomos* (Bleeker, 1856) in the northern Gulf of Mexico. *BioInvasions Rec.* **2019**, *8*, 154–166. [CrossRef]
74. Handy, S.M.; Deeds, J.R.; Ivanova, N.V.; Hebert, P.D.; Hanner, R.H.; Ormos, A.; Weigt, L.A.; Moore, M.M.; Yancy, H.F. A single-laboratory validated method for the generation of DNA barcodes for the identification of fish for regulatory compliance. *J. AOAC Int.* **2011**, *94*, 201–210. [CrossRef] [PubMed]
75. Ivanova, N.V.; Zemlak, T.S.; Hanner, R.H.; Herbert, P.D.N. Universal primer cocktails for fish DNA barcoding. *Mol. Ecol. Notes* **2007**, *7*, 544–548. [CrossRef]
76. Ward, R.D.; Zemlak, T.; Innes, B.H.; Herbert, P.D.N. DNA barcoding Australia's fish species. *Philos. Trans. B* **2005**, *360*, 1847–1857. [CrossRef] [PubMed]
77. Kryuchkov, F.; Robertson, A.; Miles, C.O.; Uhlig, S. LC-HRMS and chemical derivatization strategies for the structure elucidation of Caribbean ciguatoxins: Identification of C-CTX-3 and -4. *Mar. Drugs* **2020**, *18*, 182. [CrossRef] [PubMed]
78. FAO (Food and Agriculture Organization of the United Nations). FAO Major Fishing Areas, Indian Ocean, Western. FAO, Fisheries and Aquaculture Department. 2019. Available online: http://www.fao.org/fishery/area/Area51/en (accessed on 26 November 2019).
79. Boisier, P.; Ranaivoson, G.; Rosolofonirina, N.; Andriamahefazafy, B.; Roux, J.; Chanteau, S.; Satake, M.; Yasumoto, T. Fatal mass poisoning in Madagascar following ingestion of a shark (*Carcharhinus leucas*): Clinical and epidemiological aspects and isolation of toxins. *Toxicon* **1995**, *33*, 1359–1364. [CrossRef]
80. Jones, J.D. Observations on fish poisoning in Mauritius. *Proc. R. Soc. Arts. Sci. Mauritius* **1956**, *1*, 367–385.
81. Rajisha, R.; Kishore, P.; Panda, S.K.; Harikrishnan, G.; Ajitha, K.C.; Suresh, M.K.; Chowdhury, L.M.; Ravishankar, C.N.; Ashok Kumar, K. Incidence of ciguatoxin fish poisoning in Trivandrum, India. *Indian J. Fish* **2017**, *64*, 129–133. [CrossRef]
82. Rajisha, R.; Kishore, P.; Panda, S.K.; Jumar, K.A. Ciguatoxin—An emerging biological hazard among reef fishes of India. *Fish. Technol.* **2018**, *55*, 153–167.
83. RASFF. Ciguatera Poisoning Reports Relating to Fish Originating from India. European Commission RASFF Portal. 2016. Available online: https://webgate.ec.europa.eu/rasff-window/portal/?event=searchResultList (accessed on 25 January 2025).
84. Loeffler, C.R.; Tartaglione, L.; Friedemann, M.; Spielmeyer, A.; Kappenstein, O.; Bodi, D. Ciguatera Mini Review: 21st Century Environmental Challenges and the Interdisciplinary Research Efforts Rising to Meet Them. *Int. J. Environ. Res. Public Health* **2021**, *18*, 3027. [CrossRef] [PubMed]
85. Spielmeyer, A.; Loeffler, C.R.; Bodi, D. Extraction and LC-MS/MS analysis of ciguatoxins: A semi-targeted approach designed for fish of unknown origin. *Toxins* **2021**, *13*, 630. [CrossRef]
86. Randall, J.E.; Allen, G.R.; Anderson, W.D., Jr. Revision of the Indo-Pacific lutjanid genus *Pinjalo*, with description of a new species. *Indo-Pac. Fish* **1987**, *14*, 17.
87. Allen, G.R.; Erdmann, M.W. *Reef fishes of the East Indies*; Tropical Reef Research; University of Hawaii Press: Perth, Australia, 2012; Volume I–III.
88. Assadi, H.; Dehghani, R.P. *Atlas of the Persian Gulf and the Sea of Oman Fishes*; Iranian Fisheries Research and Training Organization: Tehran, Iran, 1997; ISBN 964-5513-15-4.
89. Sommer, C.; Schneider, W.; Poutiers, J.M. FAO species identification field guide for fishery purposes. In *The Living Marine Resources of Somalia*; FAO: Rome, Italy, 1996; 376p; Available online: http://www.fao.org/docrep/010/v8730e/v8730e00.htm (accessed on 26 November 2019).
90. Yasumoto, T.; Satake, M.; Fukui, M.; Nagai, H.; Murata, M.; LeGrand, A.-M. A turning point in ciguatera study. In *Toxic Phytoplankton Blooms in the Sea*; Smayada, T.J., Shimizu, Y., Eds.; Elsevier: Amsterdam, The Netherlands, 1993; pp. 455–461.
91. Suzuki, T.; Ha, D.V.; Uesugi, A.; Uchida, H. Analytical challenges to ciguatoxins. *Curr. Opin. Food Sci.* **2017**, *18*, 37–42. [CrossRef]
92. Yasumoto, T.; Satake, M. Chemistry, etiology and determination methods of Ciguatera Toxins. *J. Toxicol. Toxins Rev.* **1996**, *15*, 91–107. [CrossRef]
93. Vernoux, J.P. The mouse ciguatoxin bioassay: Directions for use to control fish for consumption. *Mem. Qld. Mus.* **1994**, *34*, 625–629.
94. Yasumoto, T.; Raj, U.; Bagnis, R. *Seafood Poisoning in Tropical Regions*; Laboratory of Food Hygiene, Faculty of Agriculture, Tohoku University: Miyagi, Japan, 1984; 74p.

95. Lewis, R.J.; Sellin, M.; Street, R.; Holmes, M.J.; Gillespie, N.C. Excretion of ciguatoxin from moray eel (Muraenidae) of the central Pacific. In Proceedings of the Third International Conference on Ciguatera Fish Poisoning, La Parguera, Puerto Rico, Quebec City, QC, Canada, 30 April–5 May 1991.
96. Lombet, A.; Bidard, J.N.; Lazdunski, M. Ciguatoxin and brevetoxins share a common receptor site on the neuronal voltage-dependent Na channel. *FEBS Lett.* **1987**, *219*, 355–359. [CrossRef] [PubMed]
97. Poli, M.; Lewis, R.; Dickey, R.; Musser, S.; Buckner, C.; Carpenter, L. Identification of Caribbean ciguatoxins as the cause of an outbreak of fish poisoning among U.S. soldiers in Haiti. *Toxicon* **1997**, *35*, 733–741. [CrossRef] [PubMed]
98. Manger, R.L.; Leja, L.S.; Lee, S.Y.; Hungerford, J.M.; Wekell, M.M. Tetrazolium-based cell bioassay for neurotoxins active on voltage- sensitive sodium channels: Semiautomated assay for saxitoxin, brevetoxin, and ciguatoxin. *Anal. Biochem.* **1993**, *214*, 190–194. [CrossRef] [PubMed]
99. Lewis, R.J.; Jones, A.; Vernoux, J.P. HPLC/tandem electrospray mass spectrometry for the determination of sub-ppb levels of Pacific and Caribbean ciguatoxins in crude extracts of fish. *Anal. Chem.* **1999**, *71*, 247–250. [CrossRef] [PubMed]
100. Darius, H.T.; Ponton, D.; Revel, T.; Cruchet, P.; Ung, A.; Tchou Fouc, M.; Chinain, M. Ciguatera risk assessment in two toxic sites of French Polynesia using the receptor-binding assay. *Toxicon* **2007**, *50*, 612–626. [CrossRef] [PubMed]
101. Lewis, R.J.; Yang, A.J.; Jones, A. Rapid extraction combined with LC-tandem mass spectrometry (CREM-LC/MS/MS) for the determination of ciguatoxins in ciguateric fish flesh. *Toxicon* **2009**, *54*, 62–66. [CrossRef] [PubMed]
102. Gago-Martinez, A.; Leao, J.M.; Estevez, P.; Castro, D.; Barrios, C.; Hess, P.; Sibat, M. Characterisation of ciguatoxins. EFSA external scientific report. *EFSA Support. Publ.* **2021**, *18*, 6649E. [CrossRef]
103. Dickey, R.W. Ciguatera toxins: Chemistry, toxicology, and detection. In *Seafood and Freshwater Toxins: Pharmacology, Physiology and Detection*; Botana, L.M., Ed.; CRC Press: Boca Raton, FL, USA, 2008; pp. 479–500.
104. Tsumuraya, T.; Sato, T.; Hirama, M.; Fujii, I. Highly sensitive and practical fluorescence sandwich ELISA for ciguatoxins. *Anal. Chem.* **2018**, *90*, 7318–7324. [CrossRef] [PubMed]
105. Loeffler, C.R.; Robertson, A.; Flores Quintana, H.A.; Silander, M.C.; Smith, T.B.; Olsen, D. Ciguatoxin prevalence in four commercial fish species along an oceanic gradient in the US Virgin Islands. *Env. Toxicol.* **2018**, *37*, 1852–1863. [CrossRef]
106. Robertson, A.; Garcia, A.C.; Flores Quintana, H.A.; Smith, T.B.; Castillo, B.F.; Reale-Munroe, K.; Gulli, J.A.; Olsen, D.A.; Hooe-Rollman, J.I.; Jester, E.L.E.; et al. Invasive lionfish (*Pterois volitans*): A potential human health threat for ciguatera fish poisoning in tropical waters. *Mar. Drugs* **2014**, *12*, 88–97. [CrossRef]
107. Hardison, D.R.; Holland, W.C.; Darius, H.T.; Chinain, M.; Tester, P.A.; Shea, D.; Bogdanoff, A.K.; Morris, J.A.; Quintana, H.A.F.; Loeffler, C.R.; et al. Investigation of ciguatoxins in invasive lionfish from the greater Caribbean region: Implications for fishery development. *PLoS ONE* **2018**, *13*, e0198358. [CrossRef] [PubMed]
108. Moreiras, G.; Leao, J.M.; Gago-Martinez, A. Design of experiments for the optimisation of electrospray ionisation in the LC-MS/MS analysis of ciguatoxins. *J. Mass Spec.* **2018**, *53*, 1059–1069. [CrossRef]
109. Graber, N.; Stavinsky, F.; Hoffman, R.; Button, J.; Clark, N.; Martin, S.; Robertson, A.; Hustedt, J. Ciguatera fish poisoning- New York City, 2010-2011. *Morb Mort. Wkly Rep.* **2013**, *62*, 61–65.
110. Radke, E.G.; Grattan, L.M.; Cook, R.L.; Smith, T.B.; Anderson, D.M.; Morris, J.G., Jr. Ciguatera incidence in the US Virgin Islands has not increased over a 30-year time period despite rising seawater temperatures. *Am. J. Trop. Med. Hyg.* **2013**, *88*, 908–913. [CrossRef]
111. Sibat, M.; Herrenknecht, C.; Darius, H.T.; Rou'e, M.; Chinain, M.; Hess, P. Detection of pacific ciguatoxins using liquid chromatography coupled to either low or high resolution mass spectrometry (LC-MS/MS). *J. Chromatogr. A* **2018**, *1571*, 16–28. [CrossRef] [PubMed]
112. Tudó, A.; Rambla-Alegre, M.; Flores, C.; Sagristà, N.; Aguayo, P.; Reverte, L.; Campas, M.; Gouveia, N.; Santos, C.; Andree, K.B.; et al. Identification of New CTX Analogues in Fish from the Madeira and Selvagens Archipelagos by Neuro-2a CBA and LC-HRMS. *Mar. Drugs* **2022**, *20*, 236. [CrossRef] [PubMed]
113. Chan, T.Y. Characteristic features and contributory factors in fatal ciguatera fish poisoning-implications for prevention and public education. *Am. J. Trop. Med. Hyg.* **2016**, *94*, 704–709. [CrossRef]
114. Visciano, P.; Schirone, M.; Berti, M.; Milandri, A.; Tofalo, R.; Suzzi, G. Marine Biotoxins: Occurrence, Toxicity, Regulatory Limits and Reference Methods. *Front. Microbiol.* **2016**, *7*, 1051. [CrossRef]
115. Bagnis, R.; Kuberski, T.; Laugier, S. Clinical observations on 3009 cases of ciguatera (fish poisoning) in the South Pacific. *Am. J. Trop. Med. Hyg.* **1979**, *28*, 1067–1073. [CrossRef] [PubMed]
116. Nicholson, G.M.; Lewis, R.J. Ciguatoxins: Cyclic polyether modulators of voltage-gated ion channel function. *Mar. Drugs* **2006**, *4*, 82–118. [CrossRef]
117. Perkins, R.A.; Morgan, S.S. Poisoning, envenomation and trauma from marine creatures. *Am. Fam. Physician* **2004**, *69*, 885–890. [PubMed]
118. Senthilkumaran, S.; Balamurgan, N.; Suresh, P.; Thirumalaikolundusubramanian, P. Painful ejaculation, Something fishy. *Saudi. Med. J.* **2010**, *31*, 451–452. [PubMed]

119. Anon. EC regulation no 853/2004 of the European parliament and of the council of 29 April 2004 laying down specific rules for the organisation of official controls on products of animal origin intended for human consumption. *Off. J. Eur. Union* **2003**, L139/55.
120. USFDA. *Guidance for Industry: Purchasing Reef Fish Species Associated with the Hazard of Ciguatera Fish Poisoning*; USFDA Guidance; FDA: Silver Spring, MD, USA, 2013.
121. Hossen, V.; Solino, L.; Leroy, P.; David, E.; Velge, P.; Dragacci, S.; Krys, S.; Flores Quintana, H.; Diogene, J. Contribution to the risk characterization of ciguatoxins: LOAEL estimated from eight ciguatera fish poisoning events in Guadeloupe (French West Indies). *Env. Res.* **2015**, *143*, 100–108. [CrossRef] [PubMed]
122. Spielmeyer, A.; Loeffler, C.R.; Kappenstein, O. Identical Ciguatoxin-3C group profiles in *Lutjanus bohar* from the Pacific and Indian Oceans—indicating the need to re-evaluate geographical CTX classifications. *Front. Mar. Sci.* **2022**, *9*, 937438. [CrossRef]
123. Ryu, A.H.; Eckalbar, W.L.; Kreimer, A.; Yosef, N.; Ahituv, N. Use antibiotics in cell culture with caution: Genome-wide identification of antibiotic-induced changes in gene expression and regulation. *Sci. Rep.* **2017**, *7*, 7533. [CrossRef]
124. Xiao, Y.; Xiong, T.; Meng, X.; Yu, D.; Xiao, Z.; Song, L. Different influences on mitochondrial function, oxidative stress and cytotoxicity of antibiotics on primary human neuron and cell lines. *J. Biochem. Mol. Toxicol.* **2018**, *33*, e22277. [CrossRef] [PubMed]
125. Murray, J.S.; Boundy, M.J.; Selwood, A.I.; Harwood, D.T. Development of an LC-MS/MS method to simultaneously monitor maitotoxins and selected ciguatoxins in algal cultures and P-CTX-1B in fish. *Harmful Algae* **2018**, *80*, 80–87. [CrossRef] [PubMed]
126. Turner, A.D.; Boundy, M.J.; Dhanji-Rapkova, M. Single laboratory validation of a LC-MS/MS method for quantitation of Tetrodotoxins in mussels and oysters. *J. AOAC Int.* **2017**, *100*, 1–14. [CrossRef]
127. Turner, A.D.; Dhanji-Rapkova, M.; Fong, S.Y.T.; Hungerford, J.; McNabb, P.S.; Boundy, M.J.; Harwood, T. Ultra-high performance hydrophilic interaction liquid chromatography with tandem mass spectrometry method for the determination of paralytic shellfish toxins and tetrodotoxin in mussels, oysters, clams, cockles and scallops: Collaborative study. *J. AOAC Int.* **2020**, *103*, 533–562. [CrossRef] [PubMed]

marine drugs

MDPI

Article

Monthly Variation of Tetrodotoxin Levels in Pufferfish (*Lagocephalus sceleratus*) Caught from Antalya Bay, Mediterranean Sea

Ali Rıza Kosker [1,*], Merve Karakus [2], Panagiota Katikou [3,*], İsmail Dal [2], Mustafa Durmus [1], Yılmaz Ucar [4], Deniz Ayas [5] and Fatih Özogul [1,6]

1 Department of Seafood Processing Technology, Faculty of Fisheries, Cukurova University, 01330 Adana, Turkey; mdurmus@cu.edu.tr (M.D.); fozogul@cu.edu.tr (F.Ö.)
2 Mediterranean Fisheries Research Production and Training Institute, Demre Unit, 07570 Antalya, Turkey; merve.karakus@tarimorman.gov.tr (M.K.); ismail.dal@tarimorman.gov.tr (İ.D.)
3 Veterinary Research Institute of Thessaloniki, Hellenic Agricultural Organization—DIMITRA, Ktima Thermis, 57001 Thessaloniki, Greece
4 Vocational School of Aladag, Department of Forestry, Cukurova University, 01720 Adana, Turkey; yucar@cu.edu.tr
5 Fisheries Faculty, Mersin University, 33169 Mersin, Turkey; ayasdeniz@mersin.edu.tr
6 Biotechnology Research and Application Center, Cukurova University, 01330 Adana, Turkey
* Correspondence: akosker@cu.edu.tr (A.R.K.); pkatikou@otenet.gr (P.K.)

Abstract: The silver-cheeked toadfish (*Lagocephalus sceleratus*), an invasive alien pufferfish species that has rapidly settled throughout the Mediterranean region, poses significant threats not only to native marine species and fisheries but also to public health due to the tetrodotoxin (TTX) they harbor. In this study, TTX concentrations in *L. sceleratus* from Antalya Bay in the Northeastern Mediterranean Sea were investigated using Q-TOF-LC-MS on a monthly basis over a one-year period. Pufferfish were caught by angling from May 2018 to April 2019. The TTX levels in three different tissues (gonads, liver, and muscle) of 110 pufferfish in total were determined in both male and female individuals caught for 11 months. The highest TTX mean levels generally occurred in the gonads and the lowest in the muscle samples. As regards the maximum TTX contents, the highest concentrations determined were 68.2, 34.2, and 7.8 μg/g in the gonad, liver, and muscle tissues, respectively. The highest levels were generally observed in late autumn to winter (especially in November and December) in all tissues from both genders. Female individuals were generally found to be more toxic than male individuals. The TTX levels found confirm that the consumption of *L. sceleratus* from Antalya Bay remains dangerous throughout the year, and thus *L. sceleratus* constantly constitutes an important risk source for public health.

Keywords: silver-cheeked toadfish; tetrodotoxin; *Lagocephalus sceleratus*; monthly variation

1. Introduction

Lessepsian species migrating from the Red Sea to the Mediterranean via the Suez Canal are increasing. While some of these alien species are of economic value, species such as pufferfish are considered invasive and poisonous. There are currently eleven established pufferfish species in the Mediterranean [1], among which *Lagocephalus sceleratus* is considered to be the most dangerous and invasive one due to its negative effects on autochthonous species, fishing activities and public health [2].

L. sceleratus was first reported in 2005 off the coast of Türkiye [3] and then spread rapidly. In a short time, relevant occurrence reports followed from Greece, Croatia, Spain, Libya, Tunisia, Egypt, and Italy [4–10]. The rapid spread of *L. sceleratus* has negatively affected native species in the Mediterranean ecosystem [11]. A significant part of the ecological success of *L. sceleratus* is due to the fact that it has one of the most advanced types of teeth in the animal kingdom. The 'first generation teeth' are covered with recurrent bands of teeth that are continuously renewed by stem cells [12]. These tooth bands fuse to form the upper and lower plates, which together form a beak. This strong beak-like structure allows them to crush and slice even very tough prey organisms such as decapods and bivalves [1,13]. Owing to this unique tooth structure, it can also damage the fishing gear of commercial and recreational fisheries, resulting in significant socio-economic effects on fishermen in affected countries, such as Türkiye [14].

In the wider Mediterranean area, pufferfish species have attracted remarkable attention in the last few decades due to their invasion potential as well as the economic damage and poisoning they can cause. For example, the impact of pufferfish on small-scale fisheries in Türkiye is estimated to lead to an annual loss of EUR 2 million (loss of fishing gear and labor) [15]. Similarly, in Cyprus, pufferfish expansion is estimated to generate an annual economic loss of EUR 4173.53 per small-scale fisher. It is also noteworthy that many small-scale commercial fishers in Cyprus have changed their fishing strategies (e.g., using larger nets, fishing for shorter periods of time, and relocating) due to the damage caused by pufferfish to fishing gear [16].

In addition to the ecological and economic impacts of pufferfish, the risks posed by *L. sceleratus* to public health are even more worrisome, as this species has also been implicated in poisonings in the Mediterranean, some of which have resulted in patient deaths [17–20]. This is attributed to tetrodotoxin (TTX), one of the most potent toxins known in the world, which is commonly present in many pufferfish. As such, in the context of public health protection, the EU and many other Mediterranean countries have introduced legislation to ban the placing on the market of fishery products derived from poisonous fish belonging to the families Tetraodontidae, Molidae, Diodontidae, and Canthigasteridae [2,21–23].

TTX is a compound with a low molecular weight (319.27) and non-protein structure. It is a water-soluble, colorless, odorless heat-stable toxin. In case of poisoning, symptoms begin to be observed within 2–4 h, and poisoning can result in death because there is no known antidote for this toxin [17]. The origin of TTX is still debated. It is thought to have its origins in bacteria belonging to the *Proteobacteria* phylum, which includes *Pseudomonas*, *Pseudoalteromonas*, and *Vibrio*. However, there are occasional reports of several other bacterial phyla (*Actinobacteria, Bacterioides, Firmicutes,* and *Proteobacteria*) being considered as potential sources of TTX [24,25]. These TTX-producing bacteria, such as *Vibrio, Pseudomonas, Aeromonas, Alteromonas, Nocardiopsis, Bacillus, Shewanella,* and *Roseobacter*, have been discovered in various locations within several aquatic species, including subcutaneous mucus, ovaries, and the gastrointestinal tract [26–29]. Additionally, some evidence in the literature suggests their association with specific dinoflagellate blooms, such as *Alexandrium tamarense* or *Prorocentrum cordatum* [30–32].

TTX and its analogues (TTXs) are found in a taxonomically diverse group of animals, living in both terrestrial and aquatic (marine, freshwater, and brackish) environments; however, pufferfish constitute the most prominent vectors of this toxin [17,33,34]. At the worldwide level, pufferfish are represented by 29 different genera and about 200 different species, and most of these species contain TTX [35]. Some of these pufferfish species that inhabit the Mediterranean also possess TTX at levels that are toxic and potentially deadly [17,36–39]. To date, most studies on pufferfish toxicity in the Mediterranean have focused particularly on *L. sceleratus* [2,8,33,36,38–44]. However, there are also scientific studies on the TTX levels of *T. flavimaculosus*, *L. guentheri*, *L. suezensis*, and *Sphoeroides pachygaster* [2,8,37,45], whereas poisoning cases caused by the consumption of pufferfish have also been reported in different Mediterranean countries [17,46,47]. These poisoning cases mainly stemmed from misidentification and a lack of knowledge on pufferfish. People living in the Mediterranean coastal regions often confuse the pufferfish species encountered in this area with the Japanese pufferfish, which constitute a delicacy (fugu). Nevertheless, it is worth emphasizing that the pufferfish species prevalent in the Mediterranean region do not comply with the classification endorsed by the Japanese Ministry of Health, Labor, and Welfare, designating the pufferfish that are safe for human consumption [48].

The studies conducted so far on the toxin content of pufferfish in the Mediterranean region have generally focused on investigating single-sampling or seasonal TTX contents [2,8,33,36,38–44]. Considering, however, the extremely harmful effects that pufferfish may have on public health, it would also be beneficial to determine the monthly changes in TTX levels of these organisms. Poisoning cases in many countries appear on an occasional basis, indicating the importance of monitoring the variation of TTX levels in pufferfish over time. On the other hand, clinical trials of anti-cancer and anti-pain drugs containing TTX are ongoing [42]. As such, exploitation of *L. sceleratus* for TTX purification could be of help to stabilize their population and reduce their impact on the environment. It would therefore be useful to investigate the variation in TTX levels in pufferfish throughout the year in order to target fishing efforts towards the months when TTX levels are highest. To address these considerations, the present study investigated the monthly and sex-related changes in TTX levels of the population of *L. sceleratus* in Antalya Bay.

2. Results and Discussion

Only sexually mature pufferfish were included in this study. The maximum mean weights of the female and male samples were 2062.2 g and 1665.4 g, respectively, while the maximum mean lengths of the females and males were 54.5 cm and 52.5 cm, respectively (Table 1).

Table 1. Monthly length and weight of pufferfish samples (n = 5, mean ± SD).

Months	Gender	Length (cm)	Weight (g)
February	♀	51.6 ± 2.96	1467.4 ± 253.49
	♂	49.5 ± 3.73	1309.8 ± 339.97
March	♀	52.9 ± 11.79	1702.7 ± 1058.14
	♂	52.5 ± 1.19	1665.4 ± 320.84
April	♀	54.5 ± 2.33	2062.2 ± 431.56
	♂	47.8 ± 1.63	1338.0 ± 249.61
May	♀	39.5 ± 1.95	750.9 ± 82.27
	♂	49.0 ± 6.07	1595.0 ± 539.36
June	♀	45.6 ± 1.45	1110.7 ± 110.86
	♂	40.9 ± 6.33	835.5 ± 396.92
July	♀	52.6 ± 2.99	1870.7 ± 472.26
	♂	41.1 ± 1.15	715.2 ± 331.30
August	♀	54.4 ± 4.48	1941.2 ± 305.86
	♂	44.2 ± 8.01	951.8 ± 326.46
September	♀	45.5 ± 5.69	1146.7 ± 484.61
	♂	47.3 ± 1.60	1243.3 ± 83.26
October	♀	53.0 ± 1.78	1664.1 ± 173.40
	♂	44.5 ± 3.90	1048.8 ± 281.14
November	♀	42.3 ± 0.46	844.9 ± 20.68
	♂	49.2 ± 0.44	1260.1 ± 57.30
December	♀	46.4 ± 3.26	1112.4 ± 323.96
	♂	49.4 ± 8.01	1403.99 ± 445.84
Overall (n = 110)	♀	47.8 ± 4.66	1316.0 ± 400.45
	♂	47.7 ± 4.60	1313.0 ± 399.49

The monthly and sexual variation of TTX levels in the muscle, liver, and gonads of *L. sceleratus* caught in Antalya Bay are presented in Table 2, whereas a graphical comparison of TTX contents in different tissues of male and female individuals is shown in Figure 1. The TTX levels measured were in the range of 0.8 to 68.2 µg/g. The highest TTX mean levels were present in the gonads and the lowest in the muscle samples. The maximum toxin concentrations determined per tissue were 68.2, 34.2, and 7.8 µg TTX/g in the gonad, liver, and muscle, respectively. The highest TTX concentrations for all tissues and both sexes were observed in November, with female individuals generally harboring more toxins than male ones. However, in some months, especially between October and March, liver TTX contents were higher in males than females. In terms of monthly toxin levels, the TTX levels of most female individuals were found to be higher than those of males.

Table 2. Monthly TTX levels (µg/g) in different tissues of *L. sceleratus* (mean ± SD).

Months	Gender	N	Muscle	Liver	Gonad
February	♀	5	2.8 ± 0.04 ^{e,x}	8.1 ± 0.34 ^{f,y}	21.0 ± 0.87 ^{c,x}
	♂	5	1.3 ± 0.03 ^{f,y}	21.6 ± 6.44 ^{a,x}	20.2 ± 1.86 ^{bc,x}
March	♀	5	1.5 ± 0.29 ^{f,y}	0.8 ± 0.03 ^{h,y}	6.8 ± 0.89 ^{d,x}
	♂	5	3.6 ± 1.08 ^{bc,x}	15.8 ± 4.37 ^{b,x}	6.4 ± 0.85 ^{de,x}
April	♀	5	5.6 ± 0.14 ^{b,x}	13.7 ± 0.09 ^{d,x}	25.0 ± 0.12 ^{c,x}
	♂	5	3.8 ± 0.01 ^{b,y}	7.6 ± 1.28 ^{cd,y}	2.7 ± 0.00 ^{e,y}
May	♀	5	3.5 ± 0.49 ^{d,x}	7.4 ± 0.45 ^{f,x}	16.2 ± 1.42 ^{cd,x}
	♂	5	2.6 ± 0.06 ^{de,y}	9.4 ± 1.26 ^{cd,x}	1.4 ± 0.07 ^{e,y}
June	♀	5	4.0 ± 0.36 ^{d,x}	14.4 ± 1.42 ^{cd,x}	8.6 ± 0.76 ^{d,x}
	♂	5	3.7 ± 0.37 ^{b,x}	9.2 ± 0.57 ^{cd,y}	3.5 ± 0.06 ^{e,y}
July	♀	5	2.5 ± 0.68 ^{e,x}	11.1 ± 0.64 ^{e,x}	7.5 ± 0.43 ^{d,y}
	♂	5	2.9 ± 0.32 ^{cd,x}	7.2 ± 1.09 ^{cd,y}	25.9 ± 5.70 ^{b,x}
August	♀	5	4. 8 ± 0.11 ^{c,x}	16.8 ± 1.67 ^{bc,x}	23.8 ± 0.80 ^{c,x}
	♂	5	2.6 ± 0.22 ^{de,y}	12.0 ± 1.15 ^{bc,y}	19.3 ± 6.53 ^{bc,x}
September	♀	5	3.7 ± 0.06 ^{d,x}	14.9 ± 0.55 ^{bcd,x}	38.2 ± 5.48 ^{b,x}
	♂	5	2.8 ± 0.41 ^{d,y}	11.8 ± 0.16 ^{bc,y}	9.7 ± 1.27 ^{de,y}
October	♀	5	1.5 ± 0.15 ^{f,y}	3.5 ± 0.75 ^{g,y}	9.0 ± 3.69 ^{d,x}
	♂	5	1.9 ± 0.01 ^{ef,x}	5.8 ± 0.52 ^{d,x}	7.6 ± 1.20 ^{de,y}
November	♀	5	5.5 ± 0.18 ^{b,x}	17.4 ± 3.11 ^{b,y}	65.9 ± 3.34 ^{a,x}
	♂	5	5.9 ± 0.30 ^{a,x}	25.3 ± 2.48 ^{a,x}	68.2 ± 11.40 ^{a,x}
December	♀	5	7.8 ± 0.07 ^{a,x}	34.2 ± 2.63 ^{a,x}	44.2 ± 15.61 ^{b,x}
	♂	5	6.0 ± 0.37 ^{a,y}	15.6 ± 1.22 ^{b,y}	12.9 ± 0.21 ^{cd,y}
Overall	♀	55	3.9 ± 1.92	12.9 ± 8.88	24.4 ± 19.13
	♂	55	3.4 ± 1.49	12.9 ± 6.19	15.9 ± 18.36

Different letters (a,b,c,d,e,f) in the same column represent the significant difference between TTX values in different months for each tissue ($p < 0.05$). Different letters (x,y) in the same column for each month show significant differences for both genders ($p < 0.05$). N: number of samples.

In terms of the TTX levels measured in the tissues throughout the year, the order was found to be gonad > liver > muscle. This is in agreement with other studies conducted by Christidis et al. [39], Rambla-Alegre et al. [8], Alkassar et al. [49] and Kosker et al. [2,42], where the highest toxin concentrations were also observed in the gonads and the lowest toxin levels were detected in the muscle tissue. The TTX levels in the gonads were within the range of 1.4–68.2 µg/g, similarly to the ranges reported for *L. sceleratus* individuals from the Aegean Sea [33,36,41,50], the Adriatic Sea [51], the Eastern Mediterranean [42,43], the Western Mediterranean [8] and the Southern Mediterranean [39]. The TTX levels in the liver were in the range of 0.78–34.19 µg/g, also in accordance with the levels detected in *L. sceleratus* from the Aegean Sea [36,41], the Eastern Mediterranean [40,42] and the Southern Mediterranean [39]. However, as can be seen in Table 3, the TTX levels in the livers of pufferfish caught in Antalya Bay were higher than the values reported by Reverté et al. from the Aegean Sea [33], Kosker et al. from the North East Mediterranean [2], Rambla-Alegre et al. from the West Mediterranean [8] and Acar et al. from the East Mediterranean [43], while they were lower than those reported in the livers of *L. sceleratus* from the southern Aegean by Anastasiou et al. [50]. Finally, the muscle tissue TTX concentrations (1.3–7.80 μg/g) were equivalent to those reported by Katikou et al. from the Aegean Sea [41], Kosker et al. from the North East Mediterranean [2] and Christidis et al. from the South Mediterranean [39]. On the other hand, the TTX levels determined in the present study in the gonads were lower compared to some other studies [40,43,49,50] conducted nearby. This could be attributed to differences in the sampling period, which in the other studies was only seasonal and probably involved one single sample per season, in contrast to the monthly sampling of the present study. This hypothesis is further supported by the notable monthly fluctuations in TTX concentrations, indicating that the sampling intervals' periodicity may hold a significant role. However, differences in environmental conditions in the respective sampling periods/years, such as temperature, availability of nutrients, etc., as well as specimens' sizes and ages, may well account for the variation of toxin levels observed between individual studies.

Table 3. TTX levels in different tissues of *L. sceleratus* from the Mediterranean (μg/g), reported using different liquid chromatography–mass spectrometry (LC-MS) methods.

Region	Muscle	Gonads	Liver	N	Analysis Method	Reference
Aegean Sea	<0.3–3.5	0.5–46.3	<0.3–44.2	43	LC-ESI-CID-MS/MS	Rodríguez et al. [36]
	ND	4.8	8.1	2	LC-MS/MS	Reverté et al. [33]
	0.5–2.1	NA	NA	2	LC-HRMS	Leonardo et al. [38]
	0.2–21.7	1.2–85.2	1.1–239.8	37	LC-MS/MS	Anastasiou et al. [50]
North–East Mediterranean	ND–2.8	0.4–52.1	ND–46.2	16	LC-MS/MS	Kosker et al. [42]
	0.7–5.1	0.7–35.6	0.9–21.1	20	Q-TOF LC-MS	Kosker et al. [2]
East Mediterranean	0.1–0.6	0.3–253.0	1.4–46. 7	3	LC-MS/MS	Hassoun et al. [40]
	0.1–3.4	0.2–80.0	0.1–25.4	20	LC-MS/MS	Acar et al. [43]
West Mediterranean	1.0	26.0	3.1	1	LC-MS/MS	Rambla-Alegre et al. [8]
	1.0	25.2	5.4	1	LC-HRMS	Rambla-Alegre et al. [8]
South Mediterranean	0.0–20.7	0.3–189.0	0.0–104.41	83	LC-MS/MS	Christidis et al. [39]
	7.6–36.5	2.1–1324.4	38.9–188.2	3	LC-MS/MS	Alkassar et al. [49]
Antalya Bay	1.3–7.8	1.4–68.2	0.8–34.2	110	Q-TOF LC-MS	Present Study

N, Number of individuals analyzed. LC-MS/MS, Liquid Chromatography with tandem mass spectrometry. LC-HRMS, Liquid chromatography–high-resolution mass spectrometry. LC-ESI-CID-MS/MS, liquid chromatography coupled to electrospray ionization mass spectrometry operating in the conventional mode in addition to low-energy collision dissociation tandem mass spectrometry. Q-TOF LC-MS, liquid chromatography quadrupole time-of-flight mass spectrometry. ND, not detected. NA, not Analyzed.

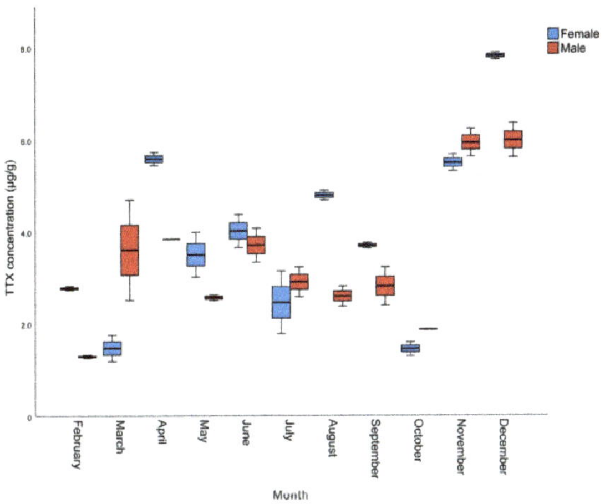

(a)

Figure 1. *Cont.*

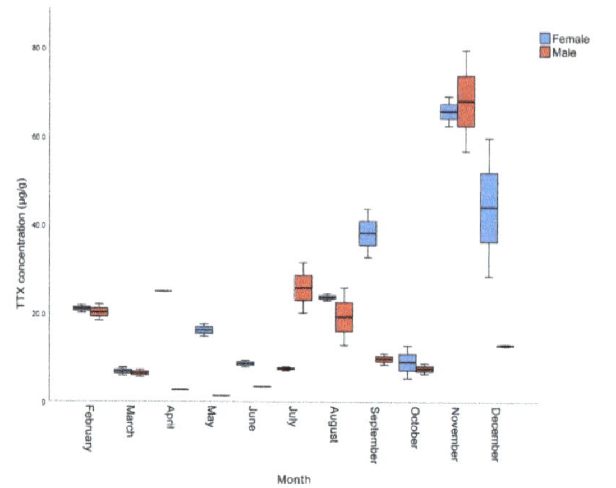

(b)

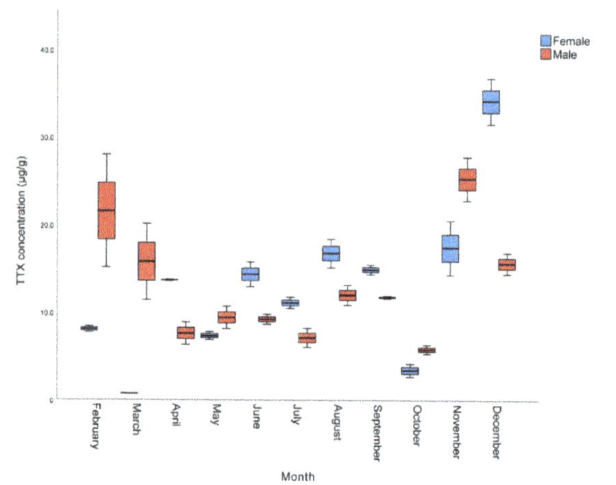

(c)

Figure 1. Monthly variation of TTX levels (µg/g) of male and female *L. sceleratus* individuals in (**a**) muscle, (**b**) liver, and (**c**) gonads (n = 5, error bars represent SD).

Considering monthly changes, a statistically significant increase was observed in the TTX levels in all tissues, especially towards the late autumn to early winter months ($p < 0.05$). The mean TTX levels in the muscle, liver, and gonads of male pufferfish in November were 5.9, 25.3 and 68.2 µg/g, respectively. In females, the mean TTX levels in the gonads was 65.9 µg/g in November, while the highest TTX levels in the muscle and liver tissues (7.8 and 34.2 µg/g), respectively, were observed in December. The spawning season of *L. sceleratus* in the Mediterranean includes a very wide period and can last from April to September [1,44,52]. It has been reported that the TTX molecule is a reproductive adaptation for pufferfish and that they show maximum toxicity in the spring and summer seasons, which are the breeding seasons of pufferfish [53,54]. However, Yu and Yu [55] reported that *Takifugu niphobles* and *Takifugu alboplumbeus* pufferfish are not toxic during their spawning period. Similarly, according to the findings of the studies conducted by Katikou et al. [41], Kosker et al. [2,42] and Acar et al. [43], *L. sceleratus* living in the Mediterranean seem to contain higher TTX in the late autumn months after the spawning period. Interestingly, in these cases, pufferfish show higher toxicity outside of their breeding season, a phenomenon that may be specific to the Mediterranean ecosystem.

Overall, the TTX levels of *L. sceleratus* caught in Antalya Bay are generally in agreement with those reported from different parts of the Mediterranean by means of analytical methods (Table 3). On the other hand, specifically where potential poisoning caused by muscle tissue consumption in humans and wild animals or pets is concerned, almost all of the pufferfish examined in this study were toxic, that is, exceeding the sole existing safety threshold of 10 mouse units (MU) TTX eq/g (equivalent to 2.2 μg TTX eq/g) established by the Japanese authorities [56]. Similarly, high concentrations of TTX have been reported in other Mediterranean pufferfish species, *L. suezensis* and *T. flavimaculosus* [2,37]. However, the main responsible species for poisoning cases in the Mediterranean is *L. sceleratus* [17]. At earlier times, these latter pufferfish were sold at fish stalls, probably due to their larger individual size and their widespread presence in almost all Mediterranean countries [57]. Currently, the trade of pufferfish is restricted by legal regulations in Türkiye, the EU, and other Mediterranean countries [22,23], but still, several poisoning cases resulting from the consumption of *L. sceleratus* have occurred in some Mediterranean countries [17,19,46]. These cases are commonly the result of accidental consumption of pufferfish caught during amateur fishing by people living in coastal areas. Nevertheless, some of the poisoning incidents have been caused by pufferfish caught professionally by fishing lines, especially in the Antalya Bay [17].

As aforementioned, *L. sceleratus* is not listed as an edible pufferfish species in Japan [28], mainly due to the fact that its muscle commonly contains TTX at levels surpassing the safety threshold of 2.2 μg TTX eq/g. This limit was established to ensure the safe consumption of pufferfish meat in Japan but has also been employed as a toxicity assessment criterion for *L. sceleratus* samples from the Mediterranean Sea [17,20]. Taking into account this threshold value, the TTX levels found in the present study generally indicate a substantial risk of foodborne poisoning associated with the potential ingestion of *L. sceleratus* edible tissues (muscle, liver, and gonads) caught by angling in Antalya Bay, regardless of the time period of sampling. As such, *L. sceleratus* should continue to be considered as an important source of risk in terms of public health in addition to its ecological and economic consequences.

It is noteworthy to mention that currently, TTX levels in seafood are a subject of debate, especially in the European Union. The European Food Safety Authority (EFSA) Panel on Contaminants in the Food Chain (CONTAM) has concluded that a concentration of 44 μg TTX/kg of shellfish meat, as regards marine bivalves and gastropods, is not expected to lead to adverse effects in humans [29]. This threshold value is notably more conservative compared to the relevant Japanese safety limit for the assessment of edible pufferfish. Subsequently, newer toxicological data obtained by administering TTX in mice through feeding, instead of gavage, but following the same calculation logic as that of the EFSA Panel, indicated that a higher concentration of 560 μg TTX/kg of shellfish meat would be expected not to lead to adverse effects in humans [58]. In addition, the EFSA opinion has also suggested exploring whether paralytic shellfish poisoning toxins (saxitoxins, STXs) and TTXs should be combined into one health-based guidance value due to their similarities as regards toxic effects and mode of action [29]. In this context, a recent study has concluded that the current regulatory threshold of 800 μg STX eq./kg [22], which is much higher than the EFSA proposed limit for TTXs alone, would be fit for purpose for the combined presence of STXs and TTXs [59]. Nevertheless, regardless of which of the above threshold values is used, almost all the pufferfish tested in our study would have been deemed unsuitable for human consumption.

3. Materials and Methods

3.1. Tetrodotoxin Standard

The TTX standard (1 mg powder, purity 99%) used for the instrumental toxin analysis was purchased from Abcam Biochemicals (Cambridge, UK). The TTX standard was diluted with methanol (Merck, Darmstadt, Germany) containing 0.01M acetic acid to obtain the stock TTX standard. Working standard solutions at concentrations of 0.05, 0.1, 0.5, and 2 µg/mL were prepared from the stock solution to obtain a calibration curve (R^2 = 0.9994). The standards were kept at −20 °C until used.

3.2. Fish Collection, Measurements, and Identification

The pufferfish were caught by angling at monthly intervals between May 2018 and April 2019, except for January 2019, where sampling was not possible due to weather conditions and technical difficulties. The fish were transported to the laboratory on ice. Length–weight measurements of the pufferfish caught in all months were obtained (Table 1), and sex determinations were made by examining the gonads of fish with the help of a microscope. A total of 110 pufferfish, 5 male and 5 female, were examined for each month. Only individuals that had reached sexual maturity and sizes likely to be caught by amateur fishermen were used.

3.3. Preparation of Samples and Toxin Extraction

Fish samples were dissected to obtain their dorsal muscle (carefully avoiding the gastrointestinal tract), gonads, and liver. TTX extractions in the liver, gonads, and muscle tissues were performed according to the method of Silva et al. [34]. Briefly, 1 g of each tissue was extracted by adding 3 mL of methanol containing 1% acetic acid, and the mixture was homogenized with an Ultra Turrax (IKA T25 Digital Ultra Turrax, Staufen, Germany) at 7200 rpm for 10 min. Then, the homogenized extracts were placed in an ultrasonic bath (Bandelin Sonorex RK 100, Berlin, Germany) at 100 Hz for 10 min. The extracts were kept at room temperature for 15 min and were then centrifuged at 4500 rpm and 4 °C, for 20 min (Hettich Zentrifugen, Universal 32R, Tuttlingen, Germany). The supernatants were collected, and the same procedure was repeated for the pellet residue. The supernatants from both extractions were combined, and the extracts were made up to 7 mL. The extracts were vortex-mixed before proceeding to solid–phase extraction (SPE). After that, 1 mL of the extract was cleaned by running it through a 500 mg/3 mL C18 solid-phase extraction (SPE) cartridge (Supelco, Bellefonte, PA, USA). The sample was eluted with 10 mL of 100% methanol and diluted with the same solvent to a final volume of 12 mL. It was then vortex-mixed and evaporated to dryness using a rotary evaporator. The dry residue was reconstituted with 1 mL of methanol, filtered through 0.45 µm membrane filters, and transferred to vials for analysis.

3.4. Tetrodotoxin Analysis

TTX analyses were performed using an Agilent 6545 Accurate-Mass Q-TOF mass spectrometer coupled to an Agilent 1260 HPLC system (Agilent Technologies, Inc., Santa Clara, CA, USA). The Q-TOF LC/MS analysis was performed according to Kosker et al. [37]. A Poroshell 120 HILIC (3.0 × 50 mm; 2.7 µm) column (Agilent Technologies, Inc., Santa Clara, CA, USA) was used for the analyses. The toxin was separated in the column using two different mobile phases. Mobile phase A was 20 mM ammonium acetate in distilled water (Sigma-Aldrich), and mobile phase B was 20 mM ammonium acetate in acetonitrile (Sigma-Aldrich). The analysis was completed in 8 min, with the retention time of TTX being 4.2 min (Figure 2). A gradient program was used as follows: the first 2.5 min were set as 3% mobile phase A, 97% mobile phase B, then for 2 min 30% mobile phase A, 70% mobile phase B, and again between 4.5 and 8 min 3% mobile phase A, 97% mobile phase B. The column temperature was 20 °C, and the injection volume was 10 µL. The LC system was operated in the positive ion mode with the ESI (electrospray ionization) interface using the following parameters: drying gas flow, 10.0 L/min; nebulizer pressure, 40 psi; gas drying temperature, 325 °C; sheath gas temperature, 400 °C; sheath gas flow, and nitrogen at 12 L/min. The scanning was performed in the range of m/z 50–500. Only the parent TTX compound was determined, as other previous studies by Christidis et al. [39] and Bane et al. [60] have confirmed that it is the most abundant among all TTX analogues in *L. sceleratus* pufferfish, whereas the limited information on the toxic potential of individual

TTX analogues, although probably present in the samples, would not allow for a solid evaluation of their contribution in overall toxicity. The Q-TOF LC/MS operating conditions were first optimized using the TTX standard. The method's limit of detection (LOD; S/N > 3) was calculated at 0.008 µg/g and the limit of quantification (LOQ; S/N > 10) at 0.026 µg/g. The working mass range was from *m/z* 50 to 500 in full-scan acquisition mode. Quantitation was performed by using the MS mode; the accurate measured mass of the detected ion was $[M + H]^+$, *m/z* 320.1088. The peaks were identified by retention time (4200–4270 min) and exact mass (mass window ±5 ppm, mass accuracy 0.93 ppm, monoisotopic pattern of the signals). Data processing was carried out using the Mass Hunter software. All Q-TOF LC/MS analyses were performed in triplicate.

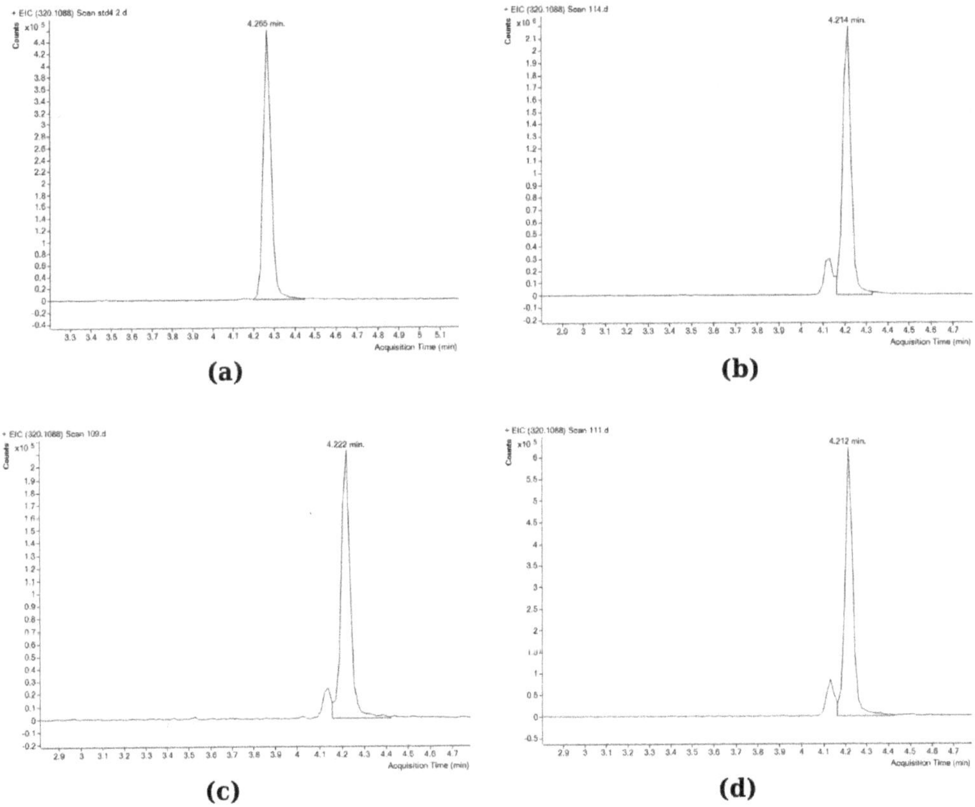

Figure 2. Q-TOF chromatograms of standard and different samples: (**a**) TTX standard (2 µg/mL); (**b**) gonad; (**c**) muscle; (**d**) liver (the smaller peak left to TTX in the tissue samples corresponds to 4-epiTTX, which was not included in the analysis). Accurate mass *m/z*: 320.1088; retention time window: 4200–4270 min; mass accuracy: 0.93 ppm; mass window: ± 5ppm.

3.5. Statistical Analysis

Results are reported as the mean and standard deviation of the measurements. SPSS version 17.0 (SPSS Inc., Chicago, IL USA) was used for the statistical evaluations of the monthly changes of TTX values among tissues and between males and females captured in the same month (Figure S1). To identify significant differences between monthly values of TTX levels in tissues, a one-way analysis of variance (ANOVA) combined with Duncan's multiple range test comparisons at $p < 0.05$ were performed. The *t*-test was applied between different genders belonging to the same month.

4. Conclusions

The present study extends our knowledge on the TTX contents of *L. sceleratus* at monthly intervals and provides further data on toxin levels required for the establishment of more efficient risk management practices for both public health and the fishing industry around Mediterranean countries. Our results demonstrate that the Mediterranean *L. sceleratus* remains poisonous and unsafe for human consumption throughout the year. As such, authorities should carefully consider the risks posed by pufferfish, especially of this species, and carry out awareness activities to prevent their consumption under any circumstances. In addition, considering that pufferfish are the most important source of purified TTX to be used as a drug [61], it is possible that Antalya Bay could constitute an appropriate potential production site for TTX with the purpose of alleviating the detrimental effects of pufferfish expansion through overfishing. In this context, this study provides further insights into the months during which TTX levels are elevated in *L. sceleratus*. To summarize, the findings highlight the toxin levels of *L. sceleratus* in Antalya Bay by offering valuable information regarding the temporal distribution of TTX in this ecosystem. Further research on the monthly variation of TTX and its analogues could also be necessary to obtain more solid information on their individual contribution to the overall toxic burden of *L. sceleratus* and their occurrence patterns, as well as on TTX metabolism.

Supplementary Materials: The following supporting information can be downloaded at: https://www.mdpi.com/article/10.3390/md21100527/s1, Figure S1: TTX spectra of different pufferfish samples.

Author Contributions: Conceptualization, A.R.K., P.K., M.K. and F.Ö.; methodology, A.R.K., M.K., İ.D., D.A. and Y.U.; writing—original draft preparation, A.R.K., P.K. and M.D.; writing—review and editing, A.R.K., P.K., M.D. and F.Ö.; supervision, P.K. and A.R.K.; funding acquisition, P.K. All authors have read and agreed to the published version of the manuscript.

Funding: This research was financially supported by the Republic of Türkiye Ministry of Agriculture and Forestry, General Directorate of Agricultural Research and Policies (Project No. TAGEM/HAYSÜD/B/17/A6/P1/507).

Institutional Review Board Statement: Not applicable.

Data Availability Statement: The data presented in this study are available on request from the corresponding authors.

Acknowledgments: We would like to thank Sedat Gündoğdu for his contribution of preparing the figures and Bahar Meryemoğlu for her contribution of the Q-TOF-LC-MS analysis.

Conflicts of Interest: The authors declare no conflict of interest.

References

1. Ulman, A.; Yildiz, T.; Demirel, N.; Canak, O.; Yemişken, E.; Pauly, D. The biology and ecology of the invasive silver-cheeked toadfish (*Lagocephalus sceleratus*), with emphasis on the Eastern Mediterranean. *NeoBiota* **2021**, *68*, 145–175. [CrossRef]
2. Kosker, A.R.; Özogul, F.; Ayas, D.; Durmus, M.; Ucar, Y.; Regenstein, J.M.; Özogul, Y. Tetrodotoxin levels of three pufferfish species (*Lagocephalus* sp.) caught in the North-Eastern Mediterranean sea. *Chemosphere* **2019**, *219*, 95–99. [CrossRef]
3. Akyol, O.; Ünal, V.; Ceyhan, T.; Bilecenoglu, M. First confirmed record of Lagocephalus sceleratus(Gmelin, 1789) in the Mediterranean Sea. *J. Fish Biol.* **2005**, *66*, 1183–1186. [CrossRef]
4. Kasapidis, P.; Peristeraki, P.; Tserpes, G.; Magoulas, A. First record of the Lessepsian migrant Lagocephalus sceleratus (Gmelin 1789) (Osteichthyes: Tetraodontidae) in the Cretan Sea (Aegean, Greece). *Aquat. Invasions* **2007**, *2*, 71–73. [CrossRef]
5. Jribi, I.; Bradai, M.N. First record of the lessepsian migrant species Lagocephalus sceleratus (Gmelin, 1789) (actinopterygii: Tetraodontidae) in the central mediterranean. *BioInvasions Rec.* **2012**, *1*, 49–52. [CrossRef]
6. Milazzo, M.; Azzurro, E.; Badalamenti, F. On the occurrence of the silverstripe blaasop Lagocephalus sceleratus (Gmelin, 1789) along the Libyan coast. *BioInvasions Rec.* **2012**, *1*, 125–127. [CrossRef]
7. Sulić Šprem, J.; Dobroslavić, T.; Kožul, V.; Kuzman, A.; Dulčić, J. First record of Lagocephalus sceleratus in the Adriatic Sea (Croatian coast), a Lessepsian migrant. *Cybium* **2014**, *38*, 147–148.
8. Rambla-Alegre, M.; Reverté, L.; del Río, V.; de la Iglesia, P.; Palacios, O.; Flores, C.; Caixach, J.; Campbell, K.; Elliott, C.T.; Izquierdo-Muñoz, A.; et al. Evaluation of tetrodotoxins in puffer fish caught along the Mediterranean coast of Spain. Toxin profile of Lagocephalus sceleratus. *Environ. Res.* **2017**, *158*, 1–6. [CrossRef]

9. Azzurro, E.; Castriota, L.; Falautano, M.; Giardina, F.; Andaloro, F. The silver-cheeked toadfish Lagocephalus sceleratus (Gmelin, 1789) reaches Italian waters. *J. Appl. Ichthyol.* **2014**, *30*, 1050–1052. [CrossRef]
10. Guardone, L.; Gasperetti, L.; Maneschi, A.; Ricci, E.; Susini, F.; Guidi, A.; Armani, A. Toxic invasive pufferfish (Tetraodontidae family) along Italian coasts: Assessment of an emerging public health risk. *Food Control* **2018**, *91*, 330–338. [CrossRef]
11. Zenetos, A.; Gofas, S.; Morri, C.; Rosso, A.; Violanti, D.; García Raso, J.E.; Çinar, M.E.; Almogi-Labin, A.; Ates, A.S.; Azzurro, E.; et al. Alien species in the Mediterranean Sea by 2012. A contribution to the application of European Union's Marine Strategy Framework Directive (MSFD). Part 2. Introduction trends and pathways. *Mediterr. Mar. Sci.* **2012**, *13*, 328–352. [CrossRef]
12. Thiery, A.P.; Shono, T.; Kurokawa, D.; Britz, R.; Johanson, Z.; Fraser, G.J. Spatially restricted dental regeneration drives pufferfish beak development. *Proc. Natl. Acad. Sci. USA.* **2017**, *114*, E4425–E4434. [CrossRef]
13. Turingan, R.G. Ecomorphological relationships among Caribbean tetraodontiform fishes. *J. Zool.* **1994**, *233*, 493–521. [CrossRef]
14. Ünal, V.; Göncüoğlu Bodur, H. The socio-economic impacts of the silver-cheeked toadfish on small-scale fishers: A comparative study from the Turkish coast. *Ege J. Fish. Aquat. Sci.* **2017**, *34*, 119–127. [CrossRef]
15. Ünal, V.; Göncüoğlu, H.; Durgun, D.; Tosunoğlu, Z.; Deval, M.C.; Turan, C. Silver-cheeked toadfish, Lagocephalus sceleratus (Actinopterygii: Tetraodontiformes: Tetraodontidae), causes a substantial economic losses in the Turkish Mediterranean coast: A call for decision makers. *Acta Ichthyol. Piscat.* **2015**, *45*, 231–237. [CrossRef]
16. Kleitou, P.; Moutopoulos, D.K.; Giovos, I.; Kletou, D.; Savva, I.; Cai, L.L.; Hall-Spencer, J.M.; Charitou, A.; Elia, M.; Katselis, G.; et al. Conflicting interests and growing importance of non-indigenous species in commercial and recreational fisheries of the Mediterranean Sea. *Fish. Manag. Ecol.* **2022**, *29*, 169–182. [CrossRef]
17. Katikou, P.; Gokbulut, C.; Kosker, A.R.; Campàs, M.; Ozogul, F. An Updated Review of Tetrodotoxin and Its Peculiarities. *Mar. Drugs* **2022**, *20*, 47. [CrossRef] [PubMed]
18. Bentur, Y.; Spanier, E. Ciguatoxin-like substances in edible fish on the eastern Mediterranean. *Clin. Toxicol.* **2007**, *45*, 695–700. [CrossRef] [PubMed]
19. Kheifets, J.; Rozhavsky, B.; Girsh Solomonovich, Z.; Marianna, R.; Soroksky, A. Severe Tetrodotoxin Poisoning after Consumption of Lagocephalus sceleratus (Pufferfish, Fugu) Fished in Mediterranean Sea, Treated with Cholinesterase Inhibitor. *Case Rep. Crit. Care* **2012**, *2012*, 1–3. [CrossRef]
20. Abd Rabou, A.F.N. On the occurrence and health risks of the silver-cheeked Toadfish (Lagocephalus sceleratus Gmelin, 1789) in the marine ecosystem of the Gaza Strip, Palestine. *Biodiversitas* **2019**, *20*, 2620–2627. [CrossRef]
21. Nader, M.; Indary, S.; Boustany, L. The Puffer Fish Lagocephalus Sceleratus (Gmelin, 1789) in the Eastern Mediterranean. Available online: http://www.fao.org/3/ap967e/ap967e.pdf (accessed on 11 August 2023).
22. European Commission Regulation (EC) No 853/2004 of the European Parlamient and of the Council. *Off. J. Eur. Union* **2004**, *L 139*, 55.
23. Republic of Türkiye Ministry of Agriculture and Forestry. Hayvansal Gidalar için Özel Hijyen Kurallari Yönetmeliği, Sayı: 28155, 27 Aralık 2011; Canlı çift kabuklu yumuşakçalar için sağlık standartları: Madde 49. Available online: https://www.resmigazete.gov.tr/eskiler/2011/12/20111227-10.htm (accessed on 11 August 2023).
24. Jal, S.; Khora, S.S. An overview on the origin and production of tetrodotoxin, a potent neurotoxin. *J. Appl. Microbiol.* **2015**, *119*, 907–916. [CrossRef] [PubMed]
25. Magarlamov, T.Y.; Melnikova, D.I.; Chernyshev, A.V. Tetrodotoxin-producing bacteria: Detection, distribution and migration of the toxin in aquatic systems. *Toxins* **2017**, *9*, 166. [CrossRef] [PubMed]
26. Bane, V.; Lehane, M.; Dikshit, M.; O'Riordan, A.; Furey, A. Tetrodotoxin: Chemistry, toxicity, source, distribution and detection. *Toxins* **2014**, *6*, 693–755. [CrossRef]
27. Wu, Z.; Yang, Y.; Xie, L.; Xia, G.; Hu, J.; Wang, S.; Zhang, R. Toxicity and distribution of tetrodotoxin-producing bacteria in puffer fish Fugu rubripes collected from the Bohai Sea of China. *Toxicon* **2005**, *46*, 471–476. [CrossRef]
28. Noguchi, T.; Arakawa, O.; Takatani, T. TTX accumulation in pufferfish. *Comp. Biochem. Physiol.—Part D Genom. Proteom.* **2006**, *1*, 145–152. [CrossRef]
29. Knutsen, H.K.; Alexander, J.; Barregård, L.; Bignami, M.; Brüschweiler, B.; Ceccatelli, S.; Cottrill, B.; Dinovi, M.; Edler, L.; Grasl-Kraupp, B.; et al. Risks for public health related to the presence of tetrodotoxin (TTX) and TTX analogues in marine bivalves and gastropods. *EFSA J.* **2017**, *15*, e04752. [CrossRef] [PubMed]
30. Kodama, M.; Sato, S.; Sakamoto, S.; Ogata, T. Occurrence of tetrodotoxin in Alexandrium tamarense, a causative dinoflagellate of paralytic shellfish poisoning. *Toxicon* **1996**, *34*, 1101–1105. [CrossRef]
31. Vlamis, A.; Katikou, P.; Rodriguez, I.; Rey, V.; Alfonso, A.; Papazachariou, A.; Zacharaki, T.; Botana, A.M.; Botana, L.M. First detection of tetrodotoxin in greek shellfish by UPLC-MS/MS potentially linked to the presence of the dinoflagellate prorocentrum minimum. *Toxins* **2015**, *7*, 1779–1807. [CrossRef]
32. Rodríguez, I.; Alfonso, A.; Alonso, E.; Rubiolo, J.A.; Roel, M.; Vlamis, A.; Katikou, P.; Jackson, S.A.; Menon, M.L.; Dobson, A.; et al. The association of bacterial C 9 -based TTX-like compounds with Prorocentrum minimum opens new uncertainties about shellfish seafood safety. *Sci. Rep.* **2017**, *7*, 40880. [CrossRef]
33. Reverté, L.; De La Iglesia, P.; Del Río, V.; Campbell, K.; Elliott, C.T.; Kawatsu, K.; Katikou, P.; Diogène, J.; Campàs, M. Detection of Tetrodotoxins in Puffer Fish by a Self-Assembled Monolayer-Based Immunoassay and Comparison with Surface Plasmon Resonance, LC-MS/MS, and Mouse Bioassay. *Anal. Chem.* **2015**, *87*, 10839–10847. [CrossRef] [PubMed]

34. Silva, M.; Azevedo, J.; Rodriguez, P.; Alfonso, A.; Botana, L.M.; Vasconcelos, V. New gastropod vectors and tetrodotoxin potential expansion in temperate waters of the Atlantic Ocean. *Mar. Drugs* **2012**, *10*, 712–726. [CrossRef] [PubMed]
35. Nelson, J.S. *Fishes of the World*; John Wiley & Sons: Hoboken, NJ, USA, 2006; ISBN 978-0-471-25031-9.
36. Rodríguez, P.; Alfonso, A.; Otero, P.; Katikou, P.; Georgantelis, D.; Botana, L.M. Liquid chromatography–mass spectrometry method to detect Tetrodotoxin and Its analogues in the puffer fish Lagocephalus sceleratus (Gmelin, 1789) from European waters. *Food Chem.* **2012**, *132*, 1103–1111. [CrossRef]
37. Kosker, A.R.; Özogul, F.; Durmus, M.; Ucar, Y.; Ayas, D.; Šimat, V.; Özogul, Y. First report on TTX levels of the yellow spotted pufferfish (*Torquigener flavimaculosus*) in the Mediterranean Sea. *Toxicon* **2018**, *148*, 101–106. [CrossRef]
38. Leonardo, S.; Kiparissis, S.; Rambla-Alegre, M.; Almarza, S.; Roque, A.; Andree, K.B.; Christidis, A.; Flores, C.; Caixach, J.; Campbell, K.; et al. Detection of tetrodotoxins in juvenile pufferfish Lagocephalus sceleratus (Gmelin, 1789) from the North Aegean Sea (Greece) by an electrochemical magnetic bead-based immunosensing tool. *Food Chem.* **2019**, *290*, 255–262. [CrossRef] [PubMed]
39. Christidis, G.; Mandalakis, M.; Anastasiou, T.I.; Tserpes, G.; Peristeraki, P.; Somarakis, S. Keeping lagocephalus sceleratus off the table: Sources of variation in the quantity of ttx, ttx analogues, and risk of tetrodotoxication. *Toxins* **2021**, *13*, 896. [CrossRef]
40. Hassoun, A.E.R.; Ujević, I.; Jemaa, S.; Roje-Busatto, R.; Mahfouz, C.; Fakhri, M.; Nazlić, N. Concentrations of Tetrodotoxin (TTX) and Its Analogue 4,9-Anhydro TTX in Different Tissues of the Silver-Cheeked Pufferfish (Lagocephalus sceleratus, Gmelin, 1789) Caught in the South-Eastern Mediterranean Sea, Lebanon. *Toxins* **2022**, *14*, 123. [CrossRef]
41. Katikou, P.; Georgantelis, D.; Sinouris, N.; Petsi, A.; Fotaras, T. First report on toxicity assessment of the Lessepsian migrant pufferfish Lagocephalus sceleratus (Gmelin, 1789) from European waters (Aegean Sea, Greece). *Toxicon* **2009**, *54*, 50–55. [CrossRef]
42. Kosker, A.R.; Özogul, F.; Durmus, M.; Ucar, Y.; Ayas, D.; Regenstein, J.M.; Özogul, Y. Tetrodotoxin levels in pufferfish (*Lagocephalus sceleratus*) caught in the Northeastern Mediterranean Sea. *Food Chem.* **2016**, *210*, 332–337. [CrossRef]
43. Acar, C.; Ishizaki, S.; Nagashima, Y. Toxicity of the Lessepsian pufferfish Lagocephalus sceleratus from eastern Mediterranean coasts of Türkiye and species identification by rapid PCR amplification. *Eur Food Res Technol* **2017**, *243*, 49–57. [CrossRef]
44. Akbora, H.D.; Kunter, İ.; Erçeti N, T.; Elagöz, A.M.; Çiçek, B.A. Determination of tetrodotoxin (TTX) levels in various tissues of the silver cheeked puffer fish (Lagocephalus sceleratus (Gmelin, 1789)) in Northern Cyprus Sea (Eastern Mediterranean). *Toxicon* **2020**, *175*, 1–6. [CrossRef] [PubMed]
45. Malloggi, C.; Rizzo, B.; Giusti, A.; Guardone, L.; Gasperetti, L.; Dall'Ara, S.; Armani, A. First Toxicological Analysis of the Pufferfish Sphoeroides pachygaster Collected in Italian Waters (Strait of Sicily): Role of Citizens Science in Monitoring Toxic Marine Species. *Animals* **2023**, *13*, 1873. [CrossRef]
46. Bentur, Y.; Ashkar, J.; Lurie, Y.; Levy, Y.; Azzam, Z.S.; Litmanovich, M.; Golik, M.; Gurevych, B.; Golani, D.; Eisenman, A. Lessepsian migration and tetrodotoxin poisoning due to Lagocephalus sceleratus in the eastern Mediterranean. *Toxicon* **2008**, *52*, 964–968. [CrossRef] [PubMed]
47. Chamandi, S.C.; Kallab, K.; Mattar, H.; Nader, E. Human Poisoning after ingestion of puffer fish caught from Mediterranean sea. *Middle East J. Anesthesiol.* **2009**, *20*, 285–288. [PubMed]
48. Ministry of Health, Labour and Welfare of Japan. Natural Toxin Risk Profile: Fish: Pufferfish Toxin. Available online: https://www.mhlw.go.jp/topics/syokuchu/poison/animal_det_01.html (accessed on 11 August 2023).
49. Alkassar, M.; Sanchez-Henao, A.; Reverté, J.; Barreiro, L.; Rambla-Alegre, M.; Leonardo, S.; Mandalakis, M.; Peristeraki, P.; Diogène, J.; Campàs, M. Evaluation of Toxicity Equivalency Factors of Tetrodotoxin Analogues with a Neuro-2a Cell-Based Assay and Application to Puffer Fish from Greece. *Mar. Drugs* **2023**, *21*, 432. [CrossRef]
50. Anastasiou, T.I.; Kagiampaki, E.; Kondylatos, G.; Tselepides, A.; Peristeraki, P.; Mandalakis, M. Assessing the Toxicity of Lagocephalus sceleratus Pufferfish from the Southeastern Aegean Sea and the Relationship of Tetrodotoxin with Gonadal Hormones. *Mar. Drugs* **2023**, *21*, 520. [CrossRef]
51. Ujević, I.; Roje-Busatto, R.; Dragičević, B.; Dulčić, J. Tetrodotoxin in Invasive Silver-cheeked Toadfish Lagocephalus sceleratus (Gmelin, 1789) in the Adriatic Sea. *Handb. Environ. Chem.* **2021**, *110*, 141–149. [CrossRef]
52. Farrag, M.M.S.; El-Haweet, A.E.A.K.; Osman, A.G.M.; Akel, E.S.K.A.; Moustafa, M.A. Reproductive behaviour of the silver-stripe blaasop; lagocephalus sceleratus (Gmelin, 1789) from the mediterranean coast, egypt. *Egypt. J. Aquat. Biol. Fish.* **2019**, *23*, 441–454. [CrossRef]
53. El-Sayed, M.; Yacout, G.A.; El-Samra, M.; Ali, A.; Kotb, S.M. Toxicity of the Red Sea pufferfish Pleuranacanthus sceleratus "El-Karad. " *Ecotoxicol. Environ. Saf.* **2003**, *56*, 367–372. [CrossRef]
54. Hwang, D.F.; Noguchi, T. Tetrodotoxin Poisoning. *Adv. Food Nutr. Res.* **2007**, *52*, 141–236. [CrossRef]
55. Yu, C.F.; Yu, P.H.F. The annual toxicological profiles of two common puffer fish, Takifugu niphobles (Jordan and Snyder) and Takifugu alboplumbeus (Richardson), collected along Hong Kong coastal waters. *Toxicon* **2002**, *40*, 313–316. [CrossRef] [PubMed]
56. Kawabata, T. *The Manual for the Methods of Food Sanitation Tests*; Japan Food Hygiene Association, Environmental Health Bureau: Tokyo, Japan, 1978.
57. Köşker, A.R.; Özoğul, F.; Ayas, D.; Durmuş, M.; Uçar, Y. The new toxin of Mediterranean:Tetrodotoxin. *Ege J. Fish. Aquat. Sci.* **2015**, *32*, 15–24. [CrossRef]
58. Finch, S.C.; Boundy, M.J.; Harwood, D.T. The acute toxicity of tetrodotoxin and tetrodotoxin–saxitoxin mixtures to mice by various routes of administration. *Toxins* **2018**, *10*, 423. [CrossRef] [PubMed]

59. Finch, S.C.; Webb, N.G.; Boundy, M.J.; Harwood, D.T.; Munday, J.S.; Sprosen, J.M.; Somchit, C.; Broadhurst, R.B. A Sub-Acute Dosing Study of Saxitoxin and Tetrodotoxin Mixtures in Mice Suggests That the Current Paralytic Shellfish Toxin Regulatory Limit Is Fit for Purpose. *Toxins* **2023**, *15*, 437. [CrossRef] [PubMed]
60. Bane, V.; Hutchinson, S.; Sheehan, A.; Brosnan, B.; Barnes, P.; Lehane, M.; Furey, A. LC-MS/MS method for the determination of tetrodotoxin (TTX) on a triple quadruple mass spectrometer. *Food Addit. Contam.—Part A Chem. Anal. Control. Expo. Risk Assess.* **2016**, *33*, 1728–1740. [CrossRef]
61. Bucciarelli, G.M.; Lechner, M.; Fontes, A.; Kats, L.B.; Eisthen, H.L.; Shaffer, H.B. From poison to promise: The evolution of tetrodotoxin and its potential as a therapeutic. *Toxins* **2021**, *13*, 517. [CrossRef]

marine drugs

MDPI

Article

Optimization of the Extraction Protocol for Pacific Ciguatoxins from Marine Products Prior to Analysis Using the Neuroblastoma Cell-Based Assay

Thomas Yon [1], Philippe Cruchet [1], Jérôme Viallon [1], J. Sam Murray [2], Emillie Passfield [2], Mireille Chinain [1], Hélène Taiana Darius [1,*] and Mélanie Roué [3,*]

[1] Institut Louis Malardé (ILM), UMR 241-SECOPOL (IFREMER, ILM, IRD, UPF), Laboratory of Marine Biotoxins, P.O. Box 30, 98713 Papeete, Tahiti, French Polynesia; tyon@ilm.pf (T.Y.); pcruchet@ilm.pf (P.C.); jviallon@ilm.pf (J.V.); mchinain@ilm.pf (M.C.)

[2] Cawthron Institute, 98 Halifax St East, PB 2, The Wood, Nelson 7010, New Zealand; sam.murray@cawthron.org.nz (J.S.M.); emillie.passfield@cawthron.org.nz (E.P.)

[3] Institut de Recherche pour le Développement (IRD), UMR 241-SECOPOL (IFREMER, ILM, IRD, UPF), P.O. Box 6570, 98702 Faa'a, Tahiti, French Polynesia

* Correspondence: tdarius@ilm.pf (H.T.D.); melanie.roue@ird.fr (M.R.)

Abstract: Ciguatera poisoning (CP) is caused by the consumption of marine products contaminated with ciguatoxins (CTXs) produced by dinoflagellates of the genus *Gambierdiscus*. Analytical methods for CTXs, involving the extraction/purification of trace quantities of CTXs from complex matrices, are numerous in the literature. However, little information on their effectiveness for nonpolar CTXs is available, yet these congeners, contributing to the risk of CP, are required for the establishment of effective food safety monitoring programs. An evaluation of six extraction/purification protocols, performed with CTX3C spiked on fish flesh and a neuroblastoma cell-based assay (CBA-N2a), revealed recoveries from 6 to 45%. This led to the development of an optimized 3-day protocol designed for a large number of samples, with CTX1B and CTX3C eluting in a single fraction and showing recoveries of 73% and 70%, respectively. In addition, a reduction in adverse matrix effects in the CBA-N2a analyses was demonstrated with naturally contaminated specimens, increasing the sensitivity of the method, which now meets the very low guidance level recommended by international agencies. However, efforts are still required to reduce the signal suppression observed in LC-MS/MS analysis. This optimized protocol contributes to the technological advancement of detection methods, promoting food safety and improving CP risk assessment in marine products.

Keywords: ciguatera poisoning; ciguatoxins; marine products; extraction protocol; CBA-N2a; LC-MS/MS; matrix effect; risk assessment; food safety

1. Introduction

Ciguatera poisoning (CP) is a foodborne illness caused by the consumption of marine products (fish or invertebrates) contaminated with ciguatoxins (CTXs), potent neurotoxins produced by certain species of dinoflagellates of the genus *Gambierdiscus* [1,2]. CP is prevalent in all tropical and subtropical regions, including islands in the Caribbean, Pacific and Indian Oceans, as well as coastal areas of the Americas, Africa and Asia [3]. While the true incidence of CP worldwide is difficult to estimate due to underreporting and occasional misdiagnosis, this poisoning is considered the most common marine toxin-related illness [3].

https://doi.org/10.3390/md23010042

The toxins responsible for CP, i.e., CTXs, are a suite of lipid-soluble polyethers that are bioaccumulated and metabolized through the food chain from herbivores to carnivores (shellfish or fish). This results in a large class of biotransformed analogues with varying degrees of oxidation, saturation and/or ring systems opening [4,5]. Depending on their structure and polarity, CTXs exert different effects on voltage-gated sodium channels (VGSCs) in excitable cells, leading to alterations in neuronal function and neurotransmitter release [6]. Hence, the symptoms and severity observed in CP cases largely depend on an individual's medical history and sensitivity, as well as the section and amount of the marine specimen consumed [7,8]. A variety of symptoms can manifest, generally starting with gastrointestinal (e.g., nausea, vomiting, diarrhea), followed by cardiovascular (e.g., bradycardia, hypotension) and then neurological (e.g., paresthesia, dysesthesia, pruritus, cold allodynia, etc.) [9]. The onset of symptoms typically occurs within hours to days after the ingestion of the contaminated marine product and may persist for weeks to years in some cases [9]. Identifying CP cases can be challenging due to their large variety of symptoms and the lack of specific clinical diagnostic tests. Diagnosis relies on a thorough history of fish consumption by the patient presenting evocative symptoms following the ingestion of reef fish [10]. Confirmation of CP requires analytical methods able to detect the presence of CTXs in leftover meal remnants; however, these are seldomly available. The very low level of CTXs in marine products, which is nonetheless sufficient to trigger symptoms in humans due to their very high toxicity (i.e., 0.01 μg kg^{-1} body weight [11]), makes their detection in leftover meal remnants very difficult [12].

Only a few approaches are sufficiently sensitive to detect CTXs in trace amounts, such as the mouse neuroblastoma cell-based assay (CBA-N2a) [13–15], the receptor binding assay [16–19], sandwich ELISA detection using monoclonal antibodies [20–22] and liquid chromatography coupled to tandem mass spectrometry (LC-MS/MS) [23–27]. In addition, the lack of certified reference materials (both analytical standards and contaminated marine products) severely hampers the development and validation of detection methods [28]. In terms of food safety issues, experts recommend that ciguatera risk management programs be conducted using a two-tiered analysis approach [11]. This typically consists of using sensitive functional methods for the screening and quantification of biological samples, such as the CBA-N2a, followed by the confirmation of CTX analogues in toxic samples using specific analytical methods, such as the LC-MS/MS.

Although these detection methods are based on very different measurement mechanisms, i.e., mode of action or chemical structure, a key step in each of these approaches is sample preparation, which must ensure the most efficient extraction of all CTXs from marine product tissues and lead to a purified fraction that can be analyzed directly without interference [29]. This represents a major challenge because only trace quantities of CTXs are accumulated in marine products (typically in the μg kg^{-1} range), which are very complex matrices rich in lipids and proteins, compounds that often co-elute with CTXs and may be responsible for matrix effects inherent to the detection method used. Therefore, numerous sample preparation protocols have been described in the literature, comprising similar main stages, i.e., extraction, liquid–liquid partitioning and solid phase extraction (SPE) clean-up steps, but each with different conditions (see Harwood et al. [28] for a review; [23,24,30]). The majority of these protocols have been developed for the more polar CTXs found primarily in carnivorous fish—such as CTX1B, 52-epi-54-deoxyCTX1B and 54-deoxyCTX1B (also called CTX2 and CTX3, respectively) or 2,3-dihydroxyCTX3C—, for which recoveries ranging from 26 to 96% have been reported, depending on the sample preparation protocol and detection method used, but also on the marine product analyzed (Table 1). In contrast, their efficiency for less polar analogues found throughout the trophic chain—such as CTX3B, CTX3C, CTX4A or CTX4B—have rarely been evaluated, and when they are, very low recoveries have been reported, ranging from 13 to 35%, except for a recent study reporting a recovery of 107% (Table 1). However, these nonpolar CTXs contribute to the risk of CP in the Pacific as several poisonings have been reported following the consumption of herbivorous fish or invertebrates, which mostly contain these nonpolar congeners [31–37]. Thus, these often-reported low recoveries imply a risk of underestimating the potentially significant quantities of CTXs in marine products with potential deleterious consequences for food safety and thus the public health of consumers. In addition, most sample preparation protocols involve multiple steps (up to five) and are, therefore, relatively time-consuming, making them incompatible with the characterization of the ciguatera risk of an area by analyzing a large number of samples [38–40].

In this context, to obtain a more reliable assessment of the actual levels of CTXs accumulated in marine products, the present study aimed to work on an extraction/purification protocol that would i) optimize the recovery of the nonpolar CTXs, ii) produce an extract with reduced adverse matrix effects observed in CBA-N2a experiments and iii) have a simple and short execution time to make it applicable for the screening of a large number of samples. First, the efficiency of six sample preparation protocols, inspired by some available in the literature, was evaluated on fish flesh spiked with a CTX3C standard and analyzed by the CBA-N2a. Second, based on the results obtained, an improved extraction/purification protocol was then developed for the subsequent detection and quantification of CTXs by the CBA-N2a, regardless of their polarity. Third, the performance of this new protocol was evaluated for the subsequent detection of CTXs by LC-MS/MS.

Table 1. CTX recoveries reported in the literature and estimated after extraction/purification of fish flesh spiked with CTX standards and subsequent analysis of the final fraction of interest considered as enriched in CTXs.

References	Detection Method	Tested Marine Products	Details of Spiking	Polar CTX Recovery (%)	Nonpolar CTX Recovery (%)
[41]	MBA	3 carnivorous fish species	500 ng CTX1B in 100 g of flesh	CTX1B = 33–85	NT [c]
[42]	LC-MS/MS	1 carnivorous fish species	0.2 ng CTX1B in 2 g of flesh	CTX1B = 85	NT
[43]	LC-MS/MS	10 carnivorous fish species	3 ng CTX1B in 2 g of flesh	CTX1B = 27–75	NT
[44]	LC-MS/MS	17 carnivorous fish species	20 ng CTX1B in 5 g of flesh	CTX1B = 49–85	NT
[45]	LC-MS/MS	4 carnivorous and 2 herbivorous fish species, 2 invertebrate species	100 pg CTX1B, 500 pg CTX2 [a], 500 pg CTX3 [b] in 2.5–5 g of flesh	CTX1B = 73–87 CTX2 = 68–83 CTX3 = 61–78	NT
[46]	LC-MS/MS	1 carnivorous fish species	Not specified	CTX1B = 26	NT
[27]	LC-MS/MS	1 carnivorous fish species	0.068, 0.341 and 0.682 μg/kg CTX1B in 5 g of flesh	CTX1B = 44	CTX3C = 13
[47]	ELISA	1 carnivorous fish species	100 pg CTX1B in 10 g of flesh	CTX1B = 32	NT

Table 1. *Cont.*

References	Detection Method	Tested Marine Products	Details of Spiking	Polar CTX Recovery (%)	Nonpolar CTX Recovery (%)
[23]	LC-MS/MS	2 carnivorous fish species	0.1 µg of CTX1B or CTX3C per kg of fish flesh	CTX1B = 87–96	CTX3C = 105–107
[24]	LC-MS/MS	2 carnivorous and 1 herbivorous fish species	Not specified	CTX1B = 39–62 CTX2 = 35–66 CTX3 = 35–69	CTX3C = 25–35
[48]	LC-MS/MS	Carnivorous fish species	0.1 ng of CTX1B, CTX2 or CTX3 in 5 g of fish flesh	CTX1B = 80–89 CTX2 = 79,80 CTX3 = 71–84	NT
[49]	CBA-N2a	2 carnivorous fish species	Not specified	CTX1B = 32–42	NT

[a] 52-epi-54-deoxyCTX1B; [b] 54-deoxyCTX1B; [c] NT: not tested.

2. Results

2.1. Evaluation of Efficiencies of Six Different Protocols for the Extraction/Purification of CTX3C

The importance of the parameters used in protocols for CTX extraction from marine products was studied using six different protocols inspired by those available in the literature (protocols #1 to #6, see Section 4.5 and Supplementary Materials). In an initial step, the six protocols were applied to a non-toxic fish sample (10 g), the Tuamotu emperor *Lethrinus atkinsoni* (Lethrinidae), to determine the matrix effects observed on N2a cell viability. Fractions of interest (i.e., the fraction expected to contain the majority of the CTXs and thus used for subsequent analysis) but also washing fractions having an extract dry mass less than 30 mg were analyzed to determine the maximum concentration of dry extract (MCE) and its equivalent in wet fish flesh (Table 2) that can be tested without causing adverse matrix effects (see Section 4.8.2).

A significant difference in MCE values was observed across the protocols: in the cases of protocols #1 and #2, for the fraction of interest, a concentration of 34 and 84 mg eq. wet weight of fish flesh, respectively, resulted in the non-specific mortality of N2a cells, whereas the maximal concentration tested for protocols #3, #4 and #5 did not induce any mortality (Table 2).

Next, three CTX3C-spiked flesh samples of the same Tuamotu emperor (5 ng CTX3C in 10 g wet weight of fish flesh) were extracted and purified according to each protocol. The fractions obtained were first screened by the CBA-N2a in a single dose at the MCE to identify and eliminate negative fractions (no mortality in OV- and OV+ conditions) from the analyses. Then, CTX quantification was performed only in fractions showing CTX-like activity (mortality in OV^+ condition only) using an appropriate range of concentrations to obtain the EC_{50} value and calculate the quantity of CTX3C actually recovered, thus allowing the evaluation of the nonpolar CTX extraction efficiency from 5 ng spiked samples.

Table 2. Results of CTX3C spiking experiments (5 ng spiked in 10 g wet weight of flesh) as estimated by CBA-N2a.

Protocol ID#	Fractions Analyzed	Dry Extract (mg)	MCE (mg Equivalent Wet Weight of Fish Flesh mL^{-1})	Recovered CTX3C (ng) [c]	CTX3C Recovery (%)	Coefficient of Variation (%)
#1	F2 [a]	2.9	34	1.08 ± 0.16	21.5 ± 3.1	14.5
	F3	10.3	39	0.53 ± 0.10	10.7 ± 1.9	17.8
	Total			1.61 ± 0.18	32.2 ± 3.5	11.0
#2	F1	3.9	NR [>244] [b]	ND [d]	-	-
	F2 [a]	1.2	83	0.32 ± 0.04	6.4 ± 0.9	13.8
	F3	1.4	NR [>343]	ND	-	-
	Total			0.32 ± 0.04	6.4 ± 0.9	13.8
#3	F2	12.9	NR [>184]	ND	-	-
	F1.1	1.4	NR [>343]	ND	-	-
	F1.2 [a]	1.2	NR [>400]	0.38 ± 0.06	7.6 ± 1.2	16.2
	F1.3	3.7	NR [>270]	0.77 ± 0.16	15.4 ± 3.2	20.7
	Total			1.15 ± 0.21	23.0 ± 4.3	18.7
#4	F4	13.5	NR [>370]	2.11 ± 0.18	42.3 ± 3.5	8.3
	F3.1	0.2	NR [>50]	0.15 ± 0.09	3.0 ± 1.7	55.9
	F3.2 [a]	0.5	NR [>400]	ND	-	-
	F3.3	0.9	NR [>444]	ND	-	-
	Total			2.27 ± 0.18	45.3 ± 3.7	8.1
#5	F4	26.2	38	0.85 ± 0.20	17.0 ± 4.0	23.3
	F3.2 [a]	3.3	NR [>145]	ND	-	-
	F3.3	5.2	NR [>92]	ND	-	-
	Total			0.85 ± 0.20	17.0 ± 4.0	23.3
#6	F1.2.1 [a]	0.4	500	2.04 ± 0.32	40.8 ± 6.3	15.5
	F1.2.2	1.7	NR [>500]	0.13 ± 0.01	2.6 ± 0.2	7.5
	Total			2.17 ± 0.32	43.4 ± 6.4	14.7

[a] Fractions of interest likely to contain majority of CTXs. [b] NR: MCE not reached. In square brackets is indicated the value corresponding to the maximum concentration tested. [c] Data represent the mean ± standard deviation (SD) of CTX3C estimated with three CTX3C-spiked fish samples tested in three independent experiments run on different days (n = 9). [d] ND: CTX3C not detected.

The CBA-N2a was able to quantify very low quantities of CTX3C from 0.13 ± 0.01 up to 2.11 ± 0.18 ng CTX3C eq. with good reproducibility. The coefficients of variation (CVs) ranged from 7.5 to 23.3%, except for the fraction F3.1 from protocol #4 tested at 0.15 ± 0.09 ng CTX3C eq., which showed a CV of 55.9% (Table 2). This result can be attributed to the difficulty of extracting/analyzing trace amounts of CTXs, which usually results in higher variability. In terms of percentage of CTX3C recovery, this corresponded to a minimum of 6.4 ± 0.9 and a maximum of 45.3 ± 1.0%. The conditions resulting in the highest recoveries of spiked CTX3C were from protocols #1, #4 and #6 (Table 2), in which the extraction solvent was either acetone, methanol (MeOH)/Hexane 73:27 or MeOH/H_2O 90:10, whereas extractions with MeOH/H_2O 60:40 or a two-step extraction with first MeOH 100% and then MeOH/H_2O 60:40 resulted in less than 25% recoveries of CTX3C (protocols #2 and #5).

In addition, using protocols #1, #3, #4 and #5, CTX3C was only partially recovered or not recovered at all in the fraction of interest (Table 2). Indeed, for protocols #1 and #3, a distribution of CTX3C in two different fractions obtained after the final SPE (C18) was observed, i.e., with 7.6 ± 1.2 and 21.5 ± 3.1% of CTX3C spike recovery in the fraction of interest, and 10.7 ± 1.9 and 15.4 ± 3.2% in the washing fraction, respectively. Furthermore, for protocols #4 and #5, the extracted CTX3C was only eluted in the additional washing phase of the C18 SPE step, resulting in the complete absence of CTX3C in the fraction of interest.

Finally, from a purely technical point of view, protocol #2 proved to be the fastest and simplest, while protocol #6 proved to be the most complicated one, with five purification steps, and protocol #1 was the most time-consuming, requiring eight days to complete.

2.2. Design of an Improved Protocol for CTX Extraction and Purification

Building on the positive attributes of each of the six protocols previously evaluated, preliminary assays on fish flesh samples spiked with CTX1B or CTX3C standards led to the development of an optimized protocol (named OP), shown in Figure 1.

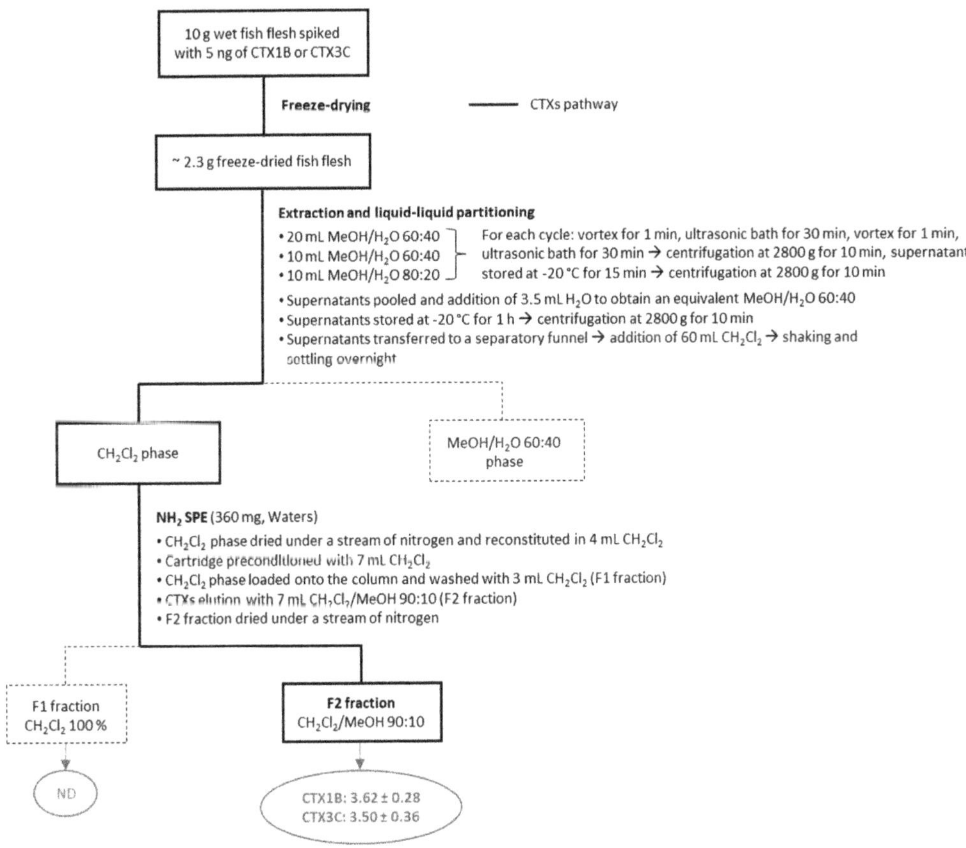

Figure 1. Schematic representation of the OP protocol with CTX1B and CTX3C amounts (ng) estimated by the CBA-N2a in testable fractions. Data represent the mean ± standard deviation (SD) (each concentration run in triplicate wells) of three independent experiments run on different days. In bold is the CTX pathway through the different steps.

Firstly, the wet fish flesh was freeze-dried to overcome differences in water content between the different samples and to facilitate the shipment and exchange of the samples between laboratories. The percentage of aqueous MeOH used in the extraction step was optimized to improve the recovery of the nonpolar CTXs. Briefly, four conditions (i.e., MeOH/H_2O 60:40, MeOH/H_2O 80:20, MeOH 100% and dichloromethane (CH_2Cl_2) 100%)

were tested on spiked fish flesh and the best compromise between CTX3C recovery and a low amount of matrix co-extractives was obtained using MeOH/H_2O 80:20.

Furthermore, to reduce the number of evaporation/redissolving steps on the extracts, and as proposed in protocol #2, a liquid/liquid partitioning step (MeOH/H_2O 60:40 vs. CH_2Cl_2 100%) was performed twice directly using the supernatant obtained from the extraction step.

Finally, based on the results previously obtained with protocol #2 (see Section 2.1), an aminopropyl (NH_2) SPE column was used to recover both polar and nonpolar CTXs in a single fraction.

This new improved protocol thus allowed CTX1B and CTX3C standards to be eluted in a single fraction of interest (F2 fraction), with recoveries of 72.5 ± 5.7 and 70.0 ± 7.2% (n = 9) for CTX1B and CTX3C, respectively, as estimated by the CBA-N2a, and with a completion time of three days.

2.3. Application of the OP Protocol to Naturally Contaminated Marine Product Samples

To confirm the efficiency of the improved OP protocol, additional tests were performed on naturally contaminated herbivorous and carnivorous marine products samples collected from French Polynesian ciguateric areas. For comparison purposes, the samples were extracted in parallel using both protocol #1 (already used in previous studies to analyze French Polynesian marine products samples) and the OP protocol. Then, for each marine product, the fraction of interest obtained from the two protocols (i.e., the F2 fraction eluted from a C18 SPE cartridge with MeOH/H_2O 90:10 for protocol #1, and the F2 fraction eluted from an NH2 SPE cartridge with CH_2Cl_2/MeOH 90:10 for the OP protocol) were tested in parallel using the CBA-N2a to determine their respective CTX contents.

The results showed that the estimated CTX levels for carnivorous fish (likely to contain mostly polar CTXs) were broadly similar for both protocols, ranging from 0.075 ± 0.006 to 0.714 ± 0.103 and from 0.067 ± 0.004 to 0.836 ± 0.036 ng eq. CTX1B g^{-1} wet fish flesh for protocol #1 (n = 9) and the OP protocol (n = 6), respectively (Figure 2). Equivalent toxicities were also calculated for the two herbivorous fish samples (likely to contain mostly nonpolar CTXs), with CTX levels quantified at 0.48 ± 0.06 and 0.09 ± 0.01 ng CTX3C eq. g^{-1} wet fish flesh for protocol #1 (n = 9) and 0.57 ± 0.09 and 0.09 ± 0.02 ng CTX3C eq. g^{-1} wet fish flesh for the OP protocol (n = 6) in the steephead parrotfish *Chlorurus microrhinos* and yellowfin surgeonfish (*Acanthurus xanthopterus*) samples, respectively (Figure 2). However, regarding the trochus *Tectus niloticus*, a 1.6-fold increase in the CTX level measured was observed using the OP protocol (16.6 ± 1.1 ng CTX3C eq. g^{-1} wet flesh) compared to protocol #1 (10.6 ± 0.9 ng CTX3C eq. g^{-1} wet flesh) (Figure 2).

The OP protocol also removed the adverse matrix effects previously observed, as demonstrated by the absence of non-specific mortality in the CBA-N2a at the maximum concentration tested (100 mg equivalent wet weight per mL) for the yellowfin surgeonfish and bluefin trevally samples (Figure 3). This reduction in matrix effects enabled the analysis of more than 100 mg eq. wet weight mL^{-1} with the new OP protocol versus. 34 mg eq. wet weight mL^{-1} with protocol #1 (Table 2). This provided improved limit of detection (LOD) and limit of quantification (LOQ) values for the CBA-N2a by 3-fold compared to the values published in Viallon et al. [13], now estimated at 0.01 and 0.02 ng CTX3C eq. g^{-1}, respectively. Of note, the improved LOD and LOQ meet the guidance safety limits recommended by the US-FDA and EFSA [50,51].

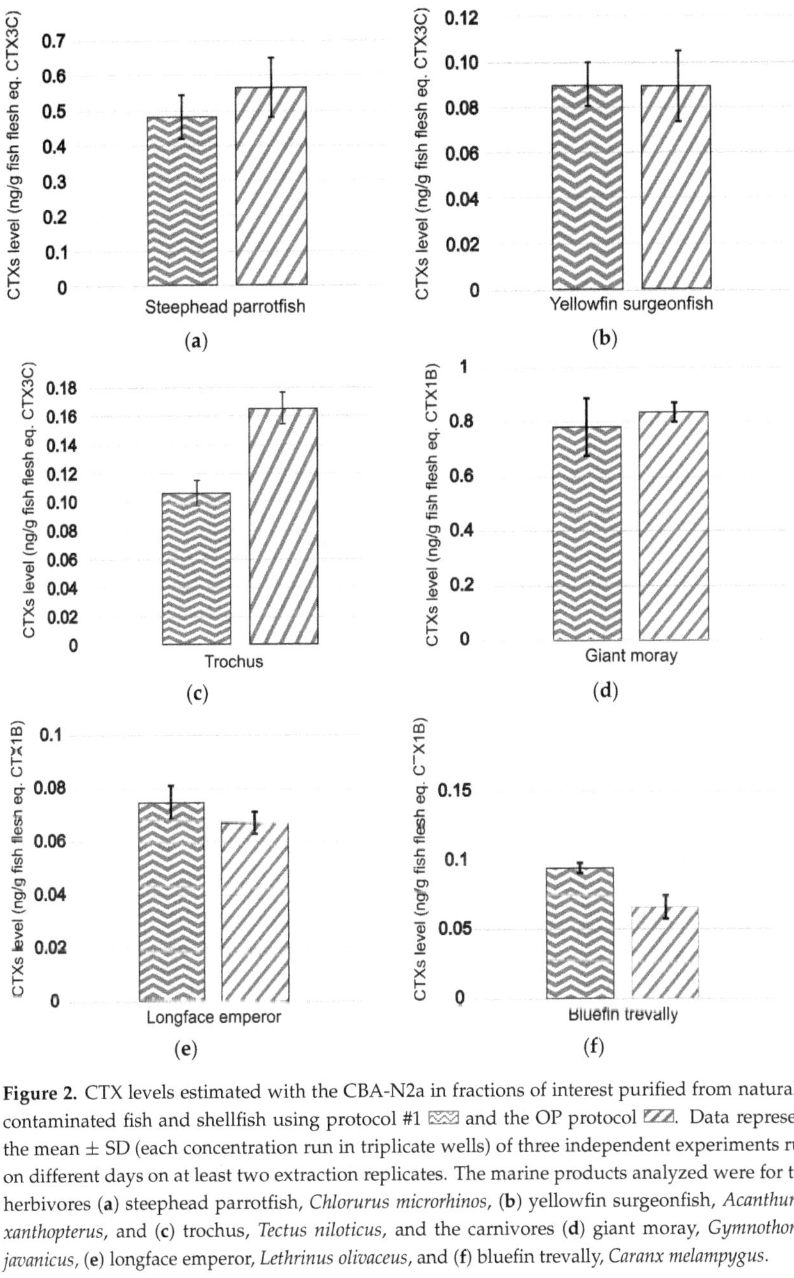

Figure 2. CTX levels estimated with the CBA-N2a in fractions of interest purified from naturally contaminated fish and shellfish using protocol #1 and the OP protocol. Data represent the mean ± SD (each concentration run in triplicate wells) of three independent experiments run on different days on at least two extraction replicates. The marine products analyzed were for the herbivores (**a**) steephead parrotfish, *Chlorurus microrhinos*, (**b**) yellowfin surgeonfish, *Acanthurus xanthopterus*, and (**c**) trochus, *Tectus niloticus*, and the carnivores (**d**) giant moray, *Gymnothorax javanicus*, (**e**) longface emperor, *Lethrinus olivaceus*, and (**f**) bluefin trevally, *Caranx melampygus*.

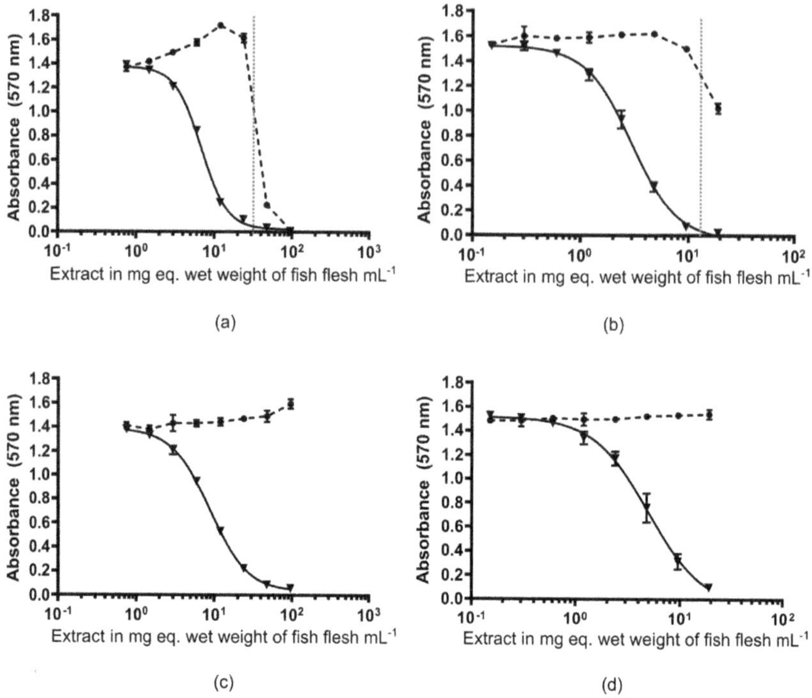

Figure 3. Cytotoxic dose–response curves under OV⁻ (black circle) and OV⁺ (black triangle) conditions obtained using the CBA-N2a on fractions of interest purified from naturally contaminated yellowfin surgeonfish and bluefin trevally samples using protocol #1 (**a**,**b**) and the OP protocol (**c**,**d**). Data represent the mean ± SD of each concentration run in triplicate wells. Non-specific mortality due to matrix effects is observed above the dotted red line.

2.4. Evaluation of the Suitability of the OP Protocol for LC-MS/MS Sample Analyses

In view of the promising results obtained with the CBA-N2a, the suitability of the OP protocol for LC-MS/MS sample analyses was subsequently evaluated.

Since the recoveries calculated from the LC-MS/MS analyses were consistently much lower than those obtained with the CBA-N2a, fortification experiments were performed to evaluate the matrix effects using two carnivorous fish and one herbivorous fish sample, i.e., a spiking of the final fraction of interest with two mixes of CTX1B, CTX3B, CTX3C and CTX4A, with each standard at a final concentration of 5 or 10 ng mL^{-1}.

Significant matrix effects were observed when analyzing the extracts of the three fishes generated using the OP protocol (Figure 4a). First, varying LC-MS/MS signal suppression was observed, which was species-specific and showed that the bluefin trevally sample induced the highest suppression. Practically, up to 95% signal suppression was evident for CTX1B for this sample, while ranging from 52 to 59% for CTX3B, CTX3C and CTX4A. Overall, CTX1B was the analogue most affected by the matrix signal suppression, with a suppression effect of 58–95% depending on the fish species. For nonpolar ciguatoxins, signal suppression was low (<25%) in the parrotfish and grouper samples.

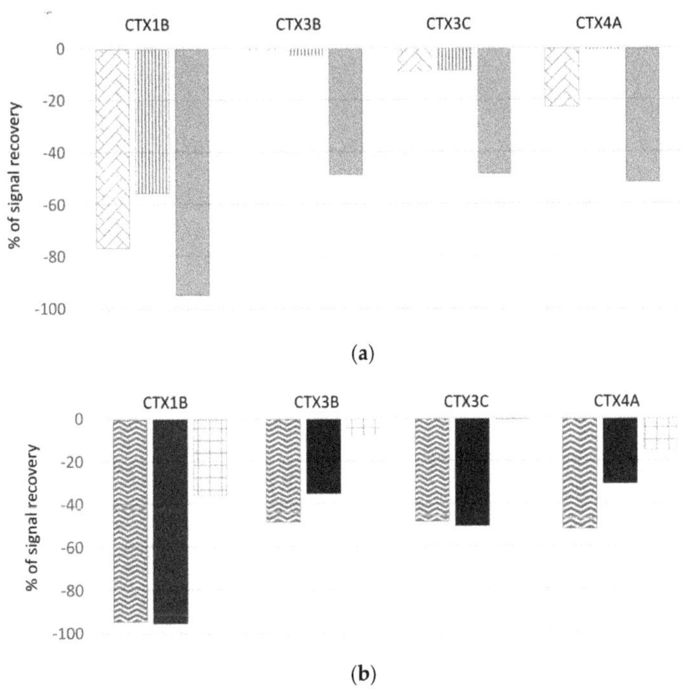

Figure 4. Percentages of signal suppression observed in LC-MS/MS analysis after fortification experiments (spiking extracts with a mix of CTX1B, CTX3B, CTX3C and CTX4A standard solutions) of the (**a**) steephead parrotfish, marbled grouper and bluefin trevally, obtained with the OP protocol, as well as of the (**b**) bluefin trevally, obtained with the OP protocol, protocol #1 and the protocol published in Murray et al. [27].

In addition, the same bluefin trevally sample was extracted using the OP protocol, protocol #1 and the protocol described in Murray et al. [27] (from which protocol #2 was inspired) (Figure 4b). Using protocol #1, equivalent signal suppressions were observed for the fractions of interest as compared to the OP protocol. In contrast, the fractions of interest purified using the Murray et al. [27] protocol, which was developed for LC-MS/MS analysis, showed much lower signal suppressions (i.e., 36% for CTX1B; 8% for CTX3B; 1% for CTX3C; and 15% for CTX4A) compared to the OP protocol.

3. Discussion

The globalization of CP has led to renewed interest from the scientific community and regulatory agencies with regard to the management of this prominent public health issue. As a result, experts have identified several priority areas for improvement, one of which is the optimization of toxin detection in marine products [11,52]. Indeed, the implementation of efficient CP risk surveillance and food safety monitoring programs is currently hampered by the critical need to improve the analytical methods used for CTXs, including sample preparation protocols [28]. When it comes to the detection and quantification of CTXs in marine products, the challenge is considerable due to (i) the trace amounts of CTXs usually present in marine product samples, though sufficient to trigger ciguatera food poisoning [12], and (ii) the complexity of these matrices, whose compositions also vary according to the part of the organism analyzed, species, size, time of year, fishing site and geographic origin [53].

Hence, achieving the development of rapid and effective protocols for extracting and purifying CTXs from marine products is critical for the development of effective monitoring programs, which will ultimately reduce the health risk to communities. Several approaches have been reported in the literature, involving sample sizes ranging from 1 to 10 g of wet flesh or a freeze-dried equivalent, as well as an extraction step (preceded or not by a cooking step or enzymatic digestion) with organic solvents or a mixture of organic and aqueous solvents, followed by one or more purification steps, either by liquid–liquid partitioning or by SPE, to reduce the matrix co-extractives among the CTXs [23,24,37,54]. However, comparing the performance of these protocols remains difficult, if not impossible, due to the lack of standardized protocols across laboratories/studies [11] and the lack of a common source of reference standards.

The aim of the present study was therefore to design a simple and rapid extraction/purification protocol applicable to the high-throughput screening of a large number of samples, while ensuring the recovery of at least 70% of all CTXs present (regardless of their polarity) and a reduction in matrix effects to improve CBA-N2a sensitivity.

3.1. CTX3C Extraction Efficiency

In this study, CTX3C recovery yields were compared between the various protocols available from the literature using triplicates of spiked fish with a single batch of standards also used for CBA-N2a analyses, enabling a direct and reliable comparison of the parameters used at each stage of these protocols. Studies using a similar approach are scarce in the literature [23,24,27] due to the limited availability and high cost of reference standards but was made possible in the present study using the standards available from the bank of CTX standards of the Institut Louis Malardé (ILM).

The results showed that the six protocols evaluated were poorly suited to the extraction and purification of nonpolar CTXs. In fact, CTX3C recoveries ranging from 0 to 41% were found in the final fractions of interest, with the remainder of the CTX3C either not extracted from the fish flesh or found in the other fractions. In protocol #2 (inspired by Murray et al. [27]), a low CTX3C recovery of 6% was observed, consistent with their study, evaluating a CTX3C recovery of 13%. In contrast, with protocol #6 (inspired by Nagae et al. [23]), a 41% CTX recovery was found in the present study while a recovery above 100% was reported by the authors. Nevertheless, this represents the best CTX3C recovery of the six protocols evaluated. The lower nonpolar CTX recovery rate observed in the present study could be due to the difficulty of reproducing the complex six-step sample preparation protocol proposed by Nagae et al. [23]. The four other protocols, from which protocols #1, #3, #4 and #5 were inspired, did not previously evaluate the recovery of nonpolar CTXs [25,45,55,56], but reported satisfactory recoveries for polar CTXs, confirming their suitability for the extraction/purification of this group of CTX analogues. For example, Mak et al. [45], from which protocol #5 was inspired, reported recoveries ranging from 61 to 87% for CTX1B, CTX2 and CTX3. Meyer et al. [56], from which protocol #3 was inspired, demonstrated a 46% improvement in CTX1B recovery compared to the protocol of Lewis et al. [42], which previously reported a CTX1B recovery of 85%. Finally, Nagae et al. [23] reported CTX1B recoveries of around 90%. However, it is important to note that some of these studies used matrix-corrected standards (i.e., considering the potential matrix effect of the extract analyzed) for their quantification compared with the solvent standards, resulting in a significant impact on their reported recovery percentages.

In addition, although most of the methods reported sensitivities meeting CTX sanitary thresholds [50,51] and could therefore detect the presence of CTXs in low-level contaminated marine products, it is likely the actual quantity of CTXs per g of wet flesh reported was greatly underestimated, which could lead to the misclassification of fish as non-toxic or safe for consumption.

All of these observations suggest that there is potential for a significant improvement to most of the currently published protocols, especially when it comes to the increased recovery of all the CTXs present in a single fraction of interest, regardless of their polarity.

3.2. Finding the Best Compromise Between Extraction Efficiencies of Ciguatoxins and Matrix

The work carried out in the present study helped provide useful insights into some of the protocol parameters likely to influence the recovery efficiency of nonpolar CTXs such as CTX3C.

One of these parameters relates to the choice of extraction solvent. The results obtained for protocols #3 and #4 highlighted that the use of a pure organic solvent such as methanol or acetone without the addition of water was not efficient for the extraction of nonpolar CTXs from freeze-dried flesh samples. Indeed, although protocols #1 and #3 shared the same extraction method (i.e., acetone, ultrasonic bath), the percentage of recovery of CTX3C with protocol #3 (23%) was lower than with protocol #1 (32.2%) because protocol #3 was carried out on freeze-dried fish flesh as opposed to the wet fish flesh (containing approximately 80% of water) in protocol #1, as previously suggested by Lewis et al. [42]. From a technical perspective, it is likely that the presence of water in the organic solvent/water mixture causes the freeze-dried flesh fibers to swell, thus increasing their contact surface with the solvent. Nevertheless, if the proportion of water in the organic solvent/water mixture selected for extraction is too high (i.e., >30%), there is a risk of reducing the extraction efficiency of nonpolar CTXs, as demonstrated for protocol #2.

The second parameter concerns the liquid–liquid partition step. Despite the different biphasic mixtures chosen (aqueous MeOH/CH_2Cl_2 for protocols #1 and #2; aqueous MeOH/Hexane for protocols #3, #4 and #6; and ACN/CH_2Cl_2 for protocol #5), low losses of CTX3C were observed at this step across all protocols, with only traces of CTXs detected in the wash phase.

The third parameter regards the SPE clean-up step to limit the amount of matrix co-extractives extracted with the CTXs [14]. The results obtained in this study highlighted the importance of the choice of the mobile phase used for CTX elution with C18 SPE. Indeed, when the concentration of H_2O in the mobile phase (i.e., a MeOH/H_2O mixture) exceeds 20%, a large amount of CTX3C remains on the solid phase, as demonstrated in protocol #4, consistent with previous findings by Spielmeyer et al. [24], whereas less CTX3C remains on the solid phase when 10% of H_2O is used, as in protocols #1 and #3. This result suggests a very narrow range of H_2O percentages should be used in an attempt to elute all CTX3C but without recovering too much matrix. Actually, when no H_2O is added to the mobile phase, such as in protocol #6, the totality of the CTX3C is eluted; however, this may considerably limit the interest of this SPE step as numerous matrix compounds are likely eluted at the same time. Alternative SPEs can also be used, as in protocols #4 and #5, in which silica was used as the stationary phase. However, for these two protocols, since the nonpolar CTXs remained on the solid phase during the previous step, it was not possible to evaluate the true benefits of this silica stationary phase. Al-though the six protocols tested were primarily developed to ensure the elution of polar CTXs in the fraction of interest, the present study was useful in characterizing the elution conditions enabling polar and nonpolar CTX recovery in a single fraction: the only stationary phase allowing the recovery of both polar (as described in Murray et al. [27]) and nonpolar CTXs in the same fraction was aminopropyl-modified silica with MeOH/CH_2Cl_2 90:10 for the elution.

Finally, another key factor to consider is the dry weight of the fraction of interest and related matrix effects. According to the CBA-N2a results, the purification steps of protocols

#3, #4, #5 and #6 significantly reduced the matrix effects to a point where no matrix effects were observed. In contrast, non-specific mortality was observed in N2a cells when exposed to concentrations of 34 and 84 mg eq. wet weight of fish flesh, respectively, obtained with protocols #1 and #2, respectively. The use of several SPEs is therefore recommended to reduce these matrix effects; however, the results obtained in this study highlight the fact that these steps are also a potential source of a significant loss of nonpolar CTXs. The careful set-up of the mobile and stationary phase conditions is therefore of utmost importance to ensure the best compromise possible.

Overall, these findings indicated that there was potential for the design of an improved extraction/purification protocol for all polar and nonpolar CTXs simultaneously.

3.3. Improved Extraction/Purification Protocol for CTXs

The second objective of this study was to develop a high-throughput protocol that would combine the pros of the previously evaluated protocols, while ensuring the recovery of at least 70% of the polar and nonpolar CTXs in a single fraction.

To this end, two solvent phases were successively used for the extraction, first 60:40 MeOH/H_2O, well suited for swelling the fish flesh fibers and extracting polar CTXs, followed by MeOH/H_2O 80:20, more suited for the extraction of nonpolar CTXs while limiting the amount of lipid co-extractives in the resulting extract. This latter supernatant was then directly subjected to 60% aqueous MeOH/CH_2Cl_2 liquid–liquid partitioning, reducing the risk of losses during evaporation cycles. This liquid–liquid partitioning is a cost-effective step commonly used for CTX purification from *Gambierdiscus* extracts [57], and it allowed a reduction in the dry mas extract by 16-fold. In addition, the SPE aminopropyl-modified silica phase also helped reduce extract complexity, as illustrated by the reduction in the final extract dry mass by 3-fold. As a result, the newly designed protocol, which required only three days to complete, allowed the recovery of > 72% of the CTX1B and 70% of the CTX3C standards spiked on fish flesh in a single fraction of interest following two purification steps.

This new protocol shares many similarities with the protocol published in Murray et al. [27], most notably the liquid–liquid partitioning and aminopropyl SPE steps. However, the OP protocol suggests an additional extraction step to improve the yield of nonpolar CTXs. Furthermore, in contrast to Murray et al. [27], the whole extract is kept throughout the protocol, which improves sensitivity and reduces the risk of false negatives when analyzing marine products with low CTX contents. In the meantime, other methods have been published, such as the study by Spielmeyer et al. [24], presenting a rapid and innovative protocol, but with a distribution of CTX3C in two different fractions and less than 35% recovery of spiked CTX3C in the fraction of interest (named CTX3C–eluate). More recently, a shortened version of this protocol was proposed by Loeffler et al. [58], but as no validation by spiking experiments was performed, the actual recovery rates remain unknown, and thus the efficiency of this protocol could not be evaluated at this time. Another protocol was also proposed by Nagae et al. [23], which provides satisfactory extraction yields for polar and nonpolar CTXs, but whose complexity and cost (i.e., it involves two liquid–liquid partitionings and four SPE steps) make it unsuitable for the screening of a large number of samples.

3.4. Applicability of the OP Protocol to Naturally Contaminated Fish Samples

The applicability of the OP protocol was further evaluated on a selection of six naturally contaminated samples—three carnivorous fish, two herbivorous fish and one gastropod—collected in French Polynesia during field trips in known ciguateric areas [35].

Unlike the previous spiking experiments in which a single CTX standard was spiked onto fresh flesh prior to freeze-drying, evaluating the OP protocol directly on naturally contaminated fish provides an opportunity to study the efficiency of the protocol on more complex samples. Indeed, naturally contaminated samples contain CTXs that are actually embedded in their flesh, and whose toxin profiles are potentially more complex due to a possible metabolization of the CTXs bioaccumulated in organisms from lower trophic levels [5,37,59,60]. Due to the limited amount of naturally contaminated material, the OP protocol was only compared to protocol #1, which was used beforehand for the selection of naturally contaminated samples.

When applied to several marine products naturally contaminated with different concentrations of CTXs (between 0.07 and 0.8 ng CTX1B eq. g^{-1} for carnivorous fish, and between 0.09 and 0.5 ng CTX3C eq. g^{-1} for herbivorous fish and shellfish), the OP protocol confirmed its good performance. This was demonstrated by the higher CTX levels measured in the steephead parrotfish and trochus samples versus similar toxicities in the three carnivorous fish and the yellowfin surgeonfish when compared to the protocol #1 toxicity data. These results align with those obtained from our previous spiking experiments. Indeed, since the OP protocol had a better recovery of nonpolar CTXs compared to protocol #1, the quantity of CTXs in CTX3C equivalence increased due to the high proportion of nonpolar CTXs contained in the steephead parrotfish and trochus shell samples [25,33]. Similarly, since the OP protocol did not significantly improve the recovery of CTX1B, the CTXs' quantity in CTX1B equivalence measured in the three carnivorous fish was similar between the two protocols.

For the herbivorous yellowfin surgeonfish sample, a similar toxicity was observed using both protocols. However, the OV- curve (i.e., without treatment sensitizing the cells to sodium channel activators such as CTXs) observed during the CBA-N2a analysis of the sample extracted with protocol #1 demonstrated an important matrix effect. Indeed, for sensitivity purposes, since the CTX amount in this sample was very low, a dry extract mass higher than the apparent MCE was analyzed for the fraction obtained with protocol #1, leading to matrix effects and thus to an overestimation of the CTX amount.

3.5. Matrix Effect Evaluation of the OP Protocol for Subsequent CBA-N2a and LC-MS/MS Analyses

One of the key considerations for a sensitive and robust analysis is to ensure that no potential matrix effect could interfere with the analysis, e.g., the non specific mortality of N2a cells in the CBA-N2a. Indeed, as demonstrated for the yellowfin surgeonfish sample in the present study, this non-specific mortality can be observed above a certain concentration of fish dry extract, consistent with observations from previous studies [13,14,61,62]. Therefore, it is of utmost importance to i) ensure the sample preparation is well suited to limit such matrix effects and increase the test sensitivity and ii) determine the MCE prior to N2a cell exposure to the fish extracts (e.g., by performing a preliminary analysis on non-toxic fish or under OV- condition). For LC-MS/MS analysis, the matrix effects can be observed by an increase in background noise [23,56], but also by signal suppression phenomena [27,63], which can differ depending on the characteristics of the sample and the CTX analogues considered. This is why, depending on the study, the detection limits using LC-MS/MS may vary for each analogue (depending on the specific CTX relative response factor and signal suppression or enhancement effects from matrix co-extractives) [25]. The OP protocol demonstrated a low matrix effect in the CBA-N2a analysis with an MCE > 100 mg equivalent wet weight mL^{-1}, which represents a true advancement for detecting low levels of CTXs. Unfortunately, it was not suitable for a subsequent detection of CTXs by LC-MS/MS, which was likely due to the residual presence of compounds that strongly interfered with CTX signals, particularly with polar CTX analogues (from 55 to 95% of signal suppression). In comparison, the protocol of Murray et al. [27] seemed to be more suitable for the detection of CTXs by LC-MS/MS, but on the other hand were less effective for their detection by the CBA-N2a. This suggests that the compounds interfering with LC-MS/MS analyses differ from those interfering with the CBA-N2a. The present results also highlight the impact of the matrix on the results of the final toxin profile obtained, and it seems very important when LC-MS/MS analyses are carried out to take into account this matrix effect, either using matrix-calibrated standards [64] or fortification experiments [27].

Risk surveillance and food safety monitoring programs of CP require the analysis of large numbers of samples in order to assess the levels of CTXs accumulated in marine products commonly consumed by local populations for a reliable assessment of the actual risk to consumers. This requires the availability of a simple and rapid sample preparation protocol providing reliable toxicity data by the CBA-N2a during the first screening assay of the two-tiered approach recommended by ciguatera experts [11]. Although it would have been ideal to have a single optimized protocol further applicable to both functional (e.g., CBA-N2a) and chemical (e.g., LC-MS/MS) screening tests, the present results highlight the difficulty, if not impossibility, to achieve such a goal given the significant differences in methodology and the related matrix effects inherent to these two methods.

4. Materials and Methods

4.1. Chemicals

Chemicals used for extraction, purification, sample resuspension and CBA-N2a were acetone (AnalaR Normapur, ACS, VWR), citric acid (ACS, VWR), acetonitrile (LiCrosolv, MERCK), ammonium acetate (LC-MS grade, Sigma Aldrich, Saint Louis, MO, USA), chloroform (HPLC-grade Fischer Scientific, Hampton, NH, USA), cyclohexane (AnalaR Normapur, ACS, VWR), dichloromethane (CH_2Cl_2) (AnalaR Normapur, ACS, VWR), dimethylsulfoxide (DMSO) (HPLC grade, PanReac Applichem (Darmstadt, Germany), VWR), ethyl acetate (EtOH) (Chromanorm HPLC, Prolabo), hexane (AnalaR, BDH), sodium chloride (NaCl) (AnalaR Normapur, ACS, VWR), methanol (MeOH) (AnalaR Normapur, ACS, VWR) and sodium carbonate (Na_2CO_3) (AnalaR Normapur, ACS, VWR).

4.2. CTX Standards

Both CTX1B and CTX3C standards were obtained from the Institut Louis Malardé (ILM) CTX standards bank. The CTX3C standard originates from in vitro cultures of the highly toxic *Gambierdiscus polynesiensis* TB92 strain, as described in Chinain et al. [57]. The CTX1B standard originates from a large stock purified from moray eels' livers, as described in previous studies [65–67]. Both standards were resuspended in MeOH to prepare stock solutions at 1 $\mu g\ mL^{-1}$.

4.3. Biological Materials

The evaluation of the CTX3C extraction/purification efficiencies of the six different protocols was performed using the same carnivorous fish specimen, the Tuamotu emperor (*Lethrinus atkinsoni*, Lethrinidae). This fish was harvested in 2017 from the lagoon of Mangareva Island (Gambier Archipelago, French Polynesia) and had previously shown no CTX-like activity with the CBA-N2a.

For the development of a new, improved extraction/purification protocol, the same batch of two carnivorous fish specimens belonging to the same species, i.e, the bluefin trevally (*Caranx melampygus*, Carangidae), mixed together was used. These two fish were harvested in 2013 and 2014 from the lagoon of Tikehau and the Tubuai Islands (Tuamotu and Australes Archipelagoes, French Polynesia) and had previously shown no CTX-like activity using the CBA-N2a.

To confirm the extraction performance of the newly improved OP protocol, six naturally contaminated fish and shellfish flesh samples (as previously confirmed by the CBA-N2a) were used: giant moray (*Gymnothorax javanicus*, Muraenidae); longface emperor (*Lethrinus olivaceus*, Lethrinidae); bluefin trevally (*Caranx melampygus*); steephead parrotfish (*Chlorurus microrhinos*, Scaridae); yellowfin surgeonfish (*Acanthurus xanthopterus*, Acanthuridae); and trochus gastropod (*Tectus niloticus*, Tegulidae). Finally, for the evaluation of the OP protocol for subsequent LC-MS/MS analyses, a marbled grouper (*Epinephelus polyphekadion*, Serranidae) fish flesh sample was used in addition to the same bluefin trevally (*Caranx melampygus*) and steephead parrotfish (*Chlorurus microrhinos*) samples used for the CBA-N2a analyses.

4.4. Spiking Procedure

Aliquots containing 10 g of ground wet non-toxic fish flesh were spiked with 5 μL of a CTX stock solution (corresponding to 5 ng of CTX) and then lyophilized for 48 hours (primary dessication: 44 h, 1 mbar, −20 °C; secondary dessication: 4 h 0.01 mbar, −76 °C) (Martin Christ, Beta 1–8 LDplus, Osterode am Harz, Germany).

4.5. Evaluation of Efficiencies of Six Different Protocols for Extraction/Purification of CTX3C

The efficiencies of the six different protocols for the extraction/purification of CTX3C were evaluated using a CBA-N2a.

For each protocol, three aliquots of 10 g wet weight of fish flesh (*Lethrinus atkinsoni*) were spiked with 5 ng of the CTX3C standard. An additional aliquot of 10 g wet weight of fish flesh (non-spiked sample) was also prepared to evaluate matrix effects using the CBA-N2a.

The samples were subjected to six different protocols inspired by some of the protocols commonly used for CTX analyses in fish matrices [23,25,27,45,56]. These six protocols are summarized in Table 3 and detailed in the Supplementary Materials. Additional washing fractions were added to each SPE purification step to ensure that all the CTXs were recovered under the conditions used in the protocols. The fractions obtained along the six chemical procedures were dried and weighted for further analyses by the CBA-N2a to assess CTX3C recovery.

Table 3. Description of the six protocols used for extraction/purification of CTX3C from fish.

Protocol	#1	#2	#3	#4	#5	#6
Reference	Darius et al. [68]	Inspired by Murray et al. [27]	Inspired by Sibat et al. [25]	Inspired by Meyer et al. [56] [a]	Inspired by Mak et al. [45]	Inspired by Nagae et al. [23] [b]
Fish flesh amount	10 g wet	10 g wet	10 g wet and then freeze-dried	10 g wet and then freeze-dried	10 g wet and then freeze-dried	10 g wet
Extraction	• Acetone 100% ×2 • Ultrasonic bath • −20 °C precipitation	• aq MeOH 60% • Blending Ultra-Turrax • Heat to 100 °C • Ice bath precipitation	• Acetone 100% ×2 • Ultrasonic bath	• MeOH/hexane 73:27 ×2 • Blending Ultra-Turrax • Washing MeOH	• MeOH 100% ×2 • aq MeOH 60% • Ultrasonic bath • Heat to 75 °C	• aq MeOH 90% • Blending Ultra-Turrax • Heat to 80 °C
Purification 1	LLP [c] aq MeOH 60%/CH_2Cl_2 ×2	LLP aq MeOH 60%/CH_2Cl_2 ×2	LLP aq MeOH 90%/hexane ×2	LLP aq MeOH 57%/Hexane	• SPE C18 • Elution aq ACN 65%	LLP aq MeOH 90% + saturated Na_2CO_3/hexane
Purification 2	LLP aq MeOH 80%/cyclohexane	• SPE NH2 • Elution CH_2Cl_2/MeOH 90:10 (F2 fraction) [d]	• SPE Florisil • Elution EtOAc/MeOH 90:10	• SPE C18 • Elution aq MeOH 80%	LLP aq ACN 65% + 1M NaCl/CH_2Cl_2	LLP aq MeOH 90% + saturated Na_2CO_3 + 5% citric acid/hexane

Table 3. *Cont.*

Protocol	#1	#2	#3	#4	#5	#6
Purification 3	• SPE C18 • Elution aq MeOH 90% (F2 fraction) [d]	/	• SPE C18 • Elution aq MeOH 90% (F1.2 fraction) [d]	• SPE silica (HILIC mode) • Elution acetone/H_2O 95:5 containing 5 mM ammonium acetate (F3.2 fraction) [d]	• SPE Silica • Elution MeOH/chloroform 10:90 (F3.2 fraction) [d]	• SPE Florisil • Elution EtOAc/MeOH 85:15
Purification 4	/	/	/	/	/	• SPE C18 • Elution 100% MeOH
Purification 5	/	/	/	/	/	• SPE PSA • Elution aq MeOH 78% (F1.2.1 fraction) [d]

[a] Previously adapted from Lewis et al. [42]. [b] Previously optimized from Yogi et al. [69]. [c] LLP = Liquid–liquid partitioning. [d] Fraction of interest to be analyzed for CTX detection/quantification.

4.6. Development of a New, Improved Extraction/Purification Protocol

Based on the advantages of each of the six protocols previously evaluated, numerous tests were performed on aliquots of 10 g wet weight of fish flesh (*Caranx melampygus*) spiked with CTX1B or CTX3C, especially to optimize the solvent mixtures used for the extraction step, leading to the development of the improved OP protocol described below.

4.6.1. Extraction

Ten grams wet weight of spiked (n = 3) and non-spiked (n = 1) fish flesh samples were aliquoted into a 50 mL plastic tube and freeze-dried for 48 h (freeze-dried dry weight approx. 2.3 g). Then, 20 mL of MeOH/H_2O 60:40 (v:v) was added to the tube. The extraction cycle was homogenization by vortex for 1 min, extraction in a 37 KHz ultrasound bath for 30 min, homogenization by vortex for 1 min before a second extraction cycle in the ultrasound bath for 30 min and finally homogenization by vortex for 1 min.

The tube was then placed in a freezer at −20 °C for 15 min to facilitate fat precipitation. The supernatant was obtained by centrifugation for 10 min at 2800× *g* and transferred to a new 50 mL plastic tube, which was placed in the freezer at −20 °C.

A second and a third extraction cycle were carried out in the same way using 10 mL of MeOH/H_2O 60:40 and 10 mL of MeOH/H_2O 80:20, respectively.

The three supernatants were combined in the same tube and 3.5 mL of water was added to bring the MeOH concentration below 60%. The combined supernatants were placed in a freezer at −20°C for 1 h.

4.6.2. Liquid–Liquid Partitioning

A centrifugation step at 2800× *g* for 10 min was then carried out to remove the fats and the resulting supernatant was transferred to a glass separating funnel. A volume of CH_2Cl_2 equivalent to twice the volume of supernatant was added to the separatory funnel (approximately 60 mL), and the mixture was mixed gently and then left to settle overnight. The CH_2Cl_2 phase was recovered and evaporated under gentle nitrogen flux.

4.6.3. SPE Fractionation

A Sep-Pak aminopropyl (NH2) classic cartridge (360 mg, Waters, Saint-Quentin-En-Yvelines, France) was conditioned with 7 mL CH_2Cl_2. The dry extract was resuspended in 2 + 1 + 1 mL of CH_2Cl_2 and loaded on the SPE phase. An additional milliliter of CH_2Cl_2 was used to rinse the sample tube and was added on the SPE phase. Washing was then performed with 2 mL CH_2Cl_2 (F1 fraction). Elution was carried out with 7 mL CH_2Cl_2/MeOH 90:10 and the resulting fraction (F2 fraction) was dried under gentle nitrogen flux and weighted for further analyses by the CBA-N2a to assess CTX1B and CTX3C recoveries.

4.7. *Application of the OP Protocol to Naturally Contaminated Marine Product Samples*

To confirm the efficiency of the new, improved OP protocol, tests were performed on six naturally contaminated shellfish and herbivorous and carnivorous fish samples collected from French Polynesian ciguateric areas. For comparison purposes, two to three aliquots of 10 g wet weight of ground marine specimen flesh were extracted using either the OP protocol or protocol #1 (inspired by Darius et al. [68]). The fractions of interest (i.e., the F2 fraction eluted from an NH2 SPE cartridge with CH_2Cl_2/MeOH 90:10 for the OP protocol, and the F2 fraction eluted from a C18 SPE cartridge with MeOH/H_2O 90:10 for protocol #1) were then analyzed with the CBA-N2a.

4.8. *Neuroblastoma Cell-Based Assay (CBA-N2a)*

All CBA-N2a experiments were conducted using N2a, a mouse neuroblastoma cell line (CCL-131) purchased from the American Type Culture Collection (ATCC, Manassas, VA, USA), according to the optimized protocol published by Viallon et al. [13].

4.8.1. CBA-N2a Procedure

Briefly, a density of 50,000 N2a cells per well (total volume 200 µL) was seeded in 5% fetal bovine serum (FBS) RPMI-1640 supplemented medium in 96-well microtiter plates and kept at 37 °C with a 5% CO_2 humidified atmosphere. The confluence of the cells was reached after 24 h of growth and the culture medium was then removed. The renewal of the culture medium consisted in the deposit of 200 µL of the 2% FBS RPMI-1640 supplemented medium in the upper half of the 96-well plate (OV- condition) and 200 µL of the 2% FBS RPMI-1640 supplemented medium containing a non-destructive treatment of ouabain–veratridine (OV) at [70/7–100/10] µM in the lower half of the 96-well plate (OV+ condition). Then, 10 µL of eight-point serial 1:2 dilutions of the dry extract or CTX standards was added to wells in triplicates for incubation overnight. Cell viability was measured using the methylthiazolyldiphenyl–tetrazolium bromide (MTT) assay after 45 min of incubation at 570 nm using a plate reader (iMark Microplate Absorbance Reader, BioRad, Marnes la Coquette, France). The net absorbance data were fitted to a sigmoidal dose–response curve (variable slope) based on a four-parameter logistic model (4PL). Half-maximal effective concentration (EC_{50}) values were obtained using Prism v9.0.2 software (GraphPad, San Diego, CA, USA) and converted in ng CTX equivalents (eq.) g^{-1} of wet tissue (flesh) by comparison with the CTX EC_{50} values.

For all experiments, the absorbance values of the viability controls such as the OV- and OV+ control wells (N2a cells alone) as well as the OV- and OV+ quality control (N2a cells exposed to VGSC standards) wells were used to verify the maintenance of N2a viability and the specific detection of CTXs, as described in Viallon et al. [13].

4.8.2. Strategy of Analyses of Fractions

Measure of Matrix Effects

Using the non-spiked samples, all the fractions obtained using the six chemical procedures and the OP protocol were weighed and resuspended in at least 100 µL and up to 520 µL of DMSO depending on the quantity of dry extract to be solubilized.

After a 10-fold dilution in culture medium, an eight-point serial 1:2 dilution in RPMI 2% FBS was prepared, and 10 µL of each dilution was added to 200 µL of the OV- and OV+ treatments (triplicate wells) to determine the MCE that did not induce unspecific effects on N2a cell viability for each fraction. The MCE value of each fraction was determined when the concentration induced more than ≥ 20% mortality and was expressed in μg of dry extract mL^{-1}. Then, this value was converted in mg eq. wet weight mL^{-1} according to the dry extract weight/fresh weight (DEW/FW) ratio of each fraction [13].

Screening of Fractions

The screening of the fractions from the spiked or naturally contaminated fish samples (n = 2 to 3 according to protocols) was undertaken by the MCE value tested in OV- and OV+ conditions (triplicate wells) to differentiate between the fractions detected with a percentage of viability ≥ 80% in both conditions (classified as negative), those found with a percentage of viability between 20% and 80% in OV- and OV+ (classified as doubtful) and those found with a percentage of viability ≤ 20% in OV+ condition only (classified as positive).

Quantification

Quantification of the doubtful fractions was undertaken by testing eight-point serial dilutions under OV- and OV+ conditions (triplicate wells). The concentration range was run from the MCE to check whether the fraction tested induced identical OV- and OV+ mortality due to an excess of dry extract at the MCE limit (false positive).

For positive fractions, the serial dilutions were performed under OV+ conditions only (triplicate wells). The concentration range was run from the MCE and then adjusted according to the toxicity of the fish to obtain a complete sigmoidal dose–response curve.

For each positive fraction, three independent CBA-N2a experiments were run on different days and were expressed in ng CTX3C equivalent g^{-1} of fresh tissue (flesh) for the spiked fish samples (see Sections 4.5 and 4.6) and herbivores (see Section 4.7) or in ng CTX1B equivalent g^{-1} of fresh tissue (flesh) for the spiked fish samples (see Section 4.6) and carnivores (see Section 4.7).

Coefficients of variation (CVs) were also calculated according to the following formula:

$$CV (\%) = (SD/mean) \times 100.$$

4.9. Evaluation of the OP Protocol for Subsequent LC-MS/MS Analyses

Fortification experiments were performed by adding two mixes of CTX1B, CTX3B, CTX3C and CTX4A, each standard at a final concentration of 5 or 10 ng mL^{-1}, to the final fractions of interest purified from two carnivorous and one herbivorous fish species according to either the OP protocol, protocol #1 or the Murray et al. [27] protocol (from which protocol #2 was inspired). Pacific CTX reference materials were obtained from the ILM CTX standards bank.

The liquid chromatography–tandem mass spectroscopy (LC-MS/MS) method previously described by Murray et al. [27] for Pacific CTXs was then performed on a Waters Xevo TQ-S triple quadrupole mass spectrometer coupled to a Waters Acquity UPLC i-Class (Waters, Milford, MA, USA). Chromatographic separation was performed using a Waters Acquity UPLC BEH phenyl column (1.7 μm, 100 × 2.1 mm) held at 50 °C. The column was eluted at 0.55 mL min^{-1} with Milli-Q water (A) and 95% acetonitrile (B) mobile phases, each containing 0.2% (*v*/*v*) of a 25% ammonium hydroxide solution (final concentration of 26.7 mM ammonia). Fresh mobile phases were prepared daily to ensure optimal sensitivity and stable retention times. The initial solvent composition was 25% B with a linear gradient to 35% B at 2.0 min, which ramped up to 50% B at 2.5 min, followed by a linear gradient to 75% B at 6.5 min, which ramped up to 95% B at 7.0 min and was maintained at 95% B between 7.0 and 8.0 min. The column was re-equilibrated with 25% B between 8.0 and 9.0 min. The autosampler rack chamber was maintained at 10 °C and the injection volume was 2 μL for the fish extracts.

For the mass spectrometer, the electrospray ionization source was operated in positive-ion mode to monitor the CTXs with the following parameters: capillary voltage 3.5 kV, cone voltage 30 V, source temperature 150 °C, nitrogen gas desolvation flowrate 1000 L h^{-1} at 600 °C, cone gas 150 L h^{-1} and collision cell operated with 0.15 mL min^{-1} argon.

Multi-reaction monitoring (MRM) transitions for the various CTXs were determined from infusion experiments of the toxin standard into the ammoniated mobile phase. Collision energies were optimized for each compound monitored. For quantitative analysis, the following MRM transitions were used, with confirmation transitions shown in parentheses: CTX3B/CTX3C: m/z1023.6 > 155.1 (m/z 1023.6 > 125.1); CTX4A/CTX4B: m/z 1061.6 > 155.1 (m/z 1061.6 > 125.1). For CTX1B quantitation, a total ion chromatogram was generated from the following MRM and pseudo-MRM transitions: m/z 1128.6 > 95.0, m/z 1128.6 > 109.0 and m/z 1133.6 > 1133.6. A dwell time of 20 ms was used. Linear five-point calibrations (0.5–10 ng mL^{-1}) of CTX1B, CTX3B, CTX3C and CTX4A were used for quantitation, with coefficients of determination >0.98.

Data acquisition and processing was performed with MassLynx and TargetLynx software V4.1 (Waters, Milford, MA, USA), respectively. Peak areas were integrated and sample concentrations calculated from linear calibration curves generated from external calibration solutions.

5. Conclusions

In conclusion, this study provides a comprehensive evaluation of the behavior of nonpolar CTXs, demonstrating large discrepancies in the recoveries in fractions obtained with six different protocols inspired by studies available in the literature. An optimized extraction protocol was then developed in this study enabling recoveries above 70% for both the polar CTX1B and the nonpolar CTX3C in a single fraction of interest, representing a significant methodological improvement in terms of accuracy and precision. This simple protocol is cost-effective, requiring only a single solid-phase extraction cartridge and three days to prepare. Moreover, this optimized protocol demonstrates reliability and efficiency, ensuring consistent results in both naturally contaminated herbivorous and carnivorous marine specimens. The matrix effects observed with CBA-N2a were reduced by a factor >3, allowing for the analysis of larger quantities of the fresh weight matrix. Most importantly, this improved the sensitivity of the detection method, which now meets the very low guidance safety limits recommended by the US-FDA and EFSA. This optimized method is suitable for processing a large number of specimens and advances the field of CTX extraction and detection. By enabling a better assessment of the risks of ciguatera poisoning in marine products, the present study represents a key step in protecting public health and supporting initiatives to monitor marine products for CP risk.

Supplementary Materials: The following supporting information can be downloaded at https://www.mdpi.com/article/10.3390/md23010042/s1. Supplementary Material: Details of the six protocols used in the present study (inspired by some of the protocols commonly used for CTX analysis from fish matrices) and a schematic representation of the protocols with CTX3C amounts (ng) estimated by CBA-N2a in the testable fractions.

Author Contributions: Conceptualization, H.T.D., M.C. and M.R.; methodology, E.P., J.V., P.C., J.S.M. and T.Y.; validation, P.C., J.V., H.T.D., M.R., J.S.M. and T.Y.; formal analysis, E.P., J.V., P.C., M.R., J.S.M. and T.Y.; data curation, E.P., J.V., M.R., J.S.M. and T.Y.; writing—original draft preparation, H.T.D., M.R. and T.Y.; writing—review and editing, H.T.D., J.V., M.C., M.R., E.P., J.S.M. and T.Y.; supervision, H.T.D., M.C. and M.R.; project administration, H.T.D. and M.R.; funding acquisition, H.T.D. and M.R. All authors have read and agreed to the published version of the manuscript.

Funding: This research was funded in the framework of the "OPEX-CTX" research program by funds from the Government of French Polynesia (no. 4402/MED/REC 21 July 2020) and from France (no. AFD CPF 1554 01 n°2113 17 April 2021).

Institutional Review Board Statement: Not applicable.

Informed Consent Statement: Not applicable.

Data Availability Statement: The original contributions presented in this study are included in the article/Supplementary Materials. Further inquiries can be directed to the corresponding authors.

Conflicts of Interest: The authors declare no conflicts of interest.

References

1. Russell, F.E.; Egen, N.B. Ciguateric fishes, ciguatoxin (CTX) and ciguatera poisoning. *J. Toxicol. Toxin Rev.* **1991**, *10*, 37–62. [CrossRef]
2. Chinain, M.; Gatti, C.; Roué, M.; Darius, H.T. Ciguatera-causing dinoflagellates in the genera *Gambierdiscus* and *Fukuyoa*: Distribution, ecophysiology and toxicology. In *Dinoflagellates*; Subba Rao, D.V., Ed.; Nova Science Publishers, Inc.: New York, NY, USA, 2020; pp. 405–457.
3. Chinain, M.; Gatti, C.M.I.; Darius, H.T.; Quod, J.P.; Tester, P.A. Ciguatera poisonings: A global review of occurrences and trends. *Harmful Algae* **2021**, *102*, 101873.
4. Soliño, L.; Costa, P.R. Differential toxin profiles of ciguatoxins in marine organisms: Chemistry, fate and global distribution. *Toxicon* **2018**, *150*, 124–143. [CrossRef]
5. Ikehara, T.; Kuniyoshi, K.; Oshiro, N.; Yasumoto, T. Biooxidation of ciguatoxins leads to species-specific toxin profiles. *Toxins* **2017**, *9*, 205. [CrossRef] [PubMed]
6. Martin, V.; Vale, C.; Rubiolo, J.A.; Roel, M.; Hirama, M.; Yamashita, S.; Vieytes, M.R.; Botana, L.M. Chronic ciguatoxin treatment induces synaptic scaling through voltage gated sodium channels in cortical neurons. *Chem. Res. Toxicol.* **2015**, *28*, 1109–1119. [CrossRef] [PubMed]
7. Château-Degat, M.L.; Chinain, M.; Cerf, N.; Gingras, S.; Hubert, B.; Dewailly, E. Seawater temperature, *Gambierdiscus* spp. variability and incidence of ciguatera poisoning in French Polynesia. *Harmful Algae* **2005**, *4*, 1053–1062. [CrossRef]
8. Oshiro, N.; Nagasawa, H.; Kuniyoshi, K.; Kobayashi, N.; Sugita-Konishi, Y.; Asakura, H.; Yasumoto, T. Characteristic distribution of ciguatoxins in the edible parts of a grouper, *Variola louti*. *Toxins* **2021**, *13*, 218. [CrossRef] [PubMed]
9. Friedman, M.A.; Fernandez, M.; Backer, L.C.; Dickey, R.W.; Bernstein, J.; Schrank, K.; Kibler, S.; Stephan, W.; Gribble, M.O.; Bienfang, P.; et al. An Updated review of ciguatera fish poisoning: Clinical, epidemiological, environmental, and public health management. *Mar. Drugs* **2017**, *15*, 72. [CrossRef] [PubMed]
10. Dickey, R.W.; Plakas, S.M. Ciguatera: A public health perspective. *Toxicon* **2010**, *56*, 123–136. [CrossRef]
11. FAO; WHO. *Report of the Expert Meeting on Ciguatera Poisoning: Rome, 19–23 November 2018*; Food safety and quality No. 9; World Health Organization: Rome, Italy, 2020.
12. Hossen, V.; Soliño, L.; Leroy, P.; David, E.; Velge, P.; Dragacci, S.; Krys, S.; Flores Quintana, H.; Diogène, J. Contribution to the risk characterization of ciguatoxins: LOAEL estimated from eight ciguatera fish poisoning events in Guadeloupe (French West Indies). *Environ. Res.* **2015**, *143 Pt B*, 100–108. [CrossRef]
13. Viallon, J.; Chinain, M.; Darius, H.T. Revisiting the neuroblastoma cell-based assay (CBA-N2a) for the improved detection of marine toxins active on voltage gated sodium channels (VGSCs). *Toxins* **2020**, *12*, 281. [CrossRef] [PubMed]
14. Castro, D.; Manger, R.; Vilariño, O.; Gago-Martínez, A. Evaluation of matrix issues in the applicability of the neuro-2a cell based assay on the detection of CTX in fish samples. *Toxins* **2020**, *12*, 308. [CrossRef] [PubMed]
15. Loeffler, C.R.; Robertson, A.; Quintana, H.A.F.; Silander, M.C.; Smith, T.B.; Olsen, D. Ciguatoxin prevalence in 4 commercial fish species along an oceanic exposure gradient in the US Virgin Islands. *Environ. Toxicol. Chem.* **2018**, *37*, 1852–1863. [CrossRef] [PubMed]
16. Yokozeki, T.; Hama, Y.; Fujita, K.; Igarashi, T.; Hirama, M.; Tsumuraya, T. Evaluation of relative potency of calibrated ciguatoxin congeners by near-infrared fluorescent receptor binding and neuroblastoma cell-based assays. *Toxicon* **2023**, *230*, 107161. [CrossRef]
17. Murata, K.; Yasumoto, T. Chemiluminescent receptor binding assay for ciguatoxins and brevetoxins using acridinium brevetoxin-B2. *Toxins* **2019**, *11*, 13. [CrossRef]
18. Hardison, D.R.; Holland, W.C.; McCall, J.R.; Bourdelais, A.J.; Baden, D.G.; Darius, H.T.; Chinain, M.; Tester, P.A.; Shea, D.; Quintana, H.A.; et al. Fluorescent receptor binding assay for detecting ciguatoxins in fish. *PLoS ONE* **2016**, *11*, e0153348. [CrossRef]

19. Darius, H.T.; Ponton, D.; Revel, T.; Cruchet, P.; Ung, A.; Tchou Fouc, M.; Chinain, M. Ciguatera risk assessment in two toxic sites of French Polynesia using the receptor-binding assay. *Toxicon* **2007**, *50*, 612–626. [CrossRef] [PubMed]
20. Tsumuraya, T.; Sato, T.; Hirama, M.; Fujii, I. Highly sensitive and practical fluorescent sandwich ELISA for ciguatoxins. *Anal. Chem.* **2018**, *90*, 7318–7324. [CrossRef] [PubMed]
21. Li, X.; Ge, A.; Li, X.; Zhang, Z.; Mei, Z.; Zhang, S. Capillary electrophoretic ELISA method for rapid detection of ciguatoxin in shellfish and fish sample. Patent CN101963614, 2 February 2011.
22. Empey Campora, C.; Dierking, J.; Tamaru, C.S.; Hokama, Y.; Vincent, D. Detection of ciguatoxin in fish tissue using sandwich ELISA and neuroblastoma cell bioassay. *J. Clin. Lab. Anal.* **2008**, *22*, 246–253. [CrossRef]
23. Nagae, M.; Igarashi, T.; Mizukoshi, K.; Kuniyoshi, K.; Oshiro, N.; Yasumoto, T. Development and validation of an LC-MS/MS method for the ultra-trace analysis of Pacific ciguatoxins in fish. *J. AOAC Int.* **2021**, *104*, 1272–1281. [CrossRef] [PubMed]
24. Spielmeyer, A.; Loeffler, C.R.; Bodi, D. Extraction and LC-MS/MS analysis of ciguatoxins: A semi-targeted approach designed for fish of unknown origin. *Toxins* **2021**, *13*, 630. [CrossRef]
25. Sibat, M.; Herrenknecht, C.; Darius, H.T.; Roué, M.; Chinain, M.; Hess, P. Detection of Pacific ciguatoxins using liquid chromatography coupled to either low or high resolution mass spectrometry (LC-MS/MS). *J. Chromatogr. A* **2018**, *1571*, 16–28. [CrossRef] [PubMed]
26. Suzuki, T.; Ha, D.V.; Uesugi, A.; Uchida, H. Analytical challenges to ciguatoxins. *Curr. Opin. Food Sci.* **2017**, *18*, 37–42. [CrossRef]
27. Murray, J.S.; Boundy, M.J.; Selwood, A.I.; Harwood, D.T. Development of an LC-MS/MS method to simultaneously monitor maitotoxins and selected ciguatoxins in algal cultures and P-CTX-1B in fish. *Harmful Algae* **2018**, *80*, 80–87. [CrossRef]
28. Harwood, D.T.; Murray, J.S.; Michael, J.B. Sample preparation prior to marine toxin analysis. In *Comprehensive Analytical Chemistry*; Elsevier: Amsterdam, The Netherlands, 2017.
29. Otero, P.; Silva, M. Emerging marine biotoxins in European waters: Potential risks and analytical challenges. *Mar. Drugs* **2022**, *20*, 199. [CrossRef] [PubMed]
30. Estevez, P.; Castro, D.; Leao, J.M.; Yasumoto, T.; Dickey, R.; Gago-Martinez, A. Implementation of liquid chromatography tandem mass spectrometry for the analysis of ciguatera fish poisoning in contaminated fish samples from Atlantic coasts. *Food Chem.* **2019**, *280*, 8–14. [CrossRef] [PubMed]
31. Chinain, M.; Gatti, C.M.I.; Ung, A.; Cruchet, P.; Revel, T.; Viallon, J.; Sibat, M.; Varney, P.; Laurent, V.; Hess, P.; et al. Evidence for the range expansion of ciguatera in French Polynesia: A revisit of the 2009 mass-poisoning outbreak in Rapa Island (Australes Archipelago). *Toxins* **2020**, *12*, 759. [CrossRef] [PubMed]
32. Mak, Y.L.; Wai, T.-C.; Murphy, M.B.; Chan, W.H.; Wu, J.J.; Lam, J.C.W.; Chan, L.L.; Lam, P.K.S. Pacific ciguatoxins in food web components of coral reef systems in the Republic of Kiribati. *Environ. Sci. Technol.* **2013**, *47*, 14070–14079. [CrossRef]
33. Darius, H.T.; Roué, M.; Sibat, M.; Viallon, J.; Gatti, C.M.I.; Vandersea, M.W.; Tester, P.A.; Litaker, R.W.; Amzil, Z.; Hess, P.; et al. *Tectus niloticus* (Tegulidae, Gastropod) as a novel vector of ciguatera poisoning: Detection of Pacific ciguatoxins in toxic samples from Nuku Hiva Island (French Polynesia). *Toxins* **2018**, *10*, 2. [CrossRef]
34. Gatti, C.M.I.; Lonati, D.; Darius, H.T.; Zancan, A.; Roué, M.; Schicchi, A.; Locatelli, C.A.; Chinain, M. *Tectus niloticus* (Tegulidae, Gastropod) as a novel vector of ciguatera poisoning: Clinical characterization and follow-up of a mass poisoning event in Nuku Hiva Island (French Polynesia). *Toxins* **2018**, *10*, 102. [CrossRef] [PubMed]
35. Chinain, M.; Gatti Howell, C.; Roué, M.; Ung, A.; Henry, K.; Revel, T.; Cruchet, P.; Viallon, J.; Darius, H.T. Ciguatera poisoning in French Polynesia: A review of the distribution and toxicity of *Gambierdiscus* spp., and related impacts on food web components and human health. *Harmful Algae* **2023**, *129*, 102525. [CrossRef] [PubMed]
36. Darius, H.T.; Roué, M.; Sibat, M.; Viallon, J.; Gatti, C.M.I.; Vandersea, M.W.; Tester, P.A.; Litaker, R.W.; Amzil, Z.; Hess, P.; et al. Toxicological investigations on the sea urchin *Tripneustes gratilla* (Toxopneustidae, Echinoid) from Anaho Bay (Nuku Hiva, French Polynesia): Evidence for the presence of Pacific ciguatoxins. *Mar. Drugs* **2018**, *16*, 122. [CrossRef] [PubMed]
37. Oshiro, N.; Tomikawa, T.; Kuniyoshi, K.; Ishikawa, A.; Toyofuku, H.; Kojima, T.; Asakura, H. LC–MS/MS Analysis of ciguatoxins revealing the regional and species distinction of fish in the tropical Western Pacific. *J. Mar. Sci. Eng.* **2021**, *9*, 299. [CrossRef]
38. Sanchez-Henao, J.A.; García-Álvarez, N.; Fernández, A.; Saavedra, P.; Silva Sergent, F.; Padilla, D.; Acosta-Hernández, B.; Martel Suárez, M.; Diogène, J.; Real, F. Predictive score and probability of CTX-like toxicity in fish samples from the official control of ciguatera in the Canary Islands. *Sci. Total Environ.* **2019**, *673*, 576–584. [CrossRef] [PubMed]
39. Oshiro, N.; Nagasawa, H.; Watanabe, M.; Nishimura, M.; Kuniyoshi, K.; Kobayashi, N.; Sugita-Konishi, Y.; Asakura, H.; Tachihara, K.; Yasumoto, T. An extensive survey of ciguatoxins on grouper *Variola louti* from the Ryukyu Islands, Japan, using liquid chromatography–tandem mass spectrometry (LC-MS/MS). *J. Mar. Sci. Eng.* **2022**, *10*, 423. [CrossRef]
40. Spielmeyer, A.; Loeffler, C.R.; Kappenstein, O. Identical ciguatoxin-3C group profiles in *Lutjanus bohar* from the Pacific and Indian Oceans—Indicating the need to re-evaluate geographical CTX classifications. *Front. Mar. Sci.* **2022**, *9*, 937438. [CrossRef]
41. Wong, C.K.; Hung, P.; Lee, K.L.; Kam, K.M. Solid-phase extraction clean-up of ciguatoxin-contaminated coral fish extracts for use in the mouse bioassay. *Food Addit. Contam. Part A Chem. Anal. Control Expo. Risk Assess.* **2009**, *26*, 236–247. [CrossRef] [PubMed]

42. Lewis, R.J.; Yang, A.J.; Jones, A. Rapid extraction combined with LC-tandem mass spectrometry (CREM-LC/MS/MS) for the determination of ciguatoxins in ciguateric fish flesh. *Toxicon* **2009**, *54*, 62–66. [CrossRef]
43. Stewart, I.; Eaglesham, G.K.; Poole, S.; Graham, G.; Paulo, C.; Wickramasinghe, W.; Sadler, R.; Shaw, G.R. Establishing a public health analytical service based on chemical methods for detecting and quantifying Pacific ciguatoxin in fish samples. *Toxicon* **2010**, *56*, 804–812. [CrossRef]
44. Wu, J.J.; Mak, Y.L.; Murphy, M.B.; Lam, J.C.W.; Chan, W.H.; Wang, M.F.; Chan, L.L.; Lam, P.K.S. Validation of an accelerated solvent extraction liquid chromatography-tandem mass spectrometry method for Pacific ciguatoxin-1 in fish flesh and comparison with the mouse neuroblastoma assay. *Anal. Bioanal. Chem.* **2011**, *400*, 3165–3175. [CrossRef]
45. Mak, Y.L.; Wu, J.J.; Chan, W.H.; Murphy, M.B.; Lam, J.C.W.; Chan, L.L.; Lam, P.K.S. Simultaneous quantification of Pacific ciguatoxins in fish blood using liquid chromatography-tandem mass spectrometry. *Anal. Bioanal. Chem.* **2013**, *405*, 3331–3340. [CrossRef]
46. Kohli, G.S.; Haslauer, K.; Sarowar, C.; Kretzschmar, A.L.; Boulter, M.; Harwood, D.T.; Laczka, O.; Murray, S.A. Qualitative and quantitative assessment of the presence of ciguatoxin, P-CTX-1B, in Spanish Mackerel (*Scomberomorus commerson*) from waters in New South Wales (Australia). *Toxicol. Rep.* **2017**, *4*, 328–334. [CrossRef] [PubMed]
47. Tsumuraya, T.; Hirama, M. Rationally designed synthetic haptens to generate anti-ciguatoxin monoclonal antibodies, and development of a practical sandwich ELISA to detect ciguatoxins. *Toxins* **2019**, *11*, 533. [CrossRef] [PubMed]
48. Zhu, J.; Lee, W.H.; Wu, J.; Zhou, S.; Yip, K.C.; Liu, X.; Kirata, T.; Chan, L.L. The Occurrence, distribution, and toxicity of high-risk ciguatera fish species (grouper and snapper) in Kiritimati Island and Marakei Island of the Republic of Kiribati. *Toxins* **2022**, *14*, 208. [CrossRef]
49. Campàs, M.; Leonardo, S.; Oshiro, N.; Kuniyoshi, K.; Tsumuraya, T.; Hirama, M.; Diogène, J. A smartphone-controlled amperometric immunosensor for the detection of Pacific ciguatoxins in fish. *Food Chem.* **2022**, *374*, 131687. [CrossRef] [PubMed]
50. EFSA. Scientific opinion on marine biotoxins in shellfish—Emerging toxins: Ciguatoxin group. *EFSA J.* **2010**, *8*, 1627.
51. FDA. Appendix 5: FDA and EPA safety levels in regulations and guidance (June 2021). In *Fish and Fishery Products Hazards and Control Guidance*; Department of Health and Human Services, Food and Drug Administration, Center for Food Safety and Applied Nutrition: Silver Spring, MD, USA, 2022; A 5-1.
52. IOC/IPHAB. Global ciguatera strategy 2015–2019. In Proceedings of the 12th Session of UNESCO-IOC's Intergovernmental Panel on Harmful Algal Blooms, Paris, France, 28–30 April 2015.
53. Gonçalves, R.M.; Petenuci, M.E.; Maistrovicz, F.C.; Galuch, M.B.; Montanher, P.F.; Pizzo, J.S.; Gualda, I.P.; Visentainer, J.V. Lipid profile and fatty acid composition of marine fish species from Northeast coast of Brazil. *J. Food Sci. Technol.* **2021**, *58*, 1177–1189. [CrossRef]
54. Estevez, P.; Oses Prieto, J.; Burlingame, A.; Gago Martinez, A. Characterization of the ciguatoxin profile in fish samples from the Eastern Atlantic Ocean using capillary liquid chromatography-high resolution mass spectrometry. *Food Chem.* **2023**, *418*, 135960. [CrossRef] [PubMed]
55. Darius, H.T.; Paillon, C.; Mou-Tham, G.; Ung, A.; Cruchet, P.; Revel, T.; Viallon, J.; Vigliola, L.; Ponton, D.; Chinain, M. Evaluating age and growth relationship to ciguatoxicity in five coral reef fish species from French Polynesia. *Mar. Drugs* **2022**, *20*, 251. [CrossRef] [PubMed]
56. Meyer, L.; Carter, S.; Capper, A. An updated ciguatoxin extraction method and silica cleanup for use with HPLC-MS/MS for the analysis of P-CTX-1, PCTX-2 and P-CTX-3. *Toxicon* **2015**, *108*, 249–256. [CrossRef]
57. Chinain, M.; Darius, H.T.; Ung, A.; Cruchet, P.; Wang, Z.; Ponton, D.; Laurent, D.; Pauillac, S. Growth and toxin production in the ciguatera-causing dinoflagellate *Gambierdiscus polynesiensis* (Dinophyceae) in culture. *Toxicon* **2010**, *56*, 739–750. [CrossRef]
58. Loeffler, C.R.; Spielmeyer, A. Faster ciguatoxin extraction methods for toxicity screening. *Sci. Rep.* **2024**, *14*, 21715. [CrossRef]
59. Costa, P.R.; Estevez, P.; Solino, L.; Castro, D.; Rodrigues, S.M.; Timoteo, V.; Leao-Martins, J.M.; Santos, C.; Gouveia, N.; Diogene, J.; et al. An update on ciguatoxins and CTX-like toxicity in fish from different trophic levels of the Selvagens Islands (NE Atlantic, Madeira, Portugal). *Toxins* **2021**, *13*, 580. [CrossRef] [PubMed]
60. Clausing, R.J.; Losen, B.; Oberhaensli, F.R.; Darius, H.T.; Sibat, M.; Hess, P.; Swarzenski, P.W.; Chinain, M.; Bottein, M.Y.D. Experimental evidence of dietary ciguatoxin accumulation in an herbivorous coral reef fish. *Aquat. Toxicol.* **2018**, *200*, 257–265. [CrossRef] [PubMed]
61. Caillaud, A.; Eixarch, H.; de la Iglesia, P.; Rodriguez, M.; Dominguez, L.; Andree, K.B.; Diogene, J. Towards the standardisation of the neuroblastoma (neuro-2a) cell-based assay for ciguatoxin-like toxicity detection in fish: Application to fish caught in the Canary Islands. *Food Addit. Contam. Part A Chem. Anal. Control Expo. Risk Assess.* **2012**, *29*, 1000–1010. [CrossRef] [PubMed]
62. Pawlowiez, R.; Darius, H.T.; Cruchet, P.; Rossi, F.; Caillaud, A.; Laurent, D.; Chinain, M. Evaluation of seafood toxicity in the Australes archipelago (French Polynesia) using the neuroblastoma cell-based assay. *Food Addit. Contam. Part A Chem. Anal. Control Expo. Risk Assess.* **2013**, *30*, 567–586. [CrossRef]
63. Spielmeyer, A.; Blaschke, V.; Loeffler, C.R. Fish extract fractionation by solid phase extraction: Investigating co-occurring ciguatoxins by LC-MS/MS and N2a-bioassay. *Separations* **2024**, *11*, 238. [CrossRef]

64. Ciminiello, P.; Dell'Aversano, C.; Dello Iacovo, E.; Fattorusso, E.; Forino, M.; Tartaglione, L.; Rossi, R.; Soprano, V.; Capozzo, D.; Serpe, L. Palytoxin in seafood by liquid chromatography tandem mass spectrometry: Investigation of extraction efficiency and matrix effect. *Anal. Bioanal. Chem.* **2011**, *401*, 1043–1050. [CrossRef] [PubMed]
65. Legrand, A.M.; Litaudon, M.; Genthon, G.N.; Bagnis, R.; Yasumoto, T. Isolation and some properties of ciguatoxin. *J. Appl. Phycol.* **1989**, *1*, 183–188. [CrossRef]
66. Murata, M.; Legrand, A.M.; Ishibashi, Y.; Fukui, M.; Yasumoto, T. Structures and configurations of ciguatoxin from the moray eel *Gymnothorax javanicus* and its likely precursor from the dinoflagellate *Gambierdiscus toxicus*. *J. Am. Chem. Soc.* **1990**, *112*, 4380–4386. [CrossRef]
67. Yasumoto, T.; Igarashi, T.; Legrand, A.-M.; Cruchet, P.; Chinain, M.; Fujita, T.; Naoki, H. Structural elucidation of ciguatoxin congeners by fast-atom bombardment tandem mass spectroscopy. *J. Am. Chem. Soc.* **2000**, *122*, 4988–4989. [CrossRef]
68. Darius, H.T.; Revel, T.; Viallon, J.; Sibat, M.; Cruchet, P.; Longo, S.; Hardison, D.R.; Holland, W.C.; Tester, P.A.; Litaker, R.W.; et al. Comparative study on the performance of three detection methods for the quantification of Pacific ciguatoxins in French Polynesian strains of *Gambierdiscus polynesiensis*. *Mar. Drugs* **2022**, *20*, 348. [CrossRef] [PubMed]
69. Yogi, K.; Sakugawa, S.; Oshiro, N.; Ikehara, T.; Sugiyama, K.; Yasumoto, T. Determination of toxins involved in ciguatera fish poisoning in the Pacific by LC/MS. *J. AOAC Int.* **2014**, *97*, 398–403. [CrossRef] [PubMed]

marine drugs

MDPI

Article

High Affinity Aptamers and Their Specificity for Azaspiracid-2 Using Capture-SELEX

Jiaping Yang †, Xinhao Li †, Weiqin Sun †, Yunyi Cui, Han Chen, Yao Yang, Mingjuan Sun * and Lianghua Wang *

Basic Medical College, Naval Medical University, Shanghai 200433, China; smmu_shenghuayjp@163.com (J.Y.); li95613@163.com (X.L.); qinjiao1225@163.com (W.S.); yunyicui@163.com (Y.C.); nmuchenhan@foxmail.com (H.C.); yangyao2017@foxmail.com (Y.Y.)

* Correspondence: sunmj@smmu.edu.cn (M.S.); lhwang@smmu.edu.cn (L.W.); Tel.: +86-021-81870967 (L.W.)

† These authors contributed equally to this work and shared the first authorship.

Abstract: Azaspiracids are a type of polyether toxin. Currently, the existing detection methods for azaspiracids all have certain drawbacks. Aptamers offer a cost-effective and convenient approach for the detection of azaspiracids. By employing the Capture-SELEX (Systematic evolution of ligands by exponential enrichment) method to screen aptamers specific to azaspiracid-2, a high-affinity aptamer can be identified for toxin detection. The bin ding affinity of the toxin is verified using biolayer interferometry (BLI) technology. Additionally, computer simulations are utilized to explore the binding sites of the aptamer and conduct molecular dynamics simulations to investigate the stability of the aptamer–toxin complex. Further optimization of the obtained aptamers is carried out to enhance their affinity for the toxin. Ultimately, two aptamers, JD2-RM3-27C28T and JD3-RMM1, are obtained, with dissociation constants (K_D) improved by two orders of magnitude (K_D = 8.7 × 10^{-8} M and K_D = 6.8 × 10^{-8} M, respectively). These aptamers have the advantage of being incorporated into a new AZA2 assay that is more accurate and ethical than biological monitoring methods, and more economical than LC-MS. In the future, this is expected to demonstrate significant advantages in the fields of food safety, environmental toxin monitoring, toxin exposure diagnosis, and public health monitoring.

Keywords: azaspiracids; nucleic acid aptamers; molecular dynamics simulation; biolayer interferometry; biosensors

1. Introduction

Azaspiracids (AZAs) are a class of lipophilic polyether and diarrheal marine toxins. They are currently known to be generated in the native organisms *Azadinium* [1,2] and *Amphidoma* [3,4] of the flagellate order, and certain species of *Protoperidium*, such as *P. crassipes* [5] (all of which belong to the phylum dinoflagellate), and can bioaccumulate in shellfish, such as oysters, scallops, clams, etc. [6–8]. It was first reported in 1996 during a food poisoning incident that occurred in Ireland in 1995, where several crew members developed acute gastrointestinal symptoms after consuming the blue mussels they had captured [9]. The toxin was initially named Killary Toxin (named after its discovery site) or KT-3, and later renamed Azaspiracids to better reflect its structural formula [10]. The physicochemical properties of AZAs toxins are significantly different from other nitrogen-containing biotoxins. As with other marine toxin groups, the toxins are relatively stable and cannot be eliminated by conventional cooking and processing. After several hours of high-temperature treatment with various organic reagents, the structure and toxicity of the azaspiracids remain unchanged, and they can be stored for a long time under refrigeration conditions [11]. At present, more than 60 types of AZAs and their derivatives have been discovered, and most of which are derived from the secondary metabolites of mussels [12,13].

There is currently no definitive experimental study on its pathogenic mechanism that can demonstrate it as a blocking agent for a specific target. However, some studies have shown that it can influence the electrical activity of cells by affecting the activity of various ion channels, including potassium ion channels [14], sodium ion channels [15], chloride ion channels [16], calcium ion channels [17], and so on. For the impact on the individual digestive system, toxins can enter the bloodstream through the human intestinal barrier, causing damage to epithelial cells [18]. It also has certain effects on inducing tumors [19], altering the cytoskeleton, and reducing human cell metabolism [20]. Experiments have shown that repeated exposure to azaspiracids can induce the expression of Jun B proto-oncogene, Jun D proto-oncogene, c-Jun proto-oncogene, FBJ murine osteosarcoma viral oncogene homolog B, and Fos-related antigen 1 in mice, leading to a significant increase in lung tumors and lymphatic necrosis of digestive system tissues [21]. In addition, there have been reports on the impact to the nervous system [22].

The European Food Safety Authority (EFSA) established an acute reference dose (ARfD) at 0.2 μg/kg body weight (b.w.) based on data from human poisoning incidents [23] and recommended a guideline limit of 30 μg/kg maximum concentration in shellfish meat [24]. However, European Law states that the maximum permitted threshold for total azaspiracids is 0.16 mg/kg AZA-1 equivalents [25].

The main detection methods currently include the following aspects:

1. Mouse bioassay (MBA) involves injecting toxins or extracts into the abdominal cavity of mice and recording the survival time of mice at different concentrations to calculate the toxin content of the sample. This method is the simplest and most commonly used method for detecting shellfish toxins, with a detection sensitivity of 1 μg level [26]. The minimum lethal doses for purified AZA-1, AZA-2, and AZA-3 are 200 μg/kg, 110 μg/kg, and 140 μg/kg, respectively [27]. This method has been listed as a routine method for toxin detection by the American Society of Analytical Chemists. The advantage is that it does not require complex equipment and has high operability. However, there are also drawbacks such as poor sensitivity, low reproducibility, high false positive rate, inability to identify toxicity, and significant influence from individual differences in mice.
2. Liquid chromatography tandem mass spectrometry (LC-MS/MS) was first used in 1999 to detect protochloroalginic acid, with a detection sensitivity of up to 50 pg [28]. This method is the European Reference Method for determination of AZAs, with advantages such as high sensitivity, high specificity, and the ability to accurately determine the composition of mixtures. However, there are still drawbacks such as expensive instruments and the need for skilled technicians to operate them.
3. Enzyme-linked immunosorbent assay (ELISA) has a detection sensitivity of 0.45–8.6 ng/mL [29]. Using new plate-coating antigen, an ELISA was developed with a working range of 0.30–4.1 ng/mL and a limit of quantification at 37 μg/kg for AZA-1 in shellfish [30]. ELISA can also be employed in conjunction with gel methods [31]. This method may have adverse effects such as false positives and cross-reactivity. And currently, there is no commercial ELISA available for purchase and use.
4. Biosensors: Current monitoring methods are limited to the detection of standard samples, including polyclonal antibody recognition [32] and monoclonal antibody recognition [33] targeting the C28–C40 site FGHI loop. However, there are still issues

with the toxicity and lethality of immune conjugates in vivo. Other biological assays, including magnetic bead based, chromatographic column based, and microsphere immunoassay [34], have provided new detection approaches.

These detection methods each have their own advantages, but most still have many obstacles in application. As a novel molecular recognition tool, nucleic acid aptamers have multiple significant advantages, with clear origins and a wide range of applications. The word comes from the Latin word "aptus" (adaptation) and the Greek word "meros" (partial) [35]. It is mainly composed of DNA and RNA, and aptamers form structured single-stranded nucleic acid secondary structures through hydrogen bonding interactions between bases, such as hair clips, loops, inner loops, multi-loops, protrusions, stems, pseudoknots, and non-standard base pairing. On the basis of the secondary structure, they can also form certain pocket structures by curling and folding themselves, or form guanine-rich triple- or quadruple-stranded structures through layer stacking [36]. Small molecules can be recognized by nucleic acid aptamers through electrostatic binding, groove surface binding, insertion binding, and other methods [37]. Electrostatic binding mainly occurs when the small molecule ligand is a positive dot molecule, and the binding occurs through charge adsorption. The electrostatic effect is very weak, and it is only meaningful for the complex to bind to the aptamer in the presence of a large distribution of positive charges. The specificity of simple small molecule binding is poor; groove surface binding mainly refers to the groove binding formed between ligands and hydrogen bond surfaces or multiple pairs of nucleic acids; The structure formed by stacking π-bonds and basal planes in compounds with smaller molecular weights using insertion binding design is the most stable [38]. After partial screening and optimization, the equilibrium dissociation constant K_D value of some aptamers can reach between nM and pM, which makes them highly accurate in molecular recognition [39–41]. Nucleic acid aptamers have demonstrated extremely high advantages in low immunogenicity, low production cost, stability after modification, small batch differences, small volume, convenient transport, resistance to pH and temperature-induced denaturation, ease of modification with nanomaterials or organic dye molecules, and flexibility in engineering performance manipulation [42]. For some target nucleic acid aptamers, they exhibit higher affinity and sensitivity than proteins while also being able to recognize hidden binding sites that are inaccessible to smaller targets or protein antibodies [43]. Nucleic acid aptamers can be combined with various signal amplification techniques to achieve high-sensitivity detection of target molecules. One of its advantages is strong adaptability, which can be applied to various monitoring scenarios such as biosensing, environmental monitoring, food safety, clinical diagnosis, and nucleic acid aptamers. After binding to the target molecule, it can be dissociated under denaturing conditions (such as high temperature, pH changes) to restore its activity, achieve reuse, and reduce detection costs. With the development of experimental technology, aptamers are expected to replace enzyme-linked immunosorbent assays as an effective method for the detection of various chemical molecules.

Azaspiracids exhibit significant cytotoxicity and cardiotoxicity, posing a direct threat to consumer health [21]. With global climate change and deterioration in the marine environment, harmful algal bloom (HAB) incidents occur frequently [44], leading to an expansion of the contamination range of AZA-2. The European Food Safety Authority (EFSA) has reported multiple shellfish recall incidents caused by AZA-2 contamination, highlighting the necessity for strengthening detection efforts [45]. Furthermore, AZA-2 is highly stable in the environment and difficult to remove through conventional water treatment processes. Its bioaccumulative properties may result in the gradual amplification of toxin concentrations within marine ecosystems [46]. The establishment of environmental benchmark values and ecological risk assessment models for AZA-2 relies on large-scale and high-frequency monitoring data, which current monitoring capabilities struggle to meet. Therefore, more novel and efficient detection methods are urgently needed.

The experimental plan is to screen for aptamers of Azaspiracid-2 using the Capture-SELEX method. To find one or more high-affinity aptamers for toxin detection, BLI technology is used to verify the affinity of the toxin. Additionally, computer simulation is used to investigate the stability of the aptamer docking site and molecular dynamics simulation to improve the affinity of the aptamer for the toxin. The results ultimately achieve specific detection of Azaspiracid-2 in the sample.

2. Results

2.1. Capture SELEX Method for Obtaining AZA-2 Aptamers

Capture-SELEX includes the following steps, which require several cycles before sequencing, as shown in Figure 1. Starting from the second round, the library was obtained through a long short-chain separation method. After high-temperature denaturation, two bands appeared on the denatured nucleic acid electrophoresis gel. The band far from the sample site was selected as the band for the next library round (Figure 2A). Successful Capture-SELEX relies on efficient library binding, and the binding rates for each round of library screening were all above 75%, as shown in Figure 2B. If the binding efficiency of the library is lower than expected, methods such as extending the incubation time, adjusting the library capture sequence ratio, and increasing the number of magnetic beads should be used to optimize the reaction conditions.

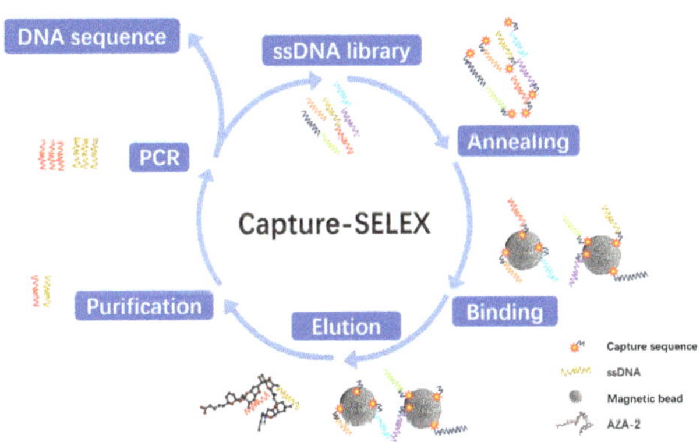

Figure 1. Schematic diagram of the Capture-SELEX screening aptamer process.

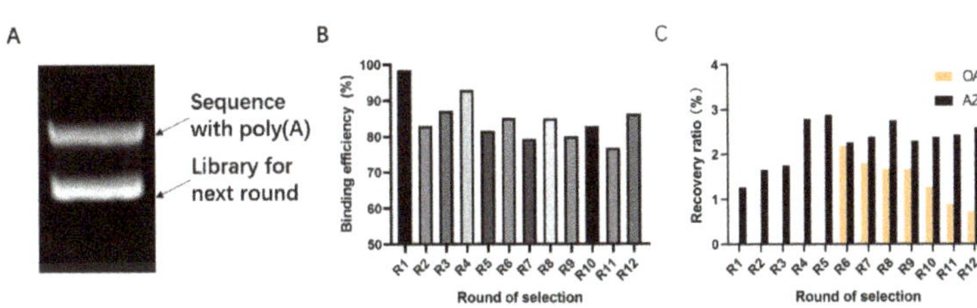

Figure 2. Monitoring data during the Capture-SELEX process. (**A**) The binding efficiency of each selection round. (**B**) Recovery ratio of ssDNA for each selection round of targets. (**C**) PCR results from the long short-chain separation method.

In the elution step, AZA-2 targets were used as forward screening targets in rounds 1–12, and okadaic acid (OA) targets were used as reverse screening targets in rounds 6–12 to increase screening pressure and improve aptamer specificity. There are also ways to increase screening pressure by reducing the number of magnetic beads used, lowering the initial library concentration, reducing the co-incubation time of library magnetic beads, and prolonging the cleaning time for the magnetic beads after incubation. The recovery rate for each library round during the screening process is shown as an important monitoring indicator in Figure 2C.

High throughput sequencing was performed on the eluate after 12 rounds of toxin mixing, and a total of 1,048,575 differential sequences were displayed. Compared with the theoretical value of the initial library, the number of library readings decreased by about 70%, from 88% to 13% (Figure 3A). The library lengths of the top 100 sequences ranked in the 12th round of high-throughput sequencing are the same as the original length of the initial library, indicating the effectiveness of high-throughput sequencing. The top ten sequences (Table 1) in the entire library increased from 0.78% in the first round to 12.60% in the 12th round (Figure 3B) (JD-1 to JD-10 represent the rankings from 1st to 10th in terms of read counts from sequences obtained after removing primers at both ends in the final round of Capture-SELEX).

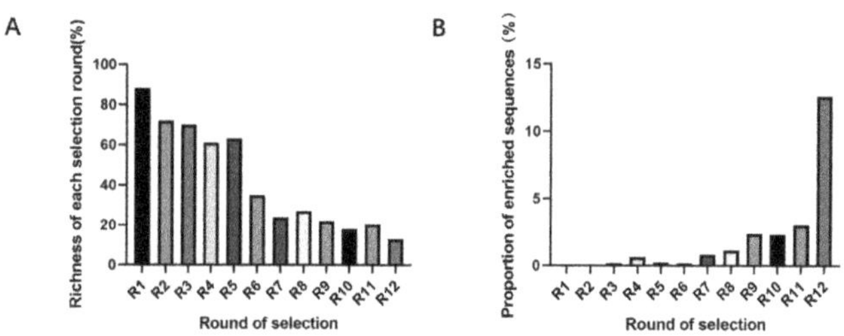

Figure 3. Richness of each selection round and proportion of enriched sequences. (**A**) Richness of each selection round. (**B**) Proportion of enriched sequences of the top 10 sequences in the pool of different rounds.

Table 1. AZA-2 aptamer sequences obtained by removing the primers at both ends.

Name	Sequence (5′-3′)
JD-1	GAATGGACCCGGTATAATTCCCTCAAGAGTGCCAATTTCA
JD-2	AAAAGACTTAGTGTTGTCCTATGTATAAGTGCCAATTTCA
JD-3	ACTAGTGCAAATCTATTCCTATGTTACAGTGCCAATTTCA
JD-4	GACATCGAGAAGAATCCTGATACGACTTGGCTTTGCTGGC
JD-5	CCAACATGATGTTCCGTCATTTTGAGGTGTGTACACCGTG
JD-6	GATGACACTTGTTTATGCCTATGATGATGTGCCAATTTCA
JD-7	GAATGGACCCGGTATAATTCCCTCGAGAGTGCCAATTTCA
JD-8	ACTAGTGCAAATCTATTCCTATGTTTCAGTGCCAATTTCA
JD-9	ACTAGTGCAAATCTATTCCTATGGTTCAGTGCCAATTTCA
JD-10	ACTAGTGCAAATCTATTCCTATGATACAGTGCCAATTTCA

Biolayer interference (BLI) technology is a label-free optical biosensing technique used for real-time monitoring and analysis of biomolecule interactions, along with quantitative analysis of their binding strength and dynamics [47]. This technology is based on the principle of light interference, using biosensors to couple nucleic acid aptamers or proteins to probes at the end of the sensor. The probes are immersed in the sample to capture the analyte and includes experimental steps such as sample loading, ligand binding and fixation, target binding, elution, and analysis (Figure 4). Preliminary affinity detection was conducted for the aptamers screened in the early stage, and the basic BLI data for the truncated aptamers are presented in Table A1. Among them, the JD-2 aptamer from the JD-2 family (with a K_D value of 3.5×10^{-6} M) and the JD-3 aptamer from the JD-3 family (with a K_D value of 3.6×10^{-6} M) (Table A1) exhibited the highest affinity. Although

JD-4 and JD-5 had high total read counts in the final round of sequencing, they showed poor abundance among the top 100 aptamer sequencing results as single-sequence family members, and their affinity performance was not outstanding.

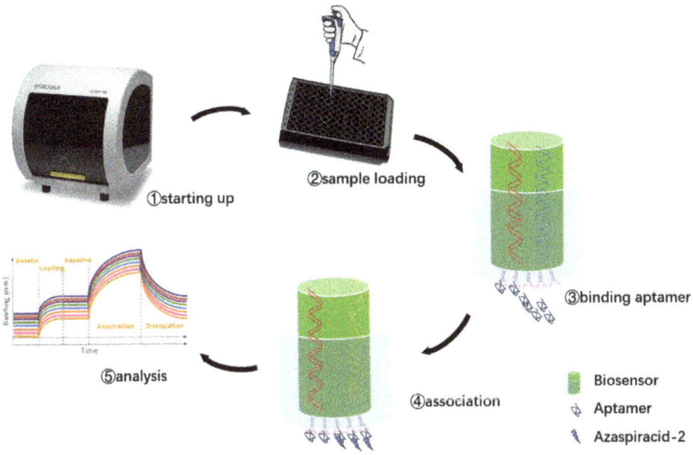

Figure 4. Basic operating procedures of BLI.

During the screening process, OA was used as a counter-screening target to enhance specificity to some extent. To verify the specificity of the aptamers against multiple marine toxins, a wider variety of toxins were employed in the affinity detection, including the small molecule Saxitoxin (STX); OA and Dinophysistoxin (DTX) with similar molecular weights to azaspiracids; the slightly larger molecular weight Palytoxin (PLTX), and azaspiracid-1, which shares a similar backbone structure with azaspiracids but differs by only one methyl group (Figure 5). Preliminary affinity results indicated that the response values of the two high-affinity aptamers and the most abundant aptamer JD-1 to OA, DTX, STX, and PLTX were all below 0.1 nm. This suggests that the initially truncated aptamers have a certain effect in distinguishing between OA, DTX, STX, and PLTX, but their ability to differentiate between AZA-1 and AZA-2 is relatively poor (Figure 6).

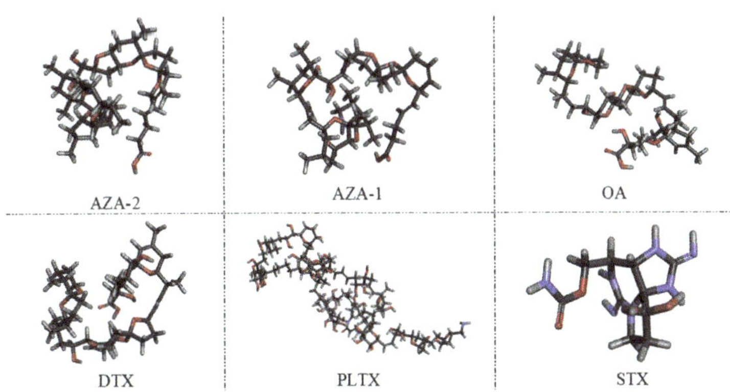

Figure 5. Three-dimensional structure of AZA-2, AZA-1, OA, DTX, PLTX, and STX.

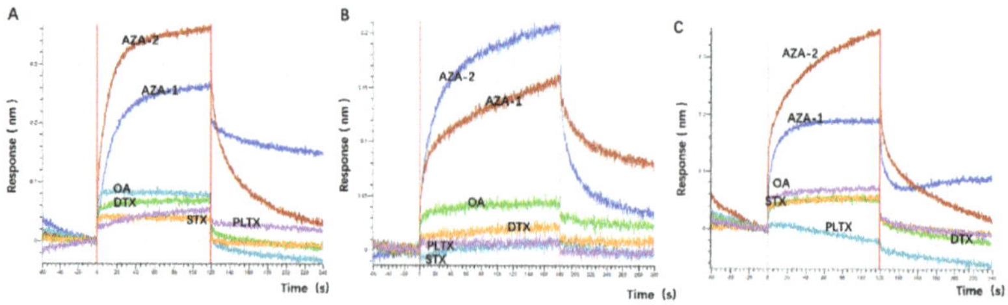

Figure 6. The specificity of some family aptamers. (**A**) BLI specificity results for aptamer JD-1. (**B**) BLI specificity results for aptamer JD-2. (**C**) BLI specificity results for aptamer JD-3.

2.2. Docking and Truncation Optimization of Aptamers

According to the results of autodock molecular docking, we can obtain binding sites with the lowest docking free energy to azaspiracids-2 for each aptamer. By removing sequences unrelated to the docking sites, including truncating the 5′-end of the aptamer (abbreviated as RM5 hereafter), removing the 3′-end of the aptamer (RM3), deleting internal sequences (RMM), and performing base substitutions based on the screening results from high-throughput sequencing, we optimized and verified the binding affinities of dozens of aptamers. We finally obtained two high affinity aptamers, JD2-RM3-27C28T (Figure 7) with $K_D = 8.7 \times 10^{-8}$ M and JD3-RMM1 with $K_D = 6.8 \times 10^{-8}$ M (Figure 8) (Table A1).

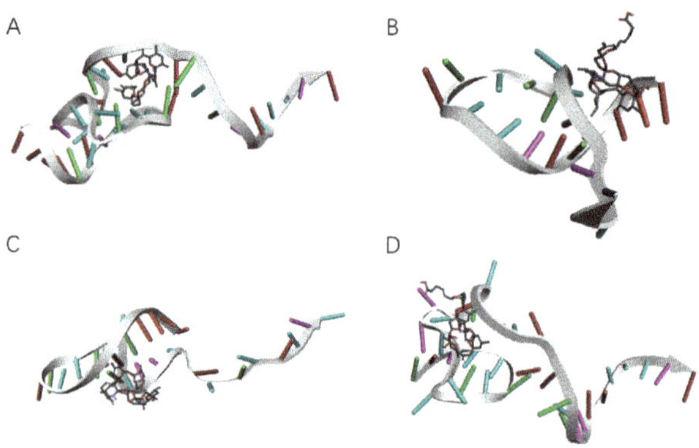

Figure 7. Docking results for aptamer JD-2 and its modifications. (**A**) Docking site between aptamer JD-2 and AZA-2. (**B**) Docking site between aptamer JD2-RM3 and AZA-2. (**C**) Docking site between aptamer JD2-RM3-27C28T and AZA-2. (**D**) Docking site between aptamer JD2-RM5 and AZA-2.

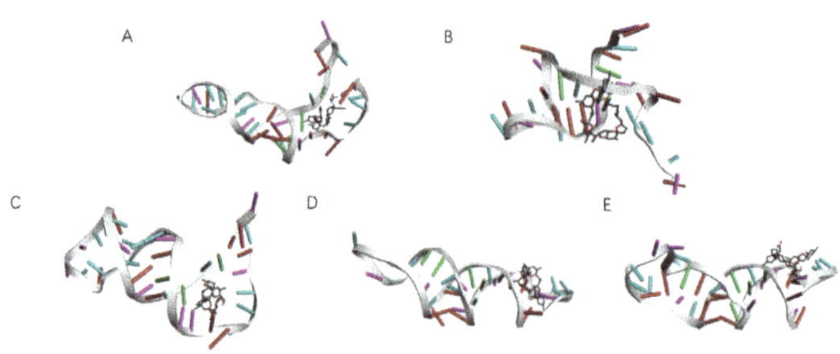

Figure 8. Docking sites of JD-3 family aptamers. (**A**) Docking site between aptamer JD-3 and AZA-2. (**B**) Docking site between aptamer JD3-RMM1 and AZA-2. (**C**) Docking site between aptamer JD-8 and AZA-2. (**D**) Docking site between aptamer JD-9 and AZA-2. (**E**) Docking site between aptamer JD-10 and AZA-2.

2.3. Molecular Dynamics Simulation of Candidate Aptamers

Subsequently, conventional dynamic simulations under an all-atom water model for 100 ns were conducted on the modified high-affinity aptamers JD2-RM3-27C28T and JD3-RMM1. The relative positional relationships between the aptamers and toxins over the 100 ns period are presented in Figures 9 and 10.

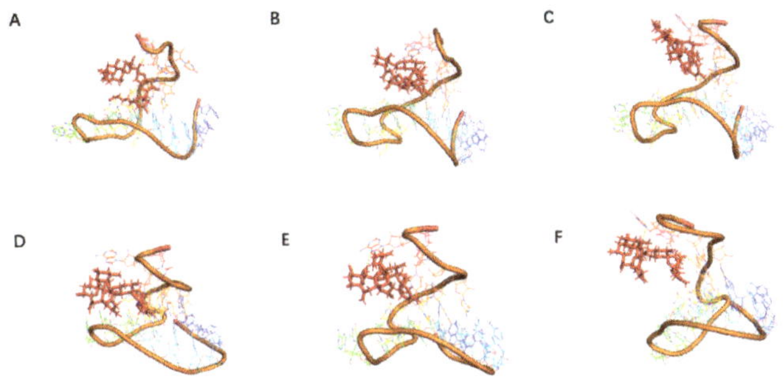

Figure 9. The positional relationship between small molecule toxins and aptamer JD2-RM3-27C28T displayed at different simulation times. (**A**) The relative position between the aptamer JD2-RM3-27C28T and the AZA-2 at 10 ns. (**B**) The relative position between the aptamer JD2-RM3-27C28T and the AZA-2 at 20 ns. (**C**) The relative position between the aptamer JD2-RM3-27C28T and the AZA-2 at 50 ns. (**D**) The relative position between the aptamer JD2-RM3-27C28T and the AZA-2 at 60 ns. (**E**) The relative position between the aptamer JD2-RM3-27C28T and the AZA-2 at 70 ns. (**F**) The relative position between the aptamer JD2-RM3-27C28T and the AZA-2 at 100 ns.

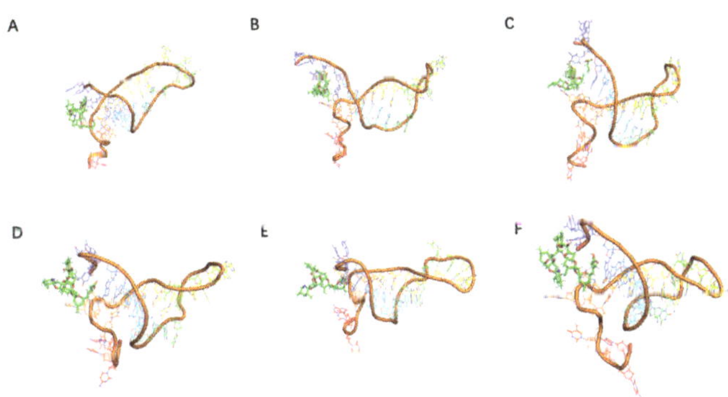

Figure 10. The positional relationship between small molecule toxins and aptamer JD3-RMM1 displayed at different simulation times. (**A**) The relative position between the aptamer JD3-RMM1 and the AZA-2 at 10 ns. (**B**) The relative position between the aptamer JD3-RMM1 and the AZA-2 at 20 ns. (**C**) The relative position between the aptamer JD3-RMM1 and the AZA-2 at 50 ns. (**D**) The relative position between the aptamer JD3-RMM1 and the AZA-2 at 60 ns. (**E**) The relative position between the aptamer JD3-RMM1 and the AZA-2 at 70 ns. (**F**) The relative position between the aptamer JD3-RMM1 and the AZA-2 at 100 ns.

The commonly used method for evaluating the stability of molecular structures in dynamic simulations is to calculate the root mean square deviation (RMSD) of the initial simulated structure (Figure 11). Although the JD-2 aptamer is in equilibrium, its RMSD value showed steep changes during some simulation time periods, indicating that the toxin ligand may have undergone unstable state changes during the simulation process. After simulation, JD2-RM3-27C28T and JD3-RMM1 reached equilibrium within a few nanoseconds and fluctuated slightly within the range of equilibrium values. Combined with the relative positions of toxin aptamers in the simulation process shown in Figures 9 and 10, the stability of the simulation state was further demonstrated. The multiple systems validated during the simulation process achieved stability within the specified simulation duration, while also verifying the rationality of the applied force field and parameters.

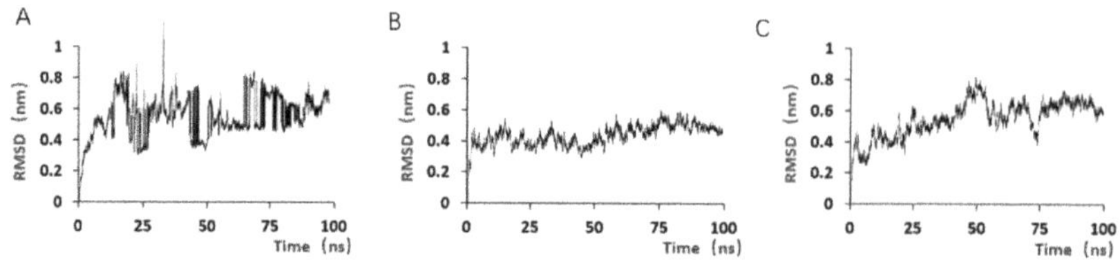

Figure 11. RMSD results for the 100 ns simulation with different modified aptamers. (**A**) RMSD of aptamer JD-2. (**B**) RMSD of aptamer JD2-RM3-27C28T. (**C**) RMSD of aptamer JD3-RMM1.

Figure 12 shows the root mean square fluctuation (RMSF) during the simulation of different aptamers and toxins. The RMSF value measures the amplitude of all atoms deviating from the average position, reflecting the flexibility and activity intensity of each nucleotide of the aptamer in the simulation. Among them, the 3′ end of the JD-2 aptamer showed a significant (2 nm) shift in some nucleotides, and other atoms also had a variation amplitude of about 1 nm (Figure 12A). However, in the aptamer JD2-RM3-27C28T, only a few free adenosine nucleotides at the 5′ end and the corresponding number of atoms at the middle nucleotides T12, G13, T14, and T15 shifted by 2 Å–6 Å, while the RMSF values of the nucleic acids at G5, A6, T17, and C18 that interact with the aptamer were all below 2 Å, indicating improved stability of the aptamer nucleotides (Figure 12B). In JD3-RMM1, the nucleotide region that binds to the target is relatively stable, while the nucleotide corresponding to the intermediate circular structure formed by non-docking exhibits significant fluctuations (Figure 12C). The binding of the aptamer to the target promotes the stability of the local region.

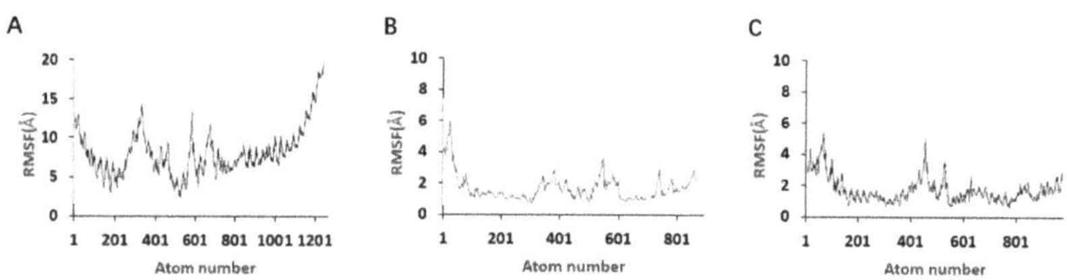

Figure 12. RMSF results for the 100 ns simulation with different modified aptamers. (**A**) RMSF of aptamer JD-2. (**B**) RMSF of aptamer JD2-RM3-27C28T. (**C**) RMSF of aptamer JD3-RMM1.

The radius of gyration (Rg) of the three further illustrates the simulated state. The Rg of the unmodified aptamer JD-2 fluctuated between 1 nm and 3 nm (Figure 13A), while JD2-RM3-27C28T remained stable between 1.9 nm and 2.0 nm after 25 ns (Figure 13B). The simulation animation revealed that at the beginning of docking, the HIJK ring of the AZA-2 was in a state far from the aptamer (Figure 9), while in the later stage, the first two ends of the AZA-2 were further approached, resulting in a decrease in the cyclotron radius at 25 ns. JD3-RMM1 remained stable between 1.2 nm and 1.5 nm (Figure 13C).

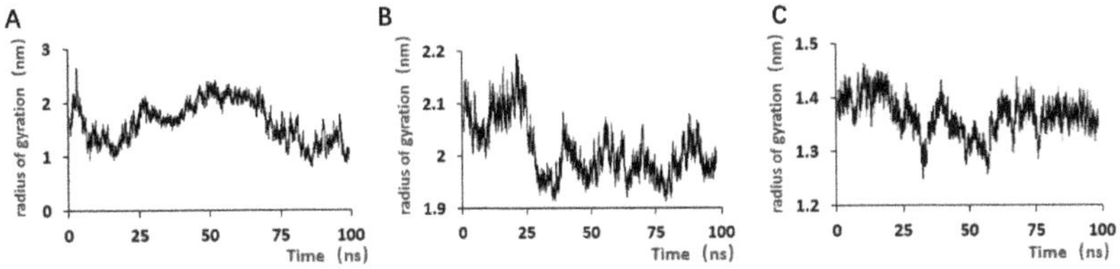

Figure 13. Radius of gyration results for the 100 ns simulation with different modified aptamers. (**A**) radius of gyration of aptamer JD-2. (**B**) radius of gyration of aptamer JD2-RM3-27C28T. (**C**) radius of gyration of aptamer JD3-RMM1.

As the simulation progresses, the solvent accessible surface area of the aptamer gradually decreases (Figure 14), indicating that the binding sites exposed by the binding of the target to the aptamer gradually decrease, and the binding stability of the aptamer to the target gradually improves.

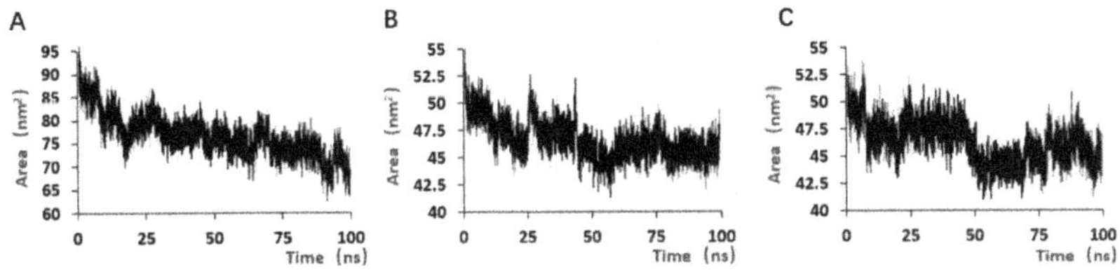

Figure 14. Solvate accessible surface area results for the 100 ns simulation with different modified aptamers. (**A**) Solvate accessible surface area of aptamer JD-2. (**B**) Solvate accessible surface area of aptamer JD2-RM3-27C28T. (**C**) Solvate accessible surface area of aptamer JD3-RMM1.

The use of graphical representation to depict changes in molecular conformational free energy can generate a free energy landscape diagram (Figure 15), in which the horizontal axis (PC1) and the vertical axis (PC2), respectively, represent RMSD (root mean square deviation) and Rg (radius of gyration) from the primary simulation results. The different dark regions in the three simulations represent distinct stable conformations. The dark region for the JD-2 aptamer is significantly larger than that for the modified aptamers (Figure 15A), and the continuous reduction in area indicates a gradual increase in aptamer stability and a decrease in conformational change values. The concentration of dark regions in the two modified aptamers, JD2-RM3-27C28T (Figure 15B) and JD3-RMM1 (Figure 15C), suggests the presence of only one stable conformation. However, the deep stable state of the unmodified JD-2 aptamer has multiple positions with similar distances, indicating the possible existence of multiple approximate conformations within the stable structure, with relatively lower difficulty in structural changes. The two modified aptamers still exhibit excellent specificity without any loss thereof (Figure 16).

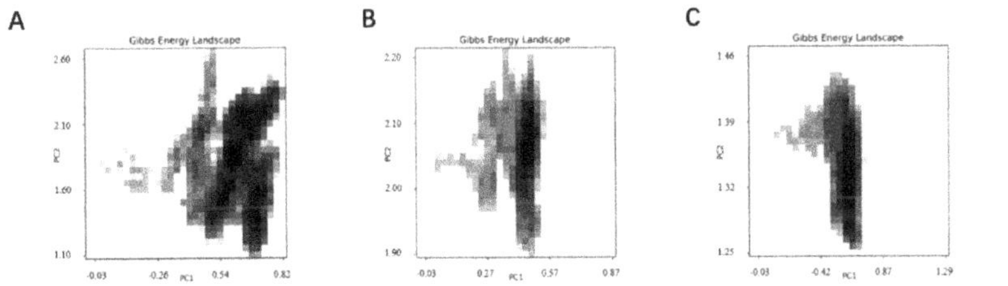

Figure 15. The 100 ns simulated free energy morphology of different modified aptamers. (**A**) Solvate accessible surface area of aptamer JD-2. (**B**) Solvate accessible surface area of aptamer JD2-RM3-27C28T. (**C**) Solvate accessible surface area of aptamer JD3-RMM1.

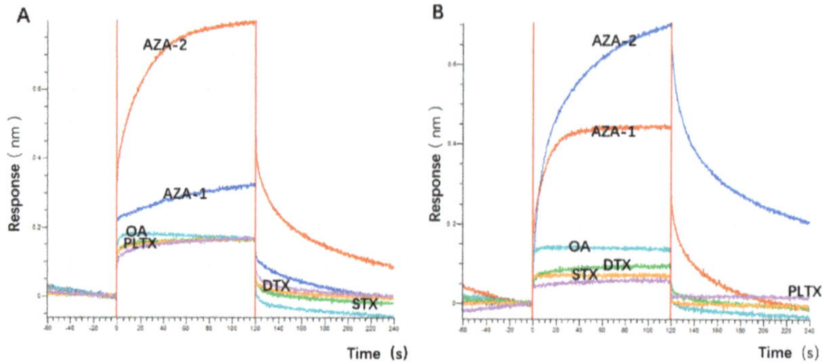

Figure 16. The specificity of some family aptamers. (**A**) BLI specificity results for aptamer JD-2-RM3-27C28T. (**B**) BLI specificity results for aptamer JD3-RMM1.

2.4. Basic Testing of Sensor Performance

After substituting the detection data from the previously validated aptamers into the sigmoidal logistic four-parameter equation, the JD2-RM3-27C28 curve and the JD3-RMM1 curve were fitted separately for the two aptamers (Figure 17). The linear regression equations for JD2-RM3-27C28 is shown in Figure 18A, its linear detection range is between 100 and ~1000 nM. For JD3-RMM1, shown in Figure 18B, its linear detection range is between 100 and 1000 nM (Figure 18). After calculation, the LOD and LOQ of the JD2-RM3-27C28 sensor are 39 nM and 131 nM, respectively. The LOD and LOQ of the JD3-RMM1 sensor are 34 nM and 116 nM, respectively.

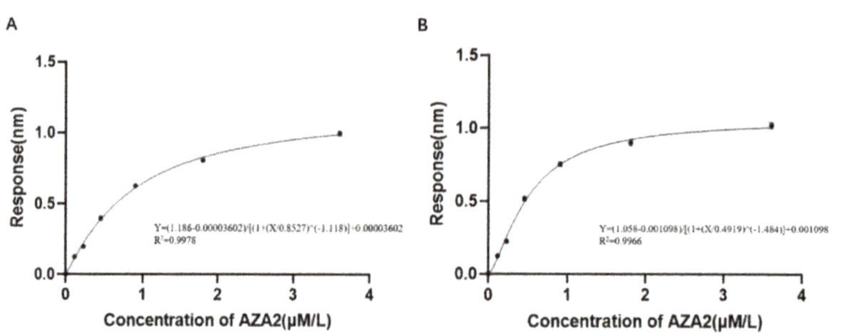

Figure 17. Fitting curves and equations of sigmoidal logistic nonlinear regression equations with different aptamers. (**A**) Fitting curve and equation of the sigmoidal logistic nonlinear regression equation for aptamer JD2-RM3-27C28T. (**B**) Fitting curve and equation of the sigmoidal logistic nonlinear regression equation for aptamer JD3-RMM1.

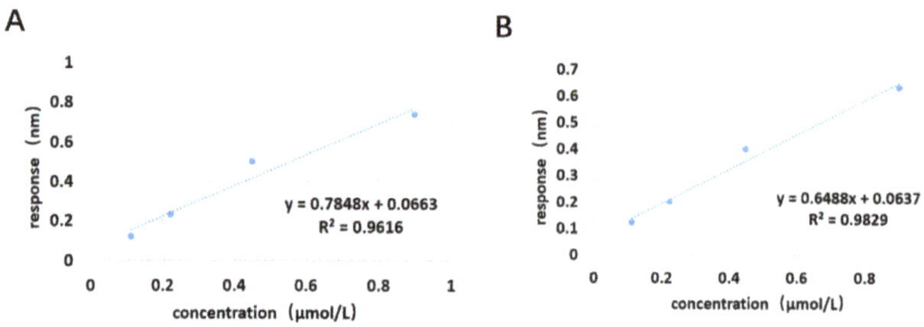

Figure 18. Linear interval fitting curve of aptamer sensor. (**A**) Linear interval fitting curve of the aptamer JD2-RM3-27C28T sensor. (**B**) Linear interval fitting curve of the aptamer JD3-RMM1 sensor.

Regarding the reusability of the sensor, it decays by about 21.2% after the first 20 cycles of use. As the number of reactions gradually increases, the degree of decay gradually decreases, and the degree of decay after the last 80 cycles is about 11.25%, indicating that this sensor can achieve stable multiple sample detections within a hundred cycles. The relative standard deviation (RSD) is 4.77%, indicating a low degree of data dispersion and good consistency (Figure 19). Moreover, for monitoring multiple samples, the detection interval only needs 60 s to be used for the next sample.

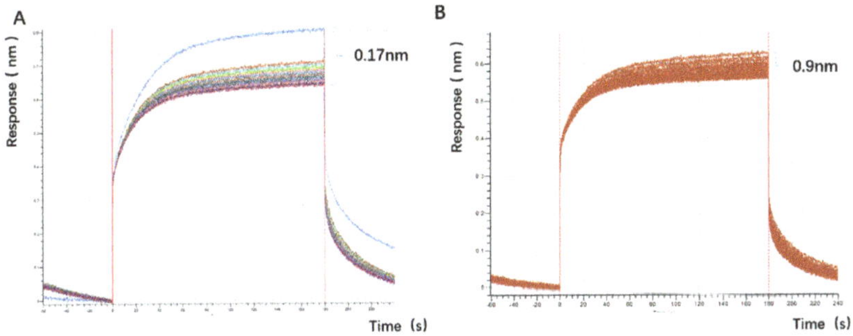

Figure 19. The degree of attenuation of sensors after multiple uses. (**A**) The degree of attenuation of the sensor in the first 20 cycles. (**B**) The degree of attenuation of the sensor between 21 and 100 times.

Regarding the specificity of these two aptamers in detecting non-azaspiracid toxins, they exhibit good specificity towards STX, DTX, OA, and PLTX. However, there is some cross-reactivity concerning distinguishing AZA-2 from similar toxins that share the same backbone but differ in the number of methyl groups (Figure 20).

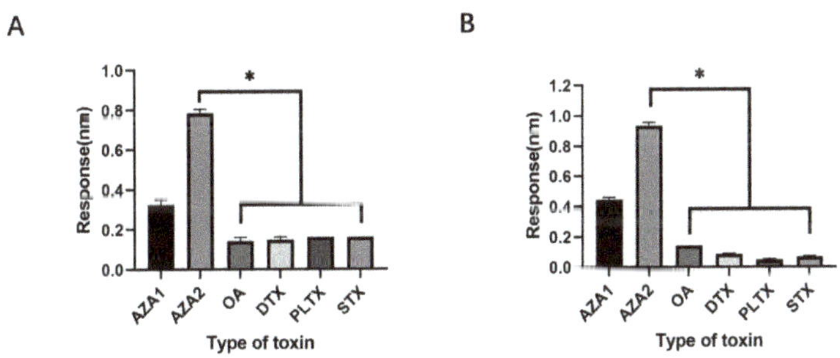

Figure 20. Specificity determination of different sensors. (**A**) Specificity of response values of aptamer JD2-RM3-27C28T to multiple toxins. (**B**) Specificity of response values of aptamer JD3-RMM1 to multiple toxins. * means $p < 0.05$ vs.AZA-2 (control).

The actual sample detection results are shown in Table 2, with a stable recovery rate between 92% and 101% and RSD ranging from 1.80% to 5.54%. The accuracy and precision meet the standard, which proves that impurities in water samples have no significant impact on the detection results in our laboratory environment.

Table 2. Concentration detection of AZA-2 in actual spiked samples.

Sample	Spiked AZA-2 (μM)	Recovery Rate (%)	RSD (%)
Tap water	0.72	100.53	3.94
	1.43	99.39	1.80
Seawater	0.72	98.11	5.44
	1.43	92.3	2.00

Therefore, both sensors have an appropriate linear range, high specificity, good repeatability, and low LOD and LOQ values, which can be applied to the preliminary detection of AZA-2 in environmental samples.

For the three environmental samples, the response values showed low or no change compared to the control response values, exceeding the minimum value of the linear detection line, indicating that the sensor did not detect any toxic components in the three hydrological samples (Table 3).

Table 3. Concentration detection of AZA-2 in hydrological samples.

Sample	Response (nm)	RSD (%)
Sea surface sample located 100 nautical miles east of Changxing Island in Shanghai	0.003	19.23
Seawater at the beach of Nanhui New City, Shanghai	0.001	0
Qiujiang Branch of the Huangpu River in Shanghai	0.000	0

3. Discussion

The choice of the Capture-SELEX method is justified as it avoids direct chemical modification of the azaspiracid (AZA-2), thereby preserving the toxin's natural structure and activity. This is crucial for ensuring that the selected aptamers possess high specificity and affinity.

This experiment experiences library loss at multiple steps. Factors such as the richness of library synthesis, the immobilization rate of the library not reaching 100%, base bias in PCR, inefficient recovery methods, qPCR prior to high-throughput sequencing, and decomposition during library collection and sample submission can all reduce library richness. These steps offer significant potential and importance for enhancing library richness during the screening process. At the same time, applying reasonable screening pressure to reduce non-specific binding is also necessary. During the screening process, methods such as reducing the number of magnetic beads, lowering library concentration, and shortening incubation time are used to increase screening pressure, which helps improve the specificity of aptamers [48]. The choice of okadaic acid (OA) as a counter-selection target is reasonable because it belongs to the same class of polyether marine toxins as AZA-2 and shares similar chemical structures and molecular weights. The impact of counter-selection target choice on aptamer specificity is discussed below.

SELEX identifies the top 10 enriched sequences from the final round of sequence analysis. In the initial rounds, some sequences already rank high in read counts. However, this does not mean that we can directly perform one-round SELEX, as there are still many complex structures that require further competitive elimination. Moreover, due to the similar structures generated by similar sequences, there are subtle differences in affinity and specificity that still require additional rounds of elimination to distinguish.

Although the selected aptamers exhibit high affinity and specificity, they may contain some non-essential nucleotides. These nucleotides not only increase the length and complexity of the aptamers but may also reduce their binding efficiency to the target [49]. Therefore, optimizing the aptamers is necessary. In this study, we truncated the aptamers by shortening the primers at both ends and further validated the effectiveness of the optimization strategy through molecular docking and dynamics simulations. The aptamer optimization in this study primarily focused on truncation. During truncation, as many existing sequences as possible were removed without disrupting existing base pairings. If changes in the docking site or a significant decrease in affinity occurred, we referred to the sequences in the selected library for backfilling or designed by adding or reducing paired base sequences. This process did not involve large-scale affinity measurements of aptamers, which may have resulted in the loss of high-affinity aptamers. Additionally, the optimization process did not employ other methods, which also reduced the likelihood of obtaining aptamers with higher affinity. Methods such as phosphate backbone modification, sugar ring modification, optimization of the binding environment, and combining aptamers with other functional molecules (such as antibody fragments, nanomaterials) can further enhance the affinity of aptamers for their targets by improving nuclease resistance and thermal stability [50].

Docking results can only provide a reference for truncation optimization. Sometimes, using several random aptamer sequences for docking can yield docking results, but these may not have practical research applications. Ultimately, it is necessary to revert to affinity detection methods such as biolayer interferometry (BLI), microscale thermophoresis (MST) [51], surface plasmon resonance (SPR) [52], isothermal titration calorimetry (ITC) [53], and fluorescence resonance energy transfer (FRET) [54] for validation. Different affinity characterization methods are not universally applicable. This experiment only used a single molecular detection method, and incorporating more affinity characterization methods could further enhance credibility.

Although the study preliminarily explored the binding mechanism between aptamers and AZA-2 using computer simulations, the research exploring the mechanism is not yet in-depth. In the future, multiple technical means, such as X-ray crystallography and nuclear magnetic resonance, should be further employed to deeply analyze the interaction mechanism between aptamers and their targets [55,56].

The application of spike detection in this experiment is highly persuasive. The sensitivity of the sensor designed in this experiment is comparable to that of ELISA, but its sensitivity cannot match that of the LC-MS detection method (Table 4). However, the detection results in environmental samples may be less convincing. 1. The distribution of toxins, especially marine biotoxins, in the environment is inherently sparse, and the distribution and occurrence of azaspiracids in China are relatively rare, reducing the likelihood of toxin detection. 2. The lack of positive control detection results based on existing detection methods such as LC-MS further reduces persuasiveness. In the future, it is necessary to learn relevant technologies and achieve controlled detection. 3. Sampling during algal blooms or harmful algal outbreaks may help improve the detection significance and expected detection rate. Additionally, the sensor developed in this experiment has not been tested in replicate environmental samples. In the future, the detection performance of the sensor under complex environments can be evaluated to further improve the affinity and specificity of aptamers, optimize the detection sensitivity and stability of the sensor, and expand its application scope.

Table 4. Existing detection methods for azaspiracids and their LOD and LOQ values.

Year	Method	LOD	LOQ
2015	SPE- HPLC-MS [57]	0.013–0.085 μg/kg	1.00 μg/kg
2015	UHPLC-HR-Orbitrap-MS [58]	0.006–0.050 ng/mL	0.018–0.227 ng/mL
2017	MB-based direct immunoassay [59]	63 μg/kg	120–2875 μg/kg
2019	ELISA [30]	0.30–4.1 ng/mL	37 μg/kg
2019	MSPE-UPLC-MS [60]	0.4–1.0 μg/kg	1.0–4.0 μg/kg
2020	SPATT-UPLC-ESI-MS [61]	0.001–0.05 μg/L	0.04 μg/mL
2020	LC-MS [62]		0.3–0.4 μg/kg

4. Materials and Methods

4.1. Chemicals and Reagents

AZA-2 was purchased from the National Research Council of Canada (Canada). OA, STX, DTX, and PLTX were purchased from Taiwan Algal Science Inc. (Taiwan, China).

All ssDNA oligonucleotides were synthesized and purified by Sangon Biotechnology Co. Ltd. (Shanghai, China). GoTaqHot® Start Colorless Master Mix was purchased from Promega Corporation (Madison, WI, USA). The QIAEX® II Gel Extraction Kit was obtained from Qiagen (Frankfurt, Germany). SA sensors were purchased from ForteBio (Shanghai, China). The selection buffer (SB) (0.9 mM $CaCl_2$, 2.7 mM KCl, 1.5 mM KH_2PO_4, 0.6 mM $MgCl_2{\cdot}6H_2O$, 0.1 M NaCl, and 20 mM Na_2HPO_4; pH = 7.4) was purchased from Tiandz (Beijing, China). The SB was used for aptamer screening, the BLI assay, and the performance evaluation of the sensor.

4.2. Aptamers Selection In Vitro

This screening uses a fixed sequence of 20 nt at the beginning and end, a random sequence of 40 nt in the middle, and a library with a total length of 80 nt for screening. The fixed nucleotides are the primer binding sites F1 and R1 for amplifying the library (Table 5). In addition, the fixed areas at both ends can form complementary feature-pairing of 10 bp. The design of R1 with a polyadenine tail (A_{20} = 20 nt A) facilitates the separation of ssDNA libraries into two different sizes of PCR products The library can be combined with the capture sequence through slow refolding, and the capture sequence (Table 5 Cap1) is bound to magnetic beads through streptavidin to achieve the binding of the three. High-affinity aptamers can be obtained by incubating with positive and negative screening targets. After PCR amplification of the library and recovery of the electrophoresis gel, the next round of library is obtained. The collected library is amplified by qPCR and subjected to high-throughput sequencing to obtain the aptamer sequence.

Table 5. List of all single stranded nucleotide sequences used.

Name	Sequence (5′-3′)
Lib	ATTGGCACTCCACGCATAGG-N40-CCTATGCGTGCTACCGTGAA [63]
F1	ATTGGCACTCCACGCATAGG
R1	A_{20}-Spacer18-TTCACGGTAGCACGCATAGG
Cap1	CCTATGCGTGGAGTGCCAAT-biotin

4.3. Affinity Determination

The aptamer sequence is labeled with biotin at the 5′ end to enable it to be immobilized on the probe of a biofilm interferometer (BLI) for affinity determination. When using BLI, the first step is to load the program: ① Baseline (1 min)—choose a 96-well plate as the buffer solution for balancing baseline measurements, and select the shaking mode when using the probe to thoroughly clean it. ② Loading (3 min)—the 96-well plate to be selected contains a final concentration of 2 μmol/L of biotin modified 5′ end aptamer that has been diluted in buffer solution. At the same time, when using the probe, the shaking mode is selected to fully bind the aptamer and remove the loosely bound aptamer. ③ Baseline (1 min)—the 96-well plate to be selected is the buffer solution to remove unbound aptamers on the sensor surface and stabilize the baseline. Next, combine the program: ① Baseline (1 min)—the 96-well plate to be selected is used as a buffer to remove unbound aptamers on the sensor surface and stabilize the baseline. ② Association (3 min)—the 96-well plate to be selected is a key step for the binding of toxins and aptamers diluted in buffer solution. ③ Dissociation (3 min)—the 96-well plate to be selected is the buffer solution for the dissociation stage of the aptamer.

4.4. Molecular Docking

Using RNAfold web server "http://rna.tbi.univie.ac.at/cgi-bin/RNAWebSuite/RNAfold.cgi (accessed on 3 May 2024)" for predicting the secondary structure of single-stranded nucleic acids, the sequence is first inputted in the order of 5′ to 3′. The prediction conditions are a folding temperature of 25 °C and ion strength equal to the concentration in the screening buffer; the generation of the tertiary structure depends on the RNAfold website for predicting the secondary structure of nucleic acid aptamers. During prediction, the DNA sequence needs to be modified to the RNA sequence. When predicting, choose the minimum free energy and allocation function while avoiding the formation of isolated base pairs. The generated dot bracket form and RNA sequence input into RNA Composer can obtain the tertiary structure of the sequence, and the file is saved in pdb format. Entering it into Discovery Studio (Ver4.5) for mutation can convert RNA into DNA. For small molecules, ChemSpider can be used "https://www.chemspider.com/ (accessed on 12 Decenmber 2023)": search for the CAS number to obtain the mol format file, obtain the aptamer and toxin pdb files, and use Autodock 1.5.7 software for molecular docking. Set the center of the small molecule in the software and detect the twist key, and then set up the docking box. After setting up the docking box, use Dejavu GUI to move the small molecule outside the box and perform grid. Before formal docking, set the docking parameters, use parameter genetic algorithm, and define the nucleic acid as a rigid structure. Ten minimum energy docking methods are used for each nucleic acid structure and small molecule, and the number of hydrogen bonds formed are counted.

4.5. Molecular Dynamic Simulation

4.5.1. Obtaining Simulated Files

Using the GROMACS 2018.8 software based on the LINUX system for molecular dynamics simulation, the top file, gro file, and itp file are first generated in the amber14sb_oL15. ff force field file. Due to software differences, there may be inconsistencies between the courtyard naming and the rtp atom naming in the force field file when using gmx, which requires manual modification in the pdb file. Use Mg^{2+} and Cl^{-} ions for cation–anion balance in charge balancing. The processing method for ligands is to use acpype to generate atomic topology under GAFF force field for 134 ligand atoms in the top file. Before using acpype, an AmberTools23 environment should be established and small molecules should be hydrogenated to calculate all atomic charges. Extract the IPT, top, and gro files with GMX suffixes and run them in the Gromacs software. Merge them with nucleic acid molecules and perform box setting, solvent addition, equilibrium charge, and other operations. Next, use MDP files for energy minimization, restrictive dynamics, and conventional dynamics. When running energy level minimization, sometimes cg (conjugate gradient method) mode is used to process complex structures that cannot be processed within a limited number of steps. To achieve equilibrium (<100 KJ/mol/nm), energy reduction can be achieved by combining the steepest descent method with cg mode.

4.5.2. Output of Dynamic Results

The root mean square deviation (RMSD) value obtained using the "gmx rms" command can reflect the stability of the system during the simulation process. Using "gmx rmsf" to obtain root mean square fluctuation (RMSF) values, RMSF calculates the fluctuation of each atom relative to its average position, characterizing the average value of structural changes over time. Using "gmx gyrate" to analyze the changes in the protein's radius of rotation, the smaller the radius, the better the density. Using "gmx sasa" to analyze the solvate accessible surface area (SASA) of solvents is an important means of describing the hydrophobicity of nucleic acids, and the hydrophobicity of nucleic acid residues may be an important physical effect affecting the conformational folding of nucleic acids. Use DulvyTools to draw free energy morphology maps and display these using "gmx sham".

4.6. Preliminary Practice of Sensors

4.6.1. Basic Testing of Sensor Performance

Using GraphPad Prism 9.5 software, we fit the data of the original concentration of polymyxin toxin and BLI response values to a four-parameter sigmoidal logistic nonlinear regression. In the equation, R_{max} and R_{min} represent the maximum and minimum response values, respectively. EC_{50} is the AZA-2 concentration corresponding to half of the maximum response value, and b is the Hill slope of the curve:

$$Y = (R_{max} - R_{min})/[(1 + (X/EC_{50})^{b})] + R_{min} \quad (1)$$

The limit of detection (LOD) refers to the lowest concentration or amount of a substance that can be identified and confirmed to be present in a sample under specific experimental conditions with a given confidence level. It does not require quantitative measurement, as long as it is indicated to be above or below the specified concentration. Calculate the response values of 20 test buffer solutions using an aptamer sensor, and obtain the LOD of the sensor according to the formula LOD = 3 Sa/b. The limit of quantification (LOQ) refers to the minimum amount of the analyte in a sample that can be quantitatively determined, usually determined by the signal-to-noise ratio method. Calculate the LOQ of the sensor based on the formula and LOQ = 10 Sa/b. Among them, Sa represents the standard deviation of the response values of 20 blank samples, and b represents the slope of the calibration curve. Under the same experimental conditions, AZA samples were repeatedly tested 100 times using the same aptamer sensor, and the response values were recorded and the RSD values were calculated. Specific detection was performed using five toxins, AZA-1, OA, DTX, PLTX, and STX, at the same concentration as the AZA-2 standard.

4.6.2. Preliminary Testing of Spiked Samples and Hydrological Samples

Firstly, take 1 mL of seawater and tap water each, remove bacteria and impurities through a 0.45 μm filter, add a bottle of toxin standard to it, evaporate at room temperature, and then add 100 μL of buffer solution to resuspend, obtaining a concentrated seawater or tap water sample containing toxins with a concentration of 7.15 μM. Store it at 4 °C for future use. To further validate the toxin detection capability of the sensor, three hydrological samples were selected for verification.

4.7. Statistical Analysis

Statistical analysis was performed using Microsoft Excel 365 or GraphPad Prism 9.5. The data were subjected to analysis by Student's *t*-test using SPSS version 22.0. The mean differences were considered significant at $p < 0.05$.

5. Conclusions

This study focuses on the screening and identification of ssDNA aptamers for the marine polyether small-molecule biotoxin AZA-2. JD2-RM3-27C28T with a dissociation constant (K_D) of 8.7×10^{-8} nM and JD3-RMM1 with a K_D of 6.8×10^{-8} nM were obtained, showing a two order-of-magnitude improvement in affinity compared to the pre-optimized aptamers. The optimized aptamers exhibited significant enhancements in both affinity and specificity, and an aptamer sensor based on BLI technology was successfully constructed. These sensors demonstrate high sensitivity, specificity, and reproducibility, enabling rapid and repeatable detection of AZA-2 in environmental samples, and its detection sensitivity surpasses that of existing enzyme-linked immunosorbent assays (ELISAs) and mouse bioassays.

In summary, this study successfully developed a BLI-based AZA-2 aptamer sensor, providing a new technical approach for the rapid detection of marine biotoxins. This innovative technology not only improves detection efficiency and accuracy but also reduces detection costs, offering strong technical support for marine food safety, environmental protection, and public health protection. The combined use of these techniques and methods not only enhances screening efficiency and accuracy but also provides robust support for aptamer optimization and validation. This study offers valuable references and insights for subsequent nucleic acid and protein research endeavors.

Author Contributions: Conceptualization, J.Y. and L.W.; methodology, J.Y.; software, X.L.; validation, J.Y., W.S., and M.S.; formal analysis, X.L.; investigation, Y.C.; resources, M.S.; data curation, Y.Y., H.C., and Y.C.; writing—original draft preparation, J.Y.; writing—review and editing, L.W.; visualization, Y.C.; supervision, M.S.; project administration, M.S.; funding acquisition, L.W. All authors have read and agreed to the published version of the manuscript.

Funding: This research was funded by National Natural Science Foundation of China (No. 82173732).

Institutional Review Board Statement: Not applicable.

Data Availability Statement: The data presented in this study are available on request from the corresponding author.

Conflicts of Interest: The authors declare no conflicts of interest.

Abbreviations

The following abbreviations are used in this manuscript:

ARfD	Acute reference dose
ASP	Amnesic shellfish poisoning
AZAs	Azaspiracids
AZP	Azaspiracid shellfish poisoning
BLI	Biolayer interferometry
DNA	Deoxyribonucleic acid
dsDNA	Double-stranded DNA
DSP	Diarrhetic shellfish poisoning
DTX	Dnophysistoxin
EFSA	European Food Safety Authority
ELISA	Enzyme-linked immunosorbent assay
HAB	Harmful algal bloom
HPLC	High-performance liquid chromatography
HTS	High-throughput sequencing
K_D	Dissociation constant
K_{dis}	Dissociation rate constant
K_{on}	Association rate constant
LC-MS	Liquid chromatograph mass spectrometer
MBA	Muse bioassay
MD	Molecular dynamics
NSP	Neurotoxic shellfish poisoning
OA	Okadaic acid
PCR	Polymerase chain reaction
PSP	Paralytic shellfish poisoning
PTX	Palytoxin
qPCR	Quantitative real-time polymerase chain reaction
RMSD	Root mean square deviation

RMSF	Root mean square fluctuation
SA	Streptavidin
SELEX	Systematic evolution of ligands by exponential
ssDNA	Single-stranded DNA
STX	Saxitoxin

Appendix A

Table A1. The docking sites and docking information of the main aptamers.

Name	Best Combination of Free Energy (kcal/mol)	Docking Site	Number of Hydrogen Bonds	K_D (μM)
JD-1	−1.82	The outer side of the complementary base structure at the 3′ and 5′ ends	2 (A3 and A40)	25.200
JD1-RMM	−1.52	The outer side of the complementary structure of the 3′ and 5′ end spaces	1 (A3)	45.600
JD-2	−3.18	Inside the pocket enclosed by the base pairing site and the linear structure at the 3′ end	2 (A25 and T12)	3.500
JD2-RM3	−3.49	The outer side of the stem structure of A6T17 and G16C7	0	2.600
JD2-RM3-27C28T	−6.49	The outer side of the stem structure of A6T17 and G16C7	1 (G5)	0.087
JD2-RM5	−1.47	Curling area of T16 and G19	0	9.750
JD-3	−2.3	Inside the grooves at the 3′ and 5′ ends	1 (C8)	3.600
JD3-RMM	−2.16	The first C G base pairing at the 3′ end	1 (G21)	56.000
JD3-RMM1	−4.18	The first C G base pairing at the 3′ end	1 (G21)	0.068
JD3-RMM2	−3.07	The first C G base pairing at the 3′ end	1 (G21)	46.120
JD-4	−2.9	The first base pairing at the 5 ′end	1 (C3)	139.400
JD4-RM3	−2.07	The first base pairing at the 5 ′end	0	237.800
JD-5	−2.32	Between the 3′ and 5′ ends	1 (T23)	39.510
JD5-RMM	−4.7	Base pairing site at the 3′ end	1 (G15)	40.290
JD-6	−2.97	The ring formed by the intermediate sequence of the aptamer	1 (A22)	41.240
JD-7	−3.04	The outer side of G11G12C32C33 base pairing	0	92.120
JD7-34G	−3.12	The outer side of C10G11C33G34 base pairing	0	91.200
JD7-34G35G	−3.58	The outer side of C10G11C33G34 base pairing	0	107.120
JD7-RM5	−2.43	The outer side of G11G12C32C33 base pairing	0	98.120
JD7-RMM	−1.52	The outer side of the complementary structure at the 3′ and 5′ ends	1 (A3)	48.980
JD-8	−0.92	The first base pairing at the 3′ end	0	18.300
JD-9	−2.5	The first base pairing at the 3′ end	0	12.800
JD9-RMM	−3.72	The first base pairing at the 3′ end	0	65.190
JD-10	−2.7	Linear curl at the 3′ end	1 (A34)	35.220
JD10-M1	−2.7	Linear curl at the 3′ end	0	38.760
JD10-RMM	−3.62	The first base pairing at the 3′ end	0	75.120
JD10-RMM-6C7A	−3.54	The first base pairing at G5-C17 at the 3′ end	1 (A19)	49.750

Table A2. Names and corresponding sequences of aptamers involved in docking.

Name	Sequence (5′ to 3′)	Count of Bases
JD1-RMM	GAATGGACCCGGTATAATAGAGTGCCAATTTCA	33
JD2-RM3	AAAAGACTTAGTGTTGTCCTAT	22
JD2-RM3-27C28T	AAAAGACTTAGTGTTGTCCTATGTATCT	28
JD2-RM5	GACTTAGTGTTGTCCTATGTATAAGTGCCAATTTCA	36
JD3-RMM	ACTAGGGCAAATCTATTACAGTGCCAATTTCA	32
JD3-RMM1	ACTAGGGCAAATCTATACAGTGCCAATTTCA	31
JD3-RMM2	ACTAGGGCAAATTATACAGTGCCAATTTCA	30
JD4-RM3	GACATCGAGAAGAATCCTGATACG	24

Table A2. *Cont.*

Name	Sequence (5′ to 3′)	Count of Bases
JD5-RMM	CCAACATGCATTTTGAGGTGTGTACACCGTG	31
JD7-RM5	GGACCCGGTATAATTCCCTCGAGAGTGCCAATTTCA	36
JD7-RMM	GAATGGACCCGGTATAATTAGAGTGCCAATTTCA	34
JD7-34G	GAATGGACCCGGTATAATTCCCTCAAGAGTGCCGATTTCA	40
JD7-34G35G	GAATGGACCCGGTATAATTCCCTCAAGAGTGCCGGTTTCA	40
JD9-RMM	ACTAGCACAAATCGTTCAGTGCCAATTTCA	30
JD10-M1	ACTAATTGGCATCTATTCCTATGATACAGTGCCAATTTCA	40
JD10-RMM	ACTAGTGCAAACAGTGCCAATTTCA	25
JD10-RMM-6C7A	ACTAGCACAAACAGTGCCAATTTCA	25

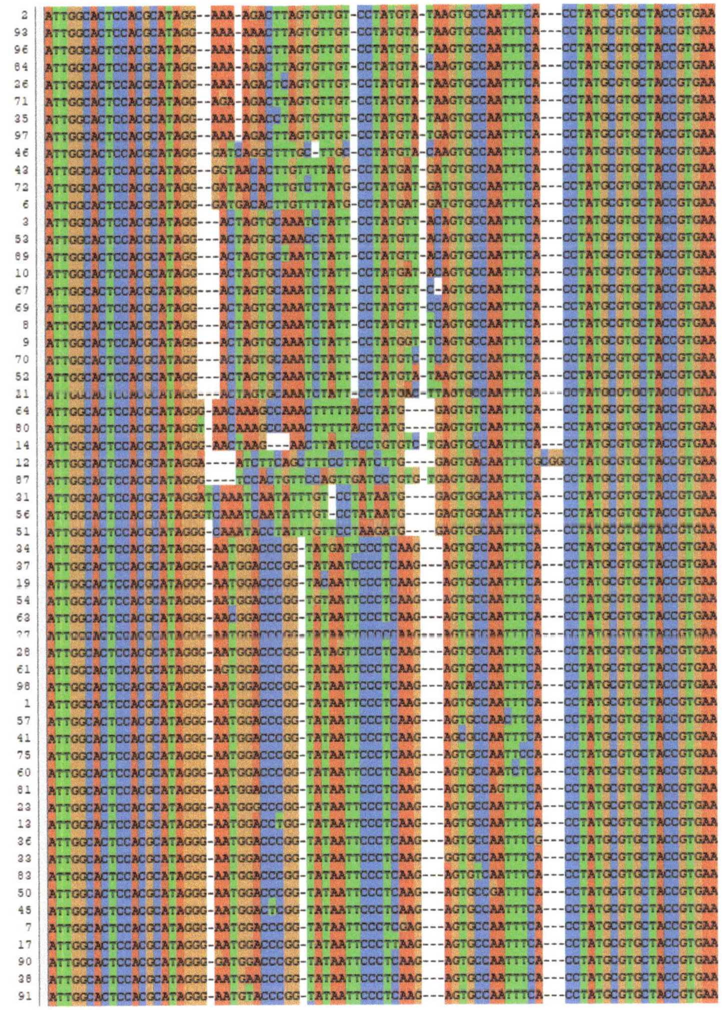

Figure A1. *Cont.*

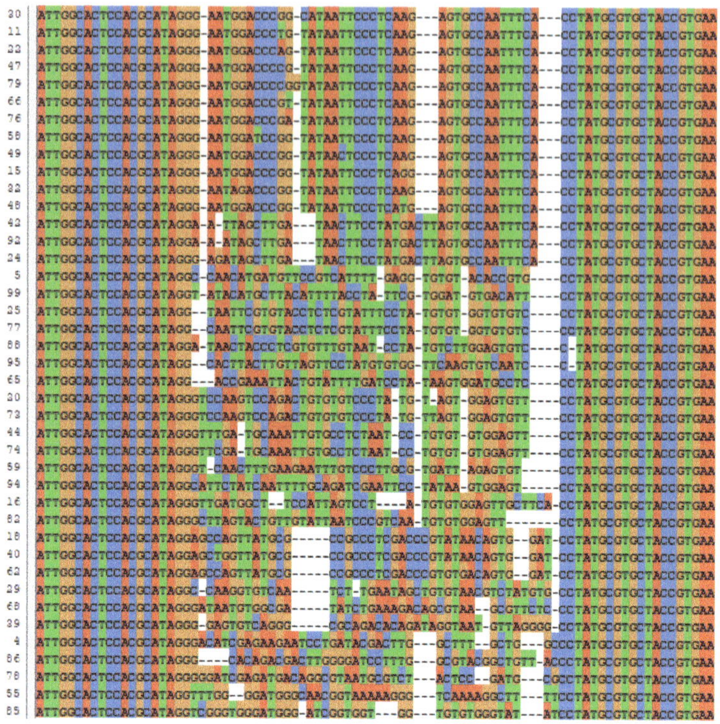

Figure A1. Cluster analysis of AZA-2 advantageous aptamer obtained by the Capture-SELEX method using Clustal X2.1.

References

1. Krock, B.; Tillmann, U.; John, U.; Cembella, A.D. Characterization of azaspiracids in plankton size-fractions and isolation of an azaspiracid-producing dinoflagellate from the North Sea. *Harmful Algae* **2009**, *8*, 254–263. [CrossRef]
2. Tillmann, U.; Elbrächter, M.; Krock, B.; John, U.; Cembella, A.D. *Azadinium spinosum* gen. et sp. nov (Dinophyceae) identified as a primary producer of azaspiracid toxins. *Eur. J. Phycol.* **2009**, *44*, 63–79. [CrossRef]
3. Tillmann, U.; Salas, R.; Gottschling, M.; Krock, B.; O'Driscoll, D.; Elbrächter, M. *Amphidoma languida* sp. nov.(Dinophyceae) reveals a close relationship between Amphidoma and Azadinium. *Protist* **2012**, *163*, 701–719. [CrossRef] [PubMed]
4. Tillmann, U.S.R.; Jauffrais, T.; Hess, P.; Silke, J. AZA: The producing organisms—Biology and trophic transfer. In *Seafood and Freshwater Toxins*; Botana, L.M., Ed.; CRC Press: Boca Raton, FL, USA, 2014; pp. 773–798.
5. James, K.J.; Moroney, C.; Roden, C.; Satake, M.; Yasumoto, T.; Lehane, M.; Furey, A. Ubiquitous 'benign' alga emerges as the cause of shellfish contamination responsible for the human toxic syndrome, azaspiracid poisoning. *Toxicon Off. J. Int. Soc. Toxinol.* **2003**, *41*, 145–151. [CrossRef]
6. Furey, A.; Braña-Magdalena, A.; Lehane, M.; Moroney, C.; James, K.J.; Satake, M.; Yasumoto, T. Determination of azaspiracids in shellfish using liquid chromatography/tandem electrospray mass spectrometry. *Rapid Commun. Mass Spectrom. RCM* **2002**, *16*, 238–242. [CrossRef] [PubMed]
7. Furey, A.; Moroney, C.; Braña-Magdalena, A.; Saez, M.J.; Lehane, M.; James, K.J. Geographical, temporal, and species variation of the polyether toxins, azaspiracids, in shellfish. *Environ. Sci. Technol.* **2003**, *37*, 3078–3084. [CrossRef]
8. Magdalena, A.B.; Lehane, M.; Moroney, C.; Furey, A.; James, K.J. Food safety implications of the distribution of azaspiracids in the tissue compartments of scallops (Pecten maximus). *Food Addit. Contam.* **2003**, *20*, 154–160. [CrossRef]
9. McMahon, T.S.J. Winter toxicity of unknown aetiology in mussels. *Harmful Algae News* **1996**, *14*, 2.
10. Satake, M.O.K.; James, K.J.; Furey, A.; Yasumoto, T. New toxic event caused by Irish mussels. *Harmful Algae* **1998**, 468–469. In *Harmful Algae: Proceedings of the VIII International Conference on Harmful Algae, Vigo, Spain, 25–29 June 1997*; Xunta de Galicia Intergovernmental Oceanographic Commission of UNESCO: Vigo, Spain, 1998.
11. James, K.J.; Sierra, M.D.; Lehane, M.; Braña Magdalena, A.; Furey, A. Detection of five new hydroxyl analogues of azaspiracids in shellfish using multiple tandem mass spectrometry. *Toxicon Off. J. Int. Soc. Toxinol.* **2003**, *41*, 277–283. [CrossRef]

12. Yang, J.; Sun, W.; Sun, M.; Cui, Y.; Wang, L. Current Research Status of Azaspiracids. *Mar. Drugs* **2024**, *22*, 79. [CrossRef]
13. Sandvik, M.; Miles, C.O.; Løvberg, K.L.E.; Kryuchkov, F.; Wright, E.J.; Mudge, E.M.; Kilcoyne, J.; Samdal, I.A. In Vitro Metabolism of Azaspiracids 1-3 with a Hepatopancreatic Fraction from Blue Mussels (Mytilus edulis). *J. Agric. Food Chem.* **2021**, *69*, 11322–11335. [CrossRef] [PubMed]
14. Twiner, M.J.; Doucette, G.J.; Rasky, A.; Huang, X.P.; Roth, B.L.; Sanguinetti, M.C. Marine algal toxin azaspiracid is an open-state blocker of hERG potassium channels. *Chem. Res. Toxicol.* **2012**, *25*, 1975–1984. [CrossRef] [PubMed]
15. Boente-Juncal, A.; Raposo-García, S.; Costas, C.; Louzao, M.C.; Vale, C.; Botana, L.M. Partial Blockade of Human Voltage-Dependent Sodium Channels by the Marine Toxins Azaspiracids. *Chem. Res. Toxicol.* **2020**, *33*, 2593–2604. [CrossRef] [PubMed]
16. Boente-Juncal, A.; Raposo-García, S.; Louzao, M.C.; Vale, C.; Botana, L.M. Targeting Chloride Ion Channels: New Insights into the Mechanism of Action of the Marine Toxin Azaspiracid. *Chem. Res. Toxicol.* **2021**, *34*, 865–879. [CrossRef]
17. Alfonso, A.; Román, Y.; Vieytes, M.R.; Ofuji, K.; Satake, M.; Yasumoto, T.; Botana, L.M. Azaspiracid-4 inhibits Ca^{2+} entry by stored operated channels in human T lymphocytes. *Biochem. Pharmacol.* **2005**, *69*, 1627–1636. [CrossRef]
18. Abal, P.; Louzao, M.C.; Fraga, M.; Vilariño, N.; Ferreiro, S.; Vieytes, M.R.; Botana, L.M. Absorption and Effect of Azaspiracid-1 Over the Human Intestinal Barrier. *Cell. Physiol. Biochem. Int. J. Exp. Cell. Physiol. Biochem. Pharmacol.* **2017**, *43*, 136–146. [CrossRef]
19. Ito, E. Toxicology of Azaspiracid-1: Acute and chronic poisoning, tumourigenicity and chemical structure relationship to toxicity in a mouse bioassay. *Seaf. Freshw. Toxins Pharmacol. Physiol. Detect.* **2008**, 775–784.
20. Bodero, M.; Hoogenboom, R.; Bovee, T.F.H.; Portier, L.; de Haan, L.; Peijnenburg, A.; Hendriksen, P.J.M. Whole genome mRNA transcriptomics analysis reveals different modes of action of the diarrheic shellfish poisons okadaic acid and dinophysis toxin-1 versus azaspiracid-1 in Caco-2 cells. *Toxicol. Vitr.* **2018**, *46*, 102–112. [CrossRef]
21. Ito, E.; Satake, M.; Ofuji, K.; Higashi, M.; Harigaya, K.; McMahon, T.; Yasumoto, T. Chronic effects in mice caused by oral administration of sublethal doses of azaspiracid, a new marine toxin isolated from mussels. *Toxicon Off. J. Int. Soc. Toxinol.* **2002**, *40*, 193–203. [CrossRef]
22. Hjørnevik, L.V.; Frøyset, A.K.; Grønset, T.A.; Rungruangsak-Torrissen, K.; Fladmark, K.E. Algal Toxin Azaspiracid-1 Induces Early Neuronal Differentiation and Alters Peripherin Isoform Stoichiometry. *Mar. Drugs* **2015**, *13*, 7390–7402. [CrossRef]
23. (CONTAM) EPoCitFC. Opinion of the Scientific Panel on Contaminants in the Food chain on a request from the European Commission on marine biotoxins in shellfish—Azaspiracids. *EFSA J.* **2008**, *723*, 1–52.
24. EFSA. Scientific Opinion of the Panel on Contaminants in the Food Chain on a request from the European Commission onMarine Biotoxins in Shellfish –Summary on regulated marine biotoxins. *EFSA J* **2009**, *1306*, 1–23.
25. Regulation (EC) No 853/2004 of the European parliament and of the council of 29 April 2004 laying down specific hygiene rules for food of animal origin. *Off. J. Eur. Union* **2004**, *139*, 55–205.
26. Draisci, R.; Palleschi, L.; Ferretti, E.; Furey, A.; James, K.J.; Satake, M.; Yasumoto, T. Development of a method for the identification of azaspiracid in shellfish by liquid chromatography-tandem mass spectrometry. *J. Chromatogr. A* **2000**, *871*, 13–21. [CrossRef]
27. Pelin, M.; Kilcoyne, J.; Nulty, C.; Crain, S.; Hess, P.; Tubaro, A.; Sosa, S. Toxic equivalency factors (TEFs) after acute oral exposure of azaspiracid 1, -2 and -3 in mice. *Toxicol. Lett.* **2018**, *282*, 136–146. [CrossRef]
28. Ofuji, K.; Satake, M.; McMahon, T.; Silke, J.; James, K.J.; Naoki, H.; Oshima, Y.; Yasumoto, T. Two analogs of azaspiracid isolated from mussels, Mytilus edulis, involved in human intoxication in Ireland. *Nat. Toxins* **1999**, *7*, 99–102. [CrossRef]
29. Samdal, I.A.; Løvberg, K.E.; Briggs, L.R.; Kilcoyne, J.; Xu, J.; Forsyth, C.J.; Miles, C.O. Development of an ELISA for the detection of Azaspiracids. *J. Agric. Food Chem.* **2015**, *63*, 7855–7861. [CrossRef]
30. Leonardo, S.; Rambla-Alegre, M.; Samdal, I.A.; Miles, C.O.; Kilcoyne, J.; Diogène, J.; O'Sullivan, C.K.; Campàs, M. Immunorecognition magnetic supports for the development of an electrochemical immunoassay for azaspiracid detection in mussels. *Biosens. Bioelectron.* **2017**, *92*, 200–206. [CrossRef]
31. Samdal, I.A.; Løvberg, K.E.; Kristoffersen, A.B.; Briggs, L.R.; Kilcoyne, J.; Forsyth, C.J.; Miles, C.O. A Practical ELISA for Azaspiracids in Shellfish via Development of a New Plate-Coating Antigen. *J. Agric. Food Chem.* **2019**, *67*, 2369–2376. [CrossRef]
32. Miles, C.O.; Kilcoyne, J.; McCarron, P.; Giddings, S.D.; Waaler, T.; Rundberget, T.; Samdal, I.A.; Løvberg, K.E. Selective Extraction and Purification of Azaspiracids from Blue Mussels (Mytilus edulis) Using Boric Acid Gel. *J. Agric. Food Chem.* **2018**, *66*, 2962–2969. [CrossRef]
33. Forsyth, C.J.; Xu, J.; Nguyen, S.T.; Samdal, I.A.; Briggs, L.R.; Rundberget, T.; Sandvik, M.; Miles, C.O. Antibodies with broad specificity to azaspiracids by use of synthetic haptens. *J. Am. Chem. Soc.* **2006**, *128*, 15114–15116. [CrossRef] [PubMed]
34. Frederick, M.O.; De Lamo Marin, S.; Janda, K.D.; Nicolaou, K.C.; Dickerson, T.J. Monoclonal antibodies with orthogonal azaspiracid epitopes. *Chembiochem. A Eur. J. Chem. Biol.* **2009**, *10*, 1625–1629. [CrossRef]
35. Rodríguez, L.P.; Vilariño, N.; Louzao, M.C.; Dickerson, T.J.; Nicolaou, K.C.; Frederick, M.O.; Botana, L.M. Microsphere-based immunoassay for the detection of azaspiracids. *Anal. Biochem.* **2014**, *447*, 58–63. [CrossRef]
36. Ellington, A.D.; Szostak, J.W. In vitro selection of RNA molecules that bind specific ligands. *Nature* **1990**, *346*, 818–822. [CrossRef] [PubMed]

37. Chen, F.; Lu, Q.; Huang, L.; Liu, B.; Liu, M.; Zhang, Y.; Liu, J. DNA Triplex and Quadruplex Assembled Nanosensors for Correlating K(+) and pH in Lysosomes. *Angew. Chem. (Int. Ed. Engl.)* **2021**, *60*, 5453–5458. [CrossRef]
38. Pheeney, C.G.; Arnold, A.R.; Grodick, M.A.; Barton, J.K. Multiplexed electrochemistry of DNA-bound metalloproteins. *J. Am. Chem. Soc.* **2013**, *135*, 11869–11878. [CrossRef]
39. Xiao, X.; Zhu, L.; He, W.; Luo, Y.; Xu, W. Functional nucleic acids tailoring and its application. *Trends Anal. Chem.* **2019**, *118*, 138–157. [CrossRef]
40. Nonaka, Y.; Yoshida, W.; Abe, K.; Ferri, S.; Schulze, H.; Bachmann, T.T.; Ikebukuro, K. Affinity improvement of a VEGF aptamer by in silico maturation for a sensitive VEGF-detection system. *Anal. Chem.* **2013**, *85*, 1132–1137. [CrossRef]
41. Singh, N.K.; Wang, Y.; Wen, C.; Davis, B.; Wang, X.; Lee, K.; Wang, Y. High-affinity one-step aptamer selection using a non-fouling porous hydrogel. *Nat. Biotechnol.* **2024**, *42*, 1224–1231. [CrossRef]
42. Xu, G.; Wang, C.; Yu, H.; Li, Y.; Zhao, Q.; Zhou, X.; Li, C.; Liu, M. Structural basis for high-affinity recognition of aflatoxin B1 by a DNA aptamer. *Nucleic Acids Res.* **2023**, *51*, 7666–7674. [CrossRef]
43. Sequeira-Antunes, B.; Ferreira, H.A. Nucleic Acid Aptamer-Based Biosensors: A Review. *Biomedicines* **2023**, *11*, 3201. [CrossRef]
44. Zhou, J.; Rossi, J. Aptamers as targeted therapeutics: Current potential and challenges. *Nat. Rev. Drug Discov.* **2017**, *16*, 181–202. [CrossRef] [PubMed]
45. Carias, J.; Vásquez-Lavín, F.; Barrientos, M.; Ponce Oliva, R.D.; Gelcich, S. Economic valuation of Harmful Algal Blooms (HAB): Methodological challenges, policy implications, and an empirical application. *J. Environ. Manag.* **2024**, *365*, 121566. [CrossRef] [PubMed]
46. EFSA. Marine biotoxins in shellfish—Azaspiracid group, scientific opinion of the panel on contaminants in the food chain. *EFSA J.* **2008**, *723*, 1–52.
47. Otero, P.; Silva, M. Emerging Marine Biotoxins in European Waters: Potential Risks and Analytical Challenges. *Mar. Drugs* **2022**, *20*, 199. [CrossRef]
48. Jiang, B.C.; Ling, Y.J.; Xu, M.L.; Gu, J.; Wu, X.B.; Sha, W.L.; Tian, T.; Bai, X.H.; Li, N.; Jiang, C.Y.; et al. Follistatin drives neuropathic pain in mice through IGF1R signaling in nociceptive neurons. *Sci. Transl. Med.* **2024**, *16*, eadi1564. [CrossRef]
49. Zhang, X.; Gao, Y.; Deng, B.; Hu, B.; Zhao, L.; Guo, H.; Yang, C.; Ma, Z.; Sun, M.; Jiao, B.; et al. Selection, Characterization, and Optimization of DNA Aptamers against Challenging Marine Biotoxin Gymnodimine-A for Biosensing Application. *Toxins* **2022**, *14*, 195. [CrossRef]
50. Zhao, L.; Guo, H.; Chen, H.; Zou, B.; Yang, C.; Zhang, X.; Gao, Y.; Sun, M.; Wang, L. A Rapid and Sensitive Aptamer-Based Biosensor for Amnesic Shellfish Toxin Domoic Acid. *Bioengineering* **2022**, *9*, 684. [CrossRef]
51. Zhou, R.; Gao, Y.; Yang, C.; Zhang, X.; Hu, B.; Zhao, L.; Guo, H.; Sun, M.; Wang, L.; Jiao, B. A Novel SELEX Based on Immobilizing Libraries Enables Screening of Saxitoxin Aptamers for BLI Aptasensor Applications. *Toxins* **2022**, *14*, 228. [CrossRef]
52. Magnez, R.; Bailly, C.; Thuru, X. Microscale Thermophoresis as a Tool to Study Protein Interactions and Their Implication in Human Diseases. *Int. J. Mol. Sci.* **2022**, *23*, 672. [CrossRef]
53. Ravindran, N.; Kumar, S.; M, Y.; S, R.; C, A.M.; Thirunavookarasu, S.N.; C, K.S. Recent advances in Surface Plasmon Resonance (SPR) biosensors for food analysis: A review. *Crit. Rev. Food Sci. Nutr.* **2023**, *63*, 1055–1077. [CrossRef] [PubMed]
54. Lima Cavalcanti, I.D.; Xavier Junior, F.H.; Santos Magalhães, N.S.; Lira Nogueira, M.C.B. Isothermal titration calorimetry (ITC) as a promising tool in pharmaceutical nanotechnology. *Int. J. Pharm.* **2023**, *641*, 123063. [CrossRef]
55. Wu, L.; Huang, C.; Emery, B.P.; Sedgwick, A.C.; Bull, S.D.; He, X.P.; Tian, H.; Yoon, J.; Sessler, J.L.; James, T.D. Förster resonance energy transfer (FRET)-based small-molecule sensors and imaging agents. *Chem. Soc. Rev.* **2020**, *49*, 5110–5139. [CrossRef] [PubMed]
56. Bijak, V.; Szczygiel, M.; Lenkiewicz, J.; Gucwa, M.; Cooper, D.R.; Murzyn, K.; Minor, W. The current role and evolution of X-ray crystallography in drug discovery and development. *Expert Opin. Drug Discov.* **2023**, *18*, 1221–1230. [CrossRef] [PubMed]
57. Caceres-Cortes, J.; Falk, B.; Mueller, L.; Dhar, T.G.M. Perspectives on Nuclear Magnetic Resonance Spectroscopy in Drug Discovery Research. *J. Med. Chem.* **2024**, *67*, 1701–1733. [CrossRef]
58. Fang, L.; Yao, X.; Wang, L.; Li, J. Solid-phase extraction-based ultra-sensitive detection of four lipophilic marine biotoxins in bivalves by high-performance liquid chromatography-tandem mass spectrometry. *J. Chromatogr. Sci.* **2015**, *53*, 373–379. [CrossRef]
59. Orellana, G.; Van Meulebroek, L.; Van Vooren, S.; De Rijcke, M.; Vandegehuchte, M.; Janssen, C.R.; Vanhaecke, L. Quantification and profiling of lipophilic marine toxins in microalgae by UHPLC coupled to high-resolution orbitrap mass spectrometry. *Anal. Bioanal. Chem.* **2015**, *407*, 6345–6356. [CrossRef]
60. Xu, F.; Liu, F.; Wang, C.; Wei, Y. Reversed-phase/weak anion exchange magnetic mesoporous microspheres for removal of matrix effects in lipophilic marine biotoxins analysis by ultrahigh-performance liquid chromatography coupled to tandem mass spectrometry. *Food Chem.* **2019**, *294*, 104–111. [CrossRef]
61. Onofrio, M.D.; Mallet, C.R.; Place, A.R.; Smith, J.L. A screening tool for the direct analysis of marine and freshwater phycotoxins in organic SPATT extracts from the chesapeake bay. *Toxins* **2020**, *12*, 322. [CrossRef]

62. Yang, L.; Singh, A.; Lankford, S.K.; Stuart, J.; Rice, D.; Wu, W.H.; Hungerford, J.M. A rapid method for the detection of diarrhetic shellfish toxins and Azaspiracid shellfish toxins in Washington State shellfish by liquid chromatography tandem mass spectrometry. *J. AOAC Int.* **2020**, *103*, 792–799. [CrossRef]
63. Guo, H.; Deng, B.; Zhao, L.; Gao, Y.; Zhang, X.; Yang, C.; Zou, B.; Chen, H.; Sun, M.; Wang, L.; et al. Programmed Aptamer Screening, Characterization, and Rapid Detection for α-Conotoxin MI. *Toxins* **2022**, *14*, 706. [CrossRef] [PubMed]

marine drugs

MDPI

Article

Comparative Analysis of Intestinal Microbiota Between Tetrodotoxin-Containing and Tetrodotoxin-Free *Takifugu rubripes*

Hanyuan Zhang [1,*], Jilun Hou [2], Youxiu Zhu [1], Biyin Wu [1], Jiangong Ren [2], Zhaohui Sun [2] and Xia Liu [2]

[1] Key Laboratory of Aquatic Genomics, Ministry of Agriculture and Rural Affairs, Beijing Key Laboratory of Fishery Biotechnology, Chinese Academy of Fishery Sciences, Beijing 100141, China
[2] Beidaihe Central Experiment Station, Chinese Academy of Fishery Sciences, Qinhuangdao 066100, China
* Correspondence: zhanghanyuan@cafs.ac.cn; Tel.: +86-010-68678587

Abstract: Tetrodotoxin (TTX) is a potent marine neurotoxin found in pufferfish, causing severe poisoning or death if consumed improperly. Studies have indicated that intestinal symbiotic microbiota are associated with the production and accumulation of TTX in pufferfish. However, the specific symbiotic microorganisms involved in these processes and their respective functions remain unclear. This study explored differences in intestinal microbiota related to the TTX content between toxic and non-toxic tiger puffer *Takifugu rubripes*. We found that the dominant phyla exhibiting significant abundance differences between the two groups were Proteobacteria and Bacteroidota, with the core bacterial genera being *Rikenella*, *Vibrio*, *Photobacterium*, and *Bacteroides*. Moreover, the genera *Marinimicrobium*, *Idomarina*, *Galbibacter*, and *Brumimicrobium* were reported for the first time to be potentially associated with TTX bioaccumulation in *T. rubripes*. In addition, an integrated analysis with our previous study indicated that the "ABC transporters" pathway may play significant roles in the production and transport of TTX in both symbiotic microorganisms and *T. rubripes*. This study preliminarily investigated the intestinal symbiotic bacteria associated with the accumulation and metabolism of TTX in *T. rubripes*, as well as screening potential microbial biomarkers for assessing the safety of pufferfish.

Keywords: *Takifugu rubripes*; tetrodotoxin; 16S rRNA sequencing; intestinal microbiota; microbial function

1. Introduction

Pufferfish is a widely distributed fish species worldwide, existing in both seawater and freshwater. Due to its tasty flavor and high nutrition, pufferfish is a well-received high-end food ingredient and an economically important breeding species in East Asian countries [1]. However, the presence of tetrodotoxin (TTX) in pufferfish has been a significant constraint on the further advancement of the pufferfish aquaculture industry. For a long time, incidents of food poisoning caused by improper consumption of pufferfish have occurred and been reported, with severe cases leading to death [2,3].

TTX is a highly potent marine bio-neurotoxin [4], which is a kind of alkaloid with a small molecular weight (chemical formula: $C_{11}H_{17}N_3O_8$, molecular weight: 319.27) and high biological activities [5]. It is a typical voltage-gated sodium ion channel (Na_V) blocker since it selectively binds to Na_V and exerts toxic effects by inhibiting nerve and muscle conduction [6]. TTX has been used in medical research for the treatment of various diseases such as analgesia, anesthesia, anti-cancer, cardiovascular disease, and neurological disorders according to its mechanism of action [7–12]. There are a variety of naturally occurring

TTX analogues [13]. So far, more than 20 TTX structural analogues have been reported with different levels of toxicity [14–17]. The significant differences in the toxicity among TTX and its analogues were found to be positively correlated with the number of hydroxyl groups [18]. TTX was firstly isolated and reported in Tetraodontidae [19], and subsequent studies have found that TTX widely exists in organisms with different evolutionary levels, including algae, parasites, mollusks, crustaceans, echinoderms, amphibians, etc. [20].

The TTX content of pufferfish varies remarkably under different habitat conditions; however, the origin and accumulation pathways were not clear yet. There are currently three main hypotheses regarding the origin of TTX in pufferfish: the exogenous hypothesis (via food chain enrichment), the endogenous hypothesis (self-transformation and accumulation of TTX by pufferfish), and the endosymbiotic origin hypothesis (the production of symbiotic microorganisms) [21]. Three hypotheses attempted to explain the possible origin of TTX in pufferfish from different perspectives [22]. Previous studies have demonstrated that tiger puffer (*Takifugu rubripes*) have the ability to convert TTX analogues or precursors enriched by the food chain or produced by symbiotic microorganisms into TTX, then transport and accumulate it to different tissues and organs [23,24]. Since the endosymbiotic origin hypothesis was proposed, researchers have successively isolated bacteria and actinomycetes that could ferment to produce TTX analogues from various TTX-containing animals [25]. Thus far, the bacterial species that can produce TTX and its analogues mainly are *Vibrio*, *Pseudomonas*, *Photobacterium*, *Aeromonas*, *Plesiomonas*, *Bacillus*, *Actinobacter*, as well as *Streptomyces* of actinomycetes [26–30]. TTX produced by marine microorganisms is typically a secondary metabolite generated during the stationary phase of bacterial growth. Microbial fermentation experiments have shown that the TTX content produced by bacterial strains is significantly lower than that found in pufferfish [31]. This difference may be attributed to the fact that pufferfish serve as hosts for TTX-producing symbiotic bacteria and can provide the necessary signaling molecules to induce TTX production [32]. In addition, the structure of TTX produced by microbial fermentation was different from that accumulated in pufferfish through food intake [33]. Studies have shown that TTX produced by *Vibrio alginolyticus* was unable to react with monoclonal antibody against TTX in *T. rubripes* [34]. These results suggested that symbiotic microorganisms in pufferfish could produce precursors or analogues of TTX, which need to be transformed and accumulated under the molecular regulation in pufferfish. Even though the molecular mechanism of TTX accumulation and transformation in *T. rubripes* has been reported before [23]. However, our understanding of the TTX-producing bacterial strains and their potential functions remains limited. Meanwhile, 16S rRNA sequencing, which is widely used in microbiota studies, can efficiently identify bacterial species, reflect the diversity and complexity of microbial communities, and compare the differences in microbiota between samples or groups [35]. Conducting a comparative study on the intestinal microbiota of toxic and non-toxic pufferfish using 16S rRNA sequencing could comprehensively identify TTX-related microorganisms.

Our previous study conducted a transcriptomic comparison between toxic and non-toxic *T. rubripes*, identifying functional genes associated with TTX accumulation, translocation, and detoxification [23]. However, the intestinal microbiota contributing to TTX accumulation and their potential functions in pufferfish remain unexplored. In the present study, we characterized the symbiotic intestinal microbiota of toxic (wild) and non-toxic (captive) *T. rubripes* to screen the intestinal microbiota involved in TTX accumulation under different habitats and to preliminarily investigate their potential roles in TTX metabolism in *T. rubripes*.

2. Results

2.1. TTX Concentration in Wild and Captive T. rubripes

The muscular TTX concentration of wild and captive *T. rubripes* was detected as the phenotypic trait in this study. The TTX concentration in the muscle tissue of wild individuals ranged from 3.9–280 μg/kg, while the TTX concentration in the muscle tissue of captive samples was all below the detectable limit (<1 μg/kg) (Table S1). Therefore, the wild individuals were considered a toxic group, and the captive individuals were considered as non-toxic group in the later comparative analyses. The TTX concentration varies significantly not only among different individuals within the toxic group (80.8 ± 80.5 μg/kg, mean ± SD, n = 15), but also between toxic and non-toxic groups under different circumstances. This reflects the TTX content in *T. rubripes* as a complex trait influenced by multiple factors.

2.2. OTU Characterization

From 16S rRNA sequencing, a total of 3,274,513 raw paired-end reads were generated (Table S2). After quality control and removal of chimeric sequences, 1,532,376 effective tags were obtained for further analysis, with an average of 51,079 tags per sample. The average sequence length was 409 bp, and the average effective tags ratio from raw reads was 46.7%. A total of 2241 OTUs were detected using all samples. The captive group detected a higher number of OTUs (1821) in total compared with the wild group (1518). There were 1098 OTUs shared between captive and wild groups, while 723 and 420 unique OTUs were detected in the captive and wild groups, respectively (Figure 1A).

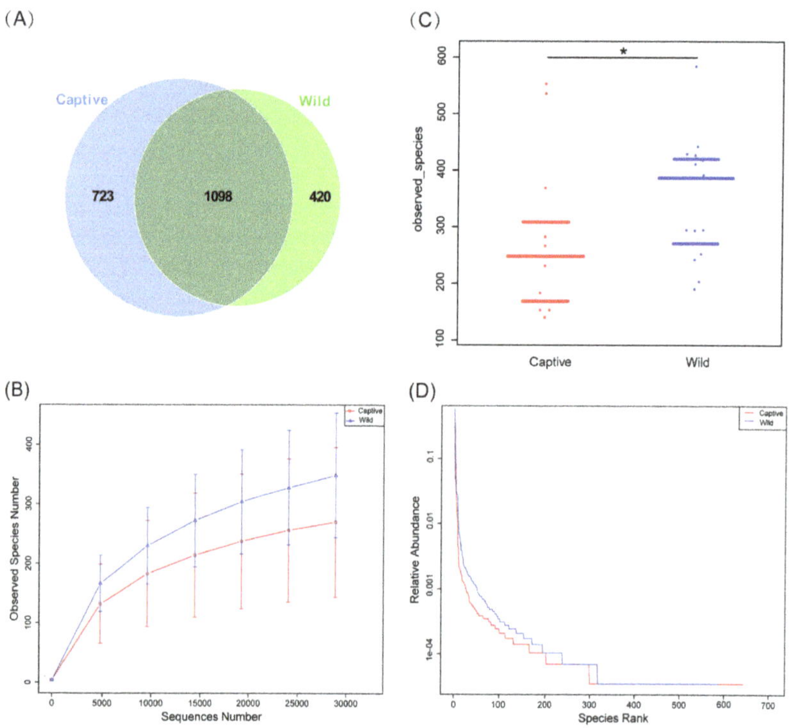

Figure 1. Observed OTUs and species in captive and wild *T. rubripes*. (**A**) Venn diagram of intestinal microbial OTUs in captive and wild groups. Captive represents non-toxic group, while wild represents toxic group. (**B**) Rarefaction curve reflected the observed species number in captive and wild groups.

(C) The beeswarm plot of observed species differences in captive and wild groups ($p = 0.038$, Wilcoxon test, $n = 15$). The asterisk indicates a statistically significant difference between wild and captive groups. **(D)** Rank abundance curve reflected the relative abundance of captive and wild samples.

2.3. Comparative Assessment of the Microbiota Diversity and Composition Between Captive and Wild *T. rubripes*

The community abundance and richness of all samples were analyzed by calculating the alpha diversity index. The rarefaction curve showed that the observed species number in each sample of the wild group was consistently higher than the captive group (Figure 1B). The observed species per sample in the wild group was 348 ± 105 (mean ± SD, $n = 15$), and the number in the captive group was 269 ± 126 (mean ± SD, $n = 15$) (Tables S3 and S4). The beeswarm plot reflected that the observed species of the wild group was significantly higher than the captive group ($p = 0.038$, Wilcoxon test, $n = 15$) (Figure 1C). Similarly, the rank abundance curve showed that the community richness in the wild group was also consistently higher than in the captive group (Figure 1D).

An average good's coverage of 99.68% indicated most of the intestinal microbiota were identified in this study. After alpha diversity analysis, there was no significant elevation in the Simpson and Shannon indices ($p > 0.05$, Wilcoxon test; Figure 2A,B), which indicated that there were no significant differences in the intestinal microbial community diversity and evenness between captive and wild groups. Whereas, a significant increase in the Chao1 and ACE indices of the community and OTU numbers per sample in the wild group were observed by comparing with the captive group ($p < 0.05$, Wilcoxon test; Figure 2C,D). These results indicated the total number of microbial community and OTUs in the wild group were significantly higher than in the captive group.

2.4. Taxonomic Classification of Microbiota

Among all intestinal samples, unidentified bacteria (mainly Campylobacteria at the class level) was the main phylum among the top 10, followed by Proteobacteria, Bacteroidota, Firmicutes, and Spirochaetota. The similar relative abundances were observed between captive and wild groups in the phyla unidentified bacteria (45.02% vs. 57.23%), Bacteroidota (7.60% vs. 5.11%), Firmicutes (4.31% vs. 4.87%), and Spirochaetota (1.39% vs. 2.52%). However, the abundance of Proteobacteria detected in the wild group (20.76%) was about five times that in the captive group (4.34%) (Figure 3A). At the genus level, a total of 563 taxa were identified. In the captive group, *Rikenella* (5.87%), *Vibrio* (3.23%), and *Brevinema* (1.38%) were the top three microbes, which account for more than 1% of total abundance. In contrast, *Vibrio* (12.07%), *Photobacterium* (5.82%), *Brevinema* (2.52%), and *Bacteroides* (1.96%) were the dominant genera in the wild group (Figure 3B). To further study the phylogenetic relationship of microbes at the genus level, the representative sequences of the top 100 genera in both groups were obtained through multi-sequence alignment, and the evolutionary tree was shown in Figure 3C. Among 13 classified phyla, Proteobacteria, Bacteroidota, and Spirochaetota had the highest abundance and the largest difference between groups. Eighteen genera were clustered to Proteobacteria, among which *Vibrio* had the highest abundance, which in the wild group was significantly higher than in the captive group. In the wild group, the second highest abundance genus in Proteobacteria was *Photobacterium*, but its abundance in the captive group was significantly lower. In addition, eighteen genera belonged to the phylum Bacteroidota. In Bacteroidota, *Rikenella* was the most abundant genus in the captive group, and its abundance was significantly higher than that in the wild group. However, *Bacteroides* and *Galbibacter*, as two other dominant taxa, were more abundant in the wild group than in the captive group. Moreover, only one genus, *Brevinema*, was classified into the phylum Spirochaetota; its abundance was relatively high in both groups, although it was somewhat higher in the wild group.

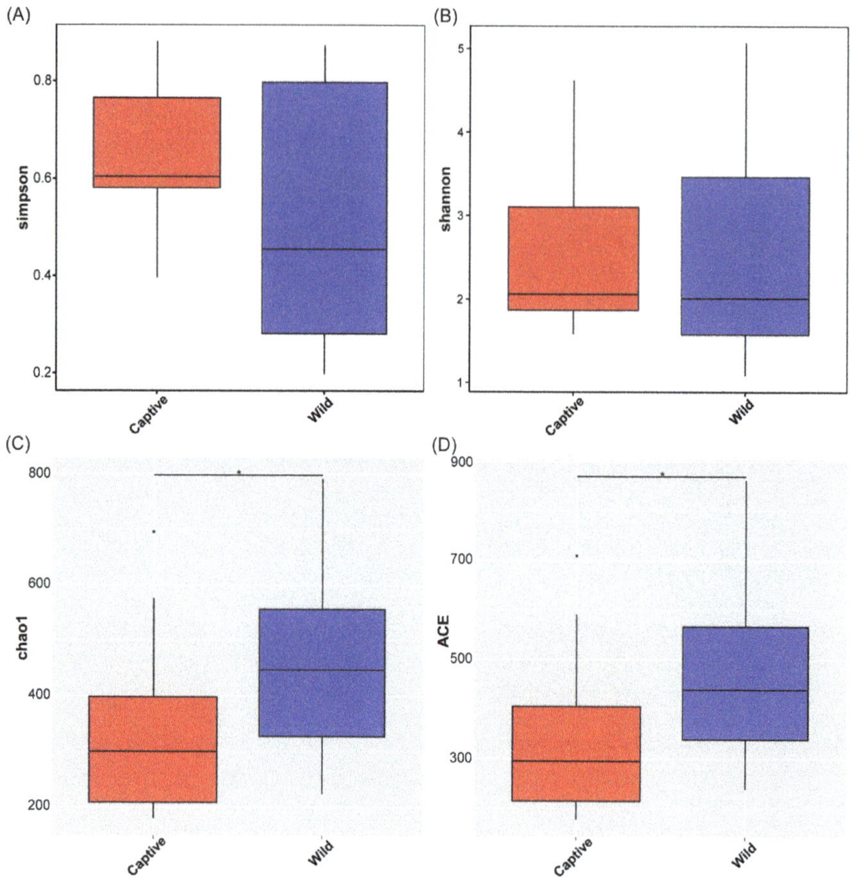

Figure 2. The alpha diversity analysis of intestinal microbiota between wild and captive groups. (**A**) Simpson's index of diversity ($p = 0.124$, Wilcoxon test). (**B**) Shannon index ($p = 0.683$, Wilcoxon test). (**C**) Chao1 index ($p = 0.029$, Wilcoxon test). The asterisk indicates a statistically significant difference between wild and captive groups. (**D**) ACE index ($p = 0.033$, Wilcoxon test). The asterisk indicates a statistically significant difference between wild and captive groups.

2.5. Beta Diversity Analysis

According to beta diversity analysis, the composition of intestinal microbiota between two groups was compared and analyzed systematically. The samples of captive and wild groups showed a partly overlapped cluster in the PCoA results (Figure 4A), indicating that there were some differences in microbial community structure between the two groups. Furthermore, Anosim analysis identified that there was an extremely significant difference in the microbial community composition between two groups ($r = 0.262$, $p = 0.001$). In Figure 4B, the notches of the captive group and the wild group did not overlap, indicating significant differences between these two groups.

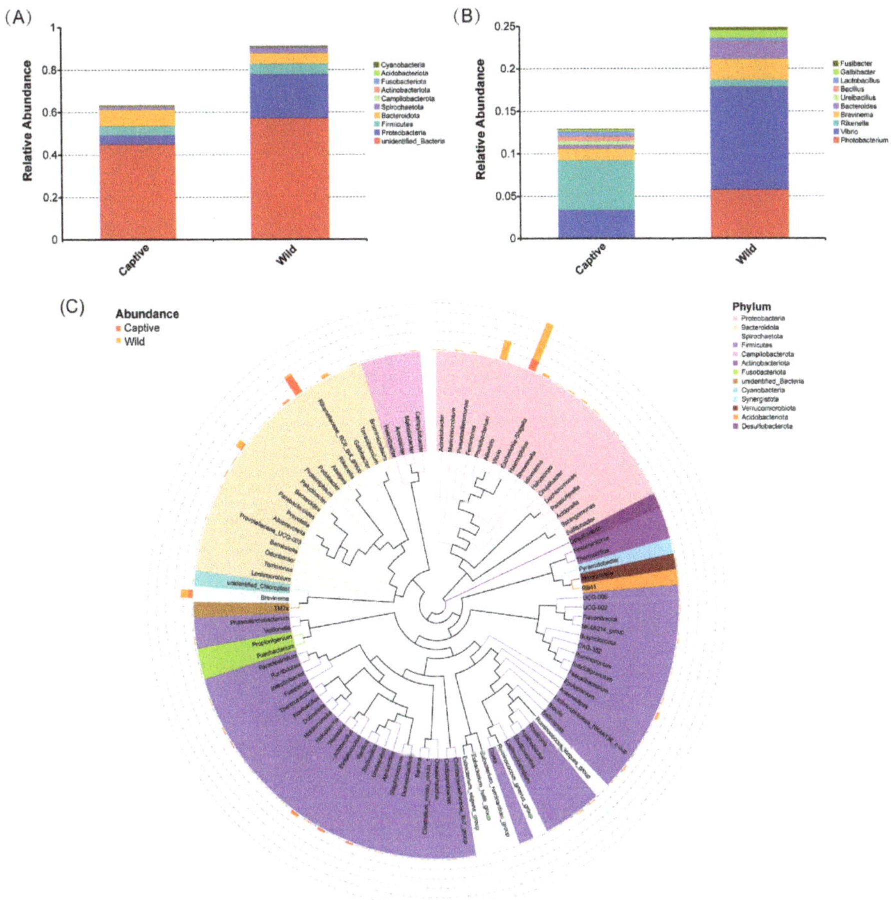

Figure 3. The proportions of intestinal microbiota in the captive and wild group of *T. rubripes*. (**A**,**B**) Relative abundance of top 10 intestinal microbiota at the phylum (**A**) and genus (**B**) level in the captive and wild groups. (**C**) Phylogenetic tree at the genus level in captive and wild group. The colors of the branches and sectors indicated their corresponding phyla, and the stacked bar diagram outside the sectors indicated the abundance distribution of the genus in different groups.

Hypothesis testing was performed on the relative abundance results between groups using Metastats software (http://metastats.cbcb.umd.edu/). Microbiota with significant differences were screened according to p values ($p < 0.05$, significant; $p < 0.01$, extremely significant). At the phylum level, Fusobacteriota and Proteobacteria showed extremely different abundance ($p < 0.01$) in captive and wild groups (Figure 5A). At the genus level, ten genera showed extremely different abundance ($p < 0.01$), and two genera showed significantly different abundance ($p < 0.05$) between two groups (Figure 5B). Among these results, *Rikenella*, *lachnospiraceae*_NK4A136_group, *Thermobacillus*, and *Ureibacillus* were extremely significantly more abundant in the captive group, while *Bacteroides*, *Escherichia-Shigella*, *Marinimicrobium*, *Idiomarina*, *Photobacterium*, and *Galbibacter* showed extremely significantly higher abundance in the wild group. T-test results at the phylum level were consistent with Metastats. Compared with the captive group, the relative abundances of Proteobacteria and Fusobacteriota were significantly higher in the wild group ($p < 0.05$; Figure 6A). At the genus level, the top 7 genera (same genera from the Metastats analysis plus *Brumimicrobium*) also exhibited significant abundant differences between the two groups in the *t*-test analysis ($p < 0.05$; Figure 6B).

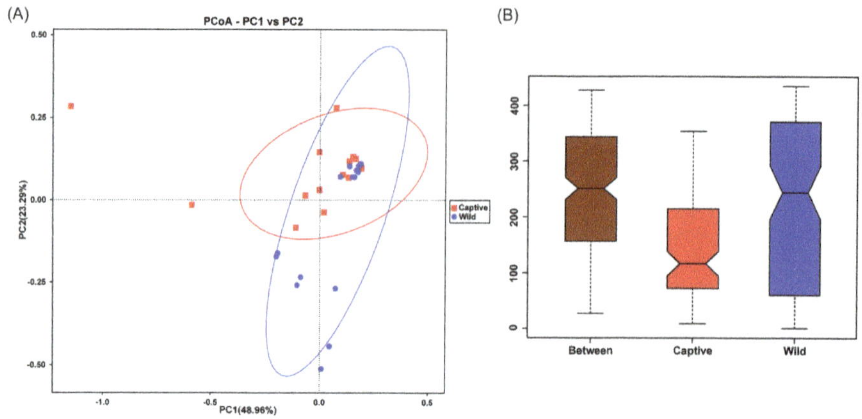

Figure 4. Beta diversity analysis results between captive and wild groups. (**A**) PCoA analysis based on weighted Unifrac distance with cluster. (**B**) The boxplot of Anosim analysis ($r = 0.262$, $p = 0.001$).

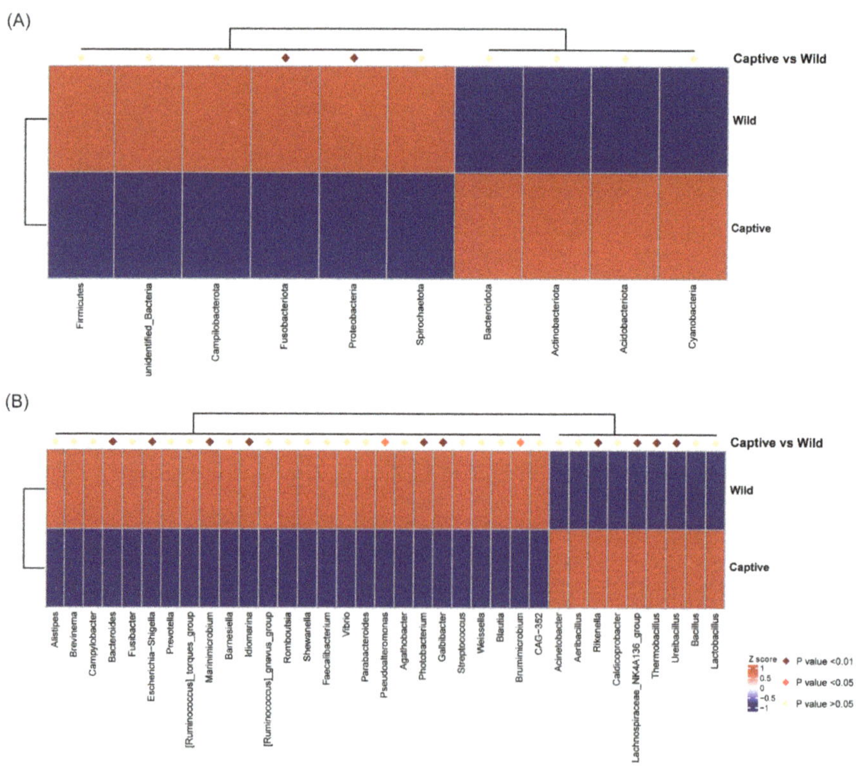

Figure 5. MetaStat analysis results reflected microbes with significant differences between two groups using clustering heatmap at phylum (**A**) and genus level (**B**).

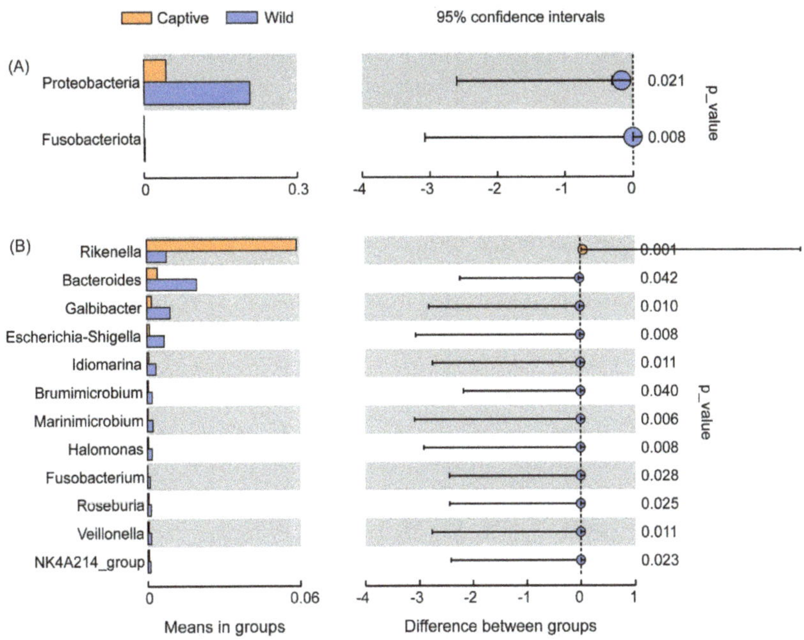

Figure 6. Analysis of intestinal microbiota with significant differences between groups. Barplots with 95% confidence intervals at phylum level (**A**) and genus level (**B**) by *t*-test with $p < 0.05$. The figure on the left represented the microbiota with significantly different abundance between groups, and the figure on the right represented the confidence of differences between groups.

LEfSe is an analytical tool for the discovery and interpretation of high-dimensional statistically different biomarkers between groups. From the histogram of LDA value distribution and phylogenetic analysis, we found that Proteobacteria at the phylum level, Gammaproteobacteria at the class level, *Photobacterium* at the genus level, and *Photobacterium_damselae* at the species level were identified as the biomarker in the wild group ($n = 15$, LDA score > 4; Figure 7A,B). Meanwhile, Bacteroidales at the order level, Rikenellaceae at the family level, and *Rikenella* at genus level were screened with an LDA score > 4 in the captive group ($n = 15$).

2.6. Function Prediction

To predict the functions and their differences in intestinal microbiota in the captive and wild *T. rubripes*, function prediction was conducted by Tax4Fun. The results showed that similar KEGG pathways were observed at level 1, 2, and 3 (Figures S1 and 8A,B), such as "Membrane transport", "Transporters", and "ABC transporters". However, the functional abundances of specific pathways were different between captive and wild populations. For instance, "Biosynthesis of other secondary metabolites", "Drug resistance", "Xenobiotics biodegradation and metabolism", "Membrane transport", and "Enzyme families" showed significant relative abundance at level 2 between captive and wild groups, while the relative abundance of "Secretion system", "Transporters", "Exosome", and "ABC transporters" was significantly different between captive and wild groups (Figure 8C,D).

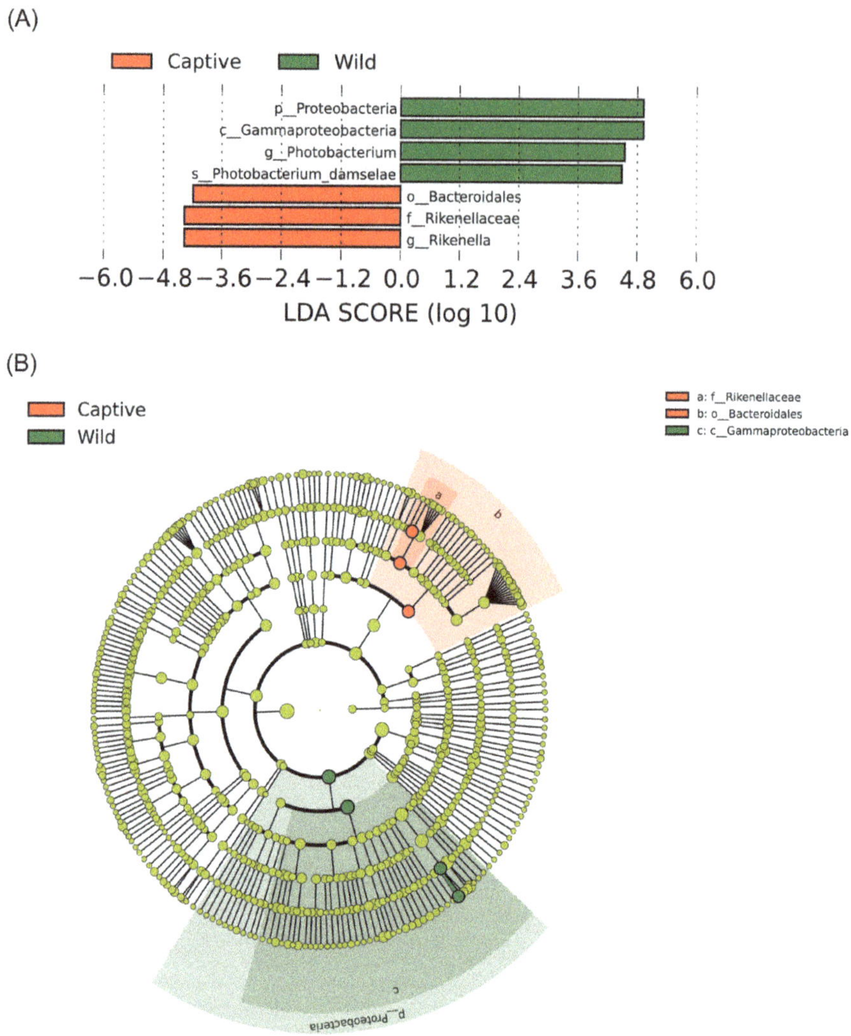

Figure 7. The LEfSe analysis results. (**A**) Histogram of LDA value distribution reflected the Biomarkers with statistical differences in the taxonomic classification (n = 15, LDA score > 4). (**B**) The evolutionary cladistic diagram. Circles radiating from the inside out represented taxonomic levels from phylum to genus (or species). Each small circle at a different classification level represented a classification at that level, and the size of the small circle diameter was proportional to the relative abundance size. Color principle: The species with no significant difference were uniformly colored yellow, and the Biomarker of different species was colored with the corresponding group, red: captive group, green: wild group.

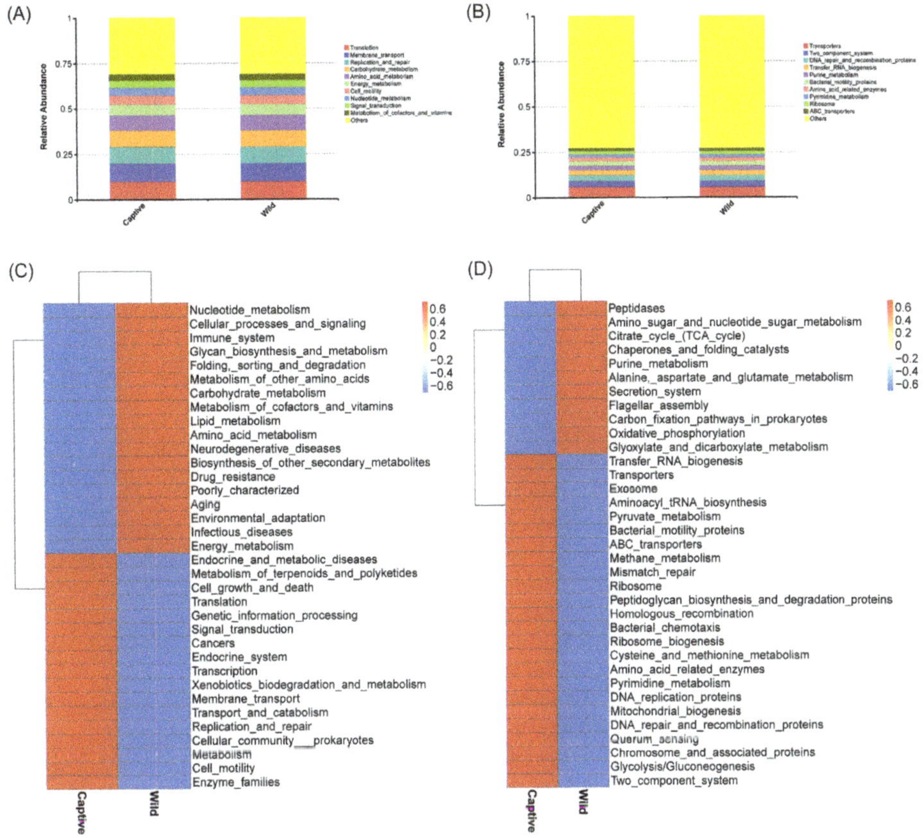

Figure 8. Predicted functions of intestinal microbiota in captive and wild *T. rubripes*. The barplots show the top 10 functional relative abundance at KEGG level 2 (**A**) and level 3 (**B**), respectively. The clustering heatmaps illustrate the KEGG level 2 (**C**) and level 3 (**D**) functional annotations using Tax4Fun.

3. Discussion

TTX is a well-known neurotoxin existing in pufferfish, as well as many other marine organisms [20,36–38]. It is distributed in multiple tissues of pufferfish, such as liver, ovary, blood, muscle, etc. [39,40]. Previous studies have shown that the accumulation of TTX in pufferfish was closely related to symbiotic bacteria [26,27]. However, the role of symbiotic bacteria in the production and accumulation of TTX is not yet clear. Therefore, this study compared the gut microbiota of wild toxic and captive non-toxic *T. rubripes* to clarify whether there are characteristic differences in the gut microbiota between toxic and non-toxic pufferfish. In addition, through functional prediction and comprehensive analysis with our previous research results, the indicative bacteria and their possible functions in the production and metabolism of TTX have been explored.

There are abundant and diverse microbial communities in the gut of fish, which is their natural defense system [41,42]. There are complex immune interactions between fish intestinal symbiotic microorganisms and their hosts [43]. The host immune system could recognize and reshape the gut microbiota, while gut microbiota could enhance host tolerance or immune response [44]. Symbiotic microorganisms co-evolve with the host immune system and play a crucial role in the physiological regulation of the host [45].

The diversity and richness of intestinal microbiota are influenced by the genetics, habitat, dietary habits, and health status of the host [46–48]. In this study, the intestinal microbiota of toxic and non-toxic *T. rubripes* were compared and analyzed systematically. The results indicated that both the observed bacterial species number and the community richness were higher in the toxic group than in the non-toxic group (Figures 1 and 2), reflecting a more complex intestinal microbiota environment in the toxic *T. rubripes*. The composition of phyla identified in the intestinal contents of toxic and non-toxic groups was almost identical, including Campylobacteria, Proteobacteria, Bacteroidota, Firmicutes, and Spirochaetota (Figure 3A). These results were consistent with recent research on the gut microbiota of juvenile *T. rubripes* [49] and among different pufferfish species [50]. Moreover, similar phyla including Proteobacteria, Firmicutes, Bacteroidota, and Spirochaetota were also found as the dominant bacterial phyla in the TTX-bearing ribbon worms; this may reflect an adaptive strategy of convergent evolution among different species under specific ecological pressures [51]. The common dominant genera identified in both toxic and non-toxic groups were *Vibrio* and *Brevinema* (Figure 3B), which was supported by earlier research related to the gut microbiome's stability and functionality in the tiger puffer [49,52]. *Brevinema* is a type of spirochete bacteria belonging to the family Brevinemataceae, primarily found in aquatic environments [53]. Based on its spiral shape and motility, it is capable of efficient movement and colonization in complex environments, which may account for its widespread presence in the gut microbiota of pufferfish [54].

Although there were many shared dominant bacteria in the intestinal microbiota between toxic and non-toxic *T. rubripes*, there were also numerous differences between them. For instance, at the phylum level, the abundance of Proteobacteria in the toxic group was approximately five times that of the non-toxic group. Previous research has demonstrated that Proteobacteria was the phylum that contained the most TTX-producing microbes among all [21]. In addition, other reported TTX-producing bacteria primarily belong to the phyla Actinobacteria, Bacteroidetes, and Firmicutes [21]. However, in our findings, differential analysis of the intestinal microbiota between toxic and non-toxic *T. rubripes* identified Proteobacteria and Fusobacteriota as the phyla with significant differences (Figures 5A and 6A). In contrast, Actinobacteria, Bacteroidetes, and Firmicutes showed no significant differences between the two groups, indicating that they may not be the primary phyla involved in TTX accumulation. However, considering the complexity of intestinal microbiota dynamics and functions, it is necessary to further study the specific roles of these microbes. At the genus level, the specific dominant genus in the non-toxic group was *Rikenella*, whereas in the toxic group, the specific dominant genera were *Photobacterium* and *Bacteroides*. Genus *Rikenella* is an important component of the gut microbiota, which plays a significant role in gut health, metabolism, and immune regulation, as well as maintaining intestinal microecological balance through microbial interactions [55]. It is worth noting that, in a previous study, genus *Photobacterium* was found as the predominant genera in the intestinal bacterial community of cultured tiger puffer by contrast [52]. Although we found that the species *Photobacterium damselae* was related to TTX content differences in *T. rubripes*, further investigations are required to verify more bacterial strains and their functions. Under the phylum Proteobacteria, *Vibrio* and *Photobacterium* were the two most abundant bacterial genera in the toxic group, and their abundances were significantly higher than those in the non-toxic group. *Vibrio*, the first isolated TTX-producing bacteria [56], has been found to be associated with TTX production in multiple organisms, such as pufferfish, starfish, xanthid crab, and nemertean worm [30,57–59]. In this study, the abundance of *Vibrio* in the wild group was nearly four times higher than that in the captive group, and it was also the most abundant genus in the wild group. However, the difference in abundance did not reach statistical significance in the differential abundance analysis. This could be due to the fact that relative abundance analysis may not fully reflect the true differences, or the differences at the genus level might be diluted by heterogeneity at the species or strain level. In future studies, metagenomic analysis and qPCR could be employed for in-depth investigation and validation of absolute abundance. In the phylum Bacteroidota, genus *Rikenella* showed a significant abundance advantage in the non-toxic group, while *Bacteroides* and *Galbibacter*, also belonging to the Bacteroidota phylum, were significantly more abundant in the toxic group. Genus *Bacteroides*, as well as *Escherichia*, were reported as the allochthonous bacteria in the *T. rubripes* [52]. However, this study is the first to find them to be related to the TTX content difference in *T. rubripes*. Furthermore, this study also discovered for the first time that genera *Marinimicrobium*, *Idomarina*, *Galbibacter*, and *Brumimicrobium* may be responsible for TTX biosynthesis in *T. rubripes*. Additionally, the phylum Fusobacteriota was also identified as a significant differential phylum between the two groups by both Metastats and T-test analyses, with its abundance being markedly higher in the toxic group. Subsequent LEfSe analysis further indicated that the phylum Proteobacteria and the genus *Photobacterium*, as well as the species *Photobacterium damselae*, could serve as the intestinal microbiota biomarker candidates for the identification of toxic *T. rubripes*. Meanwhile, the genus *Rikenella* could be considered as a biomarker candidate to identify non-toxic *T. rubripes*. Moreover, the genus *Pseudoalteromonas*, identified with significant abundant differences between groups and renowned for its ability to produce bioactive compounds and play an important role in marine ecosystems [60], may contribute to TTX bioaccumulation as well. The genus *Pseudoalteromonas* was identified as TTX-producing bacteria in *Nassarius semiplicatus* and *Hubrechtella juliae* and was also found in *Takifugu obscurus* as TTX biosynthesis-related bacteria [30,50,61].

TTX is a bioactive secondary metabolite that can be produced by marine microorganisms [31]. Therefore, based on functional enrichment analysis of the differential gut microbiota identified in this study, we screened level 2 pathways such as "biosynthesis of other secondary metabolites", "enzyme families", and "membrane transport" and level 3 pathways such as "secretion system", "transporters", and "ABC transporters", which were closely related to the production and transport of TTX in *T. rubripes*. It should be noted that the pathway "ABC transporters" was also enriched in our previous research, which included three upregulated genes (*abcb6*, *abcb7*, and *abcc5*) in the blood of toxic *T. rubripes* [23]. Likewise, "transmembrane transporter activity" was identified as the biggest pathway nodes, including more than 20 functional related genes in previous research [23]. ABC transporters are a classic membrane protein family responsible for the transport of a wide range of substances across lipid membranes [62]. Studies have reported that in aquatic animals, their function was associated with resistance to xenobiotic toxic substances [63,64]. Combined with our research findings, ABC transporters may play significant roles in the production and transport of TTX in both symbiotic microorganisms and hosts. A possible hypothesis is that TTX produced by the gut microbiota enters the host's body and reaches different tissues through the circulatory system, thereby regulating the expression of host genes [65]. In turn, the expression of host genes may alter the intestinal environment, thus affecting the growth of specific microbial communities. Additionally, dietary factors may also influence the expression of host genes and the composition of gut microbiota. Therefore, the exact correlation and co-regulatory mechanisms between host and gut microbiota in TTX production and accumulation remain to be investigated in future studies.

4. Materials and Methods

4.1. Sample Collection and TTX Detection

All animal procedures involved in this study were approved by the Animal Care and Use Committee of the Centre for Applied Aquatic Genomics at the Chinese Academy of Fishery Sciences before sampling. All procedures and animal handling were conducted in accordance with approved guidelines, with every effort made to minimize potential suffering.

Wild (toxic; body length, 16.8 ± 1.2 cm; body weight, 111.2 ± 29.0 g) and captive (non-toxic; body length, 17.1 ± 1.3 cm; body weight, 130.5 ± 22.0 g) *T. rubripes* were collected from the Bohai Bay (Tangshan, Hebei, China), then temporarily reared in the marine fish breeding laboratory of Beidaihe Central Experimental Station, Chinese Academy of Fishery Sciences. Captive *T. rubripes* were fed a combination of TTX-free natural food sources (including rotifers, brine shrimp, and fresh fish), along with a TTX-free commercial feed provided by Qingdao Saigelin Aquatic Products Technology Co., Ltd. (Qingdao, China) from the time of hatching. Wild and captive fishes were acclimated separately in the same-sized seawater tanks at 20 °C without feeding for three days. Then, fifteen individuals from each tank (approximately 30–40 fish per tank) were randomly collected for sampling. For TTX detection and DNA extraction, the intestinal contents and muscle tissues were collected from each individual. The samples used for TTX detection were taken from the dorsal muscle, with each sample weighing no less than 15 g. The samples were flash-frozen in liquid nitrogen, then stored at −80 °C until further use. The TTX detection was conducted by Meizheng Bio-Tech Co., Ltd. (Beijing, China), as our previous study described [23]. Briefly, the TTX concentration of muscle tissues was detected by the liquid chromatography-tandem mass spectrometry (LC-MS/MS) method, according to the China National Food Safety Standards GB 5009.206-2016: Determination of tetrodotoxin in aquatic products [66].

4.2. DNA Extraction, Amplification and Sequencing

Microbial DNA from the intestinal content of toxic and non-toxic *T. rubripes* was extracted using the QIAamp PowerFecal Pro DNA Kit (Qiagen, Hilden, Germany) in accordance with the manufacturer's protocols. The primers used for PCR amplification of the 16S rRNA V3–V4 region were 341F: CCTACGGGNGGCWGCAG and 806R: GGACTACHVGGGTATCTAAT [67]. The PCR amplification conditions were as follows: 95 °C for 3 min, followed by 30 cycles of 95 °C for 30 s, 50 °C for 30 s, 72 °C for 1 min, and finally 72 °C for 7 min. Each 20 µL reaction contained 10 µL premixed Taq (Takara, San Jose, CA, USA), 1 µL of forward and reverse primers each (10 µM), 2 µL gDNA, and 6 µL ddH_2O. The amplicons were purified by AMPure XP Beads (Axygen, Union City, CA, USA) and quantified by Qubit 2.0 Fluorometer (Invitrogen, Carlsbad, CA, USA). The purified amplification products were sent to the Novogene company (Tianjin, China) for 16s rRNA sequencing using the Illumina NovaSeq platform PE250 strategy.

4.3. Bioinformatics and Statistical Analysis

The raw data of each sample were separated according to barcode sequence and PCR amplification primer sequences. Paired-end clean reads were merged to raw tags using FLASH (version 1.2.7) [68]. The filtering, dereplication, denoising, and chimera sequence removing were conducted by using QIIME (version 1.9.1) [69] and Vsearch (version 1.9.6) to generate effective tags [70]. Effective tags were clustered into operational taxonomic units (OTUs) with 97% identity by Uparse software (version 7.0.1001) [71]. The sequence with the highest frequency in a cluster was selected as the representative sequence to perform species annotation. The annotation analysis was conducted using the Mothur method and SILVA138 database [72]. Taxonomic analysis was performed using MUSCLE (version 3.8.31) with a naive Bayesian classifier [73].

Alpha diversity indices (including Chao1, ACE, Shannon, and Simpson) were calculated using QIIME (version 1.9.1) [69]. QIIME was also used to calculate weighted Unifrac distance and construct a UPGMA cluster tree. The rarefaction curve and rank abundance curve were drawn using R software (version 2.15.3). PCoA (principal co-ordinate analysis) analysis based on weighted Unifrac distance was processed using R software WGCNA, stats, and ggplot2 packages. The differences analysis between toxic and non-toxic groups of alpha diversity and beta diversity indices was assessed using the Wilcoxon test in R software (version 2.15.3). The Anosim analysis, reflecting the significance of the difference between groups, was conducted using the R vegan package. To further excavate the microorganisms related to the TTX content difference between groups, T-test, MetaStat analysis, and LEfSe (LDA effect size) analysis were performed using R (version 2.15.3), Metastats (http://metastats.cbcb.umd.edu/), and LEfSe software (version 1.0) [74,75].

4.4. Functional Analysis

Tax4Fun [76] was used to predict the function of intestinal microbiota in different groups by the nearest neighbor method based on a minimum 16S rRNA sequence similarity. The specific method consisted of extracting the whole genome 16S rRNA gene sequence of prokaryotes from the KEGG database and comparing it to the SILVA SSU Ref NR database (BLAST bitscore > 1500) using the BLASTN algorithm to establish a correlation matrix. The whole genome functional information of prokaryotes in the KEGG database annotated by UProC and PAUDA corresponds to the SILVA database to achieve functional annotation of the SILVA database. The functional relative abundance distribution and clustering analysis were proceeded using R software.

5. Conclusions

In this study, we characterized the symbiotic intestinal microbiota of toxic (wild) and non-toxic (captive) *T. rubripes*. Potential intestinal microbial communities involved in the production and accumulation of TTX were identified by comparing two groups of *T. rubripes* with different TTX contents. We also preliminarily explored their potential functions in TTX accumulation and metabolism within *T. rubripes*. This study revealed that bacterial genera potentially associated with TTX production and accumulation mainly belong to the phyla Proteobacteria and Bacteroidota. The genera *Vibrio* and *Photobacterium* from Proteobacteria and the genus *Bacteroides* from Bacteroidota were significantly more abundant in the wild population than in the captive group. Combined with existing findings, we hypothesize that these genera play important biological roles in TTX accumulation within pufferfish. Meanwhile, although the genera *Marinimicrobium* and *Idomarina* from Proteobacteria and the genera *Galbibacter* and *Brumimicrobium* from Bacteroidota had relatively low abundance, they exhibited significant differences between the two groups, suggesting that these genera may also be involved in TTX accumulation. Additionally, enrichment analysis of differentially abundant microbial communities identified pathways closely related to secondary metabolite biosynthesis, secretion, and transport, implying that the potential functions of these differentially abundant communities are associated with TTX biosynthesis, secretion, and transport. However, the specific species/strains and their functions still need to be confirmed through metagenomic, metabolomic, and molecular cellular functional experiments.

Supplementary Materials: The following supporting information can be downloaded at: https://www.mdpi.com/article/10.3390/md23040140/s1, Figure S1: The barplot of KEGG 1-level top10 functional relative abundance. Table S1: Muscular TTX concentration of wild and captive *Takifugu rubripes*. Table, S2: Overall results of 16S rRNA sequencing Table S3: Summary of the Alpha diversity index Table S4: Summary of the Alpha diversity index in group.

Author Contributions: Conceptualization, H.Z.; data curation, B.W.; formal analysis, H.Z.; funding acquisition, H.Z.; investigation, H.Z. and J.H.; methodology, J.R., Z.S. and X.L.; resources, J.H.; software, Y.Z.; writing—original draft, H.Z.; writing—review and editing, H.Z. All authors have read and agreed to the published version of the manuscript.

Funding: This research was funded by the Beijing Natural Science Foundation (6202034) and the Central Public-interest Scientific Institution Basal Research Fund CAFS (2023TD25).

Institutional Review Board Statement: This study was conducted in accordance with the Animal Care and Use Committee of the Chinese Academy of Fishery Sciences (ACUC−CAFS) (ACUC-CAFS-20201012; 12 October 2020) under the established recommendations for the care and use of animals in scientific research.

Data Availability Statement: The 16S rRNA sequencing data reported in this research are available at the Sequence Read Archive (SRA) under accession number PRJNA1223122.

Conflicts of Interest: The authors declare no conflicts of interest.

References

1. Lago, J.; Rodríguez, L.P.; Blanco, L.; Vieites, J.M.; Cabado, A.G. Tetrodotoxin, an Extremely Potent Marine Neurotoxin: Distribution, Toxicity, Origin and Therapeutical Uses. *Mar. Drugs* **2015**, *13*, 6384–6406. [CrossRef] [PubMed]
2. Noguch, T.; Arakawa, O. Tetrodotoxin—Distribution and Accumulation in Aquatic Organisms, and Cases of Human Intoxication. *Mar. Drugs* **2008**, *6*, 220–242. [CrossRef]
3. Katikou, P.; Gokbulut, C.; Kosker, A.R.; Campàs, M.; Ozogul, F. An Updated Review of Tetrodotoxin and Its Peculiarities. *Mar. Drugs* **2022**, *20*, 47. [CrossRef]
4. Zhang, X.; Qiao, K.; Cui, R.; Xu, M.; Cai, S.; Huang, Q.; Liu, Z. Tetrodotoxin: The State-of-the-Art Progress in Characterization, Detection, Biosynthesis, and Transport Enrichment. *Mar. Drugs* **2024**, *22*, 531. [CrossRef]
5. Matsumura, K. A monoclonal antibody against tetrodotoxin that reacts to the active group for the toxicity. *Eur. J. Pharmacol.* **1995**, *293*, 41–45.
6. Shen, H.; Li, Z.; Jiang, Y.; Pan, X.; Wu, J.; Cristofori-Armstrong, B.; Smith, J.J.; Chin, Y.K.Y.; Lei, J.; Zhou, Q.; et al. Structural basis for the modulation of voltage-gated sodium channels by animal toxins. *Science* **2018**, *362*, eaau2596.
7. Mattei, C. Tetrodotoxin, a Candidate Drug for Nav1.1-Induced Mechanical Pain? *Mar. Drugs* **2018**, *16*, 72. [CrossRef]
8. Nieto, F.R.; Cobos, E.J.; Tejada, M.Á.; Sánchez-Fernández, C.; González-Cano, R.; Cendán, C.M. Tetrodotoxin (TTX) as a Therapeutic Agent for Pain. *Mar. Drugs* **2012**, *10*, 281–305. [CrossRef]
9. Berde, C.B.; Athiraman, U.; Yahalom, B.; Zurakowski, D.; Corfas, G.; Bognet, C. Tetrodotoxin-Bupivacaine-Epinephrine Combinations for Prolonged Local Anesthesia. *Mar. Drugs* **2011**, *9*, 2717–2728. [CrossRef]
10. Huerta, M.Á.; de la Nava, J.; Artacho-Cordón, A.; Nieto, F.R. Efficacy and Security of Tetrodotoxin in the Treatment of Cancer-Related Pain: Systematic Review and Meta-Analysis. *Mar. Drugs* **2023**, *21*, 316. [CrossRef]
11. Park, J.; Proux, C.; Ehanno, W.; Réthoré, L.; Vessières, E.; Bourreau, J.; Favre, J.; Kauffenstein, G.; Mattei, C.; Tricoire-Leignel, H.; et al. Tetrodotoxin Decreases the Contractility of Mesenteric Arteries, Revealing the Contribution of Voltage-Gated Na^+ Channels in Vascular Tone Regulation. *Mar. Drugs* **2023**, *21*, 196. [CrossRef] [PubMed]
12. Hong, B.; He, J.; Le, Q.; Bai, K.; Chen, Y.; Huang, W. Combination Formulation of Tetrodotoxin and Lidocaine as a Potential Therapy for Severe Arrhythmias. *Mar. Drugs* **2019**, *17*, 685. [CrossRef] [PubMed]
13. Yotsu-Yamashita, M.; Abe, Y.; Kudo, Y.; Ritson-Williams, R.; Paul, V.J.; Konoki, K.; Cho, Y.; Adachi, M.; Imazu, T.; Nishikawa, T.; et al. First Identification of 5,11-Dideoxytetrodotoxin in Marine Animals, and Characterization of Major Fragment Ions of Tetrodotoxin and Its Analogs by High Resolution ESI-MS/MS. *Mar. Drugs* **2013**, *11*, 2799–2813. [CrossRef] [PubMed]
14. Bane, V.; Lehane, M.; Dikshit, M.; O'Riordan, A.; Furey, A. Tetrodotoxin: Chemistry, Toxicity, Source, Distribution and Detection. *Toxins* **2014**, *6*, 693–755. [CrossRef]
15. Rossignoli, A.E.; Mariño, C.; Martín, H.; Blanco, J. First Report of Two Gymnodimines and Two Tetrodotoxin Analogues in Invertebrates from the North Atlantic Coast of Spain. *Mar. Drugs* **2023**, *21*, 232. [CrossRef]
16. Kudo, Y.; Yasumoto, T.; Konoki, K.; Cho, Y.; Yotsu-Yamashita, M. Isolation and Structural Determination of the First 8-*epi*-type Tetrodotoxin Analogs from the Newt, *Cynops ensicauda popei*, and Comparison of Tetrodotoxin Analogs Profiles of This Newt and the Puffer Fish, *Fugu poecilonotus*. *Mar. Drugs* **2012**, *10*, 655–667. [CrossRef]
17. Yu, V.C.-H.; Yu, P.H.-F.; Ho, K. C.; Lee, F.W.-F. Isolation and Identification of a New Tetrodotoxin-Producing Bacterial Species, *Raoultella terrigena*, from Hong Kong Marine Puffer Fish *Takifugu niphobles*. *Mar. Drugs* **2011**, *9*, 2384–2396. [CrossRef]

18. Arakawa, O.; Takatani, T.; Taniyama, S.; Tatsuno, R. Toxins of pufferfish—Distribution, accumulation mechanism, and physiologic functions. *Aqua-BioScience Monogr.* **2017**, *10*, 41–80. [CrossRef]
19. Mosher, H.S.; Fuhrman, F.A.; Buchwald, H.D.; Fischer, H.G. Tarichatoxin-tetrodotoxin, a potent neurotoxin. *Science* **1964**, *144*, 1100–1110. [CrossRef]
20. Ueda, H.; Itoi, S.; Sugita, H. TTX-Bearing Planocerid Flatworm (Platyhelminthes: Acotylea) in the Ryukyu Islands, Japan. *Mar. Drugs* **2018**, *16*, 37. [CrossRef]
21. Jal, S.; Khora, S.S. An overview on the origin and production of tetrodotoxin, a potent neurotoxin. *J. Appl. Microbiol.* **2015**, *119*, 907–916. [PubMed]
22. Noguchi, T.; Arakawa, O.; Takatani, T. TTX accumulation in pufferfish. *Comp. Biochem. Physiol. Part D Genom. Proteom.* **2006**, *1*, 145–152. [CrossRef] [PubMed]
23. Zhang, H.; Li, P.; Wu, B.; Hou, J.; Ren, J.; Zhu, Y.; Xu, J.; Si, F.; Sun, Z.; Liu, X. Transcriptomic analysis reveals the genes involved in tetrodotoxin (TTX) accumulation, translocation, and detoxification in the pufferfish *Takifugu rubripes*. *Chemosphere* **2022**, *303*, 134962. [CrossRef] [PubMed]
24. Yotsu-Yamashita, M.; Nagaoka, Y.; Muramoto, K.; Cho, Y.; Konoki, K. Pufferfish Saxitoxin and Tetrodotoxin Binding Protein (PSTBP) Analogues in the Blood Plasma of the Pufferfish *Arothron nigropunctatus*, *A. hispidus*, *A. manilensis*, and *Chelonodon patoca*. *Mar. Drugs* **2018**, *16*, 224. [CrossRef]
25. Chau, R.; Kalaitzis, J.A.; Wood, S.A.; Neilan, B.A. Diversity and Biosynthetic Potential of Culturable Microbes Associated with Toxic Marine Animals. *Mar. Drugs* **2013**, *11*, 2695–2712. [CrossRef]
26. Noguchi, T.; Hwang, D.F.; Arakawa, O.; Sugita, H.; Deguchi, Y.; Shida, Y.; Hashimoto, K. *Vibrio alginolyticus*, a tetrodotoxin-producing bacterium, in the intestines of the fish *Fugu vermicularis vermicularis*. *Mar. Biol.* **1987**, *94*, 625–630.
27. Cheng, C.A.; Hwang, D.F.; Tsai, Y.H.; Chen, H.C.; Jeng, S.S.; Noguchi, T.; Ohwada, K.; Hasimoto, K. Microflora and tetrodotoxin-producing bacteria in a gastropod, *Niotha clathrata*. *Food Chem. Toxicol.* **1995**, *33*, 929–934.
28. Wu, Z.; Xie, L.; Xia, G.; Zhang, J.; Nie, Y.; Hu, J.; Wang, S.; Zhang, R. A new tetrodotoxin-producing actinomycete, Nocardiopsis dassonvillei, isolated from the ovaries of puffer fish *Fugu rubripes*. *Toxicon* **2005**, *45*, 851–859.
29. Ritchie, K.B.; Nagelkerken, I.; James, S.; Smith, G.W. A tetrodotoxin-producing marine pathogen. *Nature* **2000**, *404*, 354. [CrossRef]
30. Wang, X.J.; Yu, R.C.; Luo, X.; Zhou, M.J.; Lin, X.T. Toxin-screening and identification of bacteria isolated from highly toxic marine gastropod *Nassarius semiplicatus*. *Toxicon* **2008**, *52*, 55–61. [CrossRef]
31. Miyazawa, K.; Noguchi, T. Distribution and origin of tetrodotoxin. *J. Toxicol. Toxin Rev.* **2001**, *20*, 11–33.
32. Demain, A.L. Induction of microbial secondary metabolism. *Int. Microbiol.* **1998**, *1*, 259–264. [PubMed]
33. Strand, M.; Hedström, M.; Seth, H.; McEvoy, E.G.; Jacobsson, E.; Göransson, U.; Andersson, H.S.; Sundberg, P. The Bacterial (*Vibrio alginolyticus*) Production of Tetrodotoxin in the Ribbon Worm *Lineus longissimus*—Just a False Positive? *Mar. Drugs* **2016**, *14*, 63. [CrossRef] [PubMed]
34. Matsumura, K. Reexamination of tetrodotoxin production by bacteria. *Appl. Environ. Microbiol.* **1995**, *61*, 3468–3470.
35. Johnson, J.S.; Spakowicz, D.J.; Hong, B.Y.; Petersen, L.M.; Demkowicz, P.; Chen, L.; Leopold, S.R.; Hanson, B.M.; Agresta, H.O.; Gerstein, M.; et al. Evaluation of 16S rRNA gene sequencing for species and strain-level microbiome analysis. *Nat. Commun.* **2019**, *10*, 5029.
36. Yotsu-Yamashita, M.; Mebs, D.; Flachsenberger, W. Distribution of tetrodotoxin in the body of the blue-ringed octopus (*Hapalochlaena maculosa*). *Toxicon* **2007**, *49*, 410–412.
37. Tsai, Y.H.; Hwang, D.F.; Chai, T.J.; Jeng, S.S. Occurrence of tetrodotoxin and paralytic shellfish poison in the Taiwanese crab *Lophozozymus pictor*. *Toxicon* **1995**, *33*, 1669–1673.
38. Yotsu-Yamashita, M.; Mebs, D. The levels of tetrodotoxin and its analogue 6-epitetrodotoxin in the red-spotted newt, *Notophthalmus viridescens*. *Toxicon* **2001**, *39*, 1261–1263.
39. Kosker, A.R.; Özogul, F.; Ayas, D.; Durmus, M.; Ucar, Y.; Regenstein, J.M.; Özogul, Y. Tetrodotoxin levels of three pufferfish species (*Lagocephalus* sp.) caught in the North-Eastern Mediterranean sea. *Chemosphere* **2019**, *219*, 95–99.
40. Kosker, A.R.; Özogul, F.; Durmus, M.; Ucar, Y.; Ayas, D.; Šimat, V.; Özogul, Y. First report on TTX levels of the yellow spotted pufferfish (*Torquigener flavimaculosus*) in the Mediterranean Sea. *Toxicon* **2018**, *148*, 101–106.
41. Roeselers, G.; Mittge, E.K.; Stephens, W.Z.; Parichy, D.M.; Cavanaugh, C.M.; Guillemin, K.; Rawls, J.F. Evidence for a core gut microbiota in the zebrafish. *ISME J.* **2011**, *5*, 1595–1608. [PubMed]
42. Llewellyn, M.S.; Boutin, S.; Hoseinifar, S.H.; Derome, N. Teleost microbiomes: The state of the art in their characterization, manipulation and importance in aquaculture and fisheries. *Front. Microbiol.* **2014**, *5*, 207.
43. Ruff, W.E.; Greiling, T.M.; Kriegel, M.A. Host-microbiota interactions in immune-mediated diseases. *Nat. Rev. Microbiol.* **2020**, *18*, 521–538.
44. Kiron, V. Fish immune system and its nutritional modulation for preventive health care. *Anim. Feed Sci. Technol.* **2012**, *173*, 111–133.
45. Kogut, M.H.; Lee, A.; Santin, E. Microbiome and pathogen interaction with the immune system. *Poult. Sci.* **2020**, *99*, 1906–1913.

46. Kurilshikov, A.; Wijmenga, C.; Fu, J.; Zhernakova, A. Host Genetics and Gut Microbiome: Challenges and Perspectives. *Trends Immunol.* **2017**, *38*, 633–647.
47. Zhou, Y.L.; He, G.L.; Jin, T.; Chen, Y.J.; Dai, F.Y.; Luo, L.; Lin, S.M. High dietary starch impairs intestinal health and microbiota of largemouth bass, *Micropterus salmoides*. *Aquaculture* **2021**, *534*, 736261.
48. Joe, J.T.X.; Chiou, P.P.; Kuo, C.Y.; Lin, J.H.J.; Wu, J.L.; Lu, M.W. The microbiota profile and transcriptome analysis of immune response during metamorphosis stages in orange spotted grouper (*Epinephelus coioides*). *Fish Shellfish Immunol.* **2019**, *90*, 141–149.
49. Wassel, M.A.; Makabe-Kobayashi, Y.; Iqbal, M.M.; Takatani, T.; Sakakura, Y.; Hamasaki, K. The impact of tetrodotoxin (TTX) on the gut microbiome in juvenile tiger pufferfish, *Takifugu rubripes*. *Sci. Rep.* **2024**, *14*, 16684.
50. Li, Z.; Tian, J.; Lai, Y.; Lee, C.-H.; Cai, Z.; Yu, C.-F. Puffer Fish Gut Microbiota Studies Revealed Unique Bacterial Co-Occurrence Patterns and New Insights on Tetrodotoxin Producers. *Mar. Drugs* **2020**, *18*, 278. [CrossRef]
51. Melnikova, D.I.; Magarlamov, T.Y. The Microbial Community of Tetrodotoxin-Bearing and Non-Tetrodotoxin-Bearing Ribbon Worms (Nemertea) from the Sea of Japan. *Mar. Drugs* **2020**, *18*, 177. [CrossRef] [PubMed]
52. Gao, L.; Zhang, Z.; Xing, Z.; Li, Q.; Kong, N.; Wang, L.; Song, L. The variation of intestinal autochthonous bacteria in cultured tiger pufferfish *Takifugu rubripes*. *Front. Cell. Infect. Microbiol.* **2022**, *12*, 1062512.
53. Yu, G.; Ou, W.; Liao, Z.; Xu, H.; Liang, M.; Zhang, Y.; Mai, K. Intestinal homeostasis of juvenile tiger puffer *Takifugu rubripes* was sensitive to dietary arachidonic acid in terms of mucosal barrier and microbiota. *Aquaculture* **2019**, *502*, 97–106.
54. Jin, X.; Chen, Z.; Shi, Y.; Gui, J.; Zhao, Z. Response of gut microbiota to feed-borne bacteria depends on fish growth rate: A snapshot survey of farmed juvenile *Takifugu obscurus*. *Microb. Biotechnol.* **2002**, *15*, 683–702.
55. Riazati, N.; Kable, M.E.; Stephensen, C.B. Association of intestinal bacteria with immune activation in a cohort of healthy adults. *Microbiol. Spectr.* **2023**, *11*, e0102723.
56. Noguchi, T.; Jeon, J.K.; Arakawa, O.; Sugita, H.; Deguchi, Y.; Shida, Y.; Hashimoto, K. Occurrence of tetrodotoxin and anhydrotetrodotoxin in *Vibrio* sp. isolated from the intestines of a xanthid crab (*Atergatis floridus*). *J. Biochem.* **1986**, *99*, 311–314.
57. Narita, H.; Matsubara, S.; Miwa, N.; Akahane, S.; Marukami, M.; Goto, T.; Nara, M.; Noguchi, T.; Saito, T.; Shida, Y.; et al. *Vibrio alginolyticus*, a TTX-producing bacterium isolated from the Starfish (*Astropecten polycanthus*). *Nippon Suisan Gakkaishi* **1987**, *53*, 617–621.
58. Sugita, H.; Ueda, R.; Noguchi, T.; Arakawa, O.; Hashimoto, K.; Deguchi, Y. Identification of a tetrodotoxinproducing bacterium isolated from the xanthid crab (*Atergatis floridus*). *Nippon Suisan Gakkaishi* **1987**, *53*, 1693.
59. Carroll, S.; McEvoy, E.; Gibson, R. The production of tetrodotoxin-like substances by nemertean worms in conjunction with bacteria. *J. Exp. Mar. Biol. Ecol.* **2003**, *288*, 51–63.
60. Bowman, J.P. Bioactive Compound Synthetic Capacity and Ecological Significance of Marine Bacterial Genus *Pseudoalteromonas*. *Mar. Drugs* **2007**, *5*, 220–241. [CrossRef]
61. Melnikova, D.I.; Beleneva, I.A.; Tyunin, A.P.; Magarlamov, T.Y. The taxonomic composition, characteristics, and neurotoxic activities of ribbon worm-associated bacteria from the Sea of Japan. *Russ. J. Mar. Biol.* **2017**, *43*, 383–391.
62. Zhang, D.; Yu, Z.; Zeng, B.; Liu, X. Genome-wide analysis of the ABC gene family in almond and functional predictions during flower development, freezing stress, and salt stress. *BMC Plant Biol.* **2024**, *24*, 12.
63. Romersi, R.F.; Nicklisch, S.C.T. Interactions of Environmental Chemicals and Natural Products with ABC and SLC Transporters in the Digestive System of Aquatic Organisms. *Front. Physiol.* **2022**, *12*, 767766.
64. Wang, H.; Liu, S.; Xun, X.; Li, M.; Lou, J.; Zhang, Y.; Shi, J.; Hu, J.; Bao, Z.; Hu, X. Toxin- and species-dependent regulation of ATP-binding cassette (ABC) transporters in scallops after exposure to paralytic shellfish toxin-producing dinoflagellates. *Aquat. Toxicol.* **2021**, *230*, 105697.
65. Zhu, B.; Bai, Y.; Yeo, Y.Y.; Lu, X.; Rovira-Clavé, X.; Chen, H.; Yeung, J.; Nkosi, D.; Glickman, J.; Delgado-Gonzalez, A.; et al. A multi-omics spatial framework for host-microbiome dissection within the intestinal tissue microenvironment. *Nat. Commun.* **2025**, *16*, 1230.
66. *GB 5009.206-2016*; National Standard of the People's Republic of China GB 5009.206-2016 Determination of Tetrodotoxin in Aquatic Products. China Food and Drug Administration, National Health and Family Planning Commission: Beijing, China, 2016.
67. Zhou, F.; Liu, B.; Liu, X.; Li, Y.; Wang, L.; Huang, J.; Luo, G.; Wang, X. The Impact of Microbiome and Microbiota-Derived Sodium Butyrate on *Drosophila* Transcriptome and Metabolome Revealed by Multi-Omics Analysis. *Metabolites* **2021**, *11*, 298. [CrossRef]
68. Magoč, T.; Salzberg, S.L. FLASH: Fast length adjustment of short reads to improve genome assemblies. *Bioinformatics* **2011**, *27*, 2957–2963.
69. Caporaso, J.G.; Kuczynski, J.; Stombaugh, J.; Bittinger, K.; Bushman, F.D.; Costello, E.K.; Fierer, N.; Peña, A.G.; Goodrich, J.K.; Gordon, J.I.; et al. QIIME allows analysis of high-throughput community sequencing data. *Nat. Methods* **2010**, *7*, 335–336.
70. Rognes, T.; Flouri, T.; Nichols, B.; Quince, C.; Mahé, F. VSEARCH: A versatile open source tool for metagenomics. *PeerJ* **2016**, *4*, e2584.
71. Edgar, R.C. UPARSE: Highly accurate OTU sequences from microbial amplicon reads. *Nat. Methods* **2013**, *10*, 996–998.

72. Quast, C.; Pruesse, E.; Yilmaz, P.; Gerken, J.; Schweer, T.; Yarza, P.; Peplies, J.; Glöckner, F.O. The SILVA ribosomal RNA gene database project: Improved data processing and web-based tools. *Nucleic Acids Res.* **2013**, *41*, D590–D596. [CrossRef] [PubMed]
73. Wang, Q.; Garrity, G.M.; Tiedje, J.M.; Cole, J.R. Naive Bayesian classifier for rapid assignment of rRNA sequences into the new bacterial taxonomy. *Appl. Environ. Microbiol.* **2007**, *73*, 5261–5267. [CrossRef] [PubMed]
74. White, J.R.; Nagarajan, N.; Pop, M. Statistical methods for detecting differentially abundant features in clinical metagenomic samples. *PLoS Comput. Biol.* **2009**, *5*, e1000352. [CrossRef]
75. Segata, N.; Izard, J.; Waldron, L.; Gevers, D.; Miropolsky, L.; Garrett, W.S.; Huttenhower, C. Metagenomic biomarker discovery and explanation. *Genome Biol.* **2011**, *12*, R60. [CrossRef]
76. Aßhauer, K.P.; Wemheuer, B.; Daniel, R.; Meinicke, P. Tax4Fun: Predicting functional profiles from metagenomic 16S rRNA data. *Bioinformatics* **2015**, *31*, 2882–2884. [CrossRef]

marine drugs

MDPI

Article

Morphological, Toxicological, and Biochemical Characterization of Two Species of *Gambierdiscus* from Bahía de La Paz, Gulf of California

Leyberth José Fernández-Herrera [1], Erick Julián Núñez-Vázquez [1,*], Francisco E. Hernández-Sandoval [2], Daniel Octavio Ceseña-Ojeda [1], Sara García-Davis [3], Andressa Teles [4], Marte Virgen-Félix [5] and Dariel Tovar-Ramírez [4]

[1] Laboratorio de Toxinas Marinas y Aminoácidos, Centro de Investigaciones Biológicas del Noroeste, Av. Instituto Politécnico Nacional 195, Playa Palo de Santa Rita, La Paz CP 23096, Mexico; lfernandez@pg.cibnor.mx (L.J.F.-H.); dcesena04@cibnor.mx (D.O.C.-O.)

[2] Laboratorio de Microalgas Nocivas, Centro de Investigaciones Biológicas del Noroeste, Av. Instituto Politécnico Nacional 195, Playa Palo de Santa Rita, La Paz CP 23096, Mexico; fhernan04@cibnor.mx

[3] Instituto Universitario de Bio-Orgánica Antonio González (IUBO AG), Universidad de La Laguna (ULL), Avenida Astrofísico Francisco Sánchez 2, 38206 La Laguna, Tenerife, Spain; sgdavis@ull.edu.es

[4] Laboratorio de Fisiología Comparada y Genómica Funcional, Centro de Investigaciones Biológicas del Noroeste, Av. Instituto Politécnico Nacional 195, Playa Palo de Santa Rita, La Paz CP 23096, Mexico; ateles@pg.cibnor.mx (A.T.); dtovar04@cibnor.mx (D.T.-R.)

[5] Laboratorio de Colección de Microalgas, Centro de Investigaciones Biológicas del Noroeste, Av. Instituto Politécnico Nacional 195, Playa Palo de Santa Rita, La Paz CP 23096, Mexico; marte04@cibnor.mx

* Correspondence: enunez04@cibnor.mx; Tel.: +52-612-123-8484

Abstract: We describe five new isolates of two *Gambierdiscus* species from Bahía de La Paz in the southern Gulf of California. Batch cultures of *Gambierdiscus* were established for morphological characterization using light microscopy (LM) and scanning electron microscopy (SEM). Pigment and amino acid profiles were also analyzed using high-performance liquid chromatography (HPLC-UV and HPLC-DAD). Finally, toxicity (CTX-like and MTX-like activity) was evaluated using the *Artemia salina* assay (ARTOX), mouse assay (MBA), marine fish assay (MFA), and fluorescent receptor binding assay (fRBA). These strains were identified as *Gambierdiscus* cf. *caribaeus* and *Gambierdiscus* cf. *carpenteri.* Toxicity for CTX-like and MTX-like activity was confirmed in all evaluated clones. Seven pigments were detected, with chlorophyll *a*, pyridine, Chl2, and diadinoxanthin being particularly noteworthy. For the first time, a screening of the amino acid profile of *Gambierdiscus* from the Pacific Ocean was conducted, which showed 14 amino acids for all strains except histidine, which was only present in *G.* cf. *caribeaus.* We report the presence of *Gambierdiscus* and *Fukuyoa* species in the Mexican Pacific, where ciguatera fish poisoning (CFP) cases have occurred.

Keywords: *Gambierdiscus*; amino acids; pigments; toxicity; ciguatoxins; ciguatera; Gulf of California

1. Introduction

Ciguatera fish poisoning (CFP) is the most prevalent marine toxin-related seafood poisoning worldwide, affecting up to 50,000 people annually. This illness is caused by the consumption of seafood contaminated with the lipophilic toxins ciguatoxins (CTXs) and, possibly, maitotoxins (MTXs), which are produced by benthic dinoflagellates of the genera *Gambierdiscus* and *Fukuyoa* [1,2]. Numerous fish species that are typically associated with coral reefs worldwide accumulate CTXs from these dinoflagellates through the consumption of contaminated prey, storing the toxins in their flesh and viscera [1]. In tropical and subtropical regions, CFP is a serious problem that impacts the fishing industry and human health. The true incidence rate of CFP is challenging to determine, given that CFP-related symptoms are underreported (2–10%) [1,2]. The symptoms of CFP are characterized by gastrointestinal, neurological, and cardiovascular disorders and have been described as a polymorphic syndrome, with more than 175 reported symptoms and clinical signs. Symptoms can persist for days, months, or even years, and in severe cases, CFP can lead to paralysis, coma, and death [2].

Recent reviews have described an increase in the frequency and intensity of harmful algal blooms in some regions of the world [3–6]. These trends can be partly attributed to improved monitoring and coastal eutrophication [3,7,8]. Despite being significantly underreported, cases of CFP are increasing worldwide, with reports indicating a 60% rise in the Pacific Islands over recent decades and in other regions [1,9]. Climate variability may be related to the incidence of CFP [1,9,10] and Gingold et al. [10], with some research indicating that climate change could increase the burden of this poisoning.

In Mexico, CFP primarily occurs in the Yucatan Peninsula (PYUC) in the Mexican Caribbean, the Gulf of Mexico, and the waters of the southern Baja California Peninsula (PBC) in the Eastern Tropical Pacific and Gulf of California [11]. In the PBC, CFP cases are rare and sporadic; however, at least 240 cases (some severe) have been reported in the southern region due to the consumption of carnivorous fish, such as snappers (*Lutjanus* spp.) and cabrillas and groupers (*Epinephelus* and *Mycteroperca*), from islands and islets [11–17].

Gambierdiscus is a benthic dinoflagellate that grows on a variety of seagrasses, macroalgae, corals, artificial substrates, and sediments. Until 1999, *Gambierdiscus* was considered a monotypic genus (*G. toxicus*); however, in recent decades, 19 species have been described in this genus (*G. toxicus*, *G. australes*, *G. balechii*, *G. belizeanus*, *G. caribaeus*, *G. carolinianus*, *G. carpenteri*, *G. cheloniae*, *G. excentricus*, *G. holmesii*, *G. honu*, *G. jejuelnsis*, *G. lapillus*, *G. lewissi*, *G. pacificus*, *G. polynensiensis*, *G. scabrosus*, *G. silvae*, and *G. vietnamensis*) [18]. *Fukuyoa* is closely related to *Gambierdiscus* and was originally included in this genus [18–20] before being reclassified by Gomez et al. [20]. Four species of *Fukuyoa* have been described: *F. yasumutoi*, *F. ruetzleri*, *F. paulensis*, and *F. koreensis* [20,21].

Gambierdiscus is widely distributed in coastal zones at tropical and subtropical latitudes; however, its presence has also been reported in temperate waters, including those near Australia, Japan, and the Mediterranean [22]. The breadth of species within this genus remains poorly studied, and the ability to discriminate between different *Gambierdiscus* species has only recently been developed [23]. *Gambierdiscus* produces multiple secondary metabolites, including bioactive compounds that can be either lipophilic or hydrophilic and vary in their modes of action and toxicity. These include CTXs, MTXs, gambierones, gambieroles, gambieroxide, and gambieric acids [18,22]. To date, more than 30 CTX congeners have been identified.

Ciguatoxins are lipophilic, ladder-shaped polyethers with 13 14 cyclic consecutively connected rings [22] and are classified as either CTX3C, Caribbean Sea CTXs (C-CTXs), Pacific Ocean CTXs (P-CTXs/CTX4A), or Indian Ocean CTXs (I-CTXs) based on the structural backbone of each molecule [22]. Recently, a new ciguatoxin was isolated from *G. silvae* and *G. caribeus* in the Caribbean; its structure was determined, and it was named C-CTX-5 [24,25]. Although some MTXs exhibit greater toxicity than CTXs, their role in CFP remains unclear. Six MTXs have been identified, and most of their structures have been elucidated. Like CTXs, MTXs are also polyether compounds, but their molecular masses vary significantly. Most MTXs are larger than CTXs, although they share structural similarities [22]. In addition to CTXs and MTXs, some *Gambierdiscus* species produce gambieric acids (GA), gambieroxide, gambierol, and gambierons. Some of these compounds have antifungal properties and participate in dinoflagellate defense mechanisms while also possibly regulating their growth. They also exhibit substantially lower toxicity than CTXs, opening the door for potential therapeutic and biomedical applications [22,26].

Evaluating the pigment profile, amino acids, and other metabolites in dinoflagellate species provides an opportunity to explore the chemical diversity of secondary metabolites. These metabolites can serve as biomarkers for identifying *Gambierdiscus* species (chemotaxonomy); they can also be used to evaluate possible ecotypes and to study toxin metabolism and biosynthesis, which may potentially involve allelochemical compounds [27,28].

Gambierdiscus toxicus, *Gambierdiscus* sp., *F. yasumotoi*, *G.* cf. *carolinianus*, *G.* cf. *carpenteri*, *Gambierdiscus* spp., and *Fukuyoa* sp. have been documented in the Eastern Tropical Pacific and Gulf of California around the southern PBC [29–38] (Table S1). In this work, we describe five new strains of *G.* cf. *caribaeus* and *G.* cf. *carpenteri* from Bahía de La Paz in the southern Gulf of California. We propose a morphological characterization using light microscopy (LM) and scanning electron microscopy (SEM) and report pigment and amino acid profiles using high-performance liquid chromatography and ultraviolet-visible spectrophotometry (HPLC-UV/VIS) and high-performance liquid chromatography-diode array detection (HPLC-DAD). Toxicity (CTXs-like and MTXs-like activity) was evaluated by four different assays: *Artemia salina* assay (ARTOX), mouse bioassay (MBA), marine fish assay (MFA), and fluorescent receptor binding assay (fRBA).

2. Results

2.1. Isolation and Algal Cultures

Gambierdiscus was present at all sampling sites in Bahía de La Paz (Table S2). Seventy-five percent of isolated cells were established in culture (n = 39 well plate); the lenticular cells were isolated directly from the substrate without shaking the algae or sonicating the sediment.

2.2. Morphological Identification

Gambierdiscus cf. *caribaeus* strains GbSa-2 and GbSa-7 were identified as photosynthetic, lens-shaped cells that were rounded in the apical view and compressed in the anterior-posterior axis, with golden granular chloroplasts (Figure 1A,B). The cells measured 70.31–91.61 $\pm$ 5.90 µm depth, 83.32–86.32 $\pm$ 7.90 µm width, and 61.12–64.57 $\pm$ 4.10 µm length (n = 30). The plate formula was Po, 3$'$, 7$''$, 6C, 6S, 5$'''$, 1p, and 2$''''$ (Figure 1C–F). The apical pore plate Po was in contact with apical plate 1$'$ 2$'$, and 3$'$. The rectangular hatched-shaped 2$'$ plate was the largest apical plate, with plate 4$''$ being pentagonal and symmetrical (Figure 1C,E). In the antapical view, between 2$'''$ and 4$'''$, plate 1p was the largest hypothecal plate; it was broad and pentagonal in shape. In the ventral view, minuscule plate 1$''$ was rectangular, and trapezium-shape 7$''$ was also observed (Figure 1G).

SEM micrographs (n = 15) revealed a rounded apical shape with a smooth cell surface and multiple pores scattered across the thecal plates, including symmetrical plate 4$''$ (Figure 2A). The antapical view showed a broad 1p plate and detail of postcingular plates 1$'''$, 2$'''$, 3$'''$, and 4$'''$ (Figure 2B,D). The apical pore plate, Po, was elliptical with a fishhook-shaped opening (Figure 2C). Detailed images of the sulcal zone showed minuscule plates 1$''$ and 7$''$, with the cingulum gradually descending near the sulcus (Figure 2E). The dorsal view revealed symmetrical plate 4$''$ and a narrow, stylized cingulum edge (Figure 2F). Both LM and SEM micrographs of the GbSa-2 strain are shown in Figure S3.

The maximum cell density (1835 $\pm$ 117 cells mL^{-1}) in *G.* cf. *caribaeus* was recorded during the exponential phase (day 35 for strain GbSa-2). For strain GbSa-7, the maximum cell density (2457 $\pm$ 394 cells mL^{-1}) was recorded on day 49. The calculated growth rates were 0.25 $\pm$ 0.01 and 0.26 $\pm$ 0.05 division·day^{-1}, respectively, with an average generation time of 2.71 $\pm$ 0.06 days for *G.* cf. *caribaeus* strains.

In the apical view, plates 1$'$, 2$'$, and 3$'$ were connected to the Po plate. Apical 2$'$ plate showed a hatched and large form, while plate 4$''$ was asymmetrically narrow toward plate 3$''$ and wider toward plate 5$''$. Plates 2$''$ and 6$''$ were the shortest in the precingular area (Figure 3C,E). Plate 1p was long, pentagonal, and broad, joining with small plates 1$''''$ and 2$''''$ in a ridge-like form. Plate 1p was also connected to plate 3$''$ in a quadrangular shape and to plates 2$'''$ and 4$'''$ to form the largest quadrangular plate (Figure 3D,F). In the ventral view, minuscule plate 1$''$ was rectangular and plate 7$''$ was trapezium-shaped (Figure 3G).

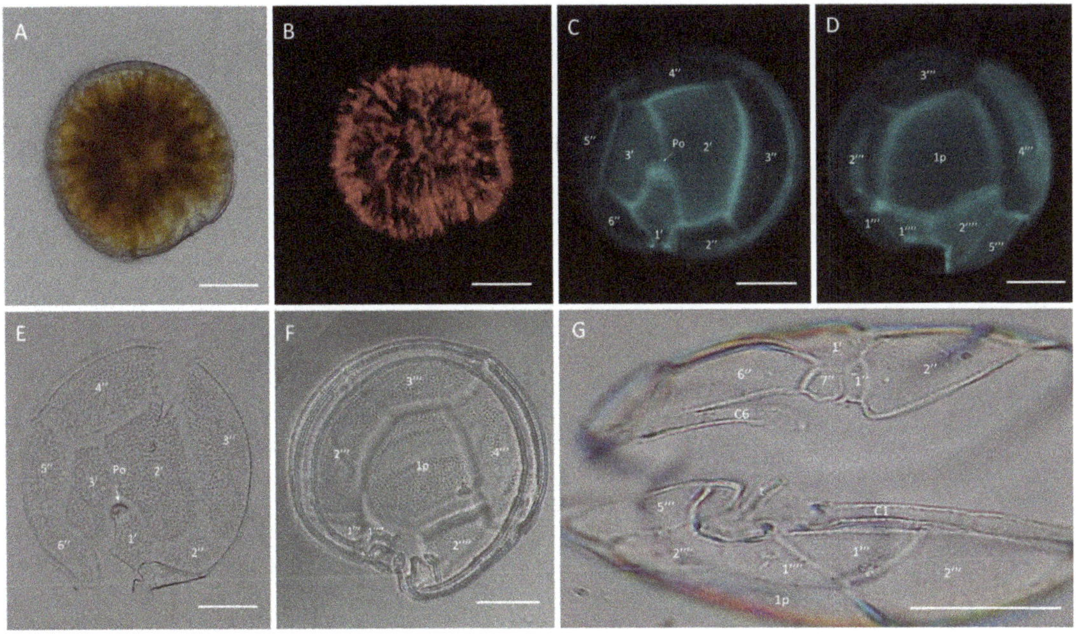

Figure 1. Light microscopy of *Gambierdiscus* cf. *caribaeus* (GbSa-7). (**A**) Live cell, (**B**) Epifluorescence of chloroplast, (**C**) Plate tabulation apical view stained with Fluorescent Brightener 28, (**D**) Antapical view, (**E**) Apical view, (**F**) Antapical view, and (**G**) Ventral view. Scale bars: (**A–F**) 50 μm and (**G**) 30 μm.

In *G.* cf. *carpenteri* strains GbSa-4, GbSa-5, and GbSa-7, the cells were photosynthetic and lens-shaped, predominantly oblong and irregular in apical view, and compressed in the anterior-posterior axis, with golden granular chloroplasts (Figure 3A,B). The cells measured 72.21–84.91 ± 5.90 μm depth, 71.72–77.62 ± 7.29 μm width, and 59.56–64.57 ± 6.15 μm length (n = 30). The plate formula was Po, 3′, 7″, 6C, 6S, 5‴, 1p, and 2⁗ (Figure 3C–F).

SEM micrographs (n = 15) of *G.* cf. *carpenteri* revealed a rounded apical shape with a smooth cell surface and multiple scattered pores. Asymmetrical plate 4″ protruded, forming an apparent rostrum where plate 3″ was articulated (Figure 4A). The antapical view revealed a long, pentagonal plate 1p (Figure 4B,D). The apical pore plate, Po, was elliptical with a fishhook-shaped opening (Figure 4C). The sulcal zone showed minuscule plate 1″ and small plate 7″, which formed ridges toward the adjacent part of the sulcus (Figure 4E). The cingulum was narrow, and the dorsal view highlighted asymmetrical 4″ plate and the apparent rostrum formed by the union of precingular 3″ and 4″ plates and postcingular 3‴ and 4‴ plates (Figure 4F). The LM and SEM micrographs of GbSa-5 and GbSa-6 strains are shown in Figures S4 and S5.

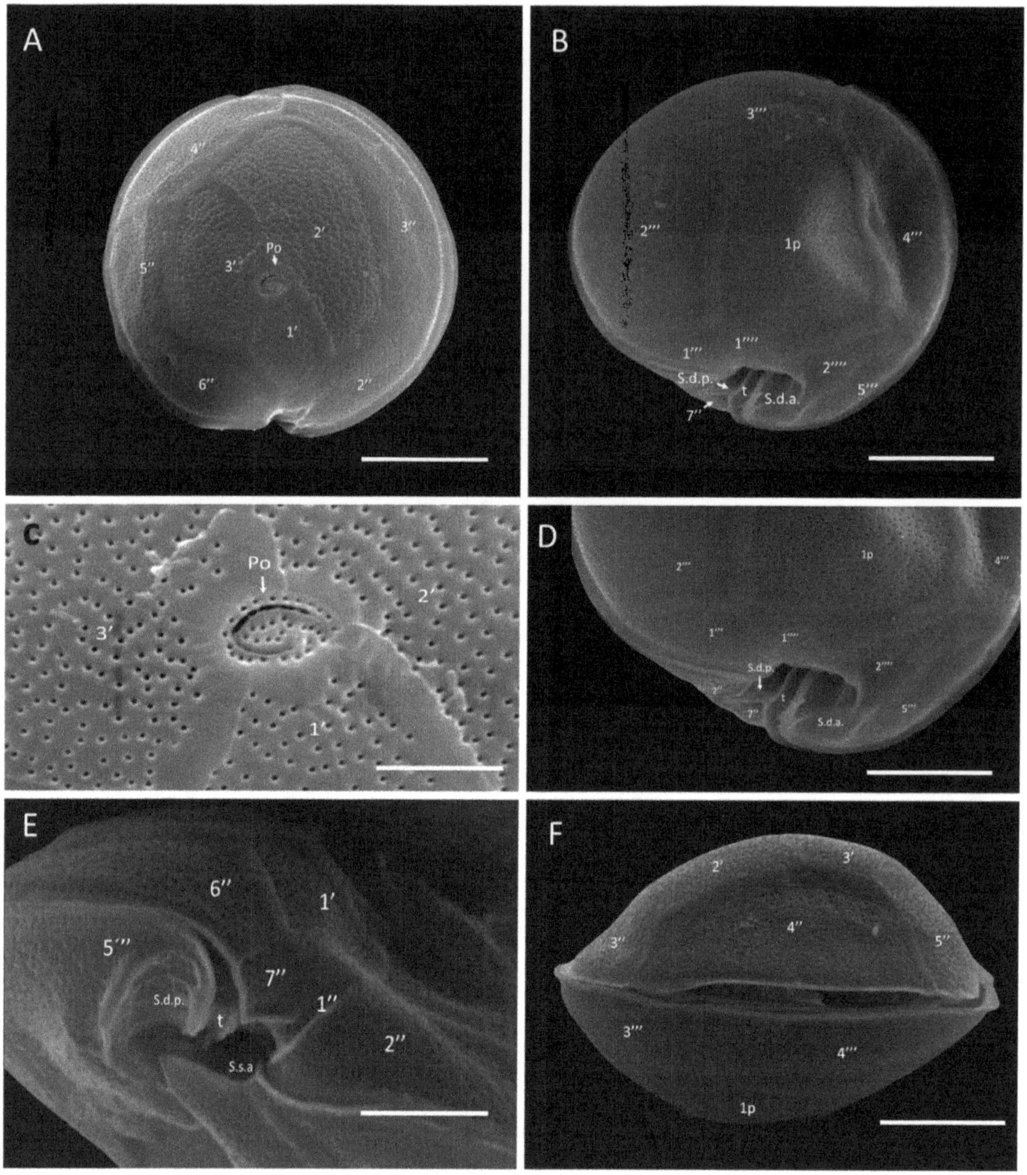

Figure 2. Scanning electron micrographs of *Gambierdiscus* cf. *caribaeus* (GbSa-7). (**A**) Apical view, (**B**) Antapical view, (**C**) Po plate, (**D**) Sulcal area in antapical view, (**E**) Sulcal area in ventral view, and (**F**) Dorsal view. Scale bars: (**A**,**B**) 50 μm, (**C**) 10 μm, (**D**) 20 μm, and (**E**,**F**) 30 μm.

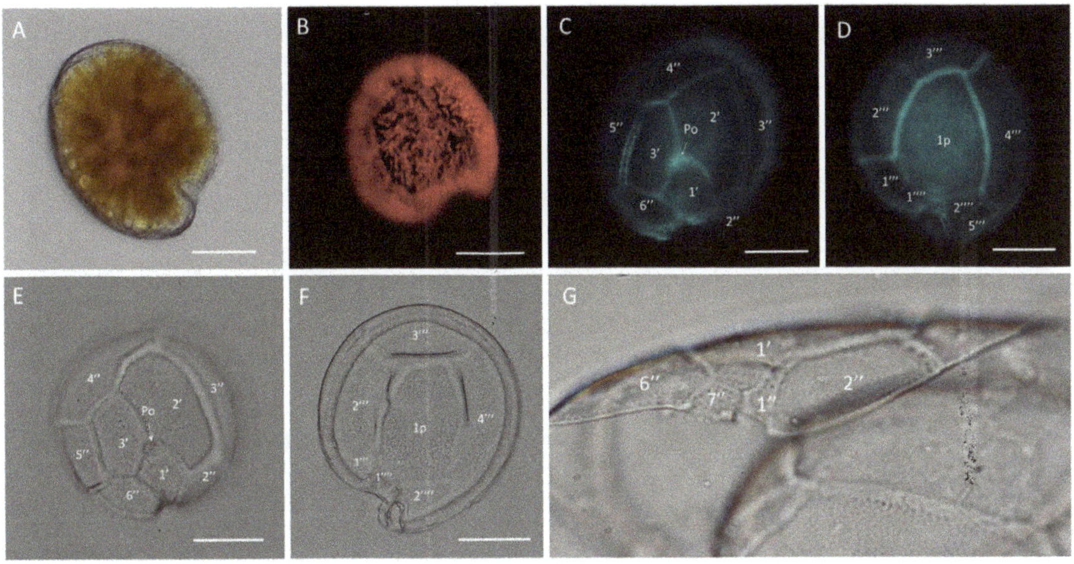

Figure 3. Light microscopy of *Gambierdiscus* cf. *carpenteri* (GbSa-4). (**A**) Live cell, (**B**) Epifluorescence of chloroplast, (**C**) Plate tabulation apical view stained with Fluorescent Brightener 28, (**D**) Antapical view, (**E**) Apical view, (**F**) Antapical view, and (**G**) Ventral view. Scale bars: (**A–F**) 50 μm and (**G**) 30 μm.

Gambierdiscus cf. *carpenteri* strains GbSa-4 and GbSa-6 presented higher cellular densities (2779 ± 424 and 1815 ± 329 cells mL^{-1}, respectively) compared to that of strain GbSa-5 (892 ± 163 cells mL^{-1}). The growth rates, based on cell counts for the strains with the highest abundance, were 0.25–0.27 ± 0.01 division·day^{-1}, with an average generation time of 2.64 ± 0.09 per day. Strain GbSa-5 showed the lowest growth rate among all strains (0.21–0 ± 0.01 division·day^{-1}, 3.36 ± 0.20 generation time). The growth curves are shown in Figure S6.

Morphological measurements revealed differences in the length-to-width ratios of *G.* cf. *caribaeus* cell size (1.05–1.13 ± 0.59 and 1.00–1.12 ± 1.29) compared to that of *G.* cf. *carpenteri*. The measurements of the apical pore plate were similar between the strains of both species, although the number of pores differed. *Gambierdiscus* cf. *caribaeus* and *G.* cf. *carpenteri* exhibited an average of 37.33 ± 3.77 pores and 40.66 ± 2.43 pores, respectively. The length-to-width ratio of plate 1p further distinguished the two species (*G.* cf. *caribaeus*: ratio of 1.28 ± 0.11; *G.* cf. *carpenteri*: 1.59 ± 0.24; Table 1).

Regarding the ratio of the $2'/4''$ suture length and $2'/2''$ suture length between the two species, no differences were observed in the measurements (n = 20). In *G.* cf. *caribaeus*, this ratio was 0.92 ± 0.01, while in *G.* cf. *carpenteri*, the ratio was 0.92 ± 0.04. After the dissociation of the apical and antapical plates, marked morphological differences were only observed in plates $1'$, $3''$, and $4''$.

In addition to the dinoflagellates characterized in this study, we also detected the presence of the potentially harmful dinoflagellates *G.* cf. *carolineanus*, *G.* cf. *toxicus*, and *F.* cf. *yasumotoi* (Figures S7–S9).

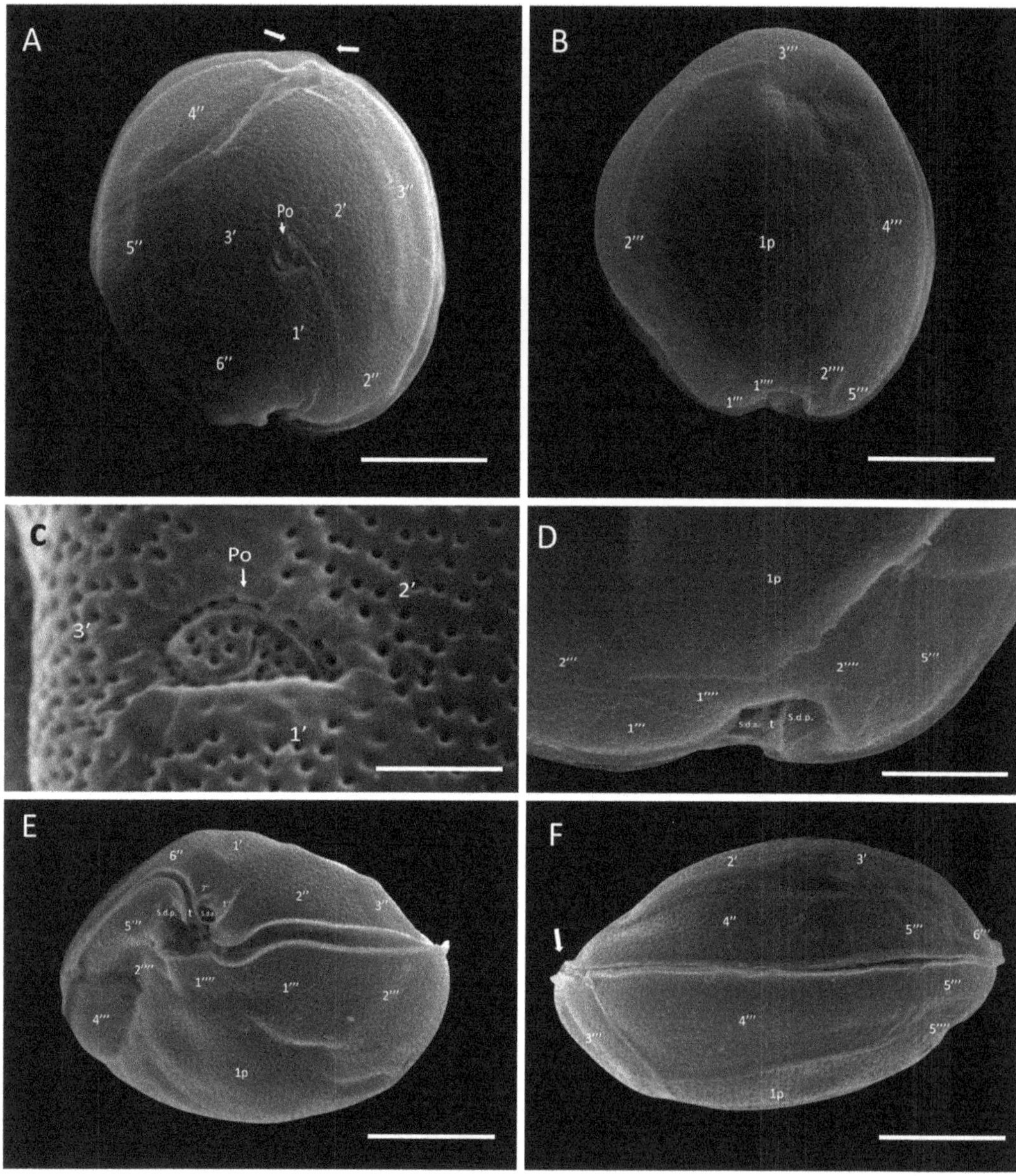

Figure 4. SEM micrographs of *Gambierdiscus* cf. *carpenteri* (GbSa-4). (**A**) Apical view arrow showing an apparent rostrum, (**B**) Antapical view, (**C**) Po plate, (**D**) Sulcal area in antapical view, (**E**) Sulcal area in ventral view, and (**F**) Dorsal view arrow points to apparent rostrum. Scale bars: (**A**,**B**) 50 µm, (**C**) 10 µm, (**D**) 20 µm, and (**E**,**F**) 30 µm.

Table 1. Morphological measurements of *Gambierdiscus* strains from El Sauzoso, Bahía de La Paz, obtained through light microscopy.

	Cell Size (μm)			Apical Pore Plate Po (μm)			Po Apical Pore (μm)		1p Plate (μm)		
Specie/Strain	**D**	**W**	**D/W**	**L**	**W**	**L/W**	**Number**	**Diameter**	**L**	**W**	**L/W**
G. cf. *caribaeus*/ GbSa-2	87.15 ± 4.73	83.32 ± 4.90	1.05 ± 0.56	10.04 ± 0.35	6.21 ± 0.31	1.62 ± 0.07	37.16 ± 2.32	0.48 ± 0.05	49.88 ± 4.24	37.55 ± 3.09	1.29 ± 0.13
G. cf. *caribaeus*/ GbSa-7	79.32 ± 8.75	70.29 ± 5.98	1.13 ± 0.63	9.91 ± 1.22	6.23 ± 0.89	1.59 ± 1.38	37.50 ± 5.23	0.41 ± 0.05	45.55 ± 3.31	36.19 ± 1.99	1.28 ± 0.10
G. cf. *carpenteri*/ GbSa-4	77.67 ± 7.12	71.31 ± 5.74	1.09 ± 1.24	10.35 ± 0.49	6.17 ± 0.37	1.68 ± 1.32	40.50 ± 3.24	0.41 ± 0.09	55.59 ± 2.95	35.99 ± 3.72	1.62 ± 0.19
G. cf. *carpenteri*/ GbSa-5	79.44 ± 6.73	71.32 ± 6.55	1.11 ± 1.03	11.03± 0.95	7.28 ± 1.71	1.52 ± 0.56	41.50 ± 1.58	0.42 ± 0.10	46.99 ± 2.26	30.90 ± 5.23	1.55 ± 0.29
G. cf. *carpenteri*/ GbSa-6	79.59 ± 8.63	71.34 ± 9.25	1.12 ± 0.93	10.47± 1.49	6.67 ± 1.29	1.57 ± 1.16	40.00 ± 2.49	0.42 ± 0.07	47.61 ± 4.33	29.86 ± 4.15	1.61 ± 0.24

Depth (D), Width (W), Length (L), and ratio for measurement plate (L/W). Data expressed as means ± standard deviation. (n = 30).

2.3. *Biochemical Characterization*

2.3.1. Pigments Composition

To investigate the composition and concentration of pigments, samples were collected between days 24–25 of culture during the early exponential growth phase. As a result of this, nine pigments were identified in *Gambierdiscus* strains (Figures 5 and 6).

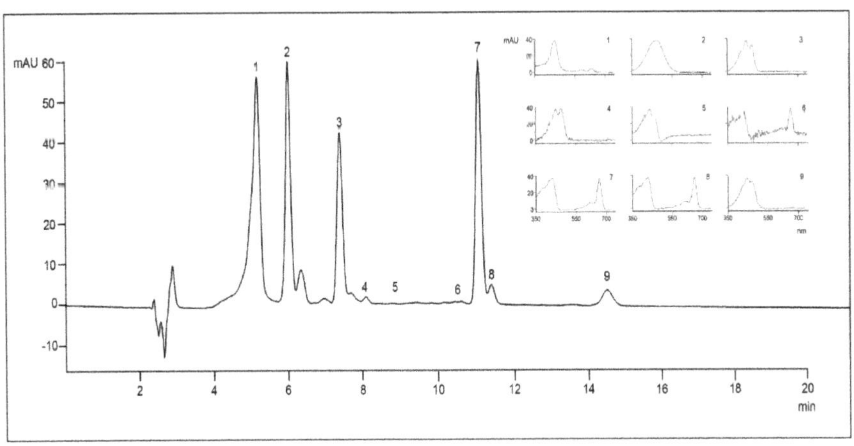

Figure 5. To investigate the composition and concentration of pigments, samples were collected between days 24 and 25 of culture during the early exponential growth phase. As a result of this, nine pigments were identified in *Gambierdiscus* strains (Figures 5 and 6).

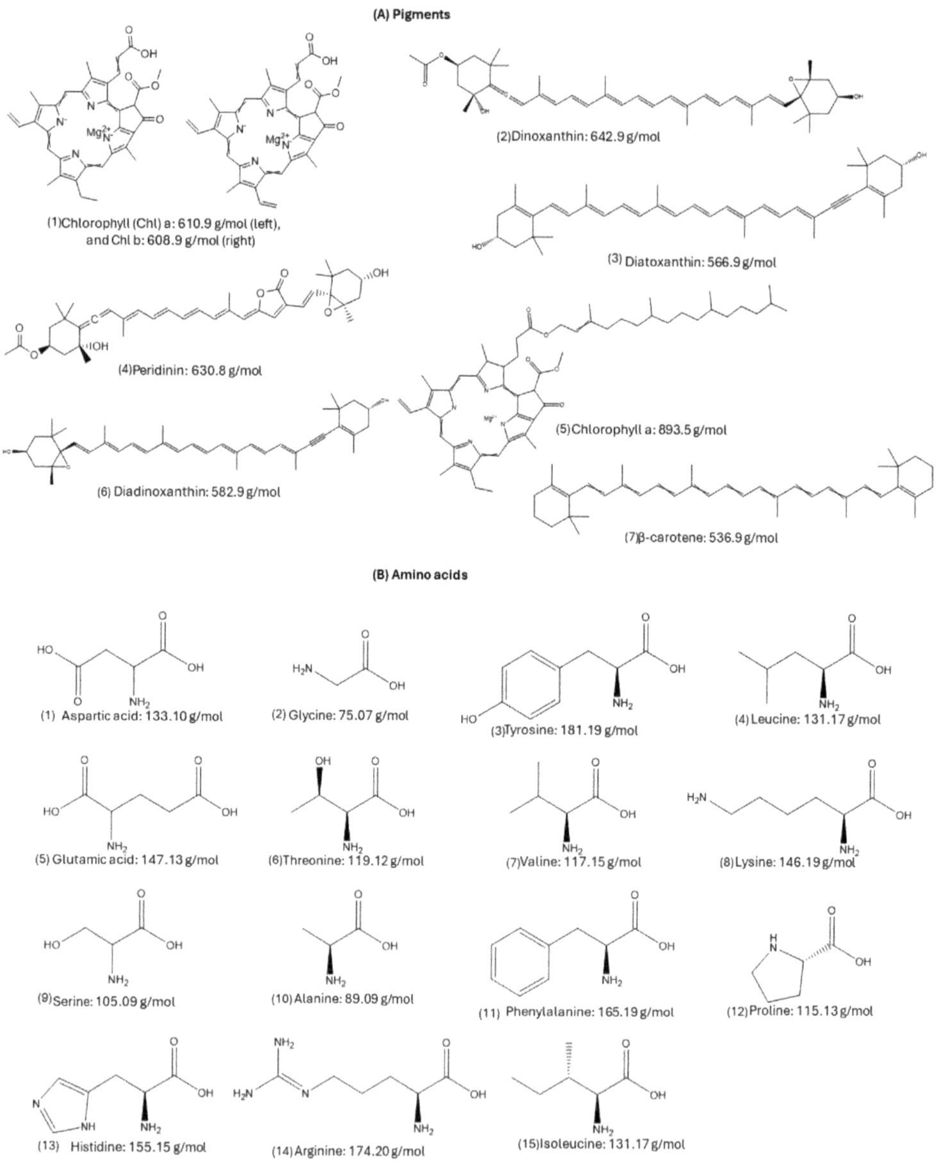

Figure 6. Principal pigments and amino acids detected in five strains of *Gambierdiscus* cf. *caribaeus* and *G.* cf. *carpenteri* in this study. (**A**) Pigments: (1) chlorophyll *c*1-*c*2, (2) Dinoxanthin, (3) diatoxanthin, (4) peridinin, (5) Chlorophyll a, (6) diadinoxanthin, and (7) β, β-carotene. (**B**) Amino acids: (1) aspartic acid, (2) glycine, (3) tyrosine, (4) leucine, (5) glutamic acid, (6) threonine, (7) valine, (8) lysine, (9) serine, (10) alanine, (11) phenylalanine, (12) proline, (13) histidine, (14) arginine, and (15) isoleucine.

Pigment production varied between the strains (Table 2). Peridinine was the main carotenoid pigment present in the *Gambierdiscus* strains (23.89–81.26 pg cell^{-1}). Chlorophyll *a* was the major pigment in *G.* cf. *carpenteri* GbSa-6 (278.9 pg·cell^{-1}), followed by chlorophyll *c*1, *c*2 (57.59 pg·cell^{-1}) in *G.* cf. *caribaeus* GbSa-2.

Table 2. Pigment content (pg·cell^{-1}) in *Gambierdiscus* species isolated from Bahía de La Paz, Gulf of California.

Species	Strain Code	Chl *a*	Chl *c2*	Diadino	Dino	Diato	Peri	ββ-Car
Gambierdiscus cf. *caribeus*	GbSa-2	149.38	57.79	26.41	2.05	26.41	81.26	6.02
	GbSa-7	203.79	37.08	25.77	4.71	3.67	55.73	6.45
Gambierdiscus cf. *carpenteri*	GbSa-4	152.41	34.24	21.10	0.00	3.90	49.47	3.99
	GbSa-5	124.95	22.88	15.70	3.28	3.26	34.42	4.10
	GbSa-6	278.69	30.96	15.99	4.74	5.67	23.89	12.85

2.3.2. Amino Acid Composition

The amino acid profile of *Gambierdiscus* was determined using HPLC-DAD. Fourteen amino acids were detected, including glutamic acid, aspartic acid, serine, glycine, threonine, alanine, arginine, tyrosine, valine, phenylalanine, isoleucine, leucine, lysine, and proline. Histidine was detected only in *G.* cf. *caribaeus* Gbsa-7 (Figure 7).

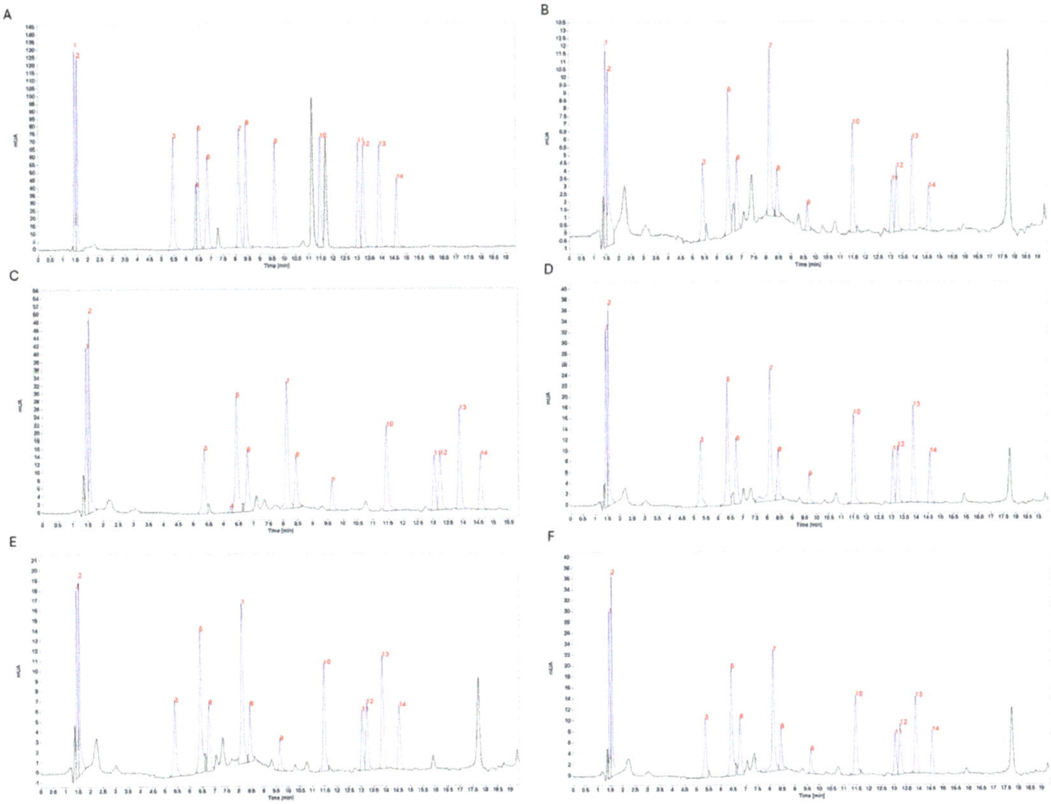

Figure 7. High-performance liquid chromatography (HPLC) chromatograms registered at 338 nm. (**A**) STD amino acids. Peak identification: (1) aspartic acid, (2) glutamic acid, (3) serine, (4) histidine, (5) glycine, (6) threonine, (7) alanine, (8) arginine, (9) tyrosine, (10) valine, (11) phenylalanine, (12) isoleucine, (13) leucine, and (14) lysine. *G.* cf. *caribaeus* strain (**B**) Gbsa-2, (**C**) Gbsa-7, *G.* cf. *carpenteri* strain (**D**) Gbsa-4, (**E**) Gbsa-5 and (**F**) Gbsa-6. The chromatograms for proline are shown in Figure S10.

The amino acid content varied between the strains (Table 3). Glutamic acid had the highest content (14.80–36.39 pg·cell^{-1}), whereas glycine had the lowest (7.72–18.66 pg·cell^{-1}). Strain GbSa-7 contained proline (<52.30 pg·cell^{-1}) and was the only strain to contain histidine.

Table 3. Concentrations of amino acids (pg·cell^{-1}) in *Gambierdiscus* species isolated from Bahía de La Paz, Gulf of California.

Species	Strain Code	Glu	Asp	Ser	His	Gly	Thr	Ala	Arg	Tyr	Val	Phe	Iso	Leu	Lys	Pro
Gambierdiscus cf. *caribeus*	Gbsa-2	17.41	15.85	13.52	Nd	8.31	14.95	12.84	17.70	18.71	15.20	17.36	16.53	16.24	15.22	17.78
	Gbsa-7	32.74	28.15	22.68	13.04	16.69	24.79	22.33	25.69	24.26	24.44	27.11	23.06	29.33	23.24	52.30
Gambierdiscus cf. *carpenteri*	Gbsa-4	36.39	30.93	26.31	Nd	18.66	28.91	27.53	30.63	29.47	28.48	30.68	27.27	32.20	26.58	18.10
	Gbsa-5	14.80	12.74	11.45	Nd	7.72	12.52	11.26	14.51	14.52	12.78	14.20	13.08	14.35	12.86	12.52
	Gbsa-6	30.94	27.14	21.12	Nd	15.04	23.83	22.13	25.23	24.93	23.26	24.86	23.13	25.26	21.99	24.14

Aspartic acid (Asp), glutamic acid (Glu), serine (Ser), histidine (His), glycine (Gly), threonine (Thr), alanine (Ala), arginine (Arg), tyrosine (Tyr), valine (Val), phenylalanine (Phe), Isoleucine (Iso) leucine (Leu), lysine (Lys), and Proline (Pro).

2.4. Toxicity Assay

2.4.1. Mouse Bioassay (MBA)

All *Gambierdiscus* strains tested by MBA were positive for CTX-like activity. The toxicity in *G.* cf. *caribaeus* GbSa-2 was 18.57 × 10^{-4} μg·kg^{-1}·1000 cells^{-1}, which was comparable to the toxicity of strain GbSa-7 (17.08 × 10^{-4} μg·kg^{-1}·1000 cells^{-1}). For *G.* cf. *carpenteri* strains, significant variations were observed. Strain GbSa-4 showed the highest toxicity (29.07 × 10^{-4} μg·kg^{-1}·1000 cells^{-1}), with CTXs-like activity three times higher compared to that of strain GbSa-5 (8.46 × 10^{-4} μg·kg^{-1}·1000 cells^{-1}) and more than twice that of strain GbSa-6 (12.54 × 10^{-4} μg·kg^{-1}·1000 cells^{-1}).

The main clinical signs related to CTXs-like activity were neurological: paralysis of the rear and front quarters, dyspnea, piloerection, spasms, loss of locomotion, body arching, arched and zig-zag tail movements, tail wriggling, blinking, and ataxia. Gastric signs included polyuria, lethargy due to peritoneal inflammation, diarrhea, and abundant defecation. Cardiovascular signs included labored breathing and vasoconstriction of the ear and other clinical signs like continuous itching and scratching, hypersalivation, closed eyes, and eventual death due to respiratory failure (Figure 8).

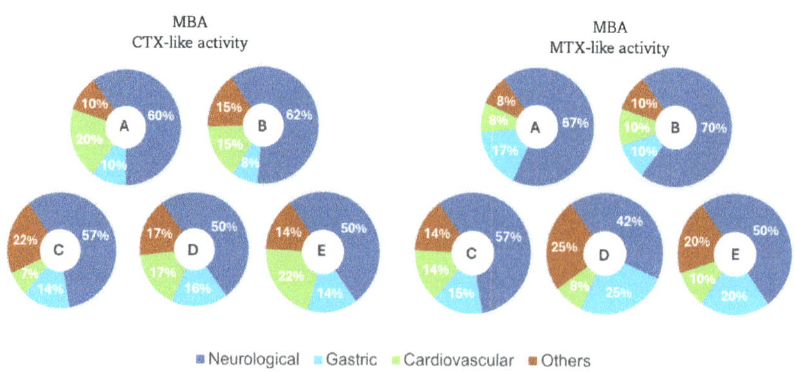

Figure 8. Left ring charts showed clinical signs recorded during the mouse bioassay (MBA) using CTX-like activity for *Gambierdiscus* cf. *caribaeus* strains (A) GbSa-2 and (B) GbSa-7 and *Gambierdiscus* cf. *carpenteri* strains (C) GbSa-4, (D) GbSa-5, and (E) GbSa-6. Right ring chart clinical signs recorded during the MBA using MTX-like activity of *Gambierdiscus* cf. *caribaeus* strains (A) GbSa-2 and (B) GbSa-7) and *Gambierdiscus* cf. *carpenteri* strains (C) GbSa-4, (D) GbSa-5, and (E) GbSa-6.

All strains of *G.* cf. *caribaeus* and *G.* cf. *carpenteri* showed toxicity related to MTX-like activity in the MBA at sublethal doses. After 24 h of observation, the LD_{50} was not recorded in any of the groups of injected mice, but several clinical signs related to MTX-like activity were observed.

The clinical signs associated with MTXs-like activity included neurological signs such as paralysis of the rear and front quarters, dyspnea, piloerection, spasms, body arching, arched tail and tail wriggling, blinking, ataxia, and convulsions. Gastric signs included polyuria, lethargy due to peritoneal inflammation, and abundant defecation. Cardiovascular signs included vasoconstriction of the ears. Other clinical signs included hypersalivation, closed eyes, and tear secretion (Figure 8).

2.4.2. *Artemia* Assay (ARTOX)

Toxicity was determined using the brine shrimp *Artemia salina*. The control group did not exhibit any mortality, whereas the organisms exposed to CTXs-like activity of *Gambierdiscus* strains showed mortality (Figure 9) and clinical signs of toxicity.

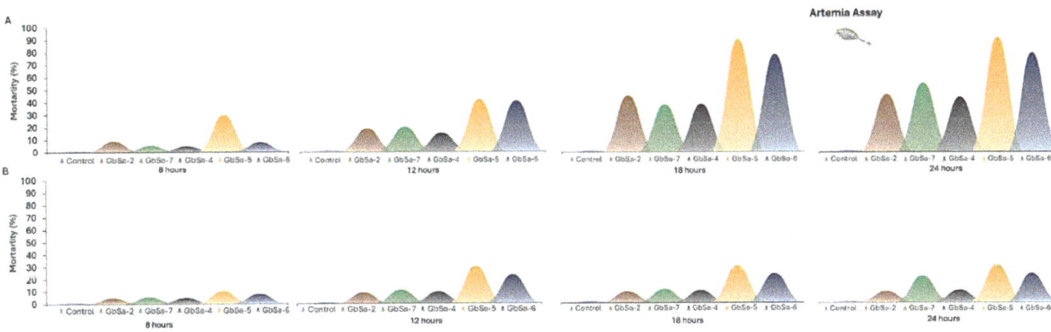

Figure 9. *Artemia salina* lethality bioassay. (**A**) Brine shrimp exposed to CTX-like activity of *Gambierdiscus* cf. *caribaeus* (GbSa-2 and GbSa-7) and *Gambierdiscus* cf. *carpenteri* (GbSa-4, GbSa-5, and GbSa-6). (**B**) Brine shrimp exposed to MTX-like activity of *Gambierdiscus* cf. *caribaeus* (GbSa-2 and GbSa-7) and *Gambierdiscus* cf. *carpenteri* (GbSa-4, GbSa-5, and GbSa-6).

At 6 h, the mortality of the brine shrimp exposed to the CTXs-like activity of *G.* cf. *caribaeus* strains GbSa-2 and GbSa-7 was 9 ± 1% and 5 ± 0.5%, respectively. The mortality caused by *G.* cf. *carpenteri* were 5 ± 1% for strain GbSa-4, 30 ± 3% for strain GbSa-5, and 8 ± 1% for strain GbSa-6.

After 12 h, *G.* cf. *caribaeus* GbSa-2 caused 18 ± 1% mortality, while strain GbSa-7 caused 14 ± 1%. The mortalities caused by *G.* cf. *carpenteri* in brine shrimp were 19 ± 1% for strain GbSa-4, 40 ± 3% for strain GbSa-5, and 39 ± 2% for strain GbSa-6.

The highest mortality in *A. salina* was recorded at 18 h of exposure to the CTXs-like activity of *G.* cf. *caribaeus* GbSa-2 (45 ± 3%) and GbSa-7 (38 ± 2%). For the *G.* cf. *carpenteri* extract, mortality was 38 ± 1% for strain GbSa-4, 90 ± 6% for strain GbSa-5, and 78 ± 5% for strain GbSa-6. This trend remained consistent until 24 h when the mortality of *A. salina* was 45 ± 3% for GbSa-2 and 43 ± 2% for GbSa-7. The highest mortality was observed for *G.* cf. *carpenteri* GbSa-5 (90 ± 12%), followed by GbSa-6 (78 ± 5%) and GbSa-4 (54 ± 3%) (Figure 9A).

Clinical signs in *A. salina* exposed to CTX-like activity were characterized by changes in swimming behavior. Starting at 6 h, *A. salina* exposed to *G.* cf. *caribaeus* and *G.* cf. *carpenteri* showed a loss of orientation while swimming, characterized by swimming in circles and rising and falling from the surface to the bottom. These clinical signs intensified between 6 and 24 h, leading to gasping, lethargy, and paralysis of the thoracic appendages.

For *A. salina* exposed to MTX-like activity, mortality was recorded at 6 h: 5 ± 0.5% and 5 ± 1% for strains GbSa-2 and GbSa-7 of *G.* cf. *caribaeus* and 5 ± 0.5%, 10 ± 1%, and

8 ± 5% for strains GbSa-4, GbSa-5, and GbSa-6 of *G.* cf. *carpenteri*, respectively. After 12 h of exposure to the MTX-like activity of *G.* cf. *caribaeus*, the mortality of *A. salina* was 9 ± 1% (GbSa-2) and 11 ± 1% (GbSa-7). At this time, the highest mortality was recorded with GbSa-5 (30 ± 4%) and GbSa-6 (23 ± 2%) from *G.* cf. *carpenteri*, while GbSa-4 showed 10 ± 2% mortality. At 18 h, *A. salina* mortality did not increase further in any of the treatments. At 24 h, *A. salina* mortality increased by 10% for those exposed to the MTXs-like activity of *G.* cf. *caribaeus* GbSa-7 (<22 ± 2%) (Figure 9B).

Although mortality due to MTX-like activity in *A. salina* was lower compared to that of CTX-like activity, several clinical signs were present from the first hour of exposure. Brine shrimp exhibited erratic swimming, lethargy, a head-up posture, reduced response to stimuli, and gasping within the first 12 h. By the end of the exposure, 80% of brine shrimp had recovered.

2.4.3. Marine Fish Assay

A toxicity test was conducted using longfin yellowtail larvae (*Seriola rivoliana*). During the first 6 h, only larvae exposed to the CTX-like activity of *G.* cf. *carpenteri* GbSa-4 presented 10 ± 1% mortality. After 12 h of treatment, *G.* cf. *caribaeus* GbSa-7 caused 10 ± 0.5% mortality. At 18 h, there was no further increase in mortality in any of the treatments. The maximum mortality was recorded at 24 h, with 20 ± 2% mortality in *S. rivoliana* larvae exposed to the CTXs-like activity of *G.* cf. *carpenteri* GbSa-4. No mortality was observed in the control group (Figure 10A).

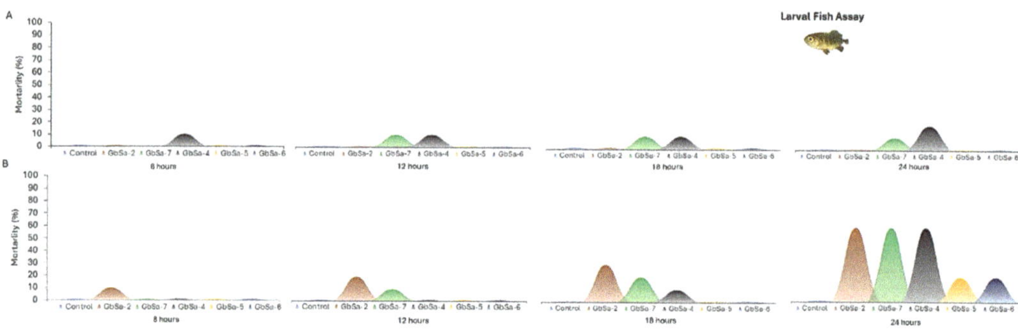

Figure 10. (**A**) Toxicity in *Seriola rivoliana* larvae exposed to the CTX-like activity of *Gambierdiscus* cf. *caribaeus* (GbSa-2 and GbSa-7) and *Gambierdiscus* cf. *carpenteri* (GbSa-4, GbSa-5, and GbSa-6). (**B**) Toxicity in larvae of *Seriola rivoliana* exposed to MTX-like activity of *Gambierdiscus* cf. *caribaeus* (GbSa-2 and GbSa-7) and *Gambierdiscus* cf. *carpenteri* (GbSa-4, GbSa-5, and GbSa-6).

Clinical signs recorded in *S. rivoliana* larvae exposed to CTXs-like activity included a loss of balance with abnormal horizontal orientation, abnormal swimming behavior, loss of buoyancy control leading to sinking to the bottom, reduced response to stimuli, erratic swimming, shaking, body arching, irregular ventilation (e.g., hyperventilation), and notable skin darkening.

Larvae exposed to the MTXs-like activity of *G.* cf. *caribaeus* GbSa-2 exhibited 10 ± 1% mortality at 6 h, which increased to 20 ± 2% at 12 h. Larvae exposed to the MTX-like activity of *G.* cf. *caribaeus* GbSa-7 showed 10 ± 1% mortality at 6 h. At 18 h, mortality due to GbSa-2 and GbSa-7 increased to 30 ± 3%. In contrast, larvae treated with the MTXs-like activity of *G.* cf. *carpenteri* GbSa-4 showed 10 ± 0.5% mortality. After 24 h, the maximum mortality of *S. rivoliana* larvae was 60 ± 4% for those exposed to the MTXs-like activity of *G.* cf. *caribaeus* GbSa-2 and GbSa-7 and *G.* cf. *carpenteri* GbSa-4. The lowest mortality was observed in larvae exposed to the MTXs-like activity of *G.* cf. *carpenteri* GbSa-5 and GbSa-6, with 20 ± 1% mortality (Figure 10B).

The exposure of larvae to dissolved MTX-like activity led to abnormal horizontal orientation, loss of balance, and loss of buoyancy control, causing the larvae to sink to the bottom. Abnormal vertical orientation with head-down posture was also observed, along with behavioral changes such as erratic swimming and pulses of hypoactivity and hyperactivity. Additional signs included shaking, body arching, abnormal respiration (e.g., hyperventilation), irregular ventilation, head shaking, and skin darkening before death.

2.5. Fluorescent Receptor Binding Assay (fRBA)

To measure the binding affinity of the CTX analogs for voltage-gated sodium channel (VGSC) receptors, we employed the fRBA assay. This assay is based on the competition for VGSC site **5** between CTXs and a brevetoxin (PbTx) labeled with fluorescent compounds. The relative fluorescence unit (RFU), defined as the fluorescence relative to the culture biomass, was adjusted to the control (ethanol 100% binding value, representing maximum fluorescence). For the highest concentration of the standard curve, the PbTx standard at 0.1 $\mu g \cdot mL^{-1}$ showed 29.8 $\pm$ 4.95% RFU. The *Karenia brevis* (Kb-3) extract (positive for PbTx-2 and Pbtx-3) resulted in 95.15 $\pm$ 2.10% RFU, with a concentration of 1.3 $\pm$ 0.07 pg eq PbTx per cell. For *G.* cf. *caribaeus* GbSa-2, the blinding value was 95.6 $\pm$ 0.89% RFU with a maximum concentration of 2.59 $\pm$ 0.25 pg eq PbTx per cell. Strain GbSa-7 showed a binding value of 92.8 $\pm$ 1.01% RFU with a concentration of 2.19 $\pm$ 0.23 pg eq PbTx per cell.

The affinity for VGCS in *G.* cf. *carpenteri* GbSa-4 was 93.9 $\pm$ 1.27% RFU, with a maximum concentration of 2.51 $\pm$ 0.30 pg eq PbTx per cell. The highest competition for VGCS was recorded for strains GbSa-5 (89.8 $\pm$ 2.50% RFU and 1.15 $\pm$ 0.31 pg eq PbTx per cell) and GbSa-6 (95.8 $\pm$ 3.49% RFU and 1.57 $\pm$ 0.05 pg eq PbTx per cell).

3. Discussion

3.1. Morphological Identification

Two out of the five isolated strains were identified as *G.* cf. *caribaeus*, and three of them were identified as *G.* cf. *carpenteri*, as both species share similarities in the plate formula Po, 3′, 7″, 6C, 6S, 5‴, 1p, and 2⁗ (Figures 1–4). However, they can be distinguished by the shape of plates 1p and 4″, as suggested by the dichotomous key of Litaker et al. [39] and the description of Diaz-Ascencio et al. [40]. *Gambierdiscus* cf. *caribaeus* (GbSa-2 and GbSa-7) is characterized by a rounded and lenticular-shaped cell, short and wide plate 1p, and symmetrical plate 4″, which is consistent with the original descriptions by Litaker et al. [39] and Soler-Onis et al. [41] in Atlantic waters and Diaz-Ascencio et al. [42] in the Caribbean region (Figure 2). Although *G. caribaeus* shares morphological features with its sister species, *G. carpenteri* and *G. jejuensis*, the presence of a symmetrical plate 4″ allows for satisfactory differentiation. However, under culture conditions, this is not always reliable, as cellular deformation occurs during the initial generations, leading to aberrant cell shapes that complicate the identification of the typical plates characterizing the genus *Gambierdiscus* [40,42,43]. In this study, the number of deformed cells in *G.* cf. *caribaeus* strains was low, and the morphological features aligned with previous descriptions, suggesting that measurements were accurate.

Strains of *G.* cf. *carpenteri* (GbSa-4, GbSa-5, and GbSa-6) from the Gulf of California exhibited clear morphological characteristics similar to those described by Litaker et al. [39]. These included oval cells, a narrower and elongated plate 1p extending toward plate 3⁗, and an asymmetric plate 4″ (Figure 3C–F). Cell size was similar to those of the strains reported for *G. carpenteri* isolated from Australia, Belize, the Mexican Caribbean, and the Philippines [39,44,45]. However, variations were observed in the shape of plate 2′, which ranged from rectangular to hatchet-shaped. Additionally, the *G.* cf. *carpenteri* strains described in this study presented several morphological characteristics similar to those described by Ramos-Santiago et al. [38] for *G. carpenteri* in the Gulf of California, with plates 2′ and 4″ corresponding to those described by Litaker et al. [39,44]. Furthermore, in the description by Litaker [39], *G. carpenteri* cells presented a rostrum resulting from the fusion of plates 3′ and 4″ and postcingular plates 3‴ and 4‴, which was similarly observed in *G.* cf. *carpenteri* (SEM GbSa-4 and GbSa-6, n = 14) (Figure 4A–F and Figure S6). However, the cellular morphology of *G.* cf. *carpenteri* (GbSa-4 and GbSa-6) was more consistent with the descriptions by Kohli et al. [44] and Ramos-Santiago et al. [38]. The variety of forms among the strains isolated from the same area supports the diversity and morphological plasticity of *G. carpenteri*, suggesting that the diverse coastal environments of the Gulf of California may harbor new ecotypes of this species.

Plate dissociation allowed for precise measurements and detailed observations using LM, which confirmed the formula tabulation [39]. Through plate dissociation, it was possible to support the morphological distinction between *G.* cf. *caribaeus* and *G.* cf. *carpenteri* and to perform more precise measurements, particularly in plate 1p. Differences in the ratio were observed between *G.* cf. *caribaeus* and *G.* cf. *carpenteri* (Table 1), although the diversity in the shape of plate 2′ did not correspond to the ratio of the 2′/4″ suture length and 2′/2″ suture length between the two species, as suggested by Diaz-Asencio et al. [40], Vacarizas et al. [45], and Litaker et al. [39]. Plate dissociation was performed to compare the shape and size of plates 1′ and 3′, leading to the suggestion of adding the shape of plate 4″ to the tabulation to distinguish *G.* cf. *caribaeus* from *G.* cf. *carpenteri* (Figure 11). Plate 1′ was more compact and robust in *G.* cf. *caribaeus*, while plate 3′ was wider in *G.* cf. *carpenteri* than in *G.* cf. *caribaeus*, including in deformed cells (Figure 11). The asymmetric plate 4″ was similar in all *G.* cf. *carpenteri* strains, and upon dissociation, it was compared with fixed cells of *G. autrales* (VGO1250), *G. excentricus* (VGO911), and *G. carolineanus* (GbEs-4), which also exhibited an asymmetrical plate 4′. In *G.* cf. *carpenteri*, the width contrast in the 3″/4″ suture was twice as small as that in the 4″/5″ suture, whereas in *G. autrales*, *G. excentricus*, and *G. carolineanus*, the proportion was lower (1.5). Regarding the forms of plate Po, only the number of pores showed differences between *G.* cf. *caribaeus* and *G.* cf. *carpenteri*. Therefore, we recommend performing plate dissociation to morphologically distinguish *G. caribaeus*, *G. carpenteri*, and *G. carolineanus* (Figure 11 and Figure S7).

Gambierdiscus caribeaus and *G. carpenteri* are two species within the genus that present a cosmopolitan distribution [46]. They exhibit a wide range of tolerance to variations in temperature, salinity, and irradiance [44,46], which contributes to their broad distribution and ability to adapt to diverse habitats, as well as their efficient growth rates [47–49]. In this study, *G.* cf. *caribaeus* and *G.* cf. *carpenteri* showed the highest cell growth under culture conditions, with average growth rates of 0.25 div/day, similar to those reported in previous studies under optimal salinity (35) and temperature (25 °C) conditions [45,50]. However, *G.* cf. *carpenteri* strain GbSa-5 exhibited a slower growth rate and lower cell abundance compared with the GbSa-4 and GbSa-6 strains.

Although the toxicity levels of *G. carpenteri* are considered medium-to-low and their contribution to CFP is uncertain [46], it is important to consider the potential implications of the presence of this species in the Gulf of California. Various factors have been identified that could increase the risk of CFP cases [1,46,51], including climate change, coastal area modifications, and increased nutrient and pollution loads. Recently, Bahía de La Paz has been impacted by tropical storms and hurricanes, leaving numerous boats at the bottom of the sea in coastal areas and nearby islands [52,53]. Some studies have suggested that the abundance of *Gambierdiscus* spp. does not depend on the presence of macroalgae and that certain species are associated with artificial substrates [49,51,54–56]. Consequently, the remains of these vessels may represent substrates that promote the potential settlement and proliferation of *Gambierdiscus* and *Fukuyoa* species associated with CFP.

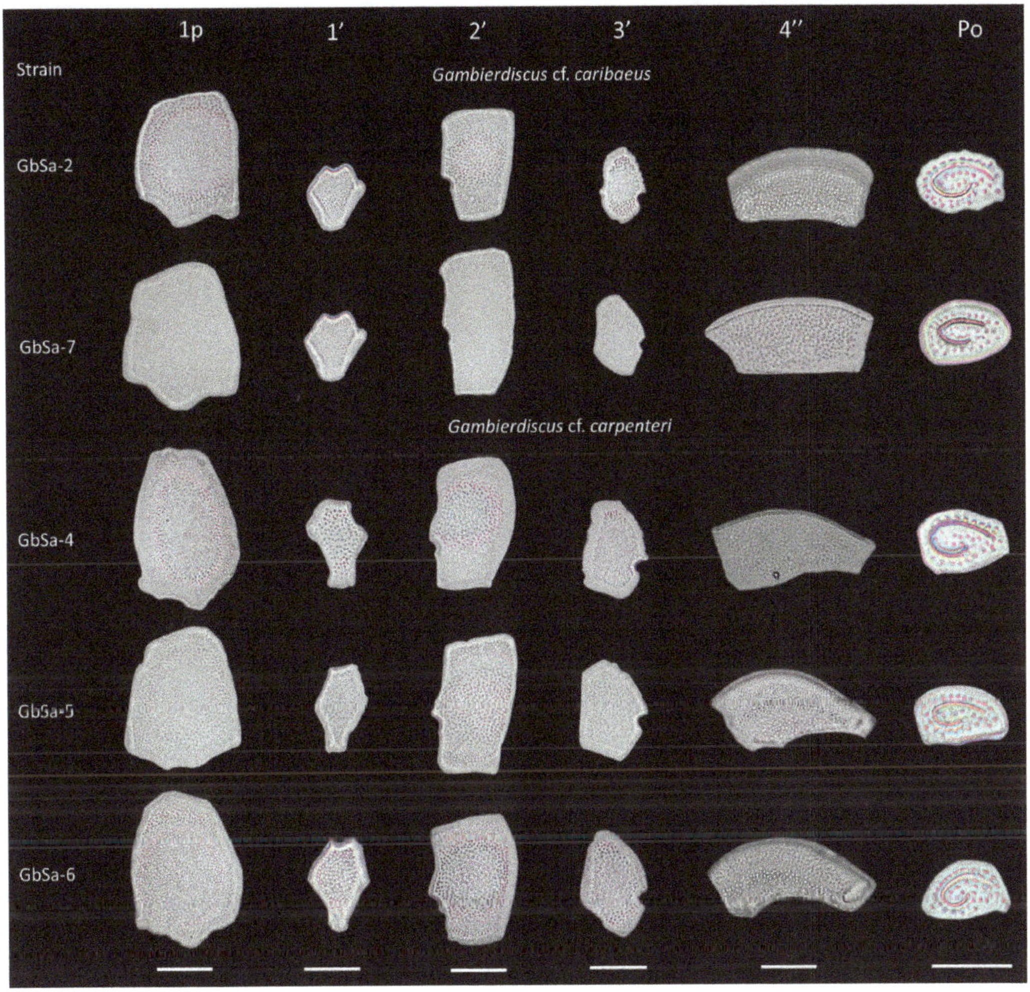

Figure 11. Light microscopy of typical morphology plate 1p, 1′, 2′, 3′, 4″, Po plate under present in *Gambierdiscus* cf. *caribaeus* (GbSa-2, GbSa-7) and *Gambierdiscus* cf. *carpenteri* (GbSa-4, GbSa-5, GbSa-6). Scale bar: 20 µm.

At sampling points near the coast, *G.* cf. *carpenteri* was the dominant species, while the highest diversity was recorded at Fang Ming, a sunken ship converted into an artificial reef near Espiritu Santo Island (Table S2, Figure 12), in addition to *G.* cf. *carpenteri*, *G.* cf. *carolineanus*, and *F.* cf. *yasumotoi* (but not *G.* cf. *caribaeus*). The Gulf of California has 244 islands, islets, and coastal areas [57]. Species of the genera *Gambierdiscus* and *Fukuyoa* are closely associated with the insular areas of the Gulf of California. Morquecho-Escamilla et al. [37] reported the presence of *Gambierdiscus* and *Fukuyoa* on San Jose Island, and Ramos-Santiago et al. [38] reported their presence on La Gaviota Island. Including this study, reports of *Gambierdiscus* and *Fukuyoa* cover only 1.2% of Gulf of California islands, suggesting that these islands are potential habitats for a great diversity of *Gambierdiscus* and *Fukuyoa* species, including novel species.

3.2. Biochemical Characterization

3.2.1. Pigments Composition by HPLC-DAD

Pigment analysis has been proposed not only for physiological studies of dinoflagellates [58–61] but also as a pigment fingerprint for monitoring harmful microalgae species during harmful algal blooms (HABs) [62–67]. Using thin-layer chromatography (TLC), Durand and Berkaloff [68] and Indelicato and Watson [69] conducted the first analyses of pigments produced by the genus *Gambierdiscus*, identifying the presence of Chl *a*, Chl *c*1, Chl *c*2, Diadino, Dino, Peri, and ββ-car. Since then, at least ten studies have been conducted to further understand pigment production and composition [63,70–73] (Table 4). Recently, Malto et al. [27] and Yon et al. [28] described the presence of these compounds using liquid chromatography coupled with mass spectrometry as part of a study on the metabolome of three strains belonging to three species of *Gambierdiscus* isolated from the Philippines and 15 strains of five species from the Atlantic Ocean. The analysis and evaluation of the pigment profile, along with other metabolites (e.g., fatty acids, sterols, carbohydrates, amino acids, and toxins), in various species of dinoflagellates allows for the chemical diversity of secondary metabolites to be explored. Biomarkers can then be identified to differentiate between *Gambierdiscus* species (chemotaxonomy), evaluate the possible presence of ecotypes, and study the metabolism and biosynthesis of toxins. In this study, we detected the presence of seven pigments in *G.* cf. *caribaeus* and *G.* cf. *carpenteri*, and the profiles obtained were compared globally with those from other isolates of species within this genus (Table 4; Figure 6).

Table 4. Pigment composition detected in isolates of the genus *Gambierdiscus*.

Specie/Strain	Pigments	Method	Reference
Gambierdiscus toxicus Gambier Island	Chl *c*1, Chl *c*2, and Peri	TLC	[68]
Gambierdiscus toxicus Southern coast Florida	Chl *a*, Chl *c*1, Chl *c*2, Diadino, Dino, Peri, and ββ-car	TLC	[69]
Gambierdiscus toxicus Fourteen clones: Bermuda Great Issacs Light, Bahamas; Drifth Algae, Gulf Stream; Gingerbreads, Bahamas; Hawaii; Marathon key, Fl; Martnique, Caribbean; Virgin Gorda, V.I.	Chl *a*, Chl *c*2, Peri, Diadino, and Dino	TLC, HPLC	[70]
Gambierdiscus toxicus Caribean clone	Chl *a*	NMR	[71]
Gambierdiscus toxicus Three clones:	Chl *a*	HPLC-DAD	[72]
Gambierdiscus excentricus Canary Islands (NE Atlantic Ocean)	Chl *a*, Chl *a* allomer, Chl *c*1, Chl *c*2, Diadchr, Diadino, Dino, Diato, (MgDVP), Peri, Perid-ol, Peri-like, Pyrrho, and ββ-car	HPLC-DAD	[73]
Gambierdiscus sp.	Chl *a*, Chl *c*1, Chl *c*2, Diadino, Dino, and Peri	HPLC-DAD	[63]
Gambierdiscus cf. *caribaeus*	Chl *a*, Chl *c*2, Diadino, Dino, Diato, Peri; and ββ-car	HPLC-DAD	This study
Gambierdiscus cf. *carpenteri* Five clones: Bahía de La Paz, Gulf of California	Chl *a*, Chl *c*2, Diadino, Dino, Diato, Peri; and ββ-car	HPLC-DAD	This study

Chlorophyll *a* (Chl *a*), Chlorophyll *c*1 (Ch *c*1), Chlorophyll *c*2, Peridine (Peri), diadinoxantin (Diadino), dinoxantina (Dino), β, β-carotene (ββ-car), Chrophyll *a* allomer (Chl *a* allomer), diatoxanthin (Diato), diadinochrome (Diadchr), divinyl protochlorophyllide (MgDVP), peridinin-like Peridininol (Perid-ol), peridinin-like (Peri-like), and pyrrhoxanthin (Pyrrho).

The predominant pigment profile for these two species included Chl *a*, Chl *c*2, Diadino, Dino, Diato, Peri, and ββ-car. Notably, we observed that strain GbSa-7 was distinct from strain GbSa-2; both corresponded to *G.* cf. *caribaeus*, as it produced a higher concentration of Chl *a* (203.79 pg·cell^{-1}) and Dino (4.71 pg·cell^{-1}). This effect was also observed in the amino acids profile, with histidine (13.04 pg·cell^{-1}) detected only in this strain, along with the highest concentration of proline (52.30 pg·cell^{-1}; Table 3 and Figure 6). Therefore, we consider that this strain may represent an ecotype. This hypothesis will be further supported by an analysis of the toxin profile and other metabolites (e.g., fatty acids, sterols, and carbohydrates) using analytical methods.

3.2.2. Amino acid Composition

Analyzing the amino acid profiles of dinoflagellates is useful for exploring the chemical diversity of secondary metabolites; these profiles can serve as biomarkers for species identification in chemotaxonomy. They also facilitate the study of ecotypes, metabolism, and toxin biosynthesis. This analysis also allows for assessments of intracellular changes in response to diurnal variations in light, nitrogen, and nutrients, as well as the physiological status of natural dinoflagellate populations [74,75].

The amino acids profile of *G.* cf. *caribaeus* and *G.* cf. *carpenteri* strains revealed the presence of 14 amino acids: glutamic acid, aspartic acid, serine, glycine, threonine, alanine, arginine, tyrosine, valine, phenylalanine, isoleucine, leucine, lysine, and proline. *Gambierdiscus.* cf. *caribaeus* GbSa-7 was the only strain that presented histidine (13.04 pg·cell^{-1}), and it also had the highest concentration of proline (52.30 pg·cell-1; Table 4 and Figure 6). Therefore, we considered that this strain may be a potential ecotype. Figure 6 provides a comparative analysis of the amino acid profile of *Gambierdiscus* detected in this study with those of other dinoflagellate species.

Studies by Okaichi et al. [76] and Hayashi [77], who used high-speed amino acid analyzers (HSAAA), in addition to the more recent analyses using high-performance liquid chromatography with fluorescence detection (HPLC-FLD) and HPLC-DAD have elucidated the amino acid profiles of various dinoflagellate species from the genera *Prorocentrum*, *Noctiluca*, *Aureodinium*, *Glenodinium*, *Heterocapsa*, *Scrippsiella*, *Alexandrium*, *Gymnodinium*, *Ansanella*, and *Takayama* [74,78]. Recently, Yon et al. [28] also described the presence of these compounds using high-performance liquid chromatography-mass spectrometry (HPLC-MS) as part of the metabolome of 15 strains from five species of *Gambierdiscus* originating from the Atlantic Ocean. Furthermore, for the first time, this study presents a screening of the amino acid profile of *Gambierdiscus* from the Pacific Ocean (Table 5).

Table 5. Amino acid composition detected in diverse species of dinoflagellates.

Dinoflagellate/Strain	Amino Acids	Method	Family	Reference
Prorocentrum triestinum	Asp, Ser, Thr, Glu, Pro, Gly, Ala, Cys, Val, Met, Iso, Leu, Tyr, Phe, Lys, His, Arg, Try		Prorocentraceae	[76]
Prorocentrum minimum, *Noctiluca scintillans*	His, Thr, Arg, Val, Phe, Iso; Leu, Lys, Asp, Glu, Ser, Gly, Ala, Tyr, Cys, Met, Try, and Pro	HSAA	Prorocentraceae, Noctilucaceae	[77]
Aureodinium pigmentosum, *Prorocentrum micans*, *Glenodinium foliaceum*, *Heterocapsa triquetra*, *Scrippsiella trochoidea*, *Alexandrium tamarense* y *Gymnodinium catenatum*	Glu, Gln, Arg, Tau, Gin/Glu	HPLC		[74]

Table 5. *Cont.*

Dinoflagellate/Strain	Amino Acids	Method	Family	Reference
Heterocapsa rotundata Shiwa Bay, Korea	His, Thr, Arg, Val, Phe, Iso; Leu, Lys, Asp, Glu, Ser, Gly, Ala, Tau, GABA, Tyr, and Pro	HPLC-FLD	Heterocapsaceae	[78]
Ansanella granifera Shiwa Bay, Korea	His, Thr, Arg, Val, Met, Phe, Iso; Leu, Lys, Asp, Glu, Ser, Gly, Ala, Tau, GABA, Tyr, and Pro		Suessiaceae	
Alexandrium andersonii Jinhae Bay, Korea	His, Thr, Arg, Val, Met, Phe, Iso; Leu, Lys, Asp, Glu, Ser, Gly, Ala, Tau, GABA, Tyr, and Pro		Ostreopsidaceae	
Takayama tasmanica Glenhaven, New Zeland	His, Thr, Arg, Val, Met, Phe, Iso; Leu, Lys, Asp, Glu, Ser, Gly, Ala, Tyr, and Pro		Brachidiniaceae	
Tanayama helix Tasman Sea, Australia	His, Thr, Arg, Val, Met, Phe, Iso; Leu, Lys, Asp, Glu, Ser, Gly, Ala, Tyr, and Pro		Brachidiniaceae	
Gymnodnium smaydae Shiwa Bay, Korea	His, Thr, Arg, Val, Met, Phe, Iso; Leu, Lys, Asp, Glu, Ser, Gly, Ala, Tau, GABA, Tyr, and Pro		Gymnodinaceae	
Gambierdiscus cf. *caribaeus*	His, Thr, Arg, Val, Phe, Iso; Leu, Lys, Asp, Glu, Ser, Gly, Ala, Tyr, and Pro	HPLC-DAD	Ostreopsidaceae	This study
Gambierdiscus cf. *carpenteri* Five clones: Bahía de La Paz, Gulf of California	His, Thr, Arg, Val, Phe, Iso; Leu, Lys, Asp, Glu, Ser, Gly, Ala, Tyr, and Pro	HPLC-DAD	Ostreopsidaceae	This study

Cysthine (Cys), histidine (His), threonine (Thr), arginine (Arg), valine (Val), methionine (Met), phenylalanine (Phe), isoleucine (Iso), leucine (Leu), lysine (Lys), aspartic acid (Asp), glutamic acid (Glu), glutamine (Gln), serine (Ser), glycine (Gly), alanine (Ala), taurine (Tau), gamma-aminobutyric acid (GABA), tryptophan (Try), tyrosine (Tyr), and proline (Pro). HSAAA: High-speed amino acid analyzer; HPLC-FLD: high-performance liquid chromatography with fluorescence detection.

3.3. Toxicity Assay

CTX-like and MTX-like activity was evaluated using four different assays: *Artemia* assay (ARTOX), mouse bioassay (MBA), marine fish assay (MFA) and fluorescent receptor binding assay (*f*RBA). Toxicity related to CTX-like and MTX-like activity was confirmed in all five clones, although the effects varied across the different toxicological assays.

3.3.1. Mouse Bioassay (MBA)

CTXs-like toxicity was evident in the MBA (18.57×10^{-4} $\mu g \cdot kg^{-1} \cdot 1000\ cell^{-1}$ and 17.08×10^{-4} $\mu g \cdot kg^{-1} \cdot 1000\ cell^{-1}$ in *G.* cf. *caribaeus* GbSa-2 and GbSa-7, respectively, and 29.07×10^{-4} $\mu g \cdot kg^{-1} \cdot 1000\ cell^{-1}$, 8.46×10^{-4} $\mu g \cdot kg^{-1} \cdot 1000\ cell^{-1}$, and 12.54×10^{-4} $\mu g \cdot kg^{-1} \cdot 1000\ cell^{-1}$ in *G.* cf. *carpenteri* GbSa-4, GbSa-5, and GbSa-6, respectively). Clinical signs were also observed for MTX-like activity; however, lethality was only associated with CTX-like activity. The predominant clinical signs were neurological (50–62%), followed by others (10–22%), cardiovascular (7–22%), and gastric (8–16%) across all analyzed strains. In the case of MTX-like activity, neurotoxicity signs predominated (42–70%), followed by gastric (10–25%), others (8–25%), and cardiovascular (8–14%) (Figure 8). These clinical signs were consistent with those described for these types of toxins [79,80]. When comparing the effects of these two types of toxins detected in strains of *G.* cf. *caribaeus* and *G.* cf. *carpenteri*, a higher percentage of neurotoxic clinical signs was observed for MTX-like activity, although it was not lethal in this assay.

3.3.2. Artemia Assay (ARTOX)

In ARTOX, toxicity was also higher for CTX-like activity compared to that of MTX-like activity (Figure 9A,B). The highest mortality was recorded at 18 h of exposure to the CTXs-like activity of *G.* cf. *caribaeus* (GbSa-2: 45 ± 3% and GbSa-7: 38 ± 2%) compared to those of the three strains of *G.* cf. *carpenteri* (GbSa-4: 38 ± 1%; GbSa-5: 90 ± 6%; GbSa-6: 78 ± 5%). Clinical signs in adult brine shrimp exposed to CTX-like activity were characterized by changes in swimming behavior, starting at 6 h in *G.* cf. *caribaeus* and *G.* cf. *carpenteri*. These changes included loss of swimming orientation characterized by circular swimming and rising and falling from the surface to the bottom, which intensified between 6 and 24 h, with gasping, lethargy, and paralysis of the thoracic appendages. These clinical signs were consistent with those described for this type of toxin in fish extracts and *Gambierdiscus* and *Fukuyoa* extracts [81,82]. Although mortality associated with MTX-like activity was lower, several clinical signs were observed, such as erratic swimming, lethargy, head-up posture, under-reactive response to stimuli, and gasping within the first 12 h. By the end of the exposure, 80% of the brine shrimp had recovered. The toxic effects of some species of benthic dinoflagellate species, such as *Prorocentrum*, *Ostreopsis*, *Gambierdiscus*, and *Fukuyoa*, have been previously demonstrated by Neves et al. [83], Heredia-Tapia et al. [84], Ajuzie [85], Faimali et al. [86], and Gong et al. [87]. Additionally, *Gambierdiscus* and *Fukuyoa* have been used as CTXs-like and MTXs-like bio-vehicles for metabolic studies of these polyether toxins in marine fish [88,89].

3.3.3. Marine Fish Assay

The MFA elucidated an opposite effect to that of the MBA, as MTX-like activity was highly ichthyotoxic for *S. rivoliana* larvae. The toxicological effects of CTXs-like activity produced by *Gambierdiscus* in fish (larvae, juveniles, and adults) have been evaluated under controlled conditions in various freshwater and marine species using different methods, such as feeding with contaminated fish pellets, dinoflagellates that produce these toxins, or a bio-vehicle like brine shrimp, as well as injections, and in some cases, dissolved toxins in seawater [89–92]. Lewis [93] described CTXs as potent ichthyotoxins. Pure ciguatoxin-1 (CTX-1) and CTX-2 added to water containing *Gambusia affinis* induced similar clinical signs, including pronounced opercular movement and uncoordinated swimming, preceding death.

Previous studies have reported the negative effects of CTXs-like activity (ranging from behavioral changes to toxicogenomic effects) on fish embryos, larvae, juveniles, and adults of marine and freshwater species (e.g., *Thalassoma bifasciatum*, *Epinephelus fulvus*, *Lutjanus apodus*, *L. mahogoni*, *Micropterus salmoiedes*, *Oryzias melastigma*, *Oryzias latipes*, *Gambusia affinis*, *Serranus cabrilla*, and *Pterois volitans*). These effects include skin color variations, hypoactivity, loss of equilibrium, erratic swimming, jerky feeding movements, loss of orientation, and death, as well as induced hyperkinetic twitching, severe spinal deformities, and an inability to feed [90–106].

Feeding cichlid fishes (*Oreochromis niloticus* and *O. mossambicus*) with *Gambierdiscus* causes balance abnormalities and erratic spiral swimming behavior [99]. Other studies under controlled laboratory conditions have also shown that feeding with different concentrations of *Gambierdiscus* causes changes in coloration, inactivity, adverse effects on swimming and orientation, an inability to eat, and, in some cases, death in reef fish species such as *Thalassoma bifasciatum*, *Acanthurus bahianus*, and *A. chirurgis* [91,92,94,98]. *Gambierdiscus* also causes color modifications and behavioral changes in *Chromis chromis* and a decrease in liver size and coloration of this organ in *Serranus cabrilla* [101]. Purified extracts of toxins from this dinoflagellate caused changes in Na^+ content in the muscle of *Chelon labrosus* [95], epithelial degeneration of the intestine, dilation of the gill filaments in *Pomacen-*

trus wardi and *Chromis nitida* [96], and mortality in *Gambusia affinis* [93]. Galarza et al. [107] described the effect of CTXs (obtained from barracuda) on the aggregation of pigment granules in the melanophores of teleost fish. A detailed pathological study using transmission electron microscopy [101] demonstrated structural alterations in the liver of the cabrilla *Serranus cabrilla* related to the ingestion of *Gambierdiscus* in the diet. Marked intercellular fibrosis, a decrease in the number and length of microvilli in the sinusoidal and canalicular areas of the hepatocytes, lipid accumulation, a high number of binucleated cells, nucleolar disorganization, loss of condensed heterochromatin in the periphery of the nucleus, nuclear pyknosis, dilated rough endoplasmic reticulum, and dilated mitochondria were observed.

Although MTXs-like toxins are among the most potent ichthyotoxins [108] and some of the largest and most powerful toxins known [109,110], their accumulation in fish and their ichthyotoxicity has not been extensively evaluated [108,111,112]. MTX was first found in the viscera of the surgeon fish *Ctenochaeus straiatus*, which bears the Tahitian name "maito" and has been occasionally associated with CFP [110]. The amphiphilic nature of MTX, derived from the uneven distribution of polar functional groups in the molecule, is thought to play an important role in its interaction with cell membranes, contributing to its potent activity [108,110,113,114].

MTXs, which are also produced by *Gambierdiscus*, have been described for their negative effects on the gills of the freshwater white cloud mountain fish *Tanichthys albonubes*. MTX caused marked vacuolization of the cytoplasm of chloride cells in the epithelium. Additionally, marked apoptosis was observed not only in the gill epithelial cells but also in all mesenchymal cells of the gill filaments, except in the cartilage in the terminal state [115].

Igarashi et al. [108] evaluated the ichthyotoxic activity of MTX added to water in *T. albonubes*. The LC_{50} of MTX in these fishes in Ca^{2+}-free media (pH 8) was 5 nM, but this value decreased significantly to 3 pMa after the pH increased to 8 and the Ca^{2+} concentration increased to 2 mM. Therefore, the authors suggested that in a marine environment, MTX is 2000 times more toxic to fish than brevetoxin-3 (PbTx-3), highlighting the need for special attention to be paid to the extraordinarily high ichthyotoxicity of MTX.

Kholi et al. [111] described the accumulation of MTXs under controlled conditions in carnivorous marine fish, using *Gambierdiscus* as a source of these toxins. Juvenile mullet (*Aldrichetta forsteri*) were fed pellets containing *G. australes*; these were also used as food for the snapper *Pagrus auratus*. The transfer of MTX-1 was confirmed by LC-MS/MS detection in viscera, liver, and muscle samples.

In this study, *S. rivoliana* larvae exposed to dissolved MTX-like activity presented clinical signs after 6 h after exposure. MTX-like activity was highly ichthyotoxic for *S. rivoliana* larvae, which exhibited changes such as abnormal horizontal position, loss of balance, loss of buoyancy, control leading to sinking to the bottom, abnormal vertical orientation with a head-down posture, behavioral changes in swimming, constant pulses of hypoactivity and hyperactivity, erratic movements, shaking, body arching, abnormal respiration activity, hyperventilation, irregular ventilation, head shaking, and notable skin darkening before death. After 24 h of exposure, the maximum mortality of *S. rivoliana* larvae was 60 ± 4% in those exposed to *G.* cf. *caribaeus* (GbSa-2 and GbSa-7) and *G.* cf. *carpenteri* (GbSa-4). The lowest mortality (20 ± 1%) was recorded in the groups of larvae exposed to MTX-like activity from *G.* cf. *carpenteri* (GbSa-5 and GbSa-6) (Figure 9B). However, after 24 h of the acute test, the surviving larvae were moribund, with most of them found at the bottom.

During epizootics and mass mortality events of reef fish (*Pomacanthus paru*, *P. arcuatus*, *Holocanthus tricolor*, *Centropyge argi*, *Scarus taeniopterus*, *Chromis cyaneus*, *Diodon holocanthus*, *Cantherhines macroceros*, *Acanthurus chirurgis*, *Chaetodon sedentarius*, and *C. capistratus*) observed in Florida and the Caribbean, Ladsberg [116] suggested a possible association with benthic HABs caused by dinoflagellates of the genera *Gambierdiscus*, *Prorocentrum*, and *Ostreopsis* and cyanobacteria of the genus *Lyngbya* as potential etiologies of these events. In Cabo Pulmo in the Gulf of California, Nuñez-Vazquez et al. [34] described epizootics and mass mortality of reef fish (tons) potentially associated with benthic HABs, highlighting the presence of *Gambierdiscus*, *Ostreopsis*, *Prorocentrum*, and *Lyngbya* species. Recently, massive die-offs of reef fish in Florida have been described, with individuals showing signs of acute poisoning similar to those observed in some trials of acute exposure to toxins from *Gambierdiscus* and other benthic dinoflagellates. Therefore, it is important to study whether *Gambierdiscus* toxins are associated with these epizootics, as has been recently suggested.

Based on existing knowledge and the results observed in the present study, new studies should be conducted that evaluate the ecotoxicological role of toxins and other related compounds from *Gambierdiscus* and other benthic species in the health of fish during different life stages in the wild and under aquaculture conditions. In the latter case, their importance as future animal models to study the metabolism and biosynthesis of CFP toxins and other syndromes of benthic origin should also be explored.

3.4. Fluorescent Receptor Binding Assay (fRBA)

Although fRBA is not specific in detecting the presence of toxins, the sensitivity of this test has demonstrated its efficiency in fish tissues, *Gambierdiscus*, and *Fukuyoa* for detecting compounds related to VGSC site 5, such as CTX and PbTx [42,117,118]. In this study, *G.* cf. *caribaeus* showed concentrations based on the VGSC competitor of 2.19–2.59 ± 0.24 pg eq PbTx $cell^{-1}$, a medium-to-low concentration similar to that reported for *G. caribeaus* from Cuba [42]. The estimated concentration for *G.* cf. *carpenteri* samples ranged between 1.15–2.51 ± 0.31 pg eq PbTX·$cell^{-1}$, which is three times lower than that reported by Vacarizas et al. [45] (7.48 ± 0.49 pg PbTx eq·$cell^{-1}$). Although PbTx was used as a standard reference to quantify the activity of toxins related to the sodium channel, previous studies have reported better sensitivity during quantification when using CTX competition as a reference standard [41,117] compared to standardized tests such as CBA-N2a and ELA [47,118]. In the present study, strains of *G.* cf. *caribaeus* and *G.* cf. *carpenteri* (Table 6) support the conclusion that these species present medium-to-low toxicity compared to *G. excentricus* and *G. polynesiensis*, which are representatives of the genus with higher CTXs production per cell. Future studies should compare the binding percentage with a CTX standard due to variations in the capacity to bind VGSC, as the extracts tested are likely to contain a mixture of analogs that could affect the sensitivity and binding signal, which could differ when quantifying using the brevetoxin standard.

Table 6. Toxicity assays performed and principal marine toxins detected in *Gambierdiscus caribaeus* and *G. carpenteri* strains.

Specie/Strain	Origin	Assays Toxicity	Toxins Detected	Method	Reference
		Gambierdiscus caribaeus			
Gam19, Jar 17 gam 20, Gam 4, SW gam 1, NCMA 1733, TC tow Gam 3, Norval Cay, SW gam 5, CBC gam1 NCMA 1651 Mexico Algae 1 gam 1 Jamaica Algae 1 gam 1 ETB Gam 6, Outfish 7-1, Outhfish 7-3 Jar 2 Tow 3, Keys jar 7 gam 7 ST1 C5 SJ3 07 BB gam 4 BRP gam 4, Coral cove gam 1 Dive 1 fa gam 1 WBHR 21 gam 2, WBHR 26 gam 1	Belize Gran Cayman) Cancun, Mexico Ocho ríos, Jamaica DryTortugas (Long key, Florida); St. Thomas St. Jhons Bathtub Beach, Florida Jupiter, Florida Ft. Pierce, Florida Flower Garden	HA	---	---	[119]

Table 6. *Cont.*

Specie/Strain	Origin	Assays Toxicity	Toxins Detected	Method	Reference
CBCGam1 CCMP1651 Dive1TA KeysJar7 MexicoAlgae1 SWGam5	Carrie Bow Cay Belize Gran Cayman Island, Caribbean Carrie Bow Cay, Belize Florida Keys, USA Cancún, Mexico Southwater Cay, Belize	CBA-N2a Average, within species toxicity (fg CTX3C eq cell^{-1} (0.66 ± 0.34, 51%))	---	---	[47]
BillHiGam8	Waikiki Beach, Honolulu, Hawaii	CBA-N2a and ELA (DSF: 1.6–1.0 fg CTX3C eq cell^{-1}) MSF: 5.3–1.0 pg MTX eq cell^{-1}	---	---	[119]
CUB4A5	Cienfuegos, Cuba	RBA	---	---	[42]
10 strains Culture Collection of Harmful Microalgae of the Spanish Institute of Oceanography (CCVIEO)	Canary Islands	Neuro-2a Cell Assay, ELA (<LOD to 90.37 ± 15.89 fg CTX1B eq cell^{-1})	---	---	[120]
1 strain	Micronesia	MBA	44-methylgambierone	LC-MS/MS	[121]
GCBG01	Weizhou Island, Beibu Gulf of China	Neuro-2a cells (0.54 fg CTX3C eq cell^{-1})	---	---	[122]
BPAug08 and USVI-08	St. Thomas, US Virgin Islands	N2a–MTT Gambierone (0.6 ± 0.2 pg cell^{-1} and 9.8 ± 0.6 pg cell^{-1} 44-methylgambierone: 7.1 ± 4.2 pg cell^{-1})	Gambierone, 44-methylgambierone C-CTX-5	LC–HRMS	[24]
DiveIFM4 NOAA STIC5 NOAA	Atlantic Ocean	---	Gambierone, 44-methylgambierone	LC-MS/MS	[28]
CAWD301	Pohnpei	---	Gambieric acid A, methylgambierone, MTX-7 trace	LC-MS/MS	[18]
BPAug08 and USVI-08	St. Thomas, US Virgin Islands	---	C-CTX-5	NMR spectroscopy	[25]
G. cf. *caribaeus*/ GbSa-2 and GbSa-7	El Sauzoso, Gulf of California, Mexico	MBA, ARTOX, MFA (CTXs and MTXs-like activity) fRBA	---	---	This study
		Gambierdiscus carpenteri			
Mixed PR gam 4NCMA 1654 GT4 PatH1jar 5 gam 3 WBHR21 ETB Exp 24 gam 1 Jamaica Algae 2 gam 1	Puerto Rico Guam Belize Oahu, Hawaii Flower Gardens Dry Tortugas Ocho Rios	Hemolytic assay	---	---	[119]
4 isolates	Cook islands, French Polynesia and Australia	MBA (LD_{50} 20–38 mg kg^{-1})	44-methylgambierone in Cook island and French Polynesia strains; Australia island negative strains	LC-MS/MS	[123]

Table 6. *Cont.*

Specie/Strain	Origin	Assays Toxicity	Toxins Detected	Method	Reference
CAWD237 CAWD364	Australia	---	MTX-6 trace Gambierone, 44-methylgambierone, Gambieric acid A	LC-MS/MS	[18]
Bill Aruba Gam15 GT4 Jamaica Algae2Gam1 Mexico Algae2Gam1 WBHR21	Aruba, Caribbean Carrie Bow Cay Belize Ocho Rios, Jamaica, Caribbean Cancun, Mexico Flower Garden Banks Nat. Mar. Sanctuary (West Bank) Gulf of Mexico, USA	CBA-N2a Average, within species toxicity (fg CTX3C eq cell^{-1}) (0.89 ± 0.41, 47%)	---	---	[48]
AWD237	New South Wales, Australia	MBA (LD_{50} 5.1–14.4 mg/Kg)	CTXs and MTXs no detected	LC-MS/MS	[123]
Tropical UTSH12C4 UTSH16C3 UTSH16A1 UTSH16D2	Heron Island Lagoon, Australia	Ca2+ Influx SH-SY5Y Cell FLIPR Bioassay MTX-like activity (3–4)	MTX-3	LC-MS/MS	[124]
Gam1BOL_080513	Bolinao, Pangasinan, Luzon Island Phillipnes	RBA (7.48 ± 0.49 pg PbTx eq cell^{-1})	---	---	[45]
NHA19 and NAH20	Anaho Bay (Nuku Hiva Island, Marquesas archipelago)	CBA-N2a Were nontoxic	---	---	[125]
Gam1BOL080513	Bolinao, Pangasinan	RBA	44-methylgambierone	UPLC-MS/MS	[27]
GCARBAPAZ3	Isla Gaviota, Gulf of California, Mexico	MBA (CTXs and MTXs-like activity) 5.9 mg kg^{-1} CTXs-like and 0.06 mg kg^{-1} MTX-like	---	---	[38]
G. c.f. *carpenteri*/ GbSa-4, GbSa-5, and GbSa-6	El Sauzoso, Gulf of California, Mexico	MBA, ARTOX, MFA (CTXs and MTXs-like activity) f -RBA	---		This study

HA: Hemolytic assay; CBA-N2a: neuroblastoma cytotoxicity assay; RBA: radioligand-receptor binding assay; ELA: erythrocyte lysis assay; MTT: 3-(4,5-dimethyl-2-thiazol)-2,5-diphenyl-2H-tetrazolium bromide (methylthiazolyldiphenyl-tetrazolium bromide) colorimetric assay; MBA: mouse bioassay; FLIPR bioassay: fluorescent imaging plate reader; ARTOX: artemia assay; MFA: marine fish assay; fRBA: fluorescent receptor binding assay; LC–HRMS; high-performance liquid chromatography/H nuclear magnetic resonance; LC-MS/MS: liquid chromatography coupled to tandem low-resolution mass spectrometry; NMR spectroscopy: nuclear magnetic resonance spectroscopy; UPLC-MS/MS: ultra-performance liquid chromatography/mass spectrometry.

Table 6 describes and compares the main toxicity assays performed at the global level and the marine toxins detected in the *G. caribaeus* and *G. carpenteri* strains. Various assays and analytical techniques have been used to determine the toxicity and chemical composition of the toxins produced by these two species in different parts of the world. Toxicity varies from nontoxic strains to those that present high toxicity. Some reports indicate 1.4 ± 0.6 fg CTX3C eq cell^{-1} in strains from Hawaii [126]), 0.3–1.4 fg CTX3C eq cell^{-1} in strains isolated from the Caribbean [39], and 7.48 ± 0.49 pg PbTx-3 eq cell^{-1} and 12.36 ± 4.38 pg PbTx-3 eq cell^{-1} in strains from the Philippines (all *G. carpenteri*) [27,44,45]. In addition, *G. carpenteri* strains from Australia have been reported to have a toxicity of 2.4 mg kg^{-1} per MBA, and Ramos-Santiago et al. [38] reported a CTX-like toxicity of 5.9 mg kg^{-1} by MBA in Isla Gaviota, Gulf of California.

In *G. caribaeus*, a toxicity of 0.66 ± 0.34 fg CTX3C eq cell^{-1} was described in strains from the Caribbean [42], 1.0–1.6 fg CTX3C eq cell^{-1} in strains from Hawaii [118,126],

15.89 fg CTX1B eq cell^{-1} in strains from the Canary Islands (127), 0.54 fg CTX3C eq cell^{-1} in strains from the Gulf of China [123], and 0.6 ± 0.2 pg cell^{-1} and 9.8 ± 0.6 pg cell^{-1} (Gambierone) and 7.1 ± 4.2 pg cell^{-1} (44-methylgambierone) in strains from St. Thomas, US Virgin Islands [24]. For the strains of *G. caribaeus* and *G. silvae*, a new ciguatoxin was isolated in the Caribbean and named C-CTX-5 [24,25]. This finding represents a significant advance in research on the origin of CFP in the Caribbean, as the precursor toxin from its algal source and its relationship with CTXs in fish that cause the disease was previously unknown, hindering the development of detection methods, diagnostics, and monitoring programs. In the same study, various in vitro and chemical techniques were applied to identify the new analog (C-CTX-5), and it was verified that the treatment of this analog using fish liver microsomes converted it to C-CTX1/2, the dominant C-CTX in ciguatoxic fish from the Caribbean. Recently, C-CTX-5 was also identified for the first time in an "amberjack" (*Seriola* sp.) involved in a CFP case in the Canary Islands, Spain [127]. Based on this, it is necessary to investigate whether *G. caribaeus* strains from other latitudes, such as those in the present study, can also produce this new ciguatoxin.

The genetic identification of the *Gambierdiscus* and *Fukuyoa* species evaluated in this study and chemical analysis of the toxins will be carried out in a subsequent study using analytical tools, given the complexity of obtaining standards and other inputs necessary to detect and analyze CTXs, MTXs, and other related compounds. The toxicity and composition of CTXs, MTXs, and other related compounds in these two species are also described in Table 6.

Over the past decades, the occurrence and geographic distribution of CFP have notably expanded due to anthropogenic activities and climate change, leading to an increase in human illness, greater public health impacts, and larger economic losses. The global spread of CFP has made *Gambierdiscus* and its toxins a concern for environmental and human health worldwide [22].

In the PBC, cases of CFP are rare and sporadic. However, at least 240 human cases have been recorded in the southern region due to the consumption of the meat, liver, and heads of carnivorous fish, such as snappers (*Lutjanus* spp.), cabrillas, and groupers (*Epinephelus* and *Mycteroperca*), caught near islands and islets [11–17]. Additionally, epizootics and the massive mortality of reef fish potentially associated with benthic HABs have been reported, with *Gambierdiscus*, *Ostreopsis*, *Prorocentrum*, and *Lyngbya* species being primarily detected [34].

The PBC is also a region of notable fishery importance, supporting artisanal fisheries that exploit numerous species of teleost fish, elasmobranchs, mollusks, and crustaceans. In the Gulf of California, coastal fisheries target around 70 species, with an annual catch of nearly 200,000 tons. Sport fishing is also prominent, targeting species such as yellowfin tuna (*Thunnus albacares*), skipjack (*Katsuwonus pelamis*), billfish striped marlin, blue marlin (*Makaira nigricans*), black marlin (*Makaira indica*), sailfish (*Istiophorus platypterus*), swordfish (*Xiphias gladius*), dorado (*Coryphaena hippurus*), yellowtail amberjack (*Seriola lalandi*), longfin yellowtail (*Seriola rivoliana*), snappers (*Lutjanus* spp.), "groupers" (*Ephinpehelus* spp. and *Mycteroperca* spp.), and around 40 species of sharks, including *Mustelus*, *Carcharhinus*, *Alopias*, *Sphyrna*, and *Squatina* [128]. The region is also important for activities such as whale watching (e.g., *Eschrichtius robustus*, *Megaptera novaeangliae*, *Balaenoptera musculus*, and *Balaenoptera physalus*) and observing dolphins (e.g., *Tursiops truncatus* and *Delphinus* spp.), sea lions (*Zalophus californianus*), sea turtles, whale sharks (*Rhincodon typus*), reef fish, a great diversity of invertebrates, and algae [128–130]. In recent decades, the southern PBC has experienced increased tourism and urban growth, increased demand for services, and the development of marine fish farming, bivalve aquaculture, and shrimp production in Bahía de La Paz.

The annual economic contribution of sport fishing in the southern PBC was estimated to be $80,801,119 USD between 2008 and 2010, equivalent to 1.4% of the Gross Domestic Product (GDP) of Mexico during that period, according to the National Institute of Geography and Statistics (INEGI) [131,132].

In this study, in addition to *G. caribaeus* and *G. carpenteri*, we report the presence *Gambierdiscus* (*G.* cf. *carolinianus* and *G.* cf. *toxicus*) and *Fukuyoa* (*Fukuyoa* cf. *yasumotoi*) species in the PBC, where cases of CFP have occurred. The Gulf of California is a potential hotspot for the biodiversity and endemism of benthic dinoflagellate species, which provides an important opportunity to research CTXs, MTXs, CFP-related compounds, other polyether toxins, and secondary metabolites. Studies must be conducted that aim to monitor and evaluate the toxicity, toxin composition, and secondary metabolites of different species of *Gambierdiscus* and *Fukuyoa* in this region. The results of these studies are essential for protecting human and animal health. These studies also represent an opportunity to advance research on chemical diversity (metabolomics), which could contribute to characterizing the species and ecotypes present and their potential biotechnological or therapeutic applications.

4. Materials and Methods

4.1. Isolated and Algal Cultures

During August 2022, samples of the macroalgae *Dyctiota* spp. and *Padina* spp. (associated with corals) were collected by snorkeling at a depth of less than 4 m in the coral reef of El Sauzoso, from artificial substrates at Playa Coromuel, from the rocky reef at El Caimancito, and from the sandy intertidal plane of El Conchalito. Additional samples were collected at the Fang Ming scuba diving site in Espiritu Santo Island (Figure 12).

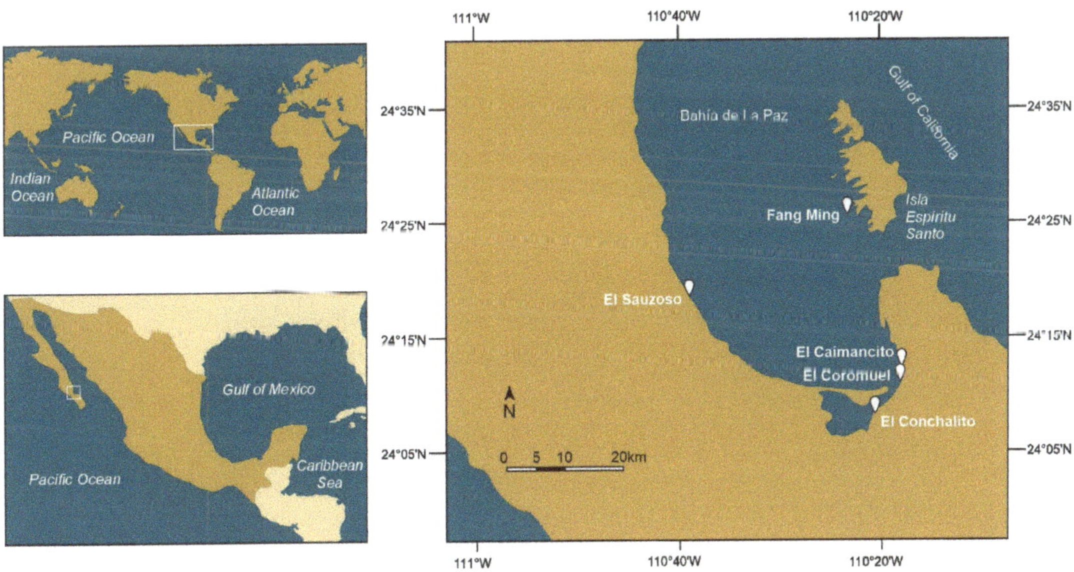

Figure 12. Sampling sites in Bahía de La Paz used to collect *Gambierdiscus* species.

Samples were carefully incubated in Nalgene 1-L bottles and immediately transported to the Marine Toxins and Amino Acids laboratory of CIBNOR. *Gambierdiscus* cells were isolated individually from macroalgae, rocks, and sandy sediments using a capillary micropipette under an inverted microscope (Labomed, TCM 400, Hicksville, NY, USA). To establish the culture, the isolated cells were transferred into 24-well flat bottom plates (Costar, Boston, MA, USA) with 1 mL of modified K medium [133] containing 1 mg/L of germanium oxide. After 45 days, the wells containing more than 20 cells were transferred to flat bottom culture tubes (Pyrex®, Sunderland, UK) containing 20 mL of medium. These cultures were maintained in the culture room of the CIBNOR microalgae collection (25 ± 2 °C, 35 salinity, a 12:12 light-dark cycle, and irradiance of 35 mmol photon m^{-2} s^{-1}

supplied by LED light (MAGGA T8 LED STD 18 W, CDMX, Mexico)). All cultures were carried out under these conditions.

Growth rates and generations per day were calculated according to the methods described by Guillard and Ryther [134] and Guillard [135]. Briefly, triplicate cultures of each species were maintained in stock cultures in 300-mL Erlenmeyer flasks with 100 mL of medium at an initial density of 50 cells·mL^{-1}. Every seven days, 2-mL samples were fixed with Lugol's solution Utermöhl for cell counts on Sedgwick-Rafter slides under a compound light microscope (Olympus BX41, Tokyo, Japan). Serial culture batches were established in 2.8-L Fernbach flasks (Pyrex®) with 1 L of culture medium to increase biomass to evaluate morphological taxonomy, pigments, and amino acids, as well as the mouse bioassay, fluorescent receptor binding assay (fRBA), artemia assay (ARTOX), and larvae fish assay.

4.2. Morphological Identification

4.2.1. Light Microscopy LM

Observations, measurements, and photomicrographs were conducted on 30 randomly selected live *Gambierdiscus* cells. Cell size, including depth (ventral/dorsal distance), width (transdiameter), and length (apical/antapical axis), was recorded under a microscope (Olympus BX41, Japan) equipped with a camera (SC-MOUNT-YW500 Sony sensor IMX335, ImageView software 2.0), and the system was calibrated using a micrometric slide (1–1000 μm) with achromatic (40×/0.65 Ph2) and 100×/1.25 Oil Ph3 ∞/−) Olympus objectives. Ten milliliters of the culture were collected in 15-mL conical tubes, and cells were fixed with 4% glutaraldehyde. To identify cell plates, a 5-mL aliquot was centrifuged at 1200× *g* for 2 min, the fixative was removed, and cells were washed three times with distilled water before being fixed with 4% sodium hypochlorite to dissociated cell plates. Plate tabulation nomenclature followed the description of Ramos-Santiago et al. [38]. Another 2 mL of fixed sample was washed again with distilled water by centrifugation and treated with Fluorescent Bright 28 (Sigma Aldrich, St. Lois, MO, USA) [136]. Fluorescent Bright 28 (blue emission) and natural autofluorescence of chloroplasts were documented using a scanning laser confocal microscope (Leica® DMi8, Wetzlar, Germany).

4.2.2. Scanning Electron Microscopy SEM

Ten milliliters of culture were fixed with osmium in 15-mL sterile conical tubes (Corning®, Corning, NY, USA). After 24 h, the cell pellet was filtered through membrane filters with a 0.2-μm pore size and 25-mm diameter (Gelman Science, Bloomfield Hills, MI, USA). Cells were washed four times with filtered distilled water and sequentially dehydrated with 30, 50, 75, 95, and 100% EtOH. After critical-point-drying, the cells were transferred and mounted on 12-mm carbon tabs PK/10 (West Chester, PA, USA), coated with gold, and examined under a scanning electron microscope (Hitachi® S300N, Tokyo, Japan) using image capture software (Quartz PCI 200).

4.3. Biochemical Characterization

4.3.1. High-Performance Liquid Chromatography (HPLC)

The pigment composition of five strains of two species of *Gambierdiscus* species was analyzed by HPLC-UV/VIS according to the methods [66,137], with some modifications. Briefly, 30 mL of *Gambierdiscus* cells were collected on a glass microfiber filter GF/F (Ahlstrom—Munksjö, Sweden). The filters were extracted with 2 mL of 100% acetone in an ice bath, hand-ground with a glass rod, and stored overnight at −40 °C for full extraction. The extract was recovered after centrifugation (3000× *g* for 5 min). A 200-μL aliquot of the extract was mixed with 100 μL of 0.5 N ammonium acetate and injected through a 100-μL loop into an HPLC system (Hewlett Packard series 1260) equipped with a photodiode array detector (DAD, 1.2 nm optical resolution). Pigments were separated and quantified using an isocratic HPLC in reverse-phase mode, as described by Vidussi et al. [137] and Bustillos et al. [66]. The mobile phase consisted of MeOH (0.5 N aqueous ammonium acetate, 70:30% v/v (solvent A), and MeOH [solvent B]), with the following gradient (minute (percent of solvent A-percent of solvent B)): 0 (75–25), 1 (50–50), 15 (0–100), and 19 (75–25). Quantification was based on the absorbance at 440 nm, and the response factor (peak area/pigment concentration) for each pigment was determined as described by Mantoura and Repeta [138]. Pigment identification was based on retention time, spectral characteristics, and chromatography using certified commercial standards (International Agency for 14C determinations, Hørsholm, Denmark).

4.3.2. High-Performance Liquid Chromatography-Diode Array Detection (HPLC-DAD) for Amino Acid Composition

The amino acid composition of five isolates of two species of *Gambierdiscus* was analyzed by HPLC-DAD according to Gratzfeld-Huesgen [139] and AOAC [140], with some modifications. A 150-mL volume of *Gambierdiscus* cultures was collected on a GF/F filter (Sartoriuos, Goettingen, Germany), and the filter was dried in an oven (Symphony™ VWR®, Radnor, PA, USA) at 60 °C. Dried samples were hydrolyzed in a mixture of 6 N hydrochloric acids/phenol at 110 °C for 24 h. After acid hydrolysis, samples were washed with deionized water and dried in a rotary evaporator (IKA®-WERKE RV05, Staufen im Breisgau, Germany) before undergoing oxidation with performic acid and alkaline hydrolysis with 4.2 M sodium hydroxide (NaOH). Finally, samples were filtered through syringe filters with 0.2–0.45 µm pore sizes before analysis on the HPLC system. An L-amino acid kit from SIGMA Chemicals Corp. (Saint Louis, MO, USA) was used. O-phthalaldehyde (OPA) solution: o-phthalaldehyde and 3-mercaptopropionic acid (Agilent Technologies Part no. 5061-3335, Santa Clara, CA, USA); 9-fluorenylmethylchloformate (FMOC) solution: 9-fluorenylmethylchloformate (Agilent Technologies Part No. 5061-3337) and Borate Buffer 0.4 M at PH 10.2 (Part Number: 5061-3339). For calibration, the following profile of L-amino acids (SIGMA Chemicals Corp., Saint Louis, MO, USA) was used: aspartic acid (Asp), glutamic acid (Glu), serine (Ser), histidine (His), glycine (Gly), threonine (Thr), alanine (Ala), arginine (Arg), tyrosine (Tyr), valine (Val), phenylalanine (Phe), Isoleucine (Iso) leucine (Leu), lysine (Lys), and Proline (Prol) at a concentration of 0.5 µmol/mL in 0.1 N HCl.

Sample derivatization method: derivatization with OPA was used for primary amino acids and FMOC for secondary amino acids. Automatic mixing of 1 µL of the sample was performed in the injector, with 5 µL borate buffer at pH 10.4, 1 µL OPA, and 1 µL FMOC. An Agilent Technologies 1100 series HPLC with a UV VIS diode array detector was used. A Hypersil 5 AA-ODS (Agilent Technologies) reversed-phase column (200 mm length, 2.1 mm diameter, and 5 µm particle size [Part No. 79916AA-572]) was used. A C18 4 × 3.0 mm Phenomenex column guard (Part No. AJ0-4287) was used to protect the column. Detection was performed by UV at 338 nm for primary amino acids and at 262 nm for secondary amino acids at 40 °C with a flow rate of 0.450 mL/min and an injection volume of 8 µL using a binary gradient program. The mobile phase was prepared based on a binary gradient (A, B), with phase A of 20 mM sodium acetate buffer with 0.018% (v/v) triethylamine and 0.03% tetrahydrofuran, and phase B consisting of 20% of 100 mM sodium acetate buffer, 40% acetonitrile, and 40% MeOH.

4.4. Toxicity Assay

4.4.1. Toxins Extraction

Batch cultures were harvested during the early exponential phase, and the number of cells was recorded. Harvesting was carried out by concentrating cells on glass microfiber filters with a pore size of 0.7 µm and a diameter of 47 mm (Whatman™, Maidstone, UK). Toxin extraction from *Gambierdiscus* cells was performed following the method described by Chinain et al. [141]. Methanol (50 mL) was added, and the filters were macerated with a glass rod for 5 min, followed by sonication for 20 min (Branson® 1210, Branson, MO, USA) three times. Methanolic extracts were dried at 40 °C using a Bücher Rotary evaporator. The extract was resuspended and partitioned in a separating funnel by adding dichloromethane (DCM) (50 mL per million cells) and MeOH:H2O (6:4. v/v) (25 mL per million cells) [142,143]. Two fractions were obtained: the methanol:water phase containing MTX and the DCM fraction containing CTX.

4.4.2. Mouse Bioassay (MBA)

The toxicity of *Gambierdiscus* was assessed using MBA for CTX-like and MTX-like toxins. The mouse model was maintained under comfortable conditions (temperature of 22 ± 1 °C, a 12:12 light:dark cycle, and a maximum humidity of 60%) in compliance with the Official Mexican Standard NOM-062-ZOO-1999. The mice were provided with bedding made of Aspen shavings from Nepco and were fed sterile formulated pellets (Zeigler, Gardners, PA, USA) and sterile acidified water provided *ad libitum*. The DCM and MeOH:H_2O phases were evaporated and resuspended in 1% Tween 60 and 0.9% sterile saline solution. A 1 mL dose was intraperitoneally (i.p.) injected into male mice weighing 18–20 g each (strain CD-1 Harlan Laboratories, CDMX, Mexico) divided in groups of two animals, with a control group injected with Tween solution. Biological activity and toxicity were determined following the recommendations described by Satake et al. [79,142,143]. The mouse unit (MU), defined as the i.p. The LD_{50} dose for a 20-g mouse was calculated according to Lewis [77]. To determine the toxicity of a mixture of ciguatoxins, the following formula was used:

$$log\ MU = 2.3log\left(1 + \frac{1}{T}\right) \quad (1)$$

To determine the toxicity of a mixture of maitotoxins, the following formula was used:

$$log\ MU = 6.7log\left(1 + \frac{1}{T}\right) \quad (2)$$

Continuous monitoring was conducted every hour for 24 h, during which the clinical signs of poisoning in mice were recorded and reported as a percentage according to the affected body system.

4.4.3. *Artemia salina* Assay (ARTOX)

The *A. salina* toxicity assay was performed to evaluate the acute toxicity of the CTXs-like and MTX-like extracts, following the procedure described by Heredia-Tapia et al. and Neves et al. [82,83]. Adult brine shrimp (*A. salina*) were cultured in plastic containers and exposed to the extracts (1 mL in 1% Tween 60 and 0. 9% sterile saline solution) of *G.* cf. *caribeaus* and *G.* cf. *carpenteri*, equivalent to 450,000 cel mL^{-1} in a glass container of brine shrimp (n = 10). Clinical signs were observed over a period of 24 h, with observation at 6, 12, 18, and 24 h, to determine the mortality rate associated with each extract.

4.4.4. Marine Fish Assay

Groups of *S. rivoliana* larvae were conditioned for 24 h in cylindrical glass tanks containing 400 mL of seawater with aeration at 25 °C, after which the water was changed. To evaluate the acute toxicity of the extracts, 3 mL of CTXs-like and MTXs-like extracts, equivalent to 450,000 cel mL^{-1} of *G.* cf. *caribaeus* and *G.* cf. *carpenteri*, were added to 0. 9% sterile saline solution. Two control groups were established, one with saline solution and the other with seawater. Mortality was recorded for each treatment at 6, 12, 18, and 24 h, and clinical signs were documented following the Organisation for Economic Co-operation and Development OECD guidelines for testing chemicals (Test No. 203: Fish, Acute Toxicity Test). Mortality was reported as a percentage, comparing the number of survivors to dead individuals in the control groups with the number of survivors to dead *S. rivoliana* larvae in the treatments with the dissolved CTXs-like and MTXs-like extracts.

4.5. Fluorescent Receptor Binding Assay (fRBA)

The SeaTox Research Inc. (Wilmington, NC, USA) Brevetoxin/Ciguatoxin RBA test kit was used to detect the presence of compounds with binding affinity to VGSC site 5, which is specific for the receptor but not for toxins (e.g., PbTx-1, PbTx-2, and PbTx-3 and ciguatoxin P-CTX-1, P-CTX-3C, and C-CTX-1) [117]. In this assay, the receptor was mixed with a fluorescent ligand and an unlabeled ligand. In the presence of a low competitor (e.g., ethanol control or 0.1 ng/mL^{-10} log in the standard brevetoxin assay), the receptor site was predominantly bound by the fluorescent ligand, resulting in high fluorescence levels in the samples. When the concentration of a competitor was high (e.g., 1 or 0.1 µg/mL standard brevetoxin; -6 and -7 log in the assay), most receptor sites were occupied by the unlabeled ligands, and low fluorescence levels were measured from the sample.

The reagents were prepared in 1× buffer (10 mL of buffer from the kit with 90 mL of sterile distilled water) and stored at 4 °C until use. Fluorescent ligands were resuspended in 60 µL of HPLC-grade ethanol to create a 100× solution in an amber glass vial, sonicated, and vortexed to ensure resuspension. A 1× working solution was prepared by diluting 55 µL of fluorescent ligand (100×) in 5.5 mL of assay buffer (1×) and kept at –20 °C until use. The brevetoxin standard (100 µg/mL) was prepared by adding 35 µL of ethanol directly to the vial, followed by sonication and vortexing to achieve a final concentration of 1 µg/mL. A series of five 1:10 dilutions (0.1–0.00001 µg/mL standard brevetoxin) were prepared in HPLC vials, starting with 10 µL of brevetoxin standard stock 1, adding 90 µL of solvent for standard 2, and continuing this process to standard 5. The samples were stored at −20 °C until use. Synaptosome receptors were refrigerated until use and were completely and gently resuspended in 5 mL of sterile distillate water. During the test, the resuspended aliquot was always kept on ice. Samples were prepared by evaporating a 3-mL aliquot at a concentration of 400,000 cell·mL^{-1} for fRBA, following the procedure described by Darius et al. [118]. The *Gambierdiscus* DCM phase, corresponding to the ciguatoxin mixture, was resuspended in 1× assay buffer. An extract of *Karenia brevis* with a known profile (PbTx-2 and PbTx-3 by LC-MS/MS) was tested, and ethanol was used as a control.

The fRBA reaction assay plate was carried out in qualitative mode: single concentration (triplicate), 5 µL added to each series of the standard dilution, 5 µL of *Gambierdiscus* extract, 5 µL of *K. brevis* extract, and 6 wells with 5 µL of ethanol control. Then, 195 µL of assay buffer and 50 µL of cold synaptosome solution were added to each well, and the plate was incubated and covered on ice for 30 min with agitation and in darkness. Afterwards, 50 µL of fluorescent ligand (1×) working solution and 200 µL of assay buffer (1×) were added to each well and incubated in darkness for 60 min on ice, with gentle mixing on a table shaker. During incubation, 250 µL of 1× assay buffer was passed through the wells of the filter plate provided in the kit assay and dried using a vacuum system until no liquid remained on the paper towel. After incubation, the reaction mixtures were transferred from the plate to the filter plate and dried in a vacuum manifold system. The filter wells were rinsed with 500 µL of assay buffer, and air was removed from the plate by vacuum for at least 30 s after the final rinse. The plate filter was read at 490–10 nm (excitation filter) and 520–10 nm (emission filter), with wavelengths determined from the extra fluorescent ligand provided and the empty filter plate to minimize background fluorescence, using a fluorescent plate reader (Thermo Scientific Varioskan Flash 3001-2158, Waltham, MA, USA).

Data were analyzed as the relative fluorescence unit (RFU) of total binding (%), using an ethanol control (100% fluorescent ligand). Background RFU values were subtracted from the final RFU data for each well, and the values were extrapolated to picograms equivalent brevetoxin per cell (pg eq PbTX) from the standard curve generated by the brevetoxin standard.

5. Conclusions

1. *Gambierdiscus* cf. *carpenteri* is the most abundant species in Bahía de La Paz and is associated with the production of toxins related to CFP. The presence of G. cf. *caribeaus*, G. cf. *carolineanus*, G. cf. *toxicus*, and F. cf. *yasumotoi* was also recorded.
2. There are differences in pigment production and in the amino acid profiles between the strains G. cf. *caribaeus* and G. cf. *carpenteri*. These results suggest the presence

of ecotypes, even among strains of the same species. The biochemical profiles could potentially establish chemotaxonomic criteria for the identification of *Gambierdiscus* species.

3. Toxicity for CTX-like and MTX-like activity was confirmed in all strains of *G.* cf. *caribeaus* and *G.* cf. *carpenteri*. CTX-like and MTX-like toxicity produced a positive result in the MBA and caused the mortality of *A. salina*. In the marine fish assay, the CTXs-like extracts presented a low-to-medium toxicity through competition for VGSC site 5, as quantified by the fluorescent receptor binding assay.
4. The differences in the responses across the various assays suggest that *G.* cf. *caribeaus* and *G.* cf. *carpenteri* from Bahía de La Paz may produce different metabolites associated with the genus *Gambierdiscus*.
5. The Gulf of California is a potential hotspot of biodiversity and endemism for benthic dinoflagellate species and a key area for research on CTXs and MTXs, CFP-related compounds, other polyether toxins, and secondary metabolites.

Supplementary Materials: The following supporting information can be downloaded at https://www.mdpi.com/article/10.3390/md22090422/s1, Table S1: Species of dinoflagellates of the genera Gambierdiscus and Fukuyoa in the Mexican Pacific and Gulf of California; Table S2. Gambierdiscus and Fukuyoa species recorded in the site collected; Figure S3. LM plate tabulation Gambierdiscus cf. caribaeus (GbSa-2); Figure S4. LM plate tabulation Gambierdiscus cf. carpenteri (GbSa-5); Figure S5. LM plate tabulation Gambierdiscus cf. carpenteri (GbSa-6); Figure S6. Growth curve of Gambierdiscus cf. caribaeus and Gambierdiscus cf. carpenteri; Figure S7. LM Gambierdiscus cf. carolineanus (GbEs-4); Figure S8. LM Gambierdiscus cf. toxicus (GbEs-3); Figure S9. LM Fukuyoa cf. yasumotoi (FyEs-1); Figure S10. Chromatogram of the amino acid proline; Figure S11. Sigmoid dose response curve as a percentage of total binding.

Author Contributions: All authors contributed significantly to this study. Conceptualization, methodology, software, formal analysis, investigation, data curation, writing—original draft preparation, visualization, review and editing: L.J.F.-H.; Conceptualization, methodology, supervision, formal analysis, investigation, data curation, writing—original draft preparation, visualization, resources, funding acquisition, review and editing: E.J.N.-V.; methodology, software, formal analysis, investigation, data curation, writing—original draft preparation, visualization, review and editing: F.E.H.-S.; methodology, software, formal analysis, investigation, data curation, writing—original draft preparation, visualization, review and editing: D.O.C.-O.; software, visualization, writing, review and editing: S.G.-D.; methodology, review and editing: M.V.-F.; methodology, writing, review and editing: A.T.; investigation, resources, funding acquisition, writing, review and editing: D.T.-R. All authors have read and agreed to the published version of the manuscript.

Funding: This research was financed by CONAHCYT (project 319865), Ciencia Básica y/o Ciencia de Frontera Modalidad Paradigmas y Controversias de la Ciencia 2022: "Toxicogenómica en peces de importancia socio-económica: establecimiento de dos nuevos modelos acuícolas para el estudio del efecto de toxinas marinas y de agua dulce" and PRONAII-CONAHCYT (Project 319104) "Atención de la problemática asociada florecimientos algales nocivos en Baja California: integración del conocimiento a necesidades socioambientales y económicas".

Institutional Review Board Statement: The experiment complied with the Mexican Government (NOM-062-ZOO-1999) for the production, care, and use of experimental animals, the ARRIVE guidelines, and the Guidelines of the European Union Council (2010/63/EU). The acute toxicity test in the mouse model was carried out based on the Mexican Standard NOM-242-SSA1-2009 (https://dof.gob.mx/normasOficiales/4295/salud2a/salud2a.htm (accessed on 11 July 2024)) of the Ministry of Health (official method in Mexico for analysis of CTXs) and in accordance with the Official Mexican Standard NOM-062-ZOO-1999 (https://www.gob.mx/senasica/documentos/nom-062-zoo-1999 (accessed on 11 July 2024)) of the Ministry of Agriculture, Livestock, Rural Development, Fisheries and Food and the technical specifications for production, care and use of laboratory animals and followed the recommendations of Heldrich, H. (2012). The laboratory mouse. 2ed. Academic Press, Elsevier, Oxford, UK.

Data Availability Statement: Data is contained within the article or Supplementary Materials.

Acknowledgments: E. J. N.-V dedicates this work to the memory of Maasaki Kodama (1943–2020) of the University of Tokyo. This study received support from Centro de Investigaciones Biológicas del Noroeste (CIBNOR) and Consejo Nacional de Humanidades, Ciencia y Tecnología (CONAHCYT). We thank A. Cruz-Villacorta (CIBNOR) for the scanning electronic microscopy service and Gabriel González-Gomez, Julian Garzon-Favela, and Atilano Ramirez Octavio (Aquaculture Genetic Laboratory and Laboratory of Microalgae of the CIBNOR). We thank Amaury Cordero-Tapia and Guadalupe Sánchez (Bioterium of the CIBNOR) for technical support related to organism maintenance, Patricia Hinojosa† (Laboratory of Comparative Physiology and Functional Genomics of the CIBNOR) for their assistance, and Adriana Landa from CIBNOR for editing the pigment chromatogram figure, and Andrea Lievana MacTavish for English editing. Leyberth J. Fernández-Herrera received a postdoctoral fellowship (CONAHCYT; project 319865 and Postdoctoral residence for Mexico 2022 (3) CVU 337101) and Andressa Teles received a postdoctoral fellowship (CONAHCYT) project 319865 and Postdoctoral residence for Mexico 2022 (3). We thank (ENI) Exploration and Diving for their help during the sampling campaign and Victor Henandez (IUBO-ULL) for the fixed samples of *G. exentricus* and *G. autrales*.

Conflicts of Interest: The authors declare no conflicts of interest.

References

1. Chinain, M.; Gatti, C.M.I.; Darius, H.T.; Quod, J.P.; Tester, P.A. Ciguatera poisonings: A global review of occurrences and trends. *Harmful Algae* **2021**, *102*, 101873. [CrossRef] [PubMed]
2. FAO; WHO. *Report of the Expert Meeting on Ciguatera Poisoning: Rome, 19–23 November 2018*; Food Safety and Quality No. 9; World Health Organization: Rome, Italy, 2020. [CrossRef]
3. Anderson, D.M.; Glibert, P.M.; Burkholder, J.M. Harmful algal blooms and eutrophication: Nutrient sources, composition, and consequences. *Estuaries* **2002**, *25*, 704–726. [CrossRef]
4. Glibert, P.M.; Anderson, D.A.; Gentien, P.; Granéli, E.; Sellner, K.G. The global, complex phenomena of harmful algal blooms. *Oceanography* **2005**, *18*, 136–147. [CrossRef]
5. Hallegraeff, G.M. Ocean climate change, phytoplankton community responses, and harmful algal blooms: A formidable predictive challenge. *J. Phycol.* **2010**, *46*, 220–235. [CrossRef]
6. Gobler, C.J.; Doherty, O.M.; Hattenrath-Lehmann, T.K.; Griffith, A.W.; Kang, Y.; Litaker, R.W. Ocean warming since 1982 has expanded the niche of toxic algal blooms in the North Atlantic and North Pacific oceans. *Proc. Natl. Acad. Sci. USA* **2017**, *114*, 4975–4980. [CrossRef]
7. Heisler, J.; Glibert, P.M.; Burkholder, J.M.; Anderson, D.M.; Cochlan, W.; Dennison, W.C.; Dortch, Q.; Gobler, C.J.; Heil, C.A.; Humphries, E.; et al. Eutrophication and harmful algal blooms: A scientific consensus. *Harmful Algae* **2008**, *8*, 3–13. [CrossRef]
8. Griffith, A.W.; Gobler, C.J. Harmful algal blooms: A climate change co-stressor in marine and freshwater ecosystems. *Harmful Algae* **2020**, *91*, 101590. [CrossRef]
9. Skinner, M.P.; Brewer, T.D.; Johnstone, R.; Fleming, L.E.; Lewis, R.J. Ciguatera fish poisoning in the Pacific Islands (1998 to 2008). *PLoS Negl. Trop. Dis.* **2011**, *5*, e1416. [CrossRef]
10. Gingold, D.B.; Strickland, M.J.; Hess, J.J. Ciguatera fish poisoning and climate change. Analysis of National Poison Center data in the United States, 2001–2011. *Environ. Health Perspect.* **2014**, *122*, 580–586. [CrossRef]
11. Núñez-Vázquez, E.J.; Almazán-Becerril, A.; López-Cortés, D.J.; Heredia-Tapia, A.; Hernández-Sandoval, F.E.; Band-Schmidt, C.J.; Bustillos-Guzmán, J.J.; Gárate-Lizárraga, I.; García-Mendoza, E.; Salinas-Zavala, C.A.; et al. Ciguatera in Mexico (1984–2013). *Mar. Drugs* **2019**, *17*, 13. [CrossRef]
12. Parrilla-Cerrillo, M.C.; Vázquez-Castellanos, J.L.; Sáldate-Castañeda, E.O.; Nava-Fernández, L.M. Brotes de toxiinfecciones alimentarias de origen microbiano y parasitario. *Salud Pública Mex.* **1993**, *35*, 456–463. [PubMed]
13. Barton, E.D.; Tanner, P.; Turchen, S.G.; Tungen, C.L.; Manoguerra, A.; Clarck, R.F. Ciguatera fish poisoning: A Southern California epidemic. *West. J. Med.* **1995**, *163*, 31–35. [PubMed]
14. Lechuga-Devéze, C.; Sierra-Beltrán, A. Documented case of ciguatera on the Mexican Pacific Coast. *Nat. Toxins* **1995**, *3*, 415–418. [CrossRef] [PubMed]
15. Núñez-Vázquez, E.; Sierra-Beltrán, A.P.; Ochoa, J.L. Presencia de Biotoxinas Tipo Ciguatoxinas en Peces Carnívoros de Baja California Sur. In Proceedings of the XI Simposium Internacional de Biología Marina, La Paz, Mexico, 18–22 November 1996.
16. Núñez-Vázquez, E.J.; Sierra-Beltrán, A.P.; and Ochoa, J.L. Fish Poisoning in Mexico. In Proceedings of the Abstracts of the VIII Conferencia Internacional sobre Algas Nocivas, Vigo, Spain, 25–29 June 1997.
17. Núñez-Vázquez, E.J.; Bustillos-Guzmán, J.; Heredia-Tapia, A.; Yasumoto, T.; Cruz-Villacorta, A.; Band-Schmidt, C.J.; Gárate-Lizárraga, I.; López-Cortés, D.; Hernández-Sandoval, F.E.; Ochoa, J.L. Múltiples toxinas marinas en el hígado de *Mycteroperca prionura, M. rosacea* y *Lutjanus colorado* asociados a la ciguatera en la isla El Pardito, B.C.S. México. In *Resúmenes del III Taller sobre Florecimientos Algales Nocivos: Integración del Conocimiento Sobre Eventos de FAN en México*; Laboratorio Estatal de Salud Pública Dr. Galo Soberón y Parra; Secretaría de Salud México: Acapulco, Mexico, 2009; p. 52.

18. Murray, J.S.; Passfield, E.M.F.; Rhodes, L.L.; Puddick, J.; Finch, S.C.; Smith, K.F.; van Ginkel, R.; Mudge, E.M.; Nishimura, T.; Funaki, H.; et al. Targeted Metabolite Fingerprints of Thirteen *Gambierdiscus*, Five *Coolia* and Two *Fukuyoa* Species. *Mar. Drugs* **2024**, *22*, 119. [CrossRef]
19. Holmes, M.J. *Gambierdiscus yasumotoi*, sp. nov. (Dinophyceae), a toxic benthic dinoflagellate from Southeastern Asia. *J. Phycol.* **1998**, *34*, 661–668. [CrossRef]
20. Gómez, F.; Qiu, D.; Lopes, R.M.; Lin, S. *Fukuyoa paulensis* gen. et sp. nov., a New Genus for the Globular Species of the Dinoflagellate *Gambierdiscus* (Dinophyceae). *PLoS ONE* **2015**, *10*, e0119676. [CrossRef]
21. Li, Z.; Park, J.S.; Kang, N.S.; Chomérat, N.; Mertens, K.N.; Gu, H.; Lee, K.-W.; Kim, K.H.; Baek, S.H.; Shin, K.; et al. A new potentially toxic dinoflagellate *Fukuyoa koreansis* sp. nov. (*Gonyaulacales*, Dinophyceae) from Korean coastal waters: Morphology, phylogeny, and effects of temperature and salinity on growth. *Harmful Algae* **2020**, *109*, 102107. [CrossRef]
22. Wang, D.Z.; Xin, Y.H.; Wang, M.H. *Gambierdiscus* and Its Associated Toxins: A Minireview. *Toxins* **2022**, *14*, 485. [CrossRef] [PubMed] [PubMed Central]
23. Kohli, G.S.; Farrell, H.; Murray, S.A. Gambierdiscus, the cause of ciguatera fish poisoning: An increased human health threat influenced by climate change. In *Climate Change and Marine and Freshwater Toxins*; Botana, L.M., Louzao, C.M., Vilariño, N., Eds.; De Gruyter: Berlin, Germany, 2015; pp. 271–312.
24. Mudge, E.M.; Miles, C.O.; Ivanova, L.; Uhlig, S.; James, K.S.; Erdner, D.L.; Fæste, C.K.; McCarron, P.; Robertson, A. Algal ciguatoxin identified as source of ciguatera poisoning in the Caribbean. *Chemosphere* **2023**, *330*, 138659. [CrossRef]
25. Miles, C.O.; Burton, I.W.; Lewis, N.I.; Robertson, A.; Giddings, S.D.; McCarron, P.; Mudge, E.M. Isolation of Caribbean Ciguatoxin-5 (C-CTX5) and confirmation of its structure by NMR spectroscopy. *Tetrahedron* **2024**, *162*, 134115. [CrossRef]
26. Flores-Holguín, N.; Salas-Leiva, J.S.; Núñez-Vázquez, E.J.; Tovar-Ramírez, D.; Glossman-Mitnik, D. Exploring marine toxins: Comparative analysis of chemical reactivity properties and potential for drug discovery. *Front. Chem.* **2023**, *11*, 1286804. [CrossRef] [PubMed]
27. Malto, Z.B.L.; Benico, G.A.; Batucan, J.D.; De la Cruz, J.; Romero, M.L.J.; Azanza, R.V.; Salvador-Reyes, L.A. Global mass spectrometric analysis reveals chemical diversity of secondary metabolites and 44-Methylgambierone production in Philippine *Gambierdiscus* strains. *Front. Mar. Sci.* **2022**, *8*, 767024. [CrossRef]
28. Yon, T.; Réveillon, D.; Sibat, M.; Holland, C.R.; Litaker, W.; Nascimento, S.M.; Rossignoli, A.E.; Riobó, P.; Hess, P.; Bertrand, S. Targeted and non-targeted mass spectrometry to explore the chemical diversity of the genus *Gambierdiscus* in the Atlantic Ocean. *Phytochemistry* **2024**, *222*, 114095. [CrossRef] [PubMed]
29. Sierra, B.A.; Cruz, A.; Núñez-Vázquez, E.J.; Del Villar, L.M.; Cerecero, J.; Ochoa, J.L. An overview of the marine food poisoning in Mexico. *Toxicon* **1998**, *36*, 1493–1502. [CrossRef]
30. Núñez-Vázquez, E.J.; Almazán-Becerril, A.; Heredia-Tapia, A.; Hernández-Becerril, D.U.; Troccoli-Ghinaglia, L.; Arredondo-Vega, B.O.; Herrera-Silveira, J.A.; Vázquez-Castellanos, J.L.; Ochoa, J.L. Incidencia del envenenamiento por Ciguatera en Mexico. In Proceedings of the 4th Reunión de Expertos en Envenenamientos por Animales Ponzoñosos, Cuernavaca, Mexico, 30–31 March 2000; pp. 56–57.
31. Okolodkov, Y.B.; Gárate-Lizárraga, I. An annotated checklist of dinoflagellates (Dinophyceae) from the Mexican Pacific. *Acta Botánica Mex.* **2006**, *74*, 1–154.
32. Gárate-Lizárraga, I.; Band-Schmidt, C.J.; Verdugo-Díaz, G.; Muñetón-Gómez, M.J.; Félix-Pico, E. Dinoflagelados (Dinophyceae) del sistema lagunar Magdalena-Almejas. In *Estudios Ecológicos en Bahía Magdalena*; Funes-Rodríguez, R., Gómez-Gutiérrez, J., Palomares-García, R., Eds.; Gobierno del Estado de B. C. S.: La Paz, Mexico; Secretaria de Turismo de B. C. S.: La Paz, Mexico; Fondo para la Protección de los Recursos Marinos de B. C. S.: La Paz, Mexico; Instituto Politécnico Nacional, Centro Interdisciplinario de Ciencias Marinas: La Paz, Mexico, 2007; pp. 145–174.
33. Cortés-Altamirano, R. Two new localities for *Gambierdiscus toxicus* in the Mexican Pacific. *Harmful Algae News* **2012**, *45*, 10–11.
34. Núñez-Vázquez, E.J.; Hernández-Sandoval, F.E.; Cordero-Tapia, A.; Almazán-Becerril, A.; Band-Schmidt, C.J.; Bustillos-Guzmán, J.; López-Cortés, D.; Salinas-Zavala, C.A.; Morales-Zárate, M.V.; Mejía-Rebollo, A.; et al. Presencia de microalgas bénticas potencialmente nocivas y mortandad de peces asociadas al Parque Marino de Cabo Pulmo, B.C.S. In Proceedings of the II Congreso Nacional de Sociedad Mexicana para el estudio de los Florecimientos Algales Nocivos, Manzanillo, Mexico, 30–31 October 2013; p. 41.
35. Garate-Lizárraga, I.; Okolodkov, Y.B.; Cortés-Altamirano, R. Microalgas formadoras de florecimientos algales en el Golfo de California. In *Florecimientos Algales Nocivos en Mexico*; García-Mendoza, E., Quijano-Scheggia, S.I., Olivos-Ortíz, A., Núñez-Vázquez, E.J., Eds.; CICESE: Ensenada, Mexico, 2016; pp. 130–145. ISBN 978-607-95688-5-6.
36. Okolodkov, Y.B.; Gárate-Lizárraga, I.; Martínez-Cruz, M.E.; Galicia-García, C. Epibenthic dinoflagellates of the southern Gulf of California: Species composition and abundance (2015–2019). In Proceedings of the 19th International Conference on Harmful Algae (ICHA 2021), La Paz, Mexico, 10–15 October 2021.
37. Morquecho-Escamilla, L.; Reyes-Salinas, A.; Okolodkov, Y.B.; Gárate-Lizárraga, I.; Mazariegos-Villarreal, A.; Galicia-García, C. Aislamiento y caracterización taxonómica de dinoflagelados epífitos en Bahía de La Paz e Isla San José, Golfo de California. In Proceedings of the Memorias del IV Congreso Somefan y II Reunión Alen, Cancun, Mexico, 23–27 October 2017.
38. Ramos-Santiago, A.E.; Band-Schmidt, C.J.; Leyva-Valencia, I.; Fernández-Herrera, L.J.; Núñez-Vázquez, E.J.; Okolodkov, Y.B. *Gambierdiscus carpenteri* (Dinophyceae) from Bahía de La Paz, Gulf of California: Morphology, genetic affinities, and mouse toxicity. *Bot. Mar.* **2024**, *67*, 309–324. [CrossRef]

39. Litaker, R.W.; Vandersea, M.W.; Faust, M.A.; Kibler, S.R.; Chinain, M.; Holmes, M.J.; Holland, W.C.; Tester, P.A. Taxonomy of *Gambierdiscus* including four new species, *Gambierdiscus caribaeus*, *Gambierdiscus carolinianus*, *Gambierdiscus carpenteri* and *Gambierdiscus ruetzleri* (Gonyaulacales, Dinophyceae). *Phycologia* **2009**, *48*, 344–390. [CrossRef]
40. Díaz-Asencio, L.; Rojas Abrahantes, G.; Pérez Avilleira, G.; Chamero, D.; Moreira González, A. Clave dicotómica para la identificación preliminar de las especies de *Gambierdiscus* y *Fukuyoa* reportadas en la región del Caribe. *Rev. De Investig. Mar.* **2023**, *43*, 25–35. Available online: https://hdl.handle.net/1834/43015 (accessed on 11 July 2024).
41. Soler-Onís, E.; Fernández-Zabala, J.; Ojeda-Rodriguez, A.; Amorim, A. Bloom of *Gambierdiscus caribaeus* in the temperate-subtropical waters of El Hierro, Canary Islands (north East Atlantic). *Harmful Algae News* **2016**, *55*, 14–17.
42. Díaz-Asencio, L.; Vandersea, M.; Chomérat, N.; Fraga, S.; Clausing, R.J.; Litaker, R.W.; Chamero-Lago, D.; Gómez-Batista, M.; Moreira-González, A.; Tester, P.; et al. Morphology, toxicity and molecular characterization of *Gambierdiscus* spp. towards risk assessment of ciguatera in south central Cuba. *Harmful Algae* **2019**, *86*, 119–127. [CrossRef] [PubMed]
43. Smith, K.F.; Rhodes, L.; Verma, A.; Curley, B.G.; Harwood, D.; Kohli, G.S.; Solomona, D.; Rongo, T.; Munday, R.; Murray, S.A. A new *Gambierdiscus* species (Dinophyceae) from Rarotonga, Cook Islands: *Gambierdiscus cheloniae* sp. nov. *Harmful Algae* **2016**, *60*, 45–56. [CrossRef]
44. Kohli, G.S.; Murray, S.A.; Neilan, B.A.; Rhodes, L.L.; Harwood, D.T.; Smith, K.F.; Meyer, L.; Capper, A.; Brett, S.; Hallegraeff, G.M. High abundance of the potentially maitotoxic dinoflagellate *Gambierdiscus carpenteri* in temperate waters of New South Wales, Australia. *Harmful Algae* **2014**, *39*, 134–145. [CrossRef]
45. Vacarizas, J.; Benico, G.; Austero, N.; Azanza, R. Taxonomy and toxin production of *Gambierdiscus carpenteri* (Dinophyceae) in a tropical marine ecosystem: The first record from the Philippines. *Mar. Pollut. Bull.* **2018**, *137*, 430–443. [CrossRef]
46. Chinain, M.; Gatti, C.M.; Roue, M.; Darius, H.T. Ciguatera-causing dinoflagellates in the genera *Gambierdiscus* and *Fukuyoa*: Distribution, ecophysiology and toxicology. In *Dinoflagellates: Classification, Evolution, Physiology and Ecological Significance*; Subba Rao, D.V., Ed.; Nova Science Publishers: Hauppauge, NY, USA, 2020; pp. 405–457.
47. Litaker, R.W.; Holland, W.C.; Hardison, D.R.; Pisapia, F.; Hess, P.; Kibler, S.R.; Tester, P.A. Ciguatoxicity of Gambierdiscus and Fukuyoa species from the Caribbean and Gulf of Mexico. *PLoS ONE* **2017**, *12*, e0185776. [CrossRef]
48. Tester, P.; Wickliffe, L.; Jossart, J.; Rhodes, L.; Enevoldsen, H.; Adachi, M.; Litaker, W. Global distribution of the genera Gambierdiscus and Fukuyoa. *Harmful Algae* **2018**, *138*, 138–143.
49. Richlen, M.L.; Horn, K.; Uva, V.; Fachon, E.; Heidmann, S.L.; Smith, T.B.; Parsons, M.L.; Anderson, D.M. Gambierdiscus species diversity and community structure in St. Thomas, USVI and the Florida Keys, USA. *Harmful Algae* **2024**, *131*, 102562. [CrossRef]
50. Xu, Y.; Richlen, M.L.; Liefer, J.D.; Robertson, A.; Kulis, D.; Smith, T.B.; Parsons, M.L.; Anderson, D.M. Influence of environmental variables on *Gambierdiscus* spp. (Dinophyceae) growth and distribution. *PLoS ONE* **2016**, *11*, e0153197. [CrossRef]
51. Loeffler, C.R.; Tartaglione, L.; Friedemann, M.; Spielmeyer, A.; Kappenstein, O.; Bodi, D. Ciguatera Mini Review: 21st Century Environmental Challenges and the Interdisciplinary Research Efforts Rising to Meet Them. *Int. J. Environ. Res. Public Health* **2021**, *18*, 3027. [CrossRef]
52. Cementerio de Yates: Negligencia y Omisiones en la Bahía de La Paz Tras Paso de Norma. Available online: https://causanaturamedia.com/periodismo-en/Cementerio-de-yates-negligencia-y-omisiones-en-la-bahia-de-La-Paz-tras-paso-de-Norma (accessed on 1 April 2024).
53. Gobierno de BCS Sin Capacidad de Retirar Embarcaciones Varadas en La Paz: VCC. Available online: https://tribunademexico.com/gobierno-de-bcs-sin-capacidad/ (accessed on 11 July 2024).
54. Parsons, M.L.; Settlemier, C.J.; Ballauer, J.M. An examination of the epiphytic nature of Gambierdiscus toxicus, a dinoflagellate involved in ciguatera fish poisoning. *Harmful Algae* **2011**, *10*, 598–605. [CrossRef]
55. Parsons, M.L.; Brandt, A.L.; Ellsworth, A.; Leynse, A.K.; Rains, L.K.; Anderson, D.M. Assessing the use of artificial substrates to monitor Gambierdiscus populations in the Florida Keys. *Harmful Algae* **2017**, *68*, 52–66. [CrossRef] [PubMed]
56. Tester, P.A.; Kibler, S.R.; Holland, W.C.; Usup, G.; Vandersea, M.W.; Leaw, C.P.; Teen, L.P.; Larsen, J.; Mohammad-Noor, N.; Faust, M.A.; et al. Sampling harmful benthic dinoflagellates: Comparison of artificial and natural substrate methods. *Harmful Algae* **2014**, *39*, 8–25. [CrossRef]
57. Islas y Áreas Protegidas del Golfo de California. Available online: https://www.gob.mx/semarnat/articulos/islas-y-areas-protegidas-del-golfo-de-california-269050 (accessed on 12 April 2021).
58. Heredia-Tapia, A.; Arredondo-Vega, B.O.; NuñezVázquez, E.J.; Ochoa, J.L. Partial biochemical characterization and toxicological evaluation of the dinoflagellate *Prorocentrum lima* (=*Exuviaella lima*) isolated from El Pardito Island in Baja California Sur, México. In Proceedings of the Memorias 9th International Conference on Algal Blooms, Hobart, Australia, 7–11 February 2000.
59. Nascimento, S.M.; Purdie, D.A.; and Morris, S. Morphology, toxin composition and pigment content of *Prorocentrum lima* strains isolated from a coastal lagoon in southern UK. *Toxicon* **2005**, *45*, 633–649. [CrossRef] [PubMed]
60. David, H.; Laza-Martínez, A.; Kromkamp, J.C.; Orive, E. Physiological response of *Prorocentrum lima* (Dinophyceae) to varying light intensities. *FEMS Microbiol. Ecol.* **2018**, *94*, 1–12. [CrossRef]
61. Wu, H.; Zhang, H.; Peng, J.; Zheng, G.; Lu, S.; Tan, Z. Adaptive responses of geographically distinct strains of the benthic dinoflagellate, *Prorocentrum lima* (Dinophyceae), to varying light intensity and photoperiod. *Harmful Algae* **2023**, *127*, 102479. [CrossRef]

62. Zapata, M.; Freire, J.; Garrido, J.L. Pigment composition of several harmful algae as determined by HPLC using pyridine-containing mobile phase polymeric octadecylsilica column. In *Harmful Algae*; Reguera, B., Blanco, J., Fernández, M.L., Wyatt, T., Eds.; UNESCO—Xunta de Galicia: Santiago de Compostela, Spain, 1998; pp. 304–307.
63. Zapata, M.; Fraga, S.; Rodríguez, F.; Garrido, J.L. Pigment-based chloroplast types in dinoflagellates. *Mar. Ecol. Prog. Ser.* **2012**, *465*, 33–52. [CrossRef]
64. Carreto, J.I.; Carignan, M.O.; Montoya, N.G. Comparative studies on mycosporine-like amino acids, paralytic shellfish toxins and pigment profiles of the toxic dinoflagellates *Alexandrium tamarense*, *Acatenella* and *A. minutum*. *Mar. Ecol. Prog. Ser.* **2001**, *223*, 49–60. [CrossRef]
65. Carreto, J.I.; Seguel, M.; Montoya, N.G.; Clément, A.; Carignan, M.O. Pigment profile of the ichthyotoxic dinoflagellate *Gymnodinium* sp. from a massive bloom in southern Chile. *J. Plank. Res.* **2001**, *23*, 1171–1175. [CrossRef]
66. Bustillos-Guzmán, J.; Garate-Lizarraga, I.; López-Cortés, D.; Hernández-Sandoval, F.E. The use of pigment "fingerprints" in the study of harmful algal blooms. *Rev. Biol. Trop.* **2004**, *52* (Suppl. S1), 17–26.
67. Liu, S.; Peng, Y.; Yu, Z.; Dong, L.I.; Deng, C.; Yu, Z. HPLC pigment Profiles of 31 Harmful Algal Bloom Species Isolated from the Coastal Sea Areas of China. *Ocean. Coast. Sea Res.* **2014**, *13*, 941–950. [CrossRef]
68. Durand, M.; Berkaloff, C. Pigments composition and chloroplast organization of *Gambierdiscus toxicus* Adachi and Fukuyo (Dinophyceae). *Phycologia* **1985**, *24*, 217–223. [CrossRef]
69. Indelicato, S.R.; Watson, D.A. Identification of the Photosynthetic Pigments of the Tropical Benthic Dinoflagellate *Gambierdiscus toxicus*. *Mar. Fish. Rev.* **1986**, *48*, 44–47.
70. Bomber, J.W.; Tindall, D.R.; Venable, C.W.; Miller, D.M. Pigment composition and low-ligth response of fourteen clones of *Gambierdiscus toxicus*. In *Toxic Marine Phytoplankton*; Graneli, E., Ed.; Elsevier Science Publishing Co., Inc.: Amsterdam, The Netherlands, 1990; pp. 263–268.
71. Miller, D.M.; Tindall, D.R.; Venable, C.W. NMR Spectroscopy of Clorophyll(s)-a isolated from *Gambierdiscus toxicus*. In *Toxic Marine Phytoplankton*; Graneli, E., Ed.; Elsevier: Amsterdam, The Netherlands, 1993; pp. 305–310.
72. Bustillos-Guzmán, J.; Diogene, J. Pigment content and toxicity in three strains of *Gambierdiscus toxicus*: Implications of cell volume differences. In *Harmful Algae*; Reguera, B., Blanco, J., Fernández, M.L., Wyatt, T., Eds.; UNESCO—Xunta de Galicia: Santiago de Compostela, Spain, 1998; pp. 372–373.
73. Fraga, S.; Rodríguez, F.; Caillaud, A.; Diogène, J.; Raho, N.; Zapata, M. *Gambierdiscus excentricus* sp. nov. (Dinophyceae), a benthic toxic dinoflagellate from the Canary Islands (NE Atlantic Ocean). *Harmful Algae* **2011**, *11*, 10–22. [CrossRef]
74. Flynn, K.; Flynn, K.J.; Jones, K.J. Changes in dinoflagellate intracellular amino acids in response to diurnal changes in light and N supply. *Mar. Ecol. Prog. Ser.* **1993**, *100*, 245–252. [CrossRef]
75. Flynn, K.J.; Jones, K.J.; Raine, R.; Richard, J.; Flynn, K. Use of intracellular amino acid analysis as an indicator of the physiological status of natural dinoflagellate populations. *Mar. Ecol. Prog. Ser.* **1994**, *103*, 175–186. [CrossRef]
76. Okaichi, T. Significance of Amino Acid Composition of Phytoplankton and suspensoid in Marine Biological Production. *Bull. Jap. Soc. Sc. Fish.* **1974**, *40*, 471–478. [CrossRef]
77. Hayashi, T.; Suitani, Y.; Murakami, M.; Yamaguchi, K.; Konosu, S.; and Noda, H. Protein and Amino Acid Compositions of Five Species of Marine Phytoplankton. *Bull. Japan Soc. Sc. Fish.* **1986**, *52*, 337–343. [CrossRef]
78. Lim, A.S.; Jeong, H.J.; Kim, S.J.; Ok, J.E. Amino acids profiles of six dinoflagellate species belonging to diverse families: Possible use as animal feeds in aquaculture. *Algae* **2018**, *33*, 279–290. [CrossRef]
79. Lewis, R.L. Detection of ciguatoxins and related benthic dinoflagellate toxins: In vivo and in vitro methods. In *Manual on Harmful Marine Microalgae*; Hallegraeff, G.M., Anderson, D.M., Cembella, A.D., Enevoldsen, H.O., Eds.; UNESCO-IOC: Paris, France, 1995; pp. 135–161.
80. FAO. *Marine Biotoxins*; Food and Nutrition Paper 80; Food and Agriculture Organozation of the United Nations: Rome, Italy, 2004; 278p.
81. Bienfang, P.; Oben, B.; DeFelice, S.; Moeller, P.; Huncik, K.; Oben, P.; Bowen, B. Ciguatera: The detection of neurotoxins in carnivorous reef fish from the coast of Cameroon, West Africa. *Afr. J. Mar. Sci.* **2008**, *30*, 533–540. [CrossRef]
82. Ley-Martínez, T.C.; Núñez-Vázquez, E.J.; Balart-Páez, E.F.; Poot-Delgado, C.A.; Band-Schmidt, C.J.; Almazán-Becerril, A. Identificación y cuantificación de ciguatoxinas en peces carnívoros de la Península de Yucatán. In Proceedings of the Memorias del IV Congreso Nacional de la SOMEFAN y II Congreso Latinoamericano de la ALEAN, Cancun, Mexico, 23–27 October 2017; p. 88.
83. Neves, R.A.F.; Fernandes, T.; Santos, L.N.d.; Nascimento, S.M. Toxicity of benthic dinoflagellates on grazing, behavior and survival of the brine shrimp *Artemia salina*. *PLoS ONE* **2017**, *12*, e0175168. [CrossRef]
84. Heredia-Tapia, A.; Arredondo-Vega, B.O.; Núñez-Vázquez, E.J.; Yasumoto, T.; Yasuda, M.; Ochoa, J.L. Isolation of *Prorocentrum lima* (Syn. *Exuviaella lima*) and diarrhetic shellfish poisoning (DSP) risk assessment in the Gulf of California, Mexico. *Toxicon* **2002**, *40*, 1121–1127. [PubMed]
85. Ajuzie, C.C. Palatability and fatality of the dinoflagellate *Prorocentrum lima* to *Artemia salina*. *J. Appl. Phycol.* **2007**, *19*, 513–519. [CrossRef]
86. Faimali, M.; Giussani, V.; Piazza, V.; Garaventa, F.; Corrà, C.; Asnaghi, V.; Privitera, D.; Gallus, L.; Cattaneo-Vietti, R.; Mangialajo, L.; et al. Toxic effects of harmful benthic dinoflagellate *Ostreopsis ovata* on invertebrate and vertebrate marine organisms. *Mar. Environ. Res.* **2012**, *76*, 97–107. [CrossRef] [PubMed]

87. Gong, Y.; Zhang, K.; Geng, N.; Wu, M.; Yi, X.; Liu, R.; Challis, J.K.; Codling, G.; Xu, E.G.; Giesy, J.P. Molecular mechanisms of zooplanktonic toxicity in the okadaic acid-producing dinoflagellate *Prorocentrum lima*. *Environ. Pollut.* **2021**, *279*, 116942. [CrossRef] [PubMed]
88. Yan, M.; Mak, M.Y.L.; Cheng, J.; Li, J.; Gu, J.R.; Leung, P.T.Y.; Lam, P.K.S. Effects of dietary exposure to ciguatoxin P-CTX-1 on the reproductive performance in marine medaka (*Oryzias melastigma*). *Mar. Pollut. Bull.* **2020**, *152*, 110837. [CrossRef]
89. Li, L.; Chen, S.; Xu, S.Y.; Li, D.W.; Li, H.Y.; Yang, W.D. Toxicity and underlying mechanism of the toxic dinoflagellate *Gambierdiscus caribaeus* to the fish *Oryzias melastigma*. *Ecotoxicol. Environ. Saf.* **2022**, *247*, 114223. [CrossRef]
90. Helfrich, P.; Bannerg, A. Experimental Induction of Ciguatera Toxicity in Fish through Diet. *Nature* **1963**, *197*, 1025–1026. [CrossRef]
91. Davin, W.T., Jr.; Kohler, C.C.; Tindall, D.R. Effects of ciguatoxins on the bluehead. *Trans. Am. Fish. Soc.* **1986**, *115*, 908–912. [CrossRef]
92. Davin, W.T., Jr.; Kohler, C.C. Ciguatera toxins adversely affect piscivorous fishes. *Trans. Am. Fish. Soc.* **1988**, *117*, 374–384. [CrossRef]
93. Lewis, R.J. Ciguatoxins are potent ichthyotoxins. *Toxicon* **1992**, *30*, 207–211. [CrossRef]
94. Magnelia, S.J.; Kohler, C.C. Acanthurids Do Not Avoid Consuming Cultured Toxic Dinoflagellates yet Do Not Become Ciguatoxic. *Trans. Am. Fish. Soc.* **1992**, *121*, 737–745. [CrossRef]
95. Quod, J. *Action de la Ciguatoxine sur Léquilibrie de Distribution Hydroionique et sur la Perméabilité au Na*$^+$ *du Muscle Blanc Dún Téléostéen, Chelon Labrosus*; University of Bordeaux: Bordeaux, France, 1983.
96. Capra, M.F.; Cameron, J.; Flowers, A.E.; Coombe, I.F.; Blanton, C.G.; Hahn, S.T. The Effects of Ciguatoxin on Teleosts. In Proceedings of the 6th International Coral Reef Congress, Townsville, Australia, 8–12 August 1988; pp. 37–41.
97. Durand-Clement, M.; Amade, P.; Puiseux-Dao, S. Induction of toxicity in fishes fed with cultures of the ciguateric dinoflagellate *Gambierdiscus toxicus*. In Proceedings of the 4th International Meeting of the SAA, Villeneuve d'Ascq, France, 15–17 September 1987; Stadier, T., Karamanos, Y., Verdus, M.C., Mollion, J., Christiaen, D., Eds.; Elsevier: Amsterdam, The Netherlands, 1988; p. 19.
98. Kohler, C.C.; Paleudis, G.A.; Tindall, D.R. Behavior abnormalities displayed by ocean surgeon following consumption of ciguatoxigenic dinoflagellates. *Proc. Assn. Island Mar. Lab. Carib.* **1989**, *22*, 34.
99. Kelly, A.M.; Kohler, C.C.; Tindall, D.R. Are crustaceans linked to the ciguatera food chain? *Environ. Biol. Fish.* **1992**, *33*, 275–286. [CrossRef]
100. Kohler, C.C.; Kholer, S. Ciguatera tropical fish poisoning: What's happening in the food chain. In Proceedings of the 26th Meeting of the Association of Marine Laboratories of the Caribbean, San Salvador, The Bahamas, 11–16 June 1994; Gerace, D.T., Ed.; UNESCO; Paris, France, 1998; pp. 112–125.
101. González, G.; Brusle, J.; Crespo, S. Ultrastructural alterations of cabrilla sea bass *Serranus cabrilla* liver related to experimental *Gambierdiscus toxicus* (dinoflagellate) ingestion. *Dis. Aqua. Org.* **1994**, *18*, 187–193. [CrossRef]
102. Edmunds, J.S.; McCarthy, R.A.; Ramsdell, J.S. Ciguatoxin reduces larval survivability in finfish. *Toxicon* **1999**, *37*, 1827–1832. [CrossRef]
103. Mak, L.Y.; Li, J.; Liu, C.N.; Cheng, S.H.; Lam, P.K.S.; Cheng, J.; Chan, L.L. Physiological and behavioural impacts of Pacific ciguatoxin-1 (P-CTX-1) on marine medaka (*Oryzias melastigma*). *J. Hazard. Mater.* **2017**, *321*, 782–790. [CrossRef]
104. Clausing, R.J.; Ben Gharbia, H.; Sdiri, K.; Sibat, M.; Rañada-Mestizo, M.L.; Lavenu, L.; Hess, P.; Chinain, M.; Bottein, M.-Y.D. Tissue Distribution and Metabolization of Ciguatoxins in an Herbivorous Fish following Experimental Dietary Exposure to *Gambierdiscus polynesiensis*. *Mar. Drugs* **2024**, *22*, 14. [CrossRef]
105. Yan, M.; Leung, P.T.; Ip, J.C.; Cheng, J.P.; Wu, J.J.; Gu, J.R.; Lam, P.K. Developmental toxicity and molecular responses of marine medaka (*Oryzias melastigma*) embryos to ciguatoxin P-CTX-1 exposure. *Aquat. Toxicol.* **2017**, *185*, 149–159. [CrossRef]
106. Leite, I.D.P.; Sdiri, K.; Taylor, A.; Viallon, J.; Gharbia, H.B.; Mafra Júnior, L.L.; Swarzenski, P.; Oberhaensli, F.; Darius, H.T.; Chinain, M.; et al. Experimental Evidence of Ciguatoxin Accumulation and Depuration in Carnivorous Lionfish. *Toxins* **2021**, *13*, 564. [CrossRef]
107. Galarza, J.R.; Edwards, R.A.; Baden, D.G.; Tosteson, T.R. Effect of Barracuda Ciguatoxins on Pigment Granule Aggregation in Teleost Melanophores. *Bull. De La Soc. De Pathol. Exot.* **1993**, *85*, 528.
108. Igarashi, T.; Aritake, S.; Yasumoto, T. Mechanisms underlying the hemolytic and ichthyotoxic activities of maitotoxin. *Nat. Toxins* **1999**, *7*, 71–79. [CrossRef]
109. Murata, M.; Naoki, H.; Iwashita, T.; Matsunaga, S.; Sasaki, M.; Yokoyama, A.; Takeshi, Y. Structure of Maitotoxin. *J. Am. Chem. Soc.* **1993**, *115*, 2060–2062. [CrossRef]
110. Nicolaou, K.C.; Frederick, M.O.; Aversa, R.J. The continuing saga of the marine polyether biotoxins. *Angew. Chem. Int. Ed. Engl.* **2008**, *47*, 7182–7225. [CrossRef]
111. Kohli, G.S.; Papiol, G.G.; Rhodes, L.L.; Harwood, D.T.; Selwood, A.; Jerrett, A.; Murray, S.A.; Neilan, B.A. A feeding study to probe the uptake of Maitotoxin by snapper (*Pagrus auratus*). *Harmful Algae* **2014**, *37*, 125–132. [CrossRef]
112. Yasumoto, T.; Bagnis, R.; Vernoux, J.P. Toxicity of the surgeonfishes. II. Properties of the principal water-soluble toxin. *Bull. Jpn. Soc. Fish.* **1976**, *42*, 359–365. [CrossRef]
113. Takahishi, M.; Tatsumi, M.; Ohizumi, Y.; Yasumoto, T. Ca^{2+} channel activation function of maitotoxin the most potent marine toxin known, in clonal rat pheochromocytoma cells. *J. Biol. Chem.* **1983**, *258*, 10944–10949. [CrossRef]

114. Murata, M.; Yasumoto, T. The structure elucidation and biological activities of high molecular weight algal toxins: Maitotoxin, prymnesins and zooxanthellatoxins. *Nat. Prod. Rep.* **2000**, *17*, 293–314. [CrossRef]
115. Terao, K.; Ito, E.; Igarashi, T.; Aritake, S.; Seki, T.; Satake, M.; Yasumoto, T. Effects of Prymnesin, Maitotoxin and Gymnomidine on the structure of Gills of small fish Akahire, *Tanichtys albonubes*. Lin. In *Harmful and Toxic Algal Blooms*; Yasumoto, T., Oshima, Y., Fukuyo, Y., Eds.; IOC-UNESCO: Paris, France, 1996; pp. 479–481.
116. Landsberg, J.H. Tropical reef fish disease outbreaks and mass mortalities in Florida: What is the role of dietary biological toxins? *Dis. Aquat. Org.* **1995**, *22*, 83–100. [CrossRef]
117. McCall, J.R.; Jacocks, H.M.; Niven, S.C.; Poli, M.A.; Baden, D.G.; Bourdelais, A.J. Development and utilization of a fluorescence-based receptor-binding assay for the site 5 voltage-sensitive sodium channel ligands brevetoxin and ciguatoxin. *J. AOAC Int.* **2014**, *97*, 307–315. [CrossRef]
118. Pisapia, F.; Holland, W.C.; Hardison, D.R.; Litaker, R.W.; Fraga, S.; Nishimura, T.; Adachi, M.; Nguyen-Ngoc, L.; Séchet, V.; Amzil, Z.; et al. Toxicity screening of 13 *Gambierdiscus* strains using neuro-2a and erythrocyte lysis bioassays. *Harmful Algae* **2017**, *63*, 173–183. [CrossRef]
119. Holland, W.C.; Litaker, R.W.; Tomas, C.R.; Kibler, S.R.; Place, A.R.; Davenport, E.D.; Tester, P.A. Differences in the toxicity of six *Gambierdiscus* (Dinophyceae) species measured using an in vitro human erythrocyte lysis assay. *Toxicon* **2013**, *65*, 15–33. [CrossRef] [PubMed]
120. Rossignoli, A.E.; Tudó, A.; Bravo, I.; Díaz, P.A.; Diogène, J.; Riobó, P. Toxicity Characterisation of *Gambierdiscus* Species from the Canary Islands. *Toxins* **2020**, *12*, 134. [CrossRef] [PubMed]
121. Murray, J.S.; Nishimura, T.; Finch, S.C.; Rhodes, L.L.; Puddick, J.; Harwood, D.T.; Larsson, M.E.; Doblin, M.A.; Leung, P.; Yan, M.; et al. The role of 44-methylgambierone in ciguatera fish poisoning: Acute toxicity, production by marine microalgae and its potential as a biomarker for *Gambierdiscus* spp. *Harmful Algae* **2020**, *97*, 101853. [CrossRef] [PubMed]
122. Xu, Y.; He, X.; Lee, W.H.; Chan, L.L.; Lu, D.; Wang, P.; Tao, X.; Li, H.; Yu, K. Ciguatoxin-Producing Dinoflagellate *Gambierdiscus* in the Beibu Gulf: First Report of Toxic *Gambierdiscus* in Chinese Waters. *Toxins* **2021**, *13*, 643. [CrossRef]
123. Munday, R.; Murray, S.; Rhodes, L.; Larsson, M.; Harwood, D. Ciguatoxins and Maitotoxins in extracts of sixteen *Gambierdiscus* isolates and one *Fukuyoa* isolate from the South Pacific and their toxicity to mice by intraperitoneal and oral administration. *Mar. Drugs* **2017**, *15*, 208. [CrossRef]
124. Larsson, M.; Laczka, O.; Harwood, D.; Lewis, R.; Himaya, S.; Murray, S.; Doblin, M. Toxicology of *Gambierdiscus* spp. (Dinophyceae) from tropical and temperate Australian waters. *Mar. Drugs* **2018**, *16*, 7. [CrossRef]
125. Darius, H.T.; Roué, M.; Sibat, M.; Viallon, J.; Gatti, C.M., II; Vandersea, M.W.; Tester, P.A.; Litaker, R.W.; Amzil, Z.; Hess, P.; et al. Toxicological Investigations on the Sea Urchin *Tripneustes gratilla* (Toxopneustidae, Echinoid) from Anaho Bay (Nuku Hiva, French Polynesia): Evidence for the Presence of Pacific Ciguatoxins. *Mar. Drugs* **2018**, *16*, 122. [CrossRef]
126. Darius, H.T.; Revel, T.; Viallon, J.; Sibat, M.; Cruchet, P.; Longo, S.; Hardison, D.R.; Holland, W.C.; Tester, P.A.; Litaker, R.W.; et al. Comparative Study on the Performance of Three Detection Methods for the Quantification of Pacific Ciguatoxins in French Polynesian Strains of *Gambierdiscus polynesiensis*. *Mar. Drugs* **2022**, *20*, 348. [CrossRef] [PubMed] [PubMed Central]
127. Estevez, P.; Oses-Prieto, J.; Castro, D.; Penin, A.; Burlingame, A.; Gago-Martinez, A. First Detection of Algal Caribbean Ciguatoxin in Amberjack Causing Ciguatera Poisoning in the Canary Islands (Spain). *Toxins* **2024**, *16*, 189. [CrossRef]
128. Lluch-Cota, S.; Aragón-Noriega, E.A.; Arreguín-Sánchez, F.; Aurioles-Gamboa, D.; Bautista-Romero, J.J.; Brusca, R.C.; Cervantes-Duarte, R.; Cortés-Altamirano, R.; Del-Monte-Luna, P.; Esquivel-Herrera, A.; et al. The Gulf of California: Review of ecosystem status and sustainability challenges. *Prog. Oceanogr.* **2007**, *73*, 1–26. [CrossRef]
129. Brusca, R.C.; Findley, L.T.; Hastings, P.A.; Hendrickx, M.E.; Torre, J.; van der Heiden, A.M. Macrofaunal diversity in the Gulf of California. In *Biodiversity, Ecosystems, and Conservation in Northern Mexico*; Cartron, J.L.E., Ceballos, G., Felger, R.S., Eds.; Oxford University Press: Oxford, UK, 2005; pp. 179–203.
130. Brusca, R.C. *The Gulf of California: Biodiversity and Conservation. Arizona-Sonora Desert Museum Studies in Natural History*; University of Arizona Press: Tucson, AZ, USA, 2012.
131. Gómez-Cabrera, I. Pesca Deportiva y Pesca Ribereña en Baja California Sur, México: Comparación del Valor Económico. Ph.D. Thesis, Universidad Autónoma de Baja California Sur, La Paz, Mexico, 2012.
132. Gómez-Cabrera, I.D.; Ivanova-Boncheva, A. Valor económico de la pesca deportiva como fuente principal de atracción turística en Los Cabos, Baja California Sur, México. *TURYDES Tur. Y Desarro. Local Sosten.* **2013**, *6*, 25.
133. Keller, M.D.; Selvin, R.C.; Claus, W.; Guillard, R.R.L. Media for the culture of oceanic ultraphytoplankton. *J. Phycol.* **1987**, *23*, 633–638. [CrossRef]
134. Guillard, R.R.L.; Ryther, J.H. Studies of marine planktonic diatoms. I. Cyclotella nana Hustedt, and Detonula confervacea (Cleve) Gran. *Can. J. Microbiol.* **1962**, *8*, 229–239. [CrossRef] [PubMed]
135. Guillard, R.R. Culture of phytoplankton for feeding marine invertebrates. In *Culture of Marine Invertebrate Animals*; Springer: Boston, MA, USA, 1975; pp. 29–60.
136. Fritz, L.; Triemer, R.E. A rapid simple technique utilizing Calcofluor White M28 for the visualization of dinoflagellate thecal plates. *J. Phycol.* **1985**, *21*, 662–664. [CrossRef]
137. Vidussi, F.; Claustre, H.; Bustillos-Guzmán, J.; Cailleau, C.; Marty, J.C. Rapid HPLC method for determination of phytoplankton chemotaxinomic pigments: Separation of chlorophyll a from divinyl-chlorophyll a and zeaxanthin from lutein. *J. Plankton Res.* **1996**, *18*, 2377–2382. [CrossRef]

138. Montoura, R.F.C.; Repeta, D. Calibration methods for HPLC. In *Phytoplankton Pigments in Oceanography: Guidelines to Modern Methods*; Jeffrey, S.W., Mantoura, R.F.C., Wright, S.W., Eds.; UNESCO: Paris, France, 1997; pp. 407–428.
139. Gratzfeld-Huesgen, A. Sensitive and Reliable Amino Acid Analysis in Protein Hydrolysates Using the Agilent 1100 Series HPLC. Agilent Technologies. Publication Number 5968–5658E. 1999. Available online: https://channel.gimitec.com/sites/default/files/59685658.pdf (accessed on 12 July 2024).
140. AOAC. *AOAC Official Method 994.12 Amino Acids in Feeds*; Association of Official Analytical Collaboration (AOAC International): Rockville, ML, USA, 2000; 1p.
141. Chinain, M.; Darius, H.T.; Ung, A.; Cruchet, P.; Wang, Z.; Ponton, D.; Laurent, D.; Pauillac, S. Growth and toxin production in the ciguatera-causing dinoflagellate *Gambierdiscus polynesiensis* (Dinophyceae) in culture. *Toxicon* **2010**, *56*, 739–750. [CrossRef]
142. Satake, M.; Murata, M.; Yasumoto, T. The structure of CTX3C, a ciguatoxin congener isolated from cultured *Gambierdiscus toxicus*. *Tetrahedr. Lett.* **1993**, *34*, 1975–1978. [CrossRef]
143. Holmes, M.J.; Lewis, R.J. Purification and characterisation of large and small maitotoxins from cultured *Gambierdiscus toxicus*. *Nat. Toxins* **1994**, *2*, 64–72. [CrossRef]

marine drugs

MDPI

Article

Comparative Analysis of Tentacle Extract and Nematocyst Venom: Toxicity, Mechanism, and Potential Intervention in the Giant Jellyfish *Nemopilema nomurai*

Xiao-Yu Geng [1,†], Ming-Ke Wang [1,†], Xiao-Chuan Hou [2,†], Zeng-Fa Wang [2,3], Yi Wang [2,3], Die-Yu Zhang [4], Blessing Danso [2], Dun-Biao Wei [5], Zhao-Yong Shou [6,*], Liang Xiao [2,*] and Ji-Shun Yang [1,*]

1 Naval Medical Center of PLA, Naval Medical University, Shanghai 200052, China; 15339625597@163.com (X.-Y.G.); wmke021@163.com (M.-K.W.)
2 Faculty of Naval Medicine, Naval Medical University, Shanghai 200433, China; xiaochuanhou@126.com (X.-C.H.); wangzengfa2022@126.com (Z.-F.W.); wyi0815@hotmail.com (Y.W.); blessingdanso@hotmail.com (B.D.)
3 College of Traditional Chinese Medicine, Jilin Agricultural University, Changchun 130118, China
4 College of Pharmacy, Bengbu Medical University, Bengbu 233030, China; 4323@outlook.com
5 Unit 92196 of the People's Liberation Army, Qingdao 266000, China; wei1026627857@163.com
6 Faculty of Health Service, Naval Medical University, Shanghai 200433, China
* Correspondence: szy@smmu.edu.cn (Z.-Y.S.); hormat830713@hotmail.com (L.X.); jasunyang@foxmail.com (J.-S.Y.)
† These authors contributed equally to this work.

Abstract: The giant jellyfish *Nemopilema nomurai* sting can cause local and systemic reactions; however, comparative analysis of the tentacle extract (TE) and nematocyst venom extract (NV), and its toxicity, mechanism, and potential intervention are still limited. This study compared venom from TE and NV for their composition, toxicity, and efficacy in vitro and in vivo used RAW264.7 cells and ICR mice. A total of 239 and 225 toxin proteins were identified in TE and NV by proteomics, respectively. Pathological analysis revealed that TE and NV caused heart and liver damage through apoptosis, necrosis, and inflammation, while TE exhibited higher toxicity ex vivo and in vivo. Biochemical markers indicated TE and NV elevated creatine kinase, lactatedehydrogenase, and aspartate aminotransferase, with the TE group showing a more significant increase. Transcriptomics and Western blotting indicated both venoms increased cytokines expression and MAPK signaling pathways. Additionally, 1 mg/kg PACOCF3 (the phospholipase A2 inhibitor) improved survival from 16.7% to 75% in mice. Our results indicate that different extraction methods impact venom activities, tentacle autolysis preserves toxin proteins and their toxicity, and PACOCF3 is a potential antidote, which establishes a good extraction method of jellyfish venom, expands our understanding of jellyfish toxicity, mechanism, and provides a promising intervention.

Keywords: jellyfish venom; toxic effects; extraction; PLA2; verapamil; EDTA; PACOCF3

1. Introduction

Venoms are usually defense mechanisms or hunting weapons evolved from various biological metabolites, and their compositions are very complex, including proteins, peptides, and small molecules [1]. Due to their potent biological activities and in-depth research, venoms as therapeutic drug candidates have gradually become a hot spot worldwide [2–6]. Marine organisms have always been an excellent source of innovative compounds and have great potential in the development of new medicines, but compared with terrestrial venom, research progress has been slow. The main limitations include the difficulty of venom extraction, the activity of which is heavily affected by the seasons, the lack of information about the venom, and so on. Establishing a method of venom extraction that maintains their activities completely is an essential foundation for the research and development of the venom [7,8].

The cnidarians are an essential component of the marine ecosystem, with about 3900 described species, mainly comprising the Anthozoa, Endocnidozoa, and Medusozoa divisions. The Medusozoa is one of the most diverse and complex taxa in the cnidarian phylum [9], and is broadly divided into four distinct lineages, including Hydrozoa, Acraspeda, Cubozoa, and Scyphozoa [10,11]. Scyphozoa is one of the most recognizable species of Medusozoa, which is widely distributed and highly toxic [12]. Meanwhile, increases in Scyphozoa jellyfish abundance have been found in some sea areas, such as the eastern Atlantic and the western Pacific [13–15], with significant ecological, economic, and human health impacts [16–19]. The *Nemopilema nomurai* (*N. nomurai*) is one of the most toxic Scyphozoa jellyfish and the main poisonous jellyfish species in China, Japan, and Korea [16,20]. The toxin components of jellyfish venom are complex and have evolved, and their compositions are diverse among species [21,22]. However, in general, jellyfish venom is usually composed of proteins and peptides used by jellyfish for predation or defense [23]. Jellyfish toxins can cause various biological effects, including hemolytic toxicity [24], cardiovascular toxicity [25], neurotoxicity [26], and skin [27] and muscle toxicity [28]. Depending on the species, jellyfish stings can induce effects ranging from localized itching and erythema to systemic shock, pulmonary edema, and even death [29–32].

Several proteins have been identified or characterized from *N. nomurai* sting venom by traditional methods, such as multilayer chromatography [25], sequential chromatography [33], matrix-assisted laser desorption/ionization time-of-flight mass spectrometry (MALDI-TOF MS) [23], gene cloning, and expression methods [34]; however, the results obtained are still limited. Recently, the emergence of high-throughput technologies combined with bioinformatics has dramatically facilitated the study of venom protein/peptide diversity in a variety of animals, which often integrates genomics, transcriptomics, and proteomics [35,36]. Recently, a tissue proteomic analysis of *N.nomurai* sting venom-induced dermatitis model was published [37], which benefited from the sequencing and annotation of the *N. nomurai* sting jellyfish proteome [20]. Proteomic methods identified metalloproteinases, proteases, and pore-forming toxins in *N.nomurai* sting venoms. Later, Li's research isolated the lethal components of *N. nomurai* sting toxins [25]. Proteomic analysis identified 13 toxin homologs, including phospholipases, potassium channel inhibitors, hemolysins, and thrombin. Yu et al. [37] also identified zinc metalloproteinases using liquid chromatograph mass spectrometer (LC-MS/MS) fractions, serine proteases, and so on. However, most of them do not provide absolute evidence that all the identified proteins are venom components, and there are slight differences in the screening criteria between these individuals, such as different fold changes in differentially expressed genes (DEGs), and some different e-values may be controversial [38]. Several transcriptomic studies have helped us to better understand the evolutionary relationships [39,40] and the mechanisms [41] by which the toxin exerts its effects. The results of these studies are discrepant [8], and the relevant genomic data still need to be improved. Therefore, there is an urgent need to explore the venom components and their effect mechanisms by applying various research techniques at the molecular level and conducting extensive analyses to accelerate the development of therapeutic drugs and the process of venom transformation.

Venom collection has been the basis for the evolution of venomous organisms [42], toxic effects, and the development of immunopharmaceuticals [43]. Relatively standardized venom collection methods have been established for snakes, scorpions, spiders, and bees [42,44–47]. This has facilitated the process of their toxic studies to a certain extent. Some studies have shown that different methods of venom extraction may affect their composition or distribution of toxin proteins in the venom [47,48]. Venom extraction from nematocysts is a prerequisite for toxicity studies of cnidarians. The currently used methods of jellyfish venom collection can be broadly categorized into two types, both of which require low-temperature autolysis of the collected jellyfish tentacles, and the difference lies in whether or not they are homogenized and the form of dialysis that follows [7,41,49,50].

We distinguished between the extraction methods of tentacle extraction (TE) and nematocyst venom extraction (NV) [51] based on whether the homogenization was performed. Unfortunately, there is no systematic comparison of the effects of different extraction methods on the venom composition, nor is there a functional assay comparing the 50% lethal dose (LD50) to determine the optimal purification method [7]. Obtaining sufficient venom without destroying the toxin proteins is of great value in fully utilizing jellyfish resources and improving the efficiency of jellyfish venom research. In addition to venom collection methods, the effective inhibitors are also controversial [30,52], and the use of inappropriate inhibitors or drugs may even worsen the clinical course of jellyfish stings [53].

In this study, we constructed an ex vivo and in vivo jellyfish toxicity model using RAW264.7 cells and ICR mice, which were exposed to naturally dissolved and mechanically crushed extracted jellyfish venom, and then compared their toxin compositions and ex vivo and in vivo toxicity effects by combining this model with proteomics analyses, histological analyses, and biochemical results. In addition, the inhibitory effects of the phospholipase A2 (PLA2) inhibitor PACOCF3, the matrix metalloproteinase (MMP) inhibitor EDTA, and the Ca^{2+} channel inhibitor verapamil on the toxicity of jellyfish venom were evaluated in vitro and in vivo. For the first time, our study compares the venom extracted from jellyfish by different methods, which is not only conducive to the establishment of a more stable and efficient method of venom extraction but also of great significance for improving the jellyfish toxin library and developing potential antidotes.

2. Results

2.1. Comparison of Toxins in TE and NV and Toxicity Evaluation

Based on label-free proteomics analysis utilizing LC-MS/MS, we conducted a comprehensive investigation of the compositions of TE and NV. By consulting the cnidarian protein database, we identified a total of 403 proteins in TE and 414 proteins in NV, with 281 proteins common to both groups (Figure 1A). Given the limited availability of toxin data within the cnidarian database, we employed BLASTp [54] to run the UniProt toxin database (Tox-Prot) in conjunction with the *N. nomurai* transcriptome database previously established by our research team. We derived the predicted toxin sequences from the *N. nomurai* transcriptome with the filtering criteria, including e-value $< 1.0 \times 10^{-P}$ and %identity > 40%. Utilizing the inferred toxin data as a reference point, we successfully identified 239 and 225 predicted toxin proteins in TE and NV, respectively (Figure 1B). The toxin characteristics of both venoms exhibited notable similarities. We classified these toxins into ten distinct families based on the expression levels and quantities of the corresponding toxin proteins, which included conotoxin-like proteins, spider neurotoxin-like proteins, Ser/Thr protein kinase, peptidase, phospholipase A2, AAA ATPase, calmodulin, PPP phosphatase, GTPase, and other toxins. It is important to note, however, that the number and relative expression levels of proteins within these toxin families may differ (Figure 1C).

First, we constructed an in vivo *N. nomurai* sting model by injecting TE into the tail vein of ICR mice at 4–6 h intervals and found that TE in the range of 12–20 mg/kg was potentially lethal, while NV in the range of 18–30 mg/kg was potentially lethal (Figure 1D,E). Quantitative analysis showed that the LD50 was 12.19 mg/kg for TE and 22.52 mg/kg for NV (Figure S1A,B). Both TE (>16 mg/kg) and NV (>18 mg/kg) mice showed an accelerated heart rate, shortness of breath, and muscle weakness at high doses, except that the TE mice showed more convulsions, which were not observed in the NV group. Pathological analysis showed that TE and NV caused heart and liver damage in the ICR mice. There was structural damage and severe congestion of cardiomyocytes, deformation of liver lobules, and necrosis of hepatocytes, accompanied by localized inflammatory cell infiltration, while the damage to renal tissues was less pronounced in the TE group. More severe damage to the cardiomyocytes was observed in the NV group, with cellular arrangement disorganization, more extensive hepatocytes necrosis, and severe lesions of glomeruli with a sizeable inflammatory infiltrate (Figure 1H). Moreover, biochemical evaluation showed that the cardiac markers creatine kinase (CK) and lactate dehydrogenase (LDH-L) were significantly elevated after both TE and NV exposures. The hepatic markers aspartate aminotransferase (AST) and alanine aminotransferase (ALT) were increased in vivo in the TE group, and even though the ALT in the NV group did not show any abnormality, its elevated AST was equally suggestive of severe liver injury, while creatinine (CREA) was also upregulated in the NV group (Figure 1I). In addition, in vitro modeling experiments also illustrated the toxicity differences between TE and NV. Cell viability decreased in both TE and NV groups, but still showed differences, with IC_{50} of 10.45 μg/mL (Figure 1F) and 32.34 μg/mL (Figure 1G), respectively.

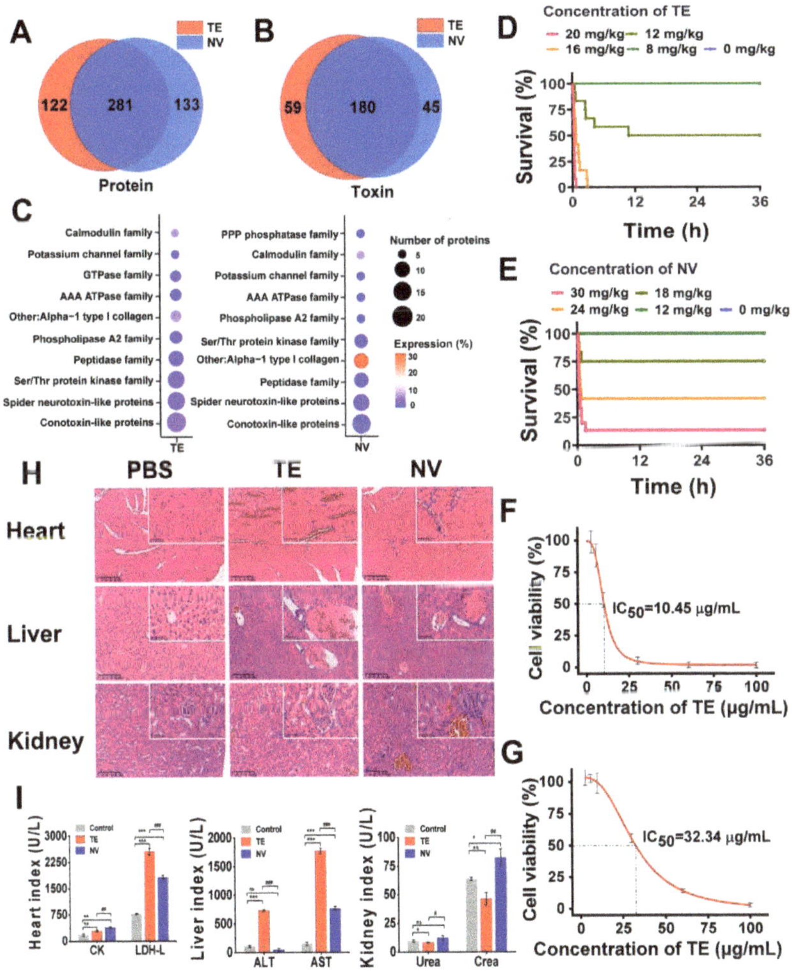

Figure 1. Comparison and toxicity evaluation of toxins in TE and NV. (**A**) Venn analysis of proteins between TE and NV groups, n = 4. (**B**) Venn analysis of toxins between TE and NV groups, n = 4. (**C**) Overview of toxin family distributions and protein number and expression in TE and NV identified in the proteomes. (**D**) Survival analysis of ICR mice after TE injections, n = 12. (**E**) Survival analysis of ICR mice after NV injections, n = 12. (**F**) Viability of RAW264.7 cells exposed to TE, n = 6. (**G**) Viability of RAW264.7 cells exposed to NV, n = 6. (**H**) HE analysis of the heart, liver, and kidneys at 400× magnification, n = 4. (**I**) Evaluation of blood biochemical indices, including creatine kinase

(CK) and lactate dehydrogenase (LDH), aspartate aminotransferase (AST), alanine aminotransferase (ALT), creatinine (CREA), and urea nitrogen (UREA), n = 3. * $p < 0.05$, ** $p < 0.01$, and *** $p < 0.001$ represent the comparison of the group with the control group, and # $p < 0.05$, ## $p < 0.01$, and ### $p < 0.001$ represents the comparison of the group with the TE group, with ns representing no significance.

2.2. TE and NV Lead to the Upregulation of the Inflammatory Signaling Pathway and the Release of Inflammatory Factors

To investigate the potential molecular mechanisms underlying the differences in the cytotoxicity of TE and NV treatments on RAW264.7, transcriptomic sequencing was performed, and differentially expressed genes (DEGs) were screened between the TE, NV, and control groups, respectively. PCA analysis showed significant differences in the transcriptomic profiles between the TE or NV groups and the control group (Figure S2A,B). Specifically, 1647 (576 upregulated and 1071 downregulated) genes were identified between the TE and control groups, and 1541 (611 upregulated and 930 downregulated) genes were identified between the NV and control groups (Figure S2). Venn diagram analysis showed that among the upregulated genes, 279 genes were common differential genes in the TE and NV groups as compared with the control group (Figure 2A). KEGG pathway analysis revealed similar enriched pathways in the TE group, including MAPK and PI3K-Akt, but the extent of gene enrichment varied. According to the degree of enrichment, the most highly enriched signaling pathways were MAPK and PI3K-Akt in the TE and NV groups, respectively (Figure 2B). It is noteworthy that although the PI3K-Akt pathway was highly enriched in the NV-treated group, the gene set enrichment analysis (GSEA) evaluation did not show statistical significance (Figure S2C,D).

Meanwhile, the screening of transcriptome upregulated genes suggested that TE and NV exposure resulted in elevated transcript levels of inflammatory factors in RAW264.7 cells. Compared with the control group, TE and NV jointly induced upregulation of transcript levels of factors, such as Ccl22, IL1β, and TNFα (Figure S2C,D). However, the induction degree might differ among the demonstrated jointly upregulated cytokines, except for Ccl7 (Figure S2C,D). NV appeared to regulate cytokines to a greater extent than TE (Figure 2D). RT-qPCR showed that the TE- and NV-treated RAW264.7 groups differentially increased the transcriptional expression levels of the cytokines IL6, TNFα, Cxcl10, Ccl2, IL1β, and Cxcl2 compared to the control group, while the stimulatory effect of NV on cellular inflammatory factors was more significant than that of TE (Figure 2E,F).

We extracted protein–protein interaction networks (PPIs) from the transcriptome database and visualized them to predict common differential genes' interaction and adhesion pathways. The PPI network of common DEGs contained 49 nodes and 233 edges (Figure 2C). Most interconnected nodes were considered hub genes in the PPI network, which revealed that Jun, Fos, and others were the most influential genes in the TE-treated group. In contrast, the NV-treated group exhibited NF-κB, Relb, Myc, and others as crucial influencers alongside the standard hub. The common pivotal genes include essential cytokines, such as IL1β and TNF, and other immune-regulating factors, such as Dusp1 and Nr4a1 (Figure 2C), which may be potential biomarkers and provide new therapeutic strategies for studying stinging injuries.

The Western blot showed that TE and NV significantly induced the phosphorylation of p65 and p38 to a comparable extent in RAW264.7 cells stimulated with TE and NV, respectively, when analyzing their total proteins and phosphorylated c-Fos, c-Jun, p65, and p38 (Figure 2G). It is worth mentioning that although the phosphorylation levels of c-Fos and c-Jun did not increase, their total protein expression was significantly increased, which was consistent with the results suggested by PPI.

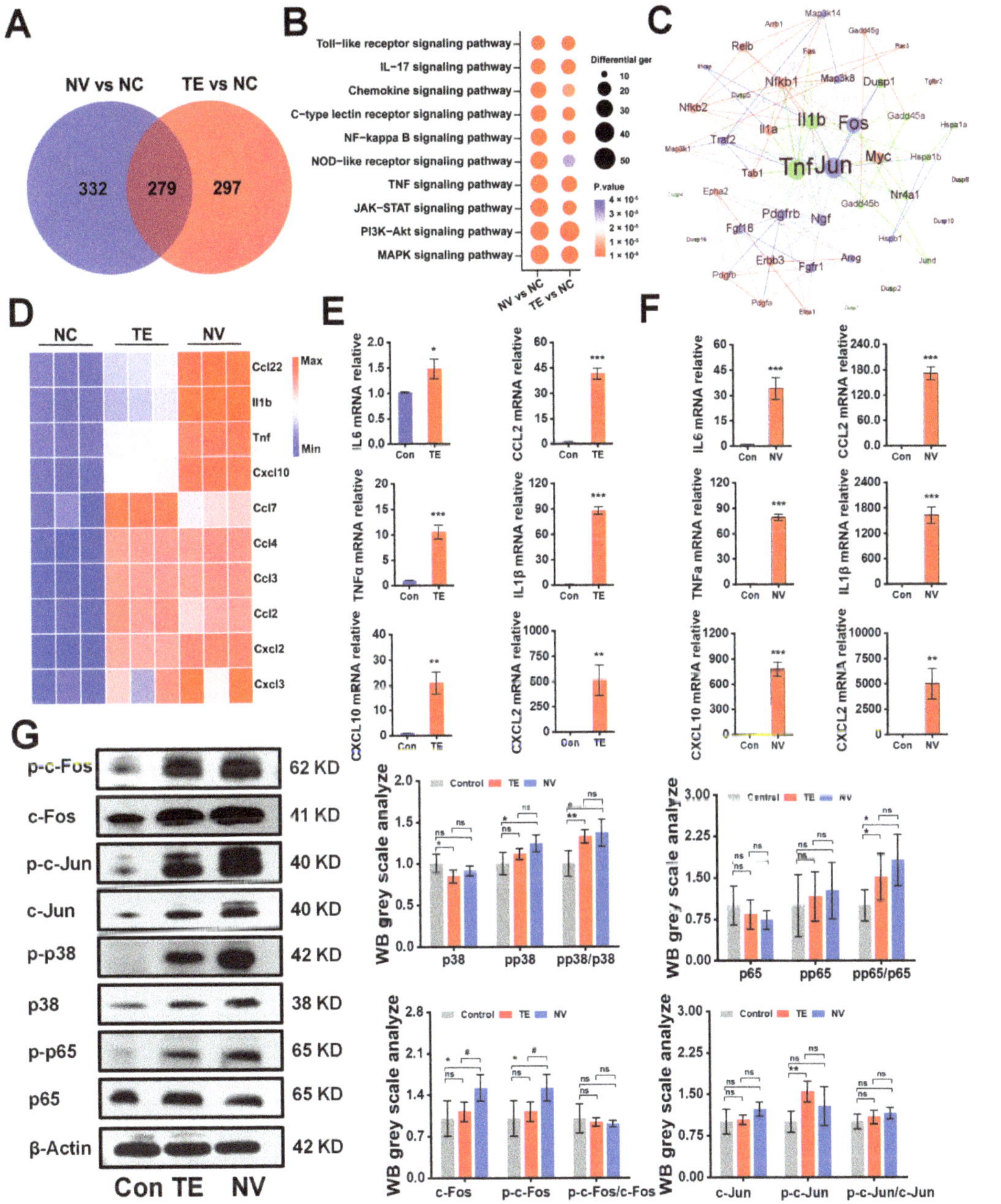

Figure 2. TE versus NV transcriptome overview and Western blot results. (**A**) Venn diagram of genes upregulated by TE versus NV in RAW264.7 compared to PBS treatment, n = 4. (**B**) Bubble plots of KEGG enrichment analysis of DEGs. (**C**) Protein–protein interaction networks of TE (10.45 μg/mL) and NV (32.34 μg/mL). (**D**) Comparative analysis of TE and NV co-induced upregulation of inflammatory factors in the transcriptome. All genes were calculated using log2 (fold change) and shown by thermography. (**E**,**F**) Relative gene expression of six inflammatory factors including IL-6,

TNF-α, Cxcl10, IL1β, Ccl2, and Cxcl2 in RAW264.7 cells treated with TE (10.45 μg/mL) or TE (32.34 μg/mL), n = 3. (**G**) Western blotting analysis of c-fos, c-Jun, p65, and p38 phosphorylated and non-phosphorylated protein, n = 4. * $p < 0.05$, ** $p < 0.01$, and *** $p < 0.001$ represent the comparison of the group with the control group, and # $p < 0.05$ represents the comparison of the group with the TE group, with ns representing no significance.

2.3. Effect of Inhibitors on TE or NV Cytotoxicity

The intervention of TE or NV with different concentrations of inhibitors, including the MMP inhibitor EDTA, the PLA2 inhibitor PACOCF3, and the calcium channel inhibitor verapamil, showed that PACOCF3 and verapamil at 10–30 μM doses had strong antagonistic action against the cytotoxicity in the TE or NV treatments, while EDTA did not have a significant action (Figure 3A,B). Additionally, PLA2 and MMP activities showed a dependence on the TE or NV concentration. Therefore, it could be inferred that the negative efficacy of EDTA was independent of the MMP activity in the venom (Figure S3A,B). RT-qPCR results showed that most of the TE-stimulated rise in the inflammatory factors TNFα, IL6, and IL1β was not decreased by PACOCF3 or verapamil, and only the transcript levels of IL6 and TNFα were significantly inhibited by PACOCF3 and verapamil, respectively (Figure 3C). Notably, even in the absence of TE intervention, PACOCF3 and verapamil increased the transcriptional levels of IL6 in RAW264.7 cells, while verapamil mainly affected the transcriptional levels of IL1β (Figure S3C).

To investigate whether PACOCF3 and verapamil could reduce the expression of NF-κB and MAPKs signaling pathways, RAW264.7 was treated with TE dissolved in PBS alone or in combination with PACOCF3 and verapamil, and the results showed that PACOCF3 significantly inhibited TE-induced phosphorylation of p38, which was also observed in the presence of verapamil but did not show statistical significance. In addition, PACOCF3 and verapamil did not show modulation of the total protein or phosphorylation of c-Jun and p65. It should be noted that although PACOCF3 and verapamil did not show any regulation of c-Fos phosphorylation, the total protein level of c-Fos was significantly downregulated by these two inhibitors (Figure 3D). Moreover, the inhibitors themselves did not affect the protein, either in terms of total protein expression or their phosphorylation (Figure S3D).

2.4. Evaluation of Antagonistic Effects of PACOCF3 and Verapamil In Vivo

A mixture of TE and PACOCF3 or verapamil was injected intravenously into ICR mice, and serum and organ samples (heart, liver, kidney) were subsequently harvested and analyzed. At 36 h post-intervention, doses of PACOCF3 ranging from 1–3 mg/kg significantly ameliorated lethality, and a higher dose (5 mg/kg) prolonged their survival time, although it did not improve lethality (Figure 4A). Doses of 0.1–1 mg/kg of Verapamil did not improve lethality associated with TE, nor did they have a positive or negative impact on survival time (Figure 4B). Biochemical tests conducted on mouse serum after treatment with 1 mg/kg of PACOCF3 or Verapamil (4–6 h) demonstrated a significant reduction in ALT and AST levels. However, there was no improvement in other parameters following treatment with either PACOCF3 or Verapamil. Notably, while Verapamil did not enhance survival rates in cases of TE, it unexpectedly showed greater efficacy in reducing ALT and AST levels compared to PACOCF3 (Figure 4C). Histopathologic examination of cardiac, hepatic, and renal tissues after treatment showed an improvement in toxin-induced injury. Specifically, denser cardiac muscle fibers, reduced hemorrhage, more regular liver lobule shape, and reduced edema were observed and were consistent with the blood biochemistry results, suggesting that verapamil was more efficacious than PACOCF3 for hepatic injury (Figure 4D). In addition, in the absence of TE, neither PACOCF3 nor verapamil caused significant damage to the liver or kidneys, and only the heart showed a slight irregularity in the arrangement of cardiomyocytes. Similar to the blood biochemistry, in the absence of TE, PACOCF3 and verapamil both upregulated CK and LDH-L expression, but significantly lowered ALT and AST levels, with no significant changes in UREA and CREA (Figure S4A,B). Thus, we hypothesized PACOCF3 may prevent jellyfish venom-induced lethality and have a hepatoprotective effect, and although verapamil did not ameliorate lethality, it may have also some hepatoprotective activity as a commonly used cardiac drug.

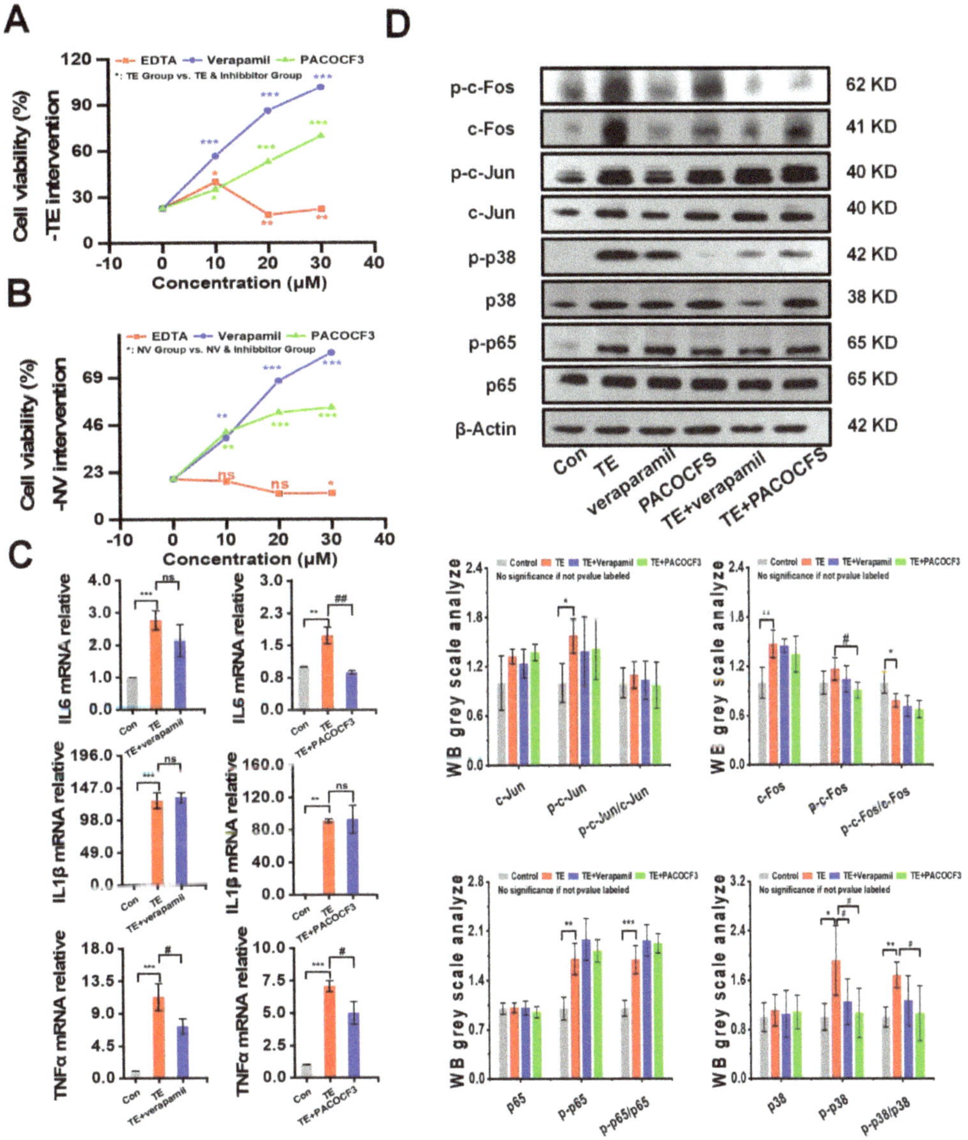

Figure 3. PACOCF3 and verapamil antagonized TE or NV toxicity in vitro. (**A**,**B**) Inhibition of TE or NV cytotoxicity in RAW264.7 cells by 10–30 μM EDTA, PACOCF3, and verapamil. n = 6. (**C**) Relative gene expression levels of three inflammatory factors, namely IL-6, IL1β, and TNF-α, in RAW264.7 cells treated with TE (10.45 μg/mL) or TE combined with PACOCF3 (20 μM) or verapamil (20 μM), n = 3. (**D**) Western blotting analysis of c-Fos, c-Jun, p65, and p38 phosphorylated and non-phosphorylated proteins, n = 4. * $p < 0.05$, ** $p < 0.01$, and *** $p < 0.001$ represent the comparison of the group with the control group, and # $p < 0.05$, and ## $p < 0.01$ represents the comparison of the group with the TE group, with ns representing no significance.

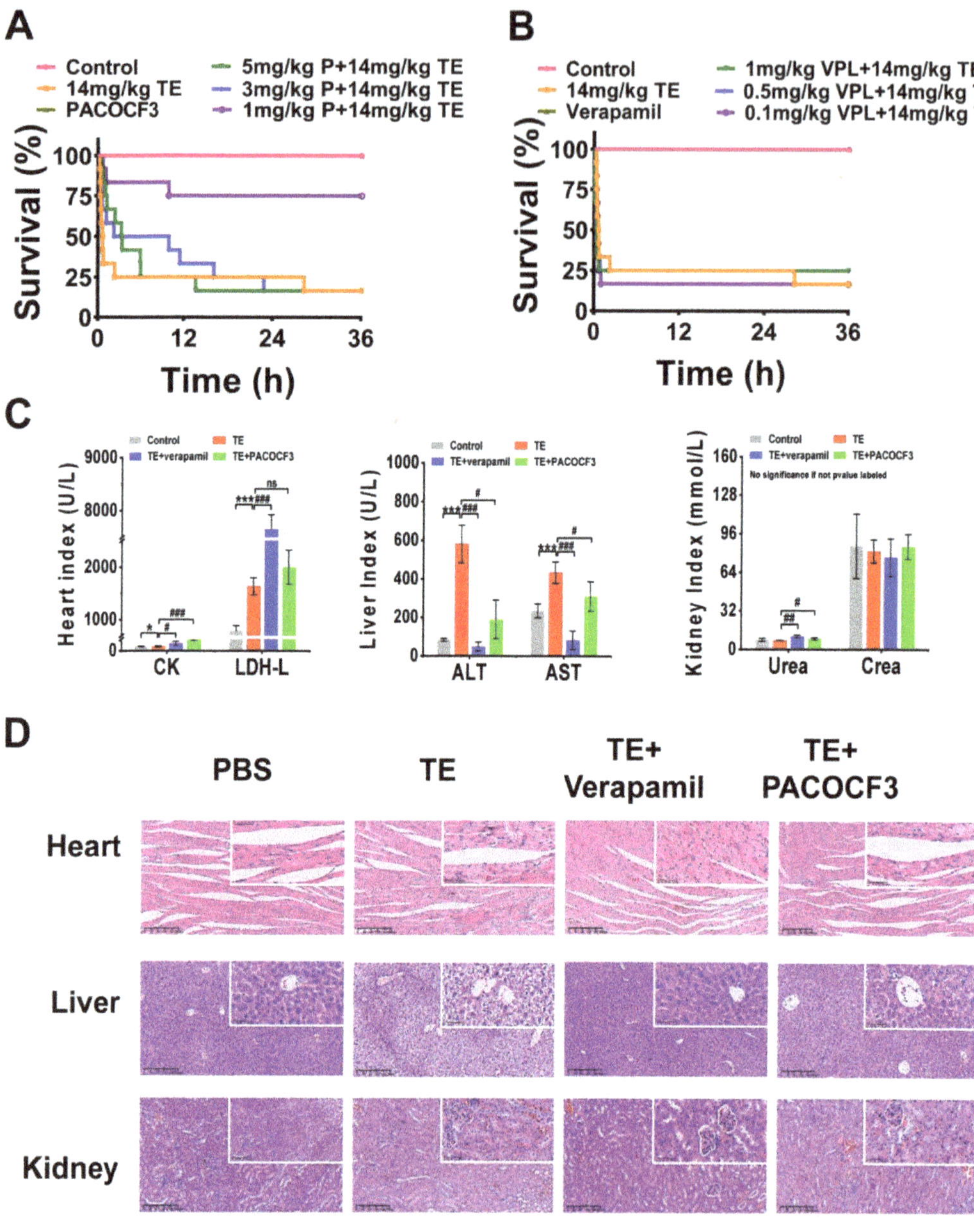

Figure 4. In vivo interventional effects of different inhibitors on TE. (**A**) Lethality analysis of TE interfered with PACOCF3, n = 12. (**B**) Lethality analysis of TE interfered with verapamil, n = 12. (**C**) Evaluation of blood biochemical parameters including creatine kinase (CK) and lactate dehydrogenase (LDH), aspartate aminotransferase (AST), alanine aminotransferase (ALT), creatinine (CREA), and urea nitrogen (UREA) in mice treated with TE (14 mg/kg) or TE combined with PACOCF3 (1 mg/kg) or verapamil (1 mg/kg), n = 3. (**D**) Detailed HE-stained analyses of the heart, liver, and kidneys. 400× magnification, n = 4. * $p < 0.05$, and *** $p < 0.001$ represent the comparison of the group with the control group, and # $p < 0.05$, ## $p < 0.01$, and ### $p < 0.001$ represents the comparison of the group with the TE group, with ns representing no significance.

3. Discussion

3.1. Effects of Different Jellyfish Venom Extraction Methods on Venom Components and Their Activities

The link between fluctuations in the composition or content of toxin proteins and their activity and the influence on subsequent symptoms by different venom extraction methods has been demonstrated in other venomous organisms [55,56]. Although this has been used as a rule for jellyfish [57], our results are consistent. We used a proteomic approach to gain insights into the proteomic differences between *N. nomurai* jellyfish venoms (TE and NV) extracted using two different methods, and the toxicity was evaluated in vitro and in vivo, respectively. The 239 and 225 toxin proteins found in TE and NV were included in 118 protein families. Among the ten highest expression protein families, most of the venom toxins were more abundant in TE than in NV, except for Alpha-1 type I collagen (COL1A1) in NV. COL1A1, as a fibrous collagen, is particularly important for skin, bone, and connective tissues, and is expressed in almost all connective tissues, suggesting that mechanical fragmentation of nematocysts still retains capsid proteins. In addition, COL1A1 has been reported to be involved in the hemorrhagic symptoms of snake venom, promoting the hemorrhagic activity of snake venom zinc-dependent metalloproteinases (SVMPs) [58], a seemingly "non-venomous" protein that has received little attention in jellyfish studies. Other toxins rarely reported in jellyfish venom, such as conotoxin-like proteins, were also identified in our study.

In summary, our results suggest that natural lysis can produce a purer and more toxic venom in jellyfish, and mechanical destruction of the nematocysts can retain a complete venom, including some structural proteins used for venom storage. These provide a new scientific rationale for using TE as a source of jellyfish toxicology, along with increased productivity, which may be helpful for drug development [59]. The extraction of NV may retain a large number of structural proteins, which makes it require higher concentrations to achieve the same cytotoxicity and lethality compared to TE, but this is not absolute. The different extraction methods of TE and NV can result in the loss of some toxin proteins, which may be critical and lethal. Identifying the two different venom extraction methods has dramatically expanded our understanding of the biological properties of jellyfish venom.

Although TE and NV did not exhibit the same cytotoxicity, similar effects at the same cytotoxic concentrations, particularly in the modulation of transcriptome-predicted total protein and phosphorylated expression levels of c-Fos, c-Jun, p65, and p38 genes, were shown in them. As predicted by the transcriptome as well as by the PPI, the mRNA levels of RAW264.7 cellular immunity-related genes (IL6, TNFα, Cxcl10, Ccl2, IL1β, and Cxcl2) were elevated by TE and NV treatments, and NV-induced cytokines were more highly upregulated than those by TE yet were less cytotoxic at the same concentrations. Therefore, the upregulation of these inflammatory factors may not be mainly involved in the venom cytotoxicity. In addition, TE and NV regulate the upregulation of phosphorylated p38, which may play a vital role in the pathogenesis of inflammation by acting on the expression of downstream regulatory factors, such as TNF-α and IL-6 [60].

3.2. Toxicological Effects of Venom and Effects on Different Organs

The heart [61,62] and liver [63,64] are the main target organs of jellyfish. A previous study reported that exposure to jellyfish venom resulted in pathologic changes in the heart and liver but not in the kidneys of mice. Consistent with the earlier study [65], mice exposed to TE (12 mg/kg) and NV (23 mg/kg) exhibited structural damage to the heart and liver; for example, cardiac tissue showed structural damage to cardiomyocytes, widening of myocardial interstitium, deformation of liver lobules, hepatocyte degeneration, necrosis, blurring, and the widening of cellular interstitial space, all accompanied by localized inflammatory cellular infiltration, which were consistent with increased activity of CK and LDH-L [66].At the same time, in the case of liver injury, AST and ALT activity enhancement occurred, as these active substances were released into the blood [67], and from the biochemical data, it seems that the TE-treated group was more seriously injured. These results suggest that the venom of *N.nomurai* jellyfish induces more severe heart and liver damage but less kidney damage, but the related toxicity mechanism is unclear. Therefore, further studies are required to investigate the molecular mechanisms of the cardiac and hepatic toxicity of jellyfish venom.

3.3. Evaluation of the Ex Vivo and In Vivo Effects of PLA2, MMP, and Calcium Channel Inhibitors

The development of jellyfish venom inhibitors has always been a research hotspot. However, due to the unstable nature of the venom [7], the presence of species, geographic variability [68], and insufficient information about the venom, it has yet to have a specific therapeutic modality, and even serum antibodies have poor efficacy [69]. It has been reported that jellyfish venom has a complex composition, in which PLA2 and MMP are essential components of jellyfish toxins [70]; PLA2 can release free fatty acids and lysophospholipids [71], MMP can induce inflammation [28] and myotoxicity [33], and calcium overload is considered to be the main cause of cardiotoxicity in jellyfish poisoning [72]. As one of the major toxins in terrestrial animals, the PLA2 inhibitor PACOCF3 has been used to inhibit PLA2 activity [73,74]. Furthermore, in jellyfish-related research, MMP inhibitors, like EDTA [51], and calcium channel blockers, including diltiazem [72] and verapamil [75], have been investigated, and have been widely demonstrated to antagonize jellyfish stings, although some of the results may be contradictory [76,77]. This contradiction also exists in our study, where EDTA was reported to be effective in [53] but showed little or no effect on our data. The antagonistic effect of calcium channel inhibitors has been contradictory, and verapamil has been reported to be effective [78] or ineffective [79] in antagonizing jellyfish venom or even exacerbating cases of toxicity [76], which may be related to the species of jellyfish. Thus, confirmation of the stinging species appears necessary when selecting a treatment. In the present study, verapamil did have an excellent antagonistic effect on toxicity at the cellular level, but this antagonistic effect was not observed in an in vivo model. In contrast, PACOCF3 showed strong antagonistic effects both for in vivo and in vitro models of jellyfish toxicity, and this antagonism was accompanied by a decrease in p38 phosphorylation levels in cells. This may suggest that PLA2 may promote MAPK activation, which is associated with jellyfish venom cytotoxicity in RAW264.7, and PLA2 inhibitors may be utilized as potential MAPK inhibitors. This conjecture has been demonstrated in snake venom [80] but remains to be tested in jellyfish

4. Materials and Methods

4.1. Jellyfish Samples Collection

In October 2023, *N. nomurai* were obtained in the East China Sea, and their tentacles were immediately cut off and stored at −80 °C for nematocyst isolation and venom extraction. No specific permit was required because the jellyfish were not an endangered or protected species. Jellyfish samples were collected from an unprotected or privately owned marine environment.

4.2. TE or NV Venom Preparation

TE and NV were prepared separately using widely accepted methods for collecting jellyfish venom. TE was obtained through the natural exsolution of the tentacles, while the venom NV was extracted from the tentacles by mechanically fragmenting the nematocysts [7]. According to the TE collection method described by Li [41], frozen tentacles were placed in a beaker and stirred continuously for 72 h at 4 °C to allow complete autolysis of the tissues. The mixture was filtered through a 200-mesh sieve to obtain the supernatant and centrifuged at 10,000× *g* for 15 min at 4 °C. The supernatant was centrifuged in PBS (0.01%) for 2 h at 0.01%. Then, the dialysate was dialyzed in PBS (0.01 mol/L, pH 7.4) for 12 h. The dialysis venom was labeled as TE. Frozen tentacles were thawed in filtered seawater at 4 °C for autolysis according to the NV collection method described by Li [28], and the seawater was changed every day until most of the visible tissue fragments were dissolved. After autolysis, the mixture was filtered through a 200-mesh sieve to remove tissue fragments. The filtrate was then centrifuged at 3000× *g* for 15 min at 4 °C, the precipitate was collected to obtain the nematocysts, and the nematocysts were washed three times with PBS, then further centrifuged at 10,000× *g* for 15 min at 4 °C. The nematocysts were fragmented using a bead grinder at 400 w. A total of 20 cycles were performed, each consisting of 10 s of sonication and 15 s resting on ice. The venom was then centrifuged at 12,000× *g* to obtain the supernatant and labeled as NV.

4.3. Label-Free LC–MS/MS

The TE and NV samples were collected and analyzed by Lianchuan Biological Co. Briefly, the proteins were hydrolyzed by trypsin and quantified. For the LC-MS/MS analysis, the enzymatic products were separated using a Thermo Ultimate 3000 nano ultra-performance liquid chromatography system at a flow rate of 300 nL/min on an Acclaim Pep Map 100 column (2 cm × 75 µm, 3 µm) with a gradient elution at a flow rate of 5 µL/min: 5 to 21% solvent B (97% ACN, pH 9.8) in 38 min, 21.5% to 40% solvent B in 20 min, 40% to 90% solvent B in 2 min, 90% solvent B for 3 min, and 5% solvent B equilibrated for 10 min. The elution peaks were monitored at 214 nm, and fractions were collected every minute. The fractions were combined according to chromatograms of the elution peaks. Ten fractions were obtained and then freeze-dried. Ionization was performed using an Orbitrap Exploris™ 480 mass spectrometer (Thermo Fisher Scientific, San Jose, CA, USA) in DDA (data-dependent acquisition) mode, followed by data-dependent MS/MS scanning. Screened proteins were identified and quantified by MaxQuant software version 2.1.4.0 (https://www.maxquant.org/, accessed on 19 February 2024) under the conditions of false-positive of the peptide-spectrum matches (PSM FDR) < 0.01 and false-positive of protein (protein FDR) < 0.01. The reference database selected for this DDA label-free mass spectrometry raw data (Raw data) was the *N. nomurai* Transcriptome Database. The reference database used for this DDA label-free mass spectrometry raw data (Raw data) was the *N. nomurai* eukaryotic non-coding transcriptome sequencing data (database of 26,066 sequences) (https://www.ncbi.nlm.nih.gov/sra/?term=Stomolophus+meleagris, accessed on 19 February 2024). The *N. nomurai* transcriptome database (database totaling 26,066 sequences) and the Uniprot biotoxin database (database totaling 1,359,673 sequences) were run using the Local Blast tool (e-value of $<1.0 \times 10^{-5}$) (https://www.uniprot.org/uniprotkb?query=toxin, accessed on 19 February 2024); then, the identified proteins were named according to the protein names in uniprot, and the families to which the proteins belonged were recorded and categorized according to their family and domain. We annotated the protein sequencing results from the reference N.nomurai eukaryotic non-coding transcriptome database using this record. Toxicity identification only screens results with a sequence alignment consistency exceeding 40%.

4.4. Survival Analysis

Male ICR mice with an average weight of 25 ± 2 g were obtained from the Laboratory Animal Center of Naval Medical University. Different concentrations of PBS diluted with TE (0–20 mg/kg, n = 12 of each treatment group) or NV (0–30 mg/kg, n = 12 of each treatment group) [65,81] were injected intravenously into each mouse through the tail vein. Subsequently, the mice were observed continuously for 3 days. Mortality data were collected to calculate the median lethal dose (LD50) for TE intervention experiments. PACOCF3 (0–5 mg/kg) or verapamil (0–1 mg/kg) diluted in PBS were pre-mixed with a dose of 14 mg/kg TE in excess of the LD50, administered through the tail vein, and death time was recorded over a 36 h period. All experiments were conducted using animal welfare principles and approved by the Naval Medical Center of PLA Ethics Committee (NMC2023011).

4.5. Histologic Examination and Blood Biochemical Analysis

Four to six hours after the injection of a median lethal dose of TE (12 mg/kg) or NV (23 mg/kg) through the tail vein, orbital blood was withdrawn for haematochemical analysis and then the mice were killed by cervical dislocation. Then, the hearts, livers, and kidneys were dissected and immediately placed in tissue fixative for histological analysis. Briefly, after routine processing, the tissue samples were embedded in paraffin, cut into thin slices, dehydrated in xylene and ethanol, and washed with clean distilled water. Subsequently, the sections were stained using a hematoxylin–eosin dye solution (B1002, Baiqiandu Biotechnology, Wuhan, China).

For blood biochemical analysis, the median lethal dose (LD50) of TE and the inhibitor dose of 1 mg/kg of PACOCF3 or verapamil, as well as 1 mg/kg of PACOCF3 or verapamil premixed with 12 mg/kg of TE, were used for the intervention experiments through the tail vein. Other operations were performed as above. The orbital blood samples were centrifuged at 2000× *g* for 10 min. The resulting supernatant was used as a template for the measurement of various biochemical parameters, including cardiac enzymes (CK, LDH-L), liver function markers (ALT, AST), and renal function markers (CREA, UREA).

4.6. Cytotoxicity Assay

RAW264.7 cells were inoculated at 10,000 cells/well in 96-well plates and incubated in DMEM containing 10% fetal bovine serum and 1% penicillin and streptomycin at 37 °C, 5% CO2, and 85% humidity under aseptic conditions for 24 h. Cells were treated with TE and NV at different concentrations (0–100 μg/mL, n = 6) for 2 h. Changes in absorbance were measured at 450 nm by using enzyme labeling instrument (BioTek, Winooski, VT, USA) to measure their absorbance at 450 nm. The wells without venom were used as a negative control and the effect was eliminated by using wells containing only DMEM. The formula for TE was calculated as follows: viability (%) = (ODTE Background)/(ODNC Background) × 100%. Nonlinear fitting was performed by logistic equations on survival curves analyzed using Origin2021 software.

In the inhibitor intervention experiments, cells were treated with a premix of 70% inhibitor dose of TE or NV (IC_{70}) and 10–30 μM doses of inhibitors (EDTA, PACOCF3, or verapamil), and the rest of the operations were performed as above. The RAW264.7 cell line was provided by QiDa Biotechnology Co., Ltd. (Shanghai, China).

4.7. Transcriptomics Analysis

4.7.1. RNA Extraction and Sequencing

RAW264.7 (2.5 × 10^4 cells/mL) was inoculated into 12-well plates and treated with TE or NV at a median inhibitory concentration (IC_{50}) for 2 h. According to the manufacturer's guidelines, total RNA was isolated from the cells using an RNeasy RNA Fast Extraction Kit (Feijie RNA fast200, Shanghai, China). Total RNA was analyzed using an enzyme marker (BioTek, Winooski, VT, USA) to assess the concentration and integrity of RNA. RNA samples were analyzed by Lianchuan Biological Co. Briefly, the RNA was quality checked and the sequencing conditions (concentration > 50 ng/μL, RIN value > 7.0, OD260/280 > 1.8, total RNA > 1 μg) were met using oligo(dT) magnetic beads (Dynabeads Oligo (dT), item 25-61005, Thermo Fisher, Waltham, MA, USA). The mRNA with PolyA (polyadenylate) was specifically captured in two rounds of purification. The captured mRNA was fragmented at high temperature, and cDNA was synthesized in the presence of reverse transcriptase (Invitrogen SuperScript™ II Reverse Transcriptase, No. 1896649, CA, USA). E. coli DNA polymerase I (NEB, No. m0209, Ipswich, MA, USA) and RNase H (NEB, No. m0297, Ipswich, MA, USA) were used for two-stranded synthesis to convert these DNA–RNA complex double strands into DNA double strands, and dUTP solution (Thermo Fisher, No. R0133, CA, Waltham, MA, USA) was mixed with the second strand to make the ends of the double-stranded DNA into flat ends. To make the ends of the double-stranded DNA flat, a base was added to each end of the double-stranded DNA to enable it to be ligated to a junction with a T base at the end, and the fragment size was screened and purified using magnetic beads. The second strand was digested with UDG enzyme (NEB, No. m0280, Ipswich, MA, USA), and PCR was used produce a library with a fragment size of 300bp±50bp. Finally, we used Illumina Novaseq™ 6000 (LC Bio Technology Co., Ltd. Hangzhou, China) to perform bipartite sequencing according to the standard operation, and the sequencing mode was PE150.

4.7.2. Differentially Expressed Genes (DEGs) Analysis

The results of comparing genomic and genomic annotation files from RSEM were used to obtain read counts of transcripts that were used for FPKM or TPM transformation to measure gene expression levels. Differentially expressed genes (DEGs) were screened (|log2(FC)| > 1 and $p < 0.05$). Functional enrichment analysis (KEGG pathway enrichment) of DEGs was performed using KOBAS. Based on our previous studies, network diagrams were generated by protein–protein interaction (PPI) network analysis to elucidate the relationships between critical DEGs and to assess the distribution trends of genes in the TE or NV gene sets in gene lists sorted by their phenotypic relevance to the PI3K-Akt pathway and MAPK pathway (GSEA).

4.8. Western Blotting

RAW 264.7 (2.5×10^4/mL) was inoculated into cell culture plates, and the cells were treated with median inhibitory concentrations of TE (10.45 μg/mL) or NV (32.34 μg/mL) for 2 h, and then the total proteins were isolated for subsequent experiments. Briefly, total protein lysates from RAW 264.7 cells were prepared using lysis buffer RIPA (Yase, pc101, Shanghai, China). A total of 10–20 μg of protein was separated by 10% SDS-PAGE (Epizyme, PG212, Shanghai, China) by adding phosphatase inhibitor (Epizyme, GRF102, Shanghai, China) and protease inhibitor (Epizyme, GRF101, Shanghai, China) to the lysate according to the manufacturer's instructions, and then transferred onto a nitrocellulose membrane (Cytiva, 10600002, Washington, DC, USA). The membranes were enclosed in a rapid containment solution (Aase Bio, China) and incubated with the first specific rabbit monoclonal antibodies c-Fos (1:1000), p-c-Fos (1:1000), c-Jun (1:1000), p-c-Jun (1:1000), p38 (1:1000), p-p38 (1:1000), p65 (1:1000), p-p65 (1:1000), and β-Actin (1:10,000), which were incubated at 4 °C for 12–16 h. After washing with TBST, each membrane was incubated with the second anti-rabbit IgG goat polyclonal antibody for 1 h at room temperature, and the protein bands were detected by chemiluminescence (BLT). Except for p-c-Fos (ImmunoWay, YP0442, Beijing, China) and p-p65 (CST, 3033T, Danvers, MA, USA), all other antibodies were obtained from HuaBio, Hangzhou, China (Antibodies c-Fos: ET1701-95; c-Jun: ET1608-3; p-c-Jun: ET1608-4; p38: ET1702-65, p-p38: ER2001-52; p65: ET1603-12; β-Actin: ET1702-67; anti-rabbit IgG goat polyclonal antibody: HA1001).

4.9. RT-qPCR

Total RNA was extracted similarly to in Section 4.7.1. Then, RNA was reversed to cDNA using the PrimeScript RT kit (AG Bio, Hunan, China), and the reverse transcription procedure was as follows: 37 °C for 2 min, 55 °C for 15 min, 85 °C for 5 min, and then maintained at 4 °C. The qpcr procedure was as follows: 94 °C for 5 min for pre-mutability, 94 °C for 20 s, 60 °C for 20 s, and 72 °C for 30 s, for amplification 40 times. The $2^{-\Delta\Delta Ct}$ method was used to calculate the gene expression using β-Actin [82] as the reference gene. The primers for all genes are listed in Table S1.

4.10. Metalloproteinase Activity

The metalloproteinase activity of TE was assayed by the azo-casein method. Different concentrations of TE or NV (0–100 μg/mL, n = 4) were added to a solution (Tris-HCl, PBS, NaCl, CaCl2, and Azocasein (Sigma, A2765, City of Saint Louis, MO, USA)) at pH 8.8, and the mixture was incubated in a water bath at 37 °C for 90 min. After incubation, 200 μL of 0.5 M trichloroacetic acid (TCA) (Ronoen, China) was added to each sample, and the mixture was kept at room temperature for 30 min. The samples were then centrifuged at

10,000× g for 10 min. 100 μL of supernatant was collected from each sample and mixed with 100 μL of 0.5 M NaOH (Amresco, 1310-73-2, Washington, DC, USA). The absorbance of these mixtures was subsequently measured at 450 nm.

4.11. PLA2 Activity

Different concentrations of TE or NV (0–100 μg/mL, n = 4) were prepared and then diluted to 250 μL with 50 mM Tris-HCl, 5 mM $CaCl_2$, and 100 mM NaCl, pH 8.0. Then, 25 μL, 1 mg/mL of 4-nitro-3-octanoyloxybenzoic acid (NOBA) was added; meanwhile, 50 mM Tris-HCl, 5 mM $CaCl_2$, 100 mM NaCl, pH 8.0, and PBS were used as controls. Subsequently, the plates were incubated at 37 °C for 1 h, and the absorbance was measured at 405 nm.

4.12. Statistical Analysis

All data were expressed as mean ± standard deviation (SD), and were analyzed by one-way ANOVA with * $p < 0.05$, ** $p < 0.01$, and *** $p < 0.001$ representing the comparison of the group with the congrol, and # p representing the comparison of the group with the TE group. All statistical analyses were performed using SPSS software 26.0 (SPSS Inc., Chicago, IL, USA). Differences between groups with $p < 0.05$ were considered significant.

5. Conclusions

The present study demonstrated that different extraction methods result in variations in the composition toxicity and effects of jellyfish venom, and different inhibitors may exhibit different antagonistic activities in vitro and in vivo. Comprehensive proteomic, transcriptomic, and biochemical analyses revealed detailed differences in toxicity and effects, including inhibitor interventions. In conclusion, the natural lysis method may result in purer venom, and the mechanical fragmentation method may result in more intact venom glands, which were largely indistinguishable from each other in modulating ex vivo and ex vivo toxicity effects, with TE being more toxic and causing more tissue damage, and NV being able to cause upregulation of inflammatory factors at a higher level. In addition, neither verapamil nor EDTA was considered as a suitable inhibitor, and PACOCF3 showed good antagonistic effects against jellyfish poisoning both in vitro and in vivo. Therefore, the present study may provide new insights for the establishment of a scientific and economical source of jellyfish sting venom and the development of therapeutic methods for jellyfish poisoning.

Supplementary Materials: The following supporting information can be downloaded at: https://www.mdpi.com/article/10.3390/md22080362/s1. Supplementary figures: Figure S1: Comparison of toxins and toxicity evaluation of TE and NV. Figure S2: TE vs. NV transcriptome overview and validation. Figure S3: PACOCF3 and verapamil antagonized the toxicity of TE or NV in vitro. Figure S4: In vivo intervention effects of inhibitors on TE and their comparison. Supplementary table: Table S1: PCR primer sequences.

Author Contributions: Conceptualization, J.-S.Y., L.X., Z.-Y.S., X.-Y.G., M.-K.W. and X.-C.H.; methodology, J.-S.Y. and L.X.; Format, Y.W.; validation, X.-Y.G., Z.-F.W., D.-Y.Z. and B.D.; resources, D.-B.W. and Z.-Y.S.; data curation, X.-Y.G.; writing—original draft preparation, X.-Y.G., M.-K.W. and X.-C.H.; writing—review and editing, B.D., M.-K.W., J.-S.Y. and L.X.; project administration, L.X.; funding acquisition, J.-S.Y.; X.-Y.G., M.-K.W. and X.-C.H. contributed equally to this work. All authors have read and agreed to the published version of the manuscript.

Funding: This study was supported by Scientific Research Project of Shanghai Municipal Health Commission (No. 202140061).

Institutional Review Board Statement: The animal study protocol was approved by the Naval Medical Center of PLA Ethics Committee (NMC2023011).

Data Availability Statement: The authors declare that all relevant data supporting the findings of this study are available within the article and its Supplementary Material files. The data related to the histology research will be uploaded to the NCBI database after the article is accepted.

Acknowledgments: We thank figdraw (https://www.figdraw.com/) for its drawing support for the graphical abstract of this article (ID: IUYWYd522f).

Conflicts of Interest: The authors declare no conflicts of interest.

References

1. Nelsen, D.R.; Nisani, Z.; Cooper, A.M.; Fox, G.A.; Gren, E.C.K.; Corbit, A.G.; Hayes, W.K. Poisons, toxungens, and venoms: Redefining and classifying toxic biological secretions and the organisms that employ them. *Biol. Rev. Camb. Philos. Soc.* **2014**, *89*, 450–465. [CrossRef]
2. Mohamed, A.E.T.; Soares, A.G.; Stockand, J.D. Snake venoms in drug discovery: Valuable therapeutic tools for life saving. *Toxins* **2019**, *11*, 564. [CrossRef]
3. Almeida, G.O.; Oliveira, I.S.; Arantes, E.C.; Sampaio, S.V. Snake venom disintegrins update: Insights about new findings. *J. Venom. Anim. Toxins* **2023**, *29*, e20230039. [CrossRef]
4. Warkentin, T.E.; Greinacher, A.; Koster, A. Bivalirudin. *J. Thromb. Haemost.* **2008**, *99*, 830–839.
5. Safavi-Hemami, H.; Brogan, S.E.; Olivera, B.M. Pain therapeutics from cone snail venoms: From Ziconotide to novel non-opioid pathways. *J. Proteom.* **2019**, *190*, 12–20. [CrossRef]
6. Yap, M.K.K.; Misuan, N. Exendin-4 from Heloderma suspectum venom: From discovery to its latest application as type II diabetes combatant. *Basic. Clin. Pharmacol. Toxicol.* **2019**, *124*, 513–527. [CrossRef]
7. Lausen, B.; Ahang, A.; Cummins, S.; Wang, T. Investigation of Best Practices for Venom Toxin Purification in Jellyfish towards Functional Characterisation. *Toxins* **2023**, *15*, 170. [CrossRef]
8. Leung, T.C.; Qu, Z.; Nong, W.; Hui, J.H.; Ngai, S.M. Proteomic Analysis of the Venom of Jellyfishes *Rhopilema esculentum* and *Sanderia malayensis*. *Mar. Drugs* **2020**, *18*, 655. [CrossRef]
9. Schnitzler, C.E. What makes a jellyfish. *Nat. Ecol. Evol.* **2019**, *3*, 724–725. [CrossRef]
10. Bridge, D.; Cunningham, C.W.; Schierwater, B.; DeSalle, R.; Buss, L.W. Class-level relationships in the phylum Cnidaria: Evidence from mitochondrial genome structure. *Proc. Natl. Acad. Sci. USA* **1992**, *89*, 8750–8753. [CrossRef]
11. Ohdera, A.; Ames, C.L.; Dikow, R.B.; Kayal, E.; Chiodin, M.; Busby, B.; La, S.; Pirro, S.; Collins, A.G.; Medina, M.; et al. Box, stalked, and upside-down? Draft genomes from diverse jellyfish (Cnidaria, Acraspeda) lineages: *Alatina alata* (Cubozoa), *Calvadosia cruxmelitensis* (Staurozoa), *and Cassiopea xamachana* (Scyphozoa). *Gigascience* **2019**, *8*, giz069. [CrossRef]
12. Remigante, A.; Costa, R.; Morabito, R.; La Spada, G.; Marino, A.; Dossena, S. Impact of Scyphozoan Venoms on Human Health and Current First Aid Options for Stings. *Toxins* **2018**, *10*, 133. [CrossRef]
13. Kim, E.; Lee, S.; Kim, J.-S.; Yoon, W.D.; Lim, D.; Hart, A.J.; Hodgson, W.C. Cardiovascular effects of *Nemopilema nomurai* (Scyphozoa: Rhizostomeae) jellyfish venom in rats. *Toxicol. Lett.* **2006**, *167*, 205–211. [CrossRef]
14. Basso, L.; Rizzo, L.; Marzano, M.; Intranuovo, M.; Fosso, B.; Pesole, G.; Piraino, S.; Stabili, L. Jellyfish summer outbreaks as bacterial vectors and potential hazards for marine animals and humans health? The case of *Rhizostoma pulmo* (Scyphozoa, Cnidaria). *Sci. Total. Environ.* **2019**, *692*, 305–318. [CrossRef]
15. Syazwan, W.M.; Rizman-Idid, M.; Low, L.B.; Then, A.Y.-H.; Chong, V.C. Assessment of scyphozoan diversity, distribution and blooms: Implications of jellyfish outbreaks to the environment and human welfare in Malaysia. *Reg. Stud. Mar. Sci.* **2020**, *39*, 101444. [CrossRef]
16. Kienberger, K.; Riera-Buch, M.; Schonemann, A.M.; Bartsch, V.; Halbauer, R.; Prieto, L. First description of the life cycle of the jellyfish *Rhizostoma luteum* (Scyphozoa: Rhizostomeae). *PLoS ONE* **2018**, *13*, e0202093. [CrossRef]
17. Ponce, D.; Brinkman, D.L.; Potriquet, J.; Mulvenna, J. Tentacle transcriptome and venom proteome of the pacific sea nettle, *Chrysaora fuscescens* (Cnidaria: Scyphozoa). *Toxins* **2016**, *8*, 102. [CrossRef]
18. Wang, X.; Jin, Q.; Yang, L.; Jia, C.; Guan, C.; Wang, H.; Guo, H. Aggregation process of two disaster-causing jellyfish species, *Nemopilema nomurai* and *Aurelia coerulea*, at the intake area of a nuclear power cooling-water system in Eastern Liaodong Bay, China. *Front. Mar. Sci.* **2023**, *9*, 1098232. [CrossRef]
19. Geng, X.Y.; Wang, M.K.; Chen, J.H.; Xiao, L.; Yang, J.S. Marine biological injuries and their medical management: A narrative review. *World J. Biol. Chem.* **2023**, *14*, 1–12. [CrossRef]
20. Li, A.; Yu, H.; Li, R.; Yue, Y.; Yu, C.; Geng, H.; Liu, S.; Xing, R.; Li, P. Jellyfish *Nemopilema nomurai* causes myotoxicity through the metalloprotease component of venom. *Biomed. Pharmacother.* **2022**, *151*, 113192. [CrossRef]
21. Rachamim, T.; Morgenstern, D.; Aharonovich, D.; Brekhman, V.; Lotan, T.; Sher, D. The dynamically evolving nematocyst content of an anthozoan, a scyphozoan, and a hydrozoan. *Mol. Biol. Evol.* **2015**, *32*, 740–753. [CrossRef]
22. Amreen Nisa, S.; Vinu, D.; Krupakar, P.; Govindaraju, K.; Sharma, D.; Vivek, R. Jellyfish venom proteins and their pharmacological potentials: A review. *Int. J. Biol. Macromol.* **2021**, *176*, 424–436. [CrossRef]
23. Choudhary, I.; Hwang, D.H.; Lee, H.; Yoon, W.D.; Chae, J.; Han, C.H.; Yum, S.; Kang, C.; Kim, E.J.T. Proteomic analysis of novel components of *Nemopilema nomurai* jellyfish venom: Deciphering the mode of action. *Toxins* **2019**, *11*, 153. [CrossRef]
24. Yu, H.; Li, C.; Li, R.; Xing, R.; Liu, S.; Li, P. Factors influencing hemolytic activity of venom from the jellyfish *Rhopilema esculentum* Kishinouye. *Food Chem. Toxicol.* **2007**, *45*, 1173–1178. [CrossRef]
25. Li, R.; Yu, H.; Li, T.; Li, P. Comprehensive proteome reveals the key lethal toxins in the venom of jellyfish *Nemopilema nomurai*. *J. Proteome. Res.* **2020**, *19*, 2491–2500. [CrossRef]

26. Yang, Y.-S.; Kang, Y.-J.; Kim, H.-J.; Kim, M.-S.; Jung, S.-C. The venom of jellyfish, *Chrysaora pacifica*, induces neurotoxicity via activating Ca^{2+}-mediated ROS signaling in HT-22 cells. *J. Appl. Biol. Chem.* **2019**, *62*, 347–353. [CrossRef]
27. Peng, X.; Liu, K.T.; Chen, J.B.; Yan, Z.H.; Danso, B.; Wang, M.K.; Peng, Z.Y.; Xiao, L. Jellyfish Stings: A Review of Skin Symptoms, Pathophysiology, and Management. *Med. Sci. Monit.* **2024**, *30*, e944265. [CrossRef]
28. Li, A.; Yu, H.; Li, R.; Liu, S.; Xing, R.; Li, P. Inhibitory effect of metalloproteinase inhibitors on skin cell inflammation induced by jellyfish *Nemopilema nomurai* nematocyst venom. *Toxins* **2019**, *11*, 156. [CrossRef]
29. Kim, J.H.; Han, S.B.; Durey, A.J.W.; Medicine, E. Fatal pulmonary edema in a child after jellyfish stings in Korea. *Wilderness Environ. Med.* **2018**, *29*, 527–530. [CrossRef]
30. Cegolon, L.; Heymann, W.; Lange, J.; Mastrangelo, G.J.M.d. Jellyfish stings and their management: A review. *Mar. Drugs* **2013**, *11*, 523–550. [CrossRef]
31. Fenner, P.J.; Hadok, J.C. Fatal envenomation by jellyfish causing Irukandji syndrome. *Med. J. Aust.* **2002**, *177*, 362–363. [CrossRef]
32. Cunha, S.A.; Dinis-Oliveira, R.J. Raising awareness on the clinical and forensic aspects of jellyfish stings: A worldwide increasing threat. *Int. J. Environ. Res. Public Health* **2022**, *19*, 8430. [CrossRef]
33. Prakash, R.L.M.; Hwang, D.H.; Asirvatham, R.D.; Hong, I.-H.; Kang, C.; Kim, E. Identification of cardiorespiratory toxic components of *Nemopilema nomurai* jellyfish venom using sequential chromatography methods. *Toxicon* **2023**, *229*, 107126. [CrossRef]
34. Hwang, D.H.; Heo, Y.; Kwon, Y.C.; Prakash, R.L.M.; Kim, K.; Oh, H.; Seyedian, R.; Munawir, A.; Kang, C.; Kim, E. Cloning of Metalloproteinase 17 Genes from Oriental Giant Jellyfish *Nemopilema nomurai* (Scyphozoa: Rhizostomeae). *Toxins* **2022**, *14*, 519. [CrossRef]
35. Slagboom, J.; Derks, R.J.; Sadighi, R.; Somsen, G.W.; Ulens, C.; Casewell, N.R.; Kool, J. High-Throughput Venomics. *J. Proteome Res.* **2023**, *22*, 1734–1746. [CrossRef]
36. von Reumont, B.M.; Anderluh, G.; Antunes, A.; Ayvazyan, N.; Beis, D.; Caliskan, F.; Crnković, A.; Damm, M.; Dutertre, S.; Ellgaard, L. Modern venomics—Current insights, novel methods, and future perspectives in biological and applied animal venom research. *GigaScience* **2022**, *11*, giac048.
37. Yu, C.; Li, R.; Yin, X.; Yu, H.; Li, P. Synergistic effect of proteinase activity by purification and identification of toxic protease from *Nemopilema nomurai*. *Front. Pharmacol.* **2021**, *12*, 791847. [CrossRef]
38. Yu, K.; Chen, J.; Bai, X.; Xiong, S.; Ye, X.; Yang, Y.; Yao, H.; Wang, F.; Fang, Q.; Song, Q. Multi-Omic Identification of Venom Proteins Collected from Artificial Hosts of a Parasitoid Wasp. *Toxins* **2023**, *15*, 377. [CrossRef]
39. Kim, H.-M.; Weber, J.A.; Lee, N.; Park, S.G.; Cho, Y.S.; Bhak, Y.; Lee, N.; Jeon, Y.; Jeon, S.; Luria, V. The genome of the giant Nomura's jellyfish sheds light on the early evolution of active predation. *BMC. Biol.* **2019**, *17*, 28. [CrossRef]
40. Ma, Y.; Li, J.; Yu, H.; Teng, L.; Geng, H.; Li, R.; Xing, R.; Liu, S.; Li, P. Comparative analysis of PacBio and ONT RNA sequencing methods for *Nemopilema nomurai* venom identification. *Genomics* **2023**, *115*, 110709. [CrossRef]
41. Li, X.; Ma, X.; Chen, X.; Wang, T.; Liu, Q.; Wang, Y.; Li, Z.; Höfer, J.; Li, F.; Xiao, L. The medusa of *Aurelia coerulea* is similar to its polyp in molecular composition and different from the medusa of *Stomolophus meleagris* in toxicity. *Toxicon* **2022**, *210*, 89–99. [CrossRef]
42. White, S.R.; Kadavakollu, S. Bradykinin in *Hemipepsis ustulata*: A novel method for safely milking wasps. *Toxicon* **2016**, *117*, 49–52. [CrossRef]
43. Tracy, J.M.; Golden, D.B. Hymenoptera venom extracts in clinical practice. *J. Allergy Clin. Immunol. Pract.* **2018**, *6*, 1856–1862. [CrossRef]
44. Rybak, M. Effect of venom collection using the method of coupled electrical and sound stimulation on honey yield in bee colonies. *J. Apic. Sci.* **2008**, *52*, 91–94.
45. Oukkache, N.; Chgoury, F.; Lalaoui, M.; Cano, A.A.; Ghalim, N. Comparison between two methods of scorpion venom milking in Morocco. *J. Venom. Anim. Toxins. Incl. Trop. Dis.* **2013**, *19*, 5. [CrossRef]
46. Nagaraj, S.K.; Dattatreya, P.; Boramuthi, T.N. Indian scorpions collected in Karnataka: Maintenance in captivity, venom extraction and toxicity studies. *J. Venom. Anim. Toxins Incl. Trop. Dis.* **2015**, *21*, 51. [CrossRef]
47. Hayes, W.K.; Fox, G.A.; Nelsen, D.R. Venom Collection from Spiders and Snakes: Voluntary and Involuntary Extractions ("Milking") and Venom Gland Extractions. *Methods Mol. Biol.* **2020**, *2068*, 53–71.
48. Li, R.; Zhang, L.; Han, B.; Fang, Y.; Feng, M.; Zhou, T.; Li, J. Proteome Comparison of Honeybee (*Apis mellifera* ligustica) Worker Venom Between Collected from Venom Glands and Electrical Stimulated. *SCI AGR* **2013**, *46*, 1448–1462.
49. Feng, J.; Yu, H.; Xing, R.; Liu, S.; Wang, L.; Cai, S.; Li, P. Partial characterization of the hemolytic activity of the nematocyst venom from the jellyfish *Cyanea nozakii* Kishinouye. *Toxicol. Vitr.* **2010**, *24*, 1750–1756. [CrossRef]
50. Kang, C.; Munawir, A.; Cha, M.; Sohn, E.T.; Lee, H.; Kim, J.S.; Yoon, W.D.; Lim, D.; Kim, E. Pharmacology, Cytotoxicity and hemolytic activity of jellyfish *Nemopilema nomurai* (Scyphozoa: Rhizostomeae) venom. *Comp. Biochem. Physiol. C Toxicol. Pharmacol.* **2009**, *150*, 85–90. [CrossRef]
51. Cantoni, J.L.; Andreosso, A.; Seymour, J. An in vitro comparison of venom recovery methods and results on the box jellyfish, *Chironex fleckeri*. *Toxicon* **2020**, *184*, 94–98. [CrossRef]
52. Kuai, W.; Wang, Y.; Xiao, L.; Wu, J. Research progress in emergency treatment of skin after jellyfish stings. *Chin. Crit. Care. Med.* **2021**, *41*, 902–906.

53. Yang, F.; Ye, R.; Ma, C.; Wang, Y.; Wang, Y.; Chen, J.; Yang, J.; Hofer, J.; Zhu, Y.; Xiao, L.; et al. Toxicity evaluation, toxin screening and its intervention of the jellyfish *Phacellophora camtschatica* based on a combined transcriptome-proteome analysis. *Ecotoxicol. Environ. Saf.* **2022**, *233*, 113315. [CrossRef]
54. Jungo, F.; Bougueleret, L.; Xenarios, I.; Poux, S. The UniProtKB/Swiss-Prot Tox-Prot program: A central hub of integrated venom protein data. *Toxicon* **2012**, *60*, 551–557. [CrossRef]
55. Herzig, V.; Cristofori-Armstrong, B.; Israel, M.R.; Nixon, S.A.; Vetter, I.; King, G.F. Animal toxins—Nature's evolutionary-refined toolkit for basic research and drug discovery. *Biochem. Pharmacol.* **2020**, *181*, 114096. [CrossRef]
56. Kristensen, C. Comments on the natural expression and artificial extraction of venom gland components from spiders. *Toxin Rev.* **2005**, *24*, 257–270. [CrossRef]
57. Carrette, T.; Seymour, J. A rapid and repeatable method for venom extraction from *Cubozoan nematocysts. Toxicon* **2004**, *44*, 135–139. [CrossRef]
58. Escalante, T.; Rucavado, A.; Fox, J.W.; Gutiérrez, J.M. Key events in microvascular damage induced by snake venom hemorrhagic metalloproteinases. *J. Proteom.* **2011**, *74*, 1781–1794. [CrossRef]
59. Song, E.; Kim, S.Y.; Chun, T.; Byun, H.-J.; Lee, Y.M. Collagen scaffolds derived from a marine source and their biocompatibility. *Biomaterials* **2006**, *27*, 2951–2961. [CrossRef]
60. Endotoxemia, D.H. Anti-inflammatory effects of a p38 mitogen-activated protein kinase inhibitor during human endotoxemia. *J. Immunol.* **2002**, *168*, 4070–4077.
61. Winter, K.L.; Isbister, G.K.; Schneider, J.J.; Konstantakopoulos, N.; Seymour, J.E.; Hodgson, W.C. An examination of the cardiovascular effects of an 'Irukandji'jellyfish, Alatina nr mordens. *Toxicol. Lett.* **2008**, *179*, 118–123. [CrossRef] [PubMed]
62. Ramasamy, S.; Isbister, G.K.; Seymour, J.E.; Hodgson, W.C. Pharmacologically distinct cardiovascular effects of box jellyfish (*Chironex fleckeri*) venom and a tentacle-only extract in rats. *Toxicol. Lett.* **2005**, *155*, 219–226. [CrossRef] [PubMed]
63. Wang, B.; Zhang, L.; Zheng, J.; Wang, Q.; Wang, T.; Lu, J.; Wen, X.; Zhang, B.; Liu, G.; Zhang, W. Multiple organ dysfunction: A delayed envenomation syndrome caused by tentacle extract from the jellyfish *Cyanea capillata. Toxicon* **2013**, *61*, 54–61. [CrossRef] [PubMed]
64. Li, R.; Yu, H.; Yue, Y.; Liu, S.; Xing, R.; Chen, X.; Wang, X.; Li, P. In depth analysis of the in vivo toxicity of venom from the jellyfish *Stomolophus meleagris. Toxicon* **2014**, *92*, 60–65. [CrossRef] [PubMed]
65. Yang, F.; Yang, K.; Wang, Y.; Yao, J.; Hua, X.; Danso, B.; Wang, Y.; Liang, H.; Wang, M.; Chen, J. Insights into the discovery and intervention of metalloproteinase in marine hazardous jellyfish. *J. Hazard Mater.* **2024**, *472*, 134526. [CrossRef] [PubMed]
66. Shaik, A.H.; Al Omar, S.Y.; Mohammad, A.; Kodidhela, L.D. Combined cardio protective ability of syringic acid and resveratrol against isoproterenol induced cardio-toxicity in rats via attenuating NF-kB and TNF-α pathways. *Sci. Rep.* **2020**, *10*, 3426.
67. Jayaraj, R.; Anand, T.; Rao, P.L. Activity and gene expression profile of certain antioxidant enzymes to microcystin-LR induced oxidative stress in mice. *Toxicology* **2006**, *220*, 136–146. [CrossRef] [PubMed]
68. Yu, C.; Yu, H.; Li, P. Highlights of animal venom research on the geographical variations of toxin components, toxicities and envenomation therapy. *Int. J. Biol. Macromol.* **2020**, *165*, 2994–3006. [CrossRef] [PubMed]
69. Winter, K.L.; Isbister, G.K.; Jacoby, T.; Seymour, J.E.; Hodgson, W.C. An in vivo comparison of the efficacy of CSL box jellyfish antivenom with antibodies raised against nematocyst-derived *Chironex fleckeri* venom. *Toxicol. Lett.* **2009**, *187*, 94–98. [CrossRef]
70. Heo, Y.; Kwon, Y.C.; Shin, K.; Yoon, W.D.; Han, C.H.; Yum, S.; Kim, E. cDNA and gene structures of two phospholipase A2 isoforms, acidic PLA2 PA4 and PLA2 PA3A/PA3B/PA5, in *Nemopilema nomurai* jellyfish venom. *Toxicon* **2016**, *122*, 160–166. [CrossRef]
71. Ferreira, F.B.; Gomes, M.S.R.; Naves de Souza, D.L.; Gimenes, S.N.C.; Castanheira, L.E.; Borges, M.H.; Rodrigues, R.S.; Yoneyama, K.A.G.; Homsi Brandeburgo, M.I.; Rodrigues, V.M. Molecular cloning and pharmacological properties of an acidic PLA2 from Bothrops pauloensis snake venom. *Toxins* **2013**, *5*, 2403–2419. [CrossRef] [PubMed]
72. Zhang, L.; He, Q.; Wang, Q.; Zhang, B.; Wang, B.; Xu, F.; Wang, T.; Xiao, L.; Zhang, L. Intracellular Ca^{2+} overload induced by extracellular Ca^{2+} entry plays an important role in acute heart dysfunction by tentacle extract from the jellyfish Cyanea capillata. *Cardiovasc. Toxicol.* **2014**, *14*, 260–274. [CrossRef] [PubMed]
73. Moreira, V.; Gutiérrez, J.M.; Soares, A.M.; Zamunér, S.R.; Purgatto, E.; Teixeira, C.d.F.P. Secretory phospholipases A2 isolated from *Bothrops asper* and from *Crotalus durissus* terrificus snake venoms induce distinct mechanisms for biosynthesis of prostaglandins E2 and D2 and expression of cyclooxygenases. *Toxicon* **2008**, *52*, 428–439. [CrossRef] [PubMed]
74. Zuliani, J.P.; Gutiérrez, J.M.; Teixeira, C. Signaling pathways involved in zymosan phagocytosis induced by two secreted phospholipases A2 isolated from *Bothrops asper* snake venom in macrophages. *Int. J. Biol. Macromol.* **2018**, *113*, 575–582. [CrossRef] [PubMed]
75. Wang, B.; Zhang, L.; He, Q.; Wang, Q.; Wang, T.; Lu, J.; Wen, X.; Ye, X.; Xiao, L.; Zhang, L. Direct cardiac toxicity of the tentacle-only extract from the jellyfish *Cyanea capillata* demonstrated in isolated rat heart. *J. Cardiovasc. Pharmacoll.* **2012**, *59*, 331–338.
76. Tibballs, J.; Williams, D.; Sutherland, S. The effects of antivenom and verapamil on the haemodynamic actions of *Chironex fleckeri* (box jellyfish) venom. *Anaesth. Intensive. Care.* **1998**, *26*, 40–45. [CrossRef] [PubMed]
77. Bloom, D.A.; Burnett, J. Effects of verapamil and CSL antivenom on *Chironex fleckeri* (box-jellyfish) induced mortality. *Toxicon* **1999**, *37*, 1621–1626. [CrossRef]

78. Kim, H.J.; Noh, J.W.; Amarsanaa, K.; Jeon, S.C.; Yang, Y.S.; Hwang, N.H.; Ko, E.A.; Kang, Y.J.; Jung, S.C. Peripheral Pain Modulation of *Chrysaora pacifica* Jellyfish Venom Requires Both Ca^{2+} Influx and TRPA1 Channel Activation in Rats. *Neurotox. Res.* **2020**, *38*, 900–913. [CrossRef]
79. Qu, X.; Xia, X.; Lai, Z.; Zhong, T.; Li, G.; Fan, L.; Shu, W. Apoptosis-like cell death induced by nematocyst venom from *Chrysaora helvola* Brandt jellyfish and an in vitro evaluation of commonly used antidotes. *Comp. Biochem. Physiol. C Toxicol. Pharmacol.* **2016**, *180*, 31–39. [CrossRef]
80. Chen, K.C.; Kao, P.H.; Lin, S.R.; Chang, L.S. p38 MAPK activation and mitochondrial depolarization mediate the cytotoxicity of Taiwan cobra phospholipase A2 on human neuroblastoma SK-N-SH cells. *Toxicol. Lett.* **2008**, *180*, 53–58. [CrossRef]
81. Li, R.; Yu, H.; Li, A.; Yu, C.; Li, P. Refinement and Neutralization Evaluation of the F (ab') 2 Type of Antivenom against the Deadly Jellyfish *Nemopilema nomurai* Toxins. *Int. J. Mol. Sci.* **2021**, *22*, 12672. [CrossRef]
82. Pfaffl, M.W. A new mathematical model for relative quantification in real-time RT–PCR. *Nucleic Acids Res.* **2001**, *29*, e45. [CrossRef] [PubMed]

marine drugs

MDPI

Article

Physiological Effects of Oxidative Stress Caused by Saxitoxin in the Nematode *Caenorhabditis elegans*

Haiyan Wu [1], Balakrishnan Prithiviraj [2] and Zhijun Tan [1,2,3,*]

[1] Key Laboratory of Testing and Evaluation for Aquatic Product Safety and Quality, Ministry of Agriculture and Rural Affairs, Yellow Sea Fisheries Research Institute, Chinese Academy of Fishery Sciences, Qingdao 266071, China; wuhy@ysfri.ac.cn

[2] Department of Plant, Food and Environmental Sciences, Faculty of Agriculture, Dalhousie University, Nova Scotia, NS B2N5E3, Canada; bprithiviraj@dal.ca

[3] State Key Laboratory of Mariculture Biobreeding and Sustainable Goods, Yellow Sea Fisheries Research Institute, Chinese Academy of Fishery Sciences, Qingdao 266071, China

* Correspondence: tanzj@ysfri.ac.cn; Tel.: +86-0532-85836348; Fax: +86-0532-85825917

Abstract: Saxitoxin (STX) causes high toxicity by blocking voltage-gated sodium channels, and it poses a major threat to marine ecosystems and human health worldwide. Our work evaluated the neurotoxicity and chronic toxicology of STX to *Caenorhabditis elegans* by an analysis of lifespan, brood size, growth ability, reactive oxygen species (ROS) and adenosine triphosphate (ATP) levels, and the overexpression of green fluorescent protein (GFP). After exposure to a series of concentrations of STX for 24 h, worms showed paralysis symptoms and fully recovered within 6 h; less than 5% of worms died at the highest concentration of 1000 ng/mL for first larval stage (L1) worms and 10,000 ng/mL for fourth larval stage (L4) worms. Declines in lifespan, productivity, and body size of *C. elegans* were observed under the stress of 1, 10, and 100 ng/mL STX, and the lifespan was shorter than that in controls. With STX exposure, the productivity declined by 32–49%; the body size, including body length and body area, declined by 13–18% and 25–27%, respectively. The levels of ROS exhibited a gradual increase over time, accompanied by a positive concentration effect of STX resulting in 1.14–1.86 times higher levels compared to the control group in L4 worms. Conversely, no statistically significant differences were observed between L1 worms. Finally, after exposure to STX for 48 h, ATP levels and GFP expression in *C. elegans* showed a significant dose-dependent increase. Our study reports the first evidence that STX is not lethal but imposes substantial oxidative stress on *C. elegans*, with a dose-responsive relationship. Our results indicated that *C. elegans* is an ideal model to further study the mechanisms underlying the fitness of organisms under the stress caused by paralytic shellfish toxins including STX.

Keywords: *Caenorhabditis elegans*; saxitoxin; oxidative stress; reproductivity; ATP

1. Introduction

Paralytic shellfish toxins (PSTs) comprise more than 57 molecules with a tetrahydropurine structure, and saxitoxin (STX) is the major component of PSTs. PSTs are produced primarily by bloom-forming algal species in both marine and freshwater ecosystems, particularly dinoflagellates belonging to the genera *Alexandrium*, *Gymnodinium*, *Pyrodinium*, and cyanobacteria [1,2]. PSTs pose major public health threats through paralytic shellfish poisoning in consumers, thus, causing symptoms including a tickling sensation or numbness around the lips, numbness of the extremities, gastrointestinal problems, and fatal respiratory paralysis [2]. Beyond humans, PSTs affects marine organisms, such as zooplankton [3–6], shellfish [7–12], finfish [13–16], crustaceans [17,18], and even marine mammals [19,20], through either direct exposure or dietary uptake. PSTs cause major negative effects on animal health, including physiological activity [6,11], reproductive development [9,12] and growth ability [21,22], and can even lead to mass mortality [14,15,23] and decreases in the biomass and yield of aquatic systems [2,9,15,19,24,25].

To our knowledge, PSTs affect organisms primarily through two mechanisms: neurotoxicity and oxidative stress. The well-known mechanism of neurotoxicity of STX is based on binding to voltage-gated sodium channels [26,27]. Acute toxicity not only inhibits physiological activity but also causes substantial mortality in contaminated organisms [28]. Generally, acute toxicity is caused by neurotoxicity, mainly at high doses of PSTs, particularly during red tides caused by dinoflagellates [14,15]. In natural seawater, the dose of PSTs is insufficient to directly kill marine organisms. These PSTs consequently accumulate in other marine organisms along the food chain, and are metabolized to low- or non-toxic metabolites [29–31].

The accumulation and metabolism of PSTs cause the generation of excessive ROS in marine organisms, and subsequently perturb the redox balance and cause oxidative stress in contaminated organisms. To date, oxidative stress due to PSTs has been reported in zooplankton [6,32], shellfish [11,33–35], finfish [22,36–38] and marine mammals [39,40]. Similar to other toxins, PSTs lead to ROS attacks on lipids, proteins, and DNA; damage to the structures of cells, tissues and organs; inhibition of the functions of physiological, growth and reproductivity; and, finally, a decrease in marine biomass.

Despite many years of investigation, numerous questions surround the mechanisms of oxidative stress caused by PSTs, particularly regarding STX. Prior experiments have been performed in zooplankton, shellfish, finfish, mammals, and even cell lines, all of which have cell membranes with voltage-gated sodium channels. As described in previous studies [41,42], STX blocking channels not only alters the influx of sodium ions into cells but also has adverse effects on physiological function, similar to responses to oxidative stress. Consequently, experiments with organisms that have Na^+ channels are not able to accurately assess oxidative stress caused by STX due to possible neurotoxic effects. Therefore, more work is needed on the oxidative stress caused by PSTs to provide a clearer explanation for the marine biomass decline in natural seawater.

Caenorhabditis elegans is used as the test animal primarily because this worm has no voltage-gated sodium channels in its cell membranes [43]; therefore, PSTs do not cause neurotoxicity but instead cause oxidative stress in *C. elegans*. *C. elegans* has many advantages as a model in biological research, such as its small size, rapid generation time, and consistent brood size [44,45]. Importantly, the *C. elegans* nervous system has been mapped, and *C. elegans* was the first multicellular animal for which the genome sequence was completed [46]. In this study, we used an alternative approach based on the model organism *C. elegans* to study the oxidative stress caused by PSTs. We exposed *C. elegans* to solutions of STX, the basic component of PSTs. Several different endpoints of *C. elegans*, including lethality, lifespan, brood size, body length, body size, and reactive oxygen species (ROS) and adenosine triphosphate (ATP) levels, were tested. We also used green fluorescent protein (GFP) markers for specific neuronal populations to assess the response of protein expression under the oxidative stress caused by STX.

2. Results

2.1. STX Caused no Lethal Paralytic Reaction in C. elegans

First larval stage (L1) and fourth larval stage (L4) worms were used to assess the toxicity of STX to *C. elegans*. *C. elegans* showed abnormal behaviors, such as body bends and slow head thrashes. Some worms were paralyzed completely and continued ankylosing, showing no response to touch. However, most worms had fully recovered and appeared normal within 6 h; less than 5% worms died in total, and these deaths were as a result of regular culturing losses that have no statistical significance.

L1 worms were much more sensitive than L4 worms under the same STX treatment level, as shown in Figure 1. Approximately 59% of worms were completely paralyzed under the highest concentration of 1000 ng/mL in the L1 worms, as compared with 20% under the same concentration in the L4 worms. A serious negative effect on *C. elegans* was observed at the highest concentrations of 500, 1000 ng/mL for L1 worms and 10,000 ng/mL for L4 worms with the lowest motility below 50%. Generally, worms in the same stage showed dose–response trends under different concentrations of STX.

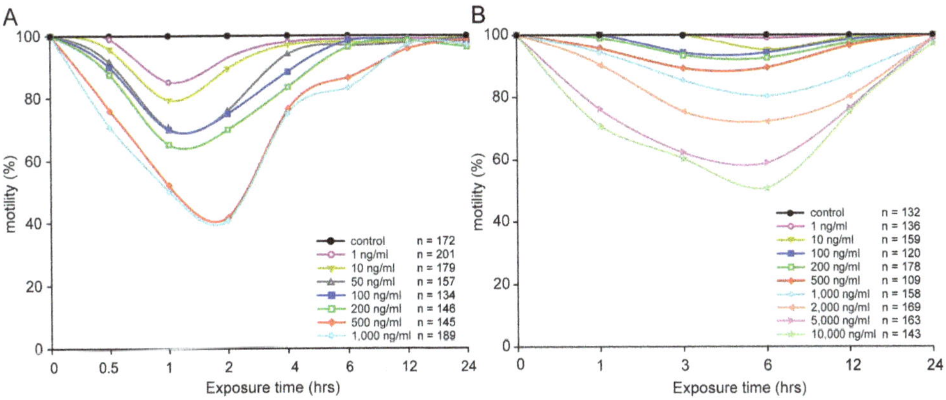

Figure 1. Influence of different concentrations of STX on L1 (**A**) and L4 (**B**) worms' motility for 24 h. All data in the curve are mean values.

2.2. *STX Induces the Production of ROS in Worms*

We assessed the degree of oxidative stress in *C. elegans* under STX treatment by measuring the dynamic variation in general ROS produced by the worms. During the experiments, the ROS level increased with time, and the highest level was approximately four to six times greater than the lowest level in L1 worms (Figure 2A), and three to four times greater than that in L4 worms (Figure 2B). A positive concentration effect of STX on ROS production was observed in the L4 worms; the ROS levels at 1, 10, and 100 ng/mL were 1.14, 1.56, and 1.86 times higher than those of controls, respectively, at the end of the experiment. During the experiment, the ROS levels in L4 worms were much higher than those in L1 worms, mainly because of the body sizes of the worms.

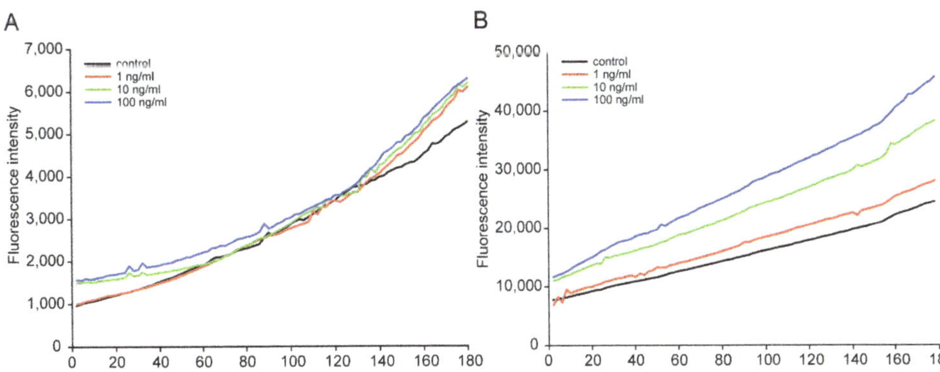

Figure 2. Dynamic variation in the general ROS in L1 (**A**) and L4 (**B**) worms after exposure to STX. The fluorescence intensity represents the relative concentrations of general ROS in worms, measured immediately after exposure.

2.3. *STX Induced Body Size Decrease in* C. elegans

Body size directly contributes to the biomass of animals and is also closely associated with lifespan. In this study, worm growth was significantly impaired after STX exposure.

The body length and body area of worms treated with STX were smaller than those in the control ($p < 0.05$) (Figure 3A). Even at the lowest concentration (1 ng/mL) of STX exposure, the body length and body area were approximately 87% and 75% of the control groups, respectively (Figure 3). STX did not exhibit a dose–response relationship on the growth of worms, and no significant difference was observed among treatment groups ($p > 0.05$).

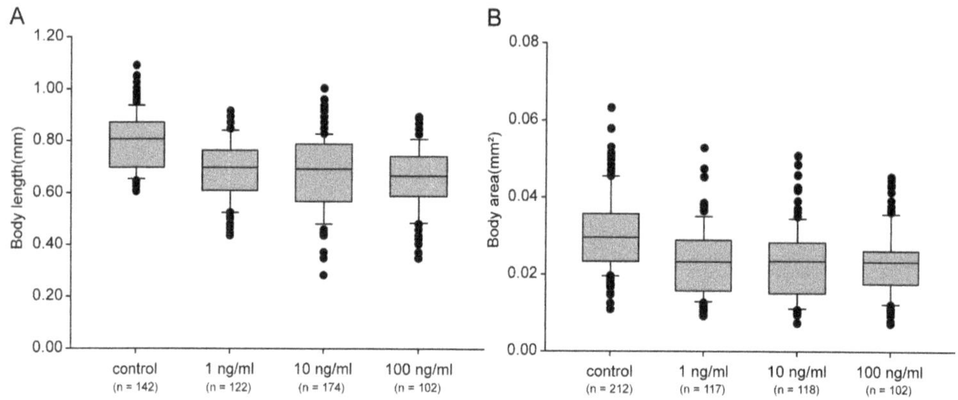

Figure 3. Body sizes, measured as length (**A**) and area (**B**), clearly decreased after treatment with STX for 48 h. Bars represent means ± SD, and different superscript letters indicate significant differences ($p < 0.05$). The horizontal lines represent the median, the hinges of the box plots represent 1st and the 3rd quartile, the whiskers are plotted at a distance of 1.5 times the inter-quartile range, the thick dots represent outliers and the thin dots represent the raw data.2.4. STX Shortens the Lifespan of *C. elegans*.

After exposure to STX for 48 h in 96-well plates, the worms were mainly at the young adult stage and were transferred to NGM plates to observe the lifespan. Although no worms died during the exposure experiment, the lifespan of the worms appeared to be shortened by STX. Compared with that of the controls, the mean lifespans of worms stressed by STX was 1 or 2 days shorter (Figure 4). A significant difference was observed between the 1 ng/mL and 10 ng/mL groups, although the maximum lifespans of the worms in these two groups were both 20 days. To validate the trend, we further investigated the survival time (mean ± SD) with the Kaplan–Meier survival function log-rank test. The mean survival time showed a dose-dependent trend and was 13.88 ± 0.34, 12.20 ± 0.37, 11.33 ± 0.36, and 10.38 ± 0.36 days in the control, 1, 10, and 100 ng/mL groups, respectively (Table 1). Significant differences were observed between the control and treatment groups with $p < 0.05$.

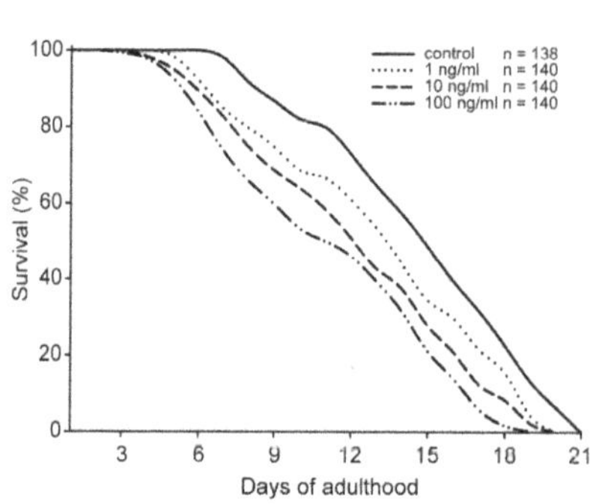

Figure 4. Lifespan of *C. elegans* exposed to different concentrations of STX. All data are mean values.

Table 1. Comparison of the mean lifespans for *C. elegans* exposed to STX.

Treatment	Number of *C. elegans*		Survival Time (Days)
	Total	Censored	
Control	138	0	13.88 ± 0.34 [a]
1 ng/mL	140	0	12.20 ± 0.37 [b]
10 ng/mL	140	0	11.33 ± 0.36 [c]
100 ng/mL	140	0	10.38 ± 0.36 [d]

Different superscript letters indicate significant differences ($p < 0.05$).

2.4. STX Decreases the Reproductive Function of C. elegans

Brood size usually reflects the reproductive capacity of animals. Reproductive organs may be important secondary targeted organs for toxicants in *C. elegans*. In this study, we investigated the effects of STX on the reproductive capacity of *C. elegans*. After being exposed to a series of concentrations of STX for 48 h, most worms in the treatment or control groups began to lay eggs soon after being transferred to NGM plates. The total generation trend was not perturbed by STX, and all worms laid the most eggs on the second day (Figure 5A). However, all STX treatment groups stopped breeding one day in advance. In addition, the reproductive speed (mean eggs laid each day) in the treatment group was significantly slower than that in the control group (Figure 5A). Therefore, the productivity of worms was substantially altered by STX, and the brood size was nearly 50% lower in the 100 ng/mL treatment group than in the control group (Figure 5B). No significant difference in reproductive speed was observed between 1 ng/mL and 10 ng/mL. However, the brood sizes in these two groups were approximately 67% of the control group.

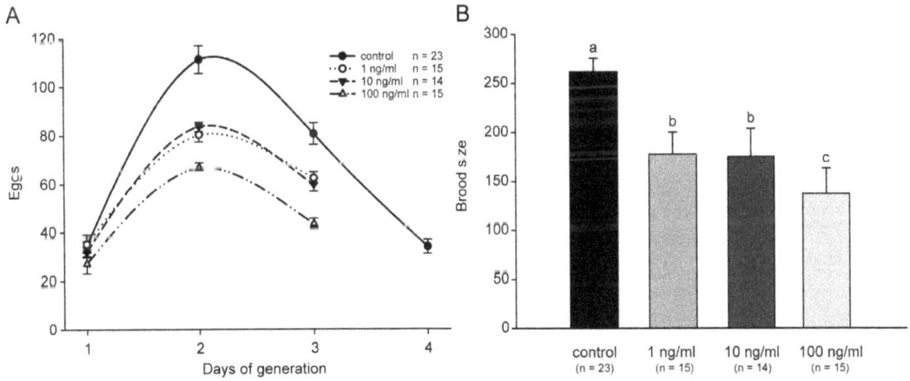

Figure 5. STX decreases the generation period (**A**) and the brood size (**B**). All treatments showed statistically significant differences vs. control. Bars indicate means ± SD, and different superscript letters indicate significant differences ($p < 0.05$).

2.5. Fluorescent Protein Expression in GFP Worms

By testing several physiological endpoints, we observed the oxidative stress caused by STX in *C. elegans*. All these endpoints reflected potential changes in protein content. Therefore, we further performed fluorescent protein intensity quantification in the whole body in *C. elegans* to gain a general understanding of the protein expression after STX stress exposure. Figure 6A clearly indicated a difference in the fluorescence intensity. The mean fluorescence intensity in the 1, 10, and 100 ng/mL groups were 1.19, 1.62, and 2.18 times greater than that in control group, respectively, with the 10 and 100 ng/mL groups indicating a significant increase in intensity with STX concentration (Figure 6B).

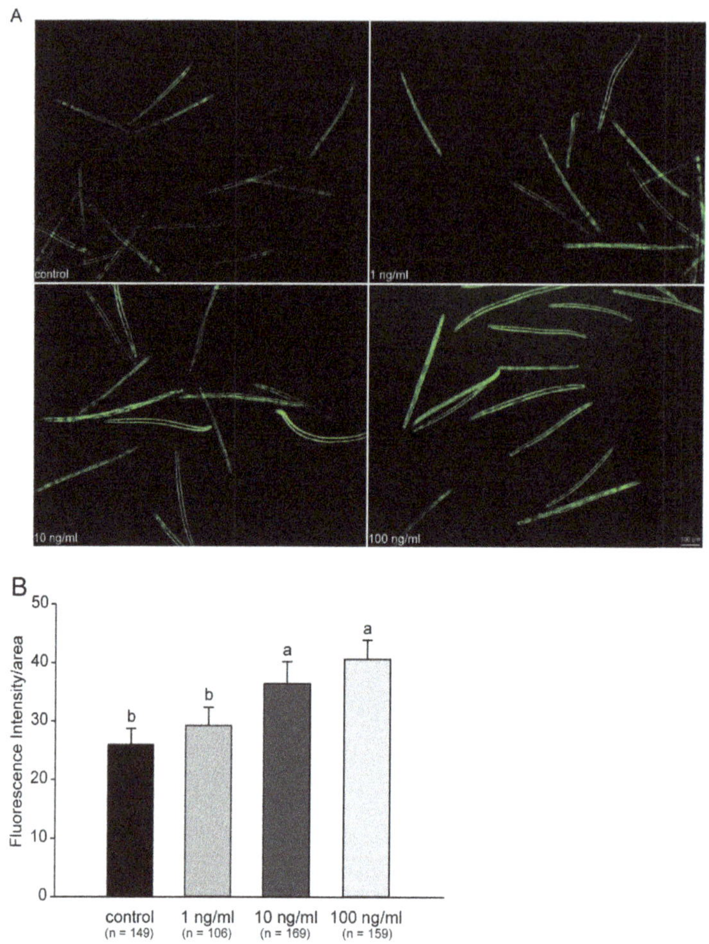

Figure 6. Fluorescent protein expression in GFP worms exposed to STX for 48 h (**A**) and fluorescent intensity (**B**). Intensity quantification of fluorescent protein calculated in ImageJ software (1.46r) with background removal. Bars represent means ± SD, and different superscript letters indicate significant differences ($p < 0.05$). Scale bar: 100 μm.

2.6. ATP Levels Significantly Increase after Exposure to STX

Variations in ATP levels indicate a response in cellular energy production and utilization in *C. elegans*. In this study, the ATP levels showed a significant increase with a clear dose-dependent relationship after exposure to STX for 48 h. The elevation in ATP reflected induction of energy production in *C. elegans*, because more energy was produced to defend against the stress. The average ATP level in the 1, 10 and 100 ng/mL treatment groups was 1.16, 1.54, and 1.91 times that in the control group at the end of the experiment, respectively (Figure 7). All treatment groups showed a significant difference with respect to one another.

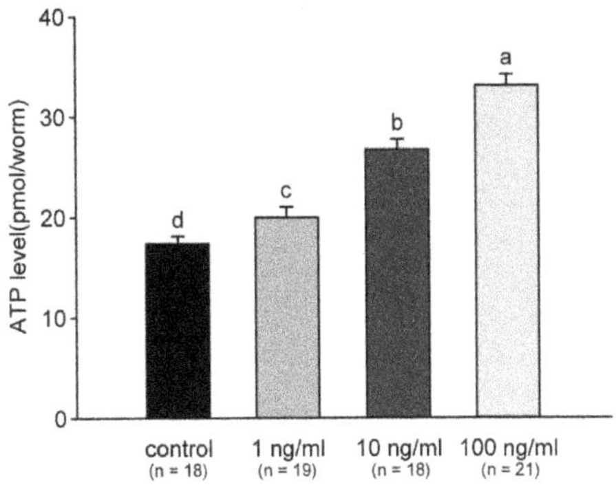

Figure 7. Average concentrations of ATP per worm across three independent experiments. ATP in worms significantly increased with the concentration of STX. Bars represent means ± SD, and superscript letters indicate significant differences ($p < 0.05$).

3. Discussion

On the basis of the available information, PSTs are considered among the most toxic phycotoxins. According to the previous data, the mouse intraperitoneal 50% lethal dose (LD50) of STX is 10 µg/kg b.w., while the intravenous LD50 is only 3 µg/kg b.w. [1]. Compared with other toxins, for example the cobrotoxin (CBT B with an LD50 of 400 mg/kg in mice) [47], the toxicity of STX is much higher. Therefore, the occurrence of HABs associated with PST producers often leads to the mass mortality of marine organisms [14,15,23]. However, we observed that treated worms entered a state of temporary paralysis and recovered within several hours, showing high tolerance during acute exposure to STX. Gao and Zhen [43] have indicated that *C. elegans* lacks voltage-gated sodium channels. The high selective binding of Na^+ channels is the major molecular mechanism of action of STX [26,27]. Therefore, *C. elegans* has no receptor for STX in the cell membrane, explaining the absence of neurotoxic effects of STX in *C. elegans*. On the other hand, the potassium (K^+) channels of *C. elegans* are highly homologous to humans, controlling important behaviors, such as motility and spawning [44]. Studies have unraveled the interaction of STX with K^+ channels. While the interaction is highly different from Na^+ channels, it shows 100~10,000 lower binding affinities, corroborating a relatively slight toxicological impact [45]. Paralysis symptoms in *C. elegans* may be caused through K^+ channels.

C. elegans is an ideal animal model to study fitness under the stress of many environmental toxins [46,48]. Although substantial research has focused on the oxidative stress caused by STX and found that this toxin exerts adverse effects in tested animals [1], limited data are available on the chronic effects of STX in animals. In our study, even though neurotoxicity was not fatal for *C. elegans*, STX caused significant oxidative stress, which could result in chronic adverse effects. Therefore, we further investigated the responses of *C. elegans* under STX-causing oxidative stress. To validate our hypothesis, we first evaluated the changes in the lifespan of *C. elegans*. Generally, the lifespan of *C. elegans* is altered according to the stress caused by toxins, which means the aging status of this animal [49,50]. In our study, the shortened lifespan revealed an accelerated aging process for *C. elegans*, and the adult generation exposed to STX showed poorer survival than the control group. Under the stress of STX in natural seawater, it is possible that the biomass of adult organisms consistently decreases over time.

A decrease in animal lifespan indicates an intensification of aging, which may result in parental losses in the population unless more offspring are produced to compensate [51]. However, no reports have indicated that animals generate more offspring in crises; instead, parents may sacrifice themselves by decreasing their lifespan [49,51]. Many studies have found that STX producers inhibit the offspring size of marine organisms, and the parental generation experiences major damage from stress [47,52]. A decline in reproductive output caused by STX producers results primarily from the inhibition of embryonic and larval development [47,52], decreased quality of gametes [53], and decreased energy status and spermatozoa motility in marine organisms [54]. As described before, PSTs or STX cause acute toxicity by blocking voltage-gated sodium channels [26,27], thereby threatening the reproductivity of nematodes. In this study, we first indicated a decline in brood size of *C. elegans*, and observed that the offspring size was suppressed by the oxidative toxicity of STX. The generation of behavior in *C. elegans* was not significantly perturbed by STX, unlike in other animals, such as *Neomysis awatschensis* [55]. Regardless of the decline in the reproductive ability of *C. elegans*, the smaller offspring size decreased the chance of the population overcoming the adverse environment caused by the oxidative stress of STX.

Beyond the lifespan and offspring size, biomass also depends on the body size and body weight of single organisms. Biomass directly correlates with aquaculture output. Previous studies have found that organisms might alter their body sizes under environmental stress; most tested organisms have shown a decline in body size, for example, benthic macroinvertebrates under temperature stress [56], mussels under oiling stress [57], and crabs under temperature stress [58]. In marine fish, both smaller and larger species have shown declines in size under the stresses of a century of climate change and fishing [59]. To our knowledge, only a few studies have reported the growth decline in marine organisms under the stress of PSTs and its causative dinoflagellate [10,60]. Li et al. [21] have found that *A. tamarense* depresses the growth rate of juvenile *Ruditapes philippinarum* clams, primarily because this toxic alga depresses the energy budget of clams, thus, decreasing their scope for growth. Similar results have been reported in rock lobsters fed toxic mussels contaminated with 6 mg STX.2HCL eq./kg [61].

To date, studies on the decline in biomass have focused on the energy intake depression in marine organisms under the stress of pollution, primarily the dynamic energy budget and scope for growth [62,63]. Biomass size depends not only on energy intake but also on the cellular energy status of organisms. As a result of evolution, there are strict expenditure rules, and cellular energy is distributed among three functions: basic metabolism, adaptive thermogenesis, and physical activity. [64,65]. Many studies have revealed that the energy distribution balance is perturbed, and more cellular energy is allocated to defend against stress, thus, decreasing utilization for physical activity, such as survival, growth, and reproduction [66–69]. Therefore, more research is needed to further indicate the mechanism of cellular energy distribution and to obtain a clear explanation for the marine biomass decline under the oxidative stress caused by PSTs in natural seawater.

ROS, a commonly used indicator of oxidative stress caused by environmental stress in organisms, lead to many subsequent changes, such as damage to lipids, proteins, and DNA [67], thereby influencing physical activities including basic behaviors, and declines in lifespan, reproductive ability and body size, to adapt to the environment. Prior studies have found that additional ROS are produced in marine organisms after exposure to PSTs; toxic dinoflagellates also elicit changes in DNA/RNA damage [35,70] and antioxidant enzymes [34,35]. On the basis of our results, the ROS in *C. elegans* rose significantly during exposure, and showed clear STX dose- and body size-dependent relationships. STX enhanced the production of ROS in worms, indicating oxidative stress caused by STX. Moreover, the larger bodies of worms resulted in greater ROS production. The decline in lifespan, brood size and body size also presumably resulted from oxidative stress caused by STX, because of the absence of neurotoxicity of this toxin through binding of voltage-gated sodium channels [26,27].

ATP and ROS are simultaneously produced by mitochondria [71]. ATP is the basic energy currency not only for all physical activities in organisms but also for the reduction in ROS through the induction of the expression of functional proteins, such as antioxidative enzymes, to maintain the redox balance within cells and help organisms perform normal physical activities [72,73]. In our study, the total ATP level in *C. elegans* showed a dynamic rising trend during exposure, with a clearly dose-dependent relationship with STX. Additionally, ROS varieties showed a similar trend. Therefore, our results demonstrated that the increased ATP enhanced the ability of *C. elegans* to remove more ROS and thereby modulate the redox balance. The removal of ROS is implemented primarily by antioxidant enzymes, e.g., superoxide dismutase, catalase, thioredoxin, glutathione peroxidase and others [35,72,73]. The occurrence of oxidative stress or the removal of ROS both require ATP to provide sufficient energy for the antioxidant system and protect the physiological activities of organisms [74]. Under the stress of ROS, the increases in ATP also provide opportunities for organisms to adapt to environmental effects by providing more energy for reproduction, survival, or growth [67,75]. However, our study indicated that the lifespan, body size and brood size of *C. elegans* all declined under the stress of STX, while we did not analyze the potential effects of changes in antioxidant enzymes, which is one limitation of this study. As an adaptation to STX stress, these phenomena stimulate ATP allocation between survival, growth, or reproduction.

GFP is sensitive to oxidative stress, and its variation represents the oxidative damage of proteins [76]. The increasing of expression of GFP in this study indicated that stress caused protein damage in *C. elegans*. Similar results were reported by previous studies [6], and the expression of GFP in yeast has been found to significantly increase after exposure to STX. STX is believed to bind a copper transporter that is mechanistically and structurally similar to ion channels [6], the major target of STX for its neurotoxicity [26,27]. Our results contrast with those of Cusick et al. [6] in that no voltage-gated sodium channels exist in the cell membrane of *C. elegans* [43]; therefore, the protein damage in *C. elegans* was caused mainly by oxidative stress due to STX, not by the neurotoxicity through ion channel binding. At present, the mechanism through which STX induces the high expression of GFP is unclear, but protein was clearly disrupted and might have been responsible for the decline in lifespan, body size, and brood size observed in this study. In addition, the long-term presence of STX in seawater has been reported worldwide; consequently, persistent oxidative stress on marine organisms may lead to biome instability, explaining the decline in catches recent years. In the future, more research is needed to reveal the mechanisms by which marine organisms adapt to STX oxidative stress, and emphasis should be placed on cellular energy metabolism, distribution, and application.

4. Materials and Methods

4.1. Chemical and Toxin Preparation

All organic solvents used were of analytical grade and were purchased from Sigma Chemical Company (Oakville, ON, Canada) unless otherwise stated. STX standard (NRC CRM-STX-f), initially at a concentration of 65 μmol/L was purchased from the National Research Council Canada, Halifax NS, Canada. The toxin was diluted with sterile water to 10 μg/mL as a stock solution and stored at 4 °C, for use within 2 weeks.

4.2. Strains

The wild type *C. elegans* strain Bristol N2 and the GFP worm strain NL5901, as well as the OP50 *Escherichia coli* strain, were originally provided by the Caenorhabditis Genetics Center (University of Minnesota, Minneapolis, MN, USA). The worms were maintained at 20 °C on nematode growth medium (NGM) agar plates seeded with *Escherichia coli* OP50. The *C. elegans* population was treated with hypochlorite/NaOH solution to synchronize the population at day 0 of the lifespan, and larval stage 1 (L1) worms were synchronized by hatching eggs overnight. Some L1 worms were transferred to freshly spotted plates to develop into L4 worms, whereas others were directly used for treatment with STX. All treatments were performed in 96-well plates with a total volume of 100 μL containing approximately 30–40 larvae in each well. During the treatment, worms were fed with heat-inactivated bacteria (OP50 HIT) to avoid interference from the xenobiotic metabolizing activity of *E. coli* [77]. All lifespan assays were performed at 20 °C according to standard protocols, as previously described [78]. Subsequently, the worms were washed and placed on OP50-containing NGM plates to observe brood size, lifespan, and other parameters. All treatments in each experiment were performed on at least three biological replicates, and three independent experiments were performed.

4.3. Toxicity Assessment

The synchronized L1 and L4 larvae stages of N2 worms were exposed to different concentrations of STX to compare the toxicity of STX on different stages of *C. elegans* development. On the basis of the prior results, the stock solution of STX was added to freshly autoclaved S-medium to a final concentration of 0, 1, 10, 50, 100, 200, 500, or 1000 ng/mL for L1 worms, and 0, 10, 100, 200, 500, 1000, 2000, 5000, or 10,000 ng/mL for L4 worms before the exposure treatment. Both treatments lasted for 24 h, and the dead worms were recorded at 0.5, 1, 2, 4, 6, 12 h, and 24 h for the L1 worms' group and 1, 3, 6, 12, and 24 h for the L4 worms' group. At least 100 worms were subjected to each treatment for statistical analysis. Worms' motilities were evaluated with repeated gentle mechanical prodding, and a transient peristalsis reaction was recorded as shown paralytic symptoms. Survival was scored throughout the entire test process.

4.4. Measurement of ROS Production

The generation of ROS was compared between L1 and L4 larvae stages of NS worms, which were treated with 1, 10, or 100 ng/mL STX for 180 min. More than 100 animals per sample were transferred into the wells of a black 96-well plate containing 100 μL M9 buffer. Dichlorofluorescein diacetate (DCFDA; 50 mM in M9 buffer, 100 mL; Invitrogen) was added to each well. ROS-associated fluorescence levels were measured kinetically with a fluorescence plate reader (Spectra-Max Gemini EM; Molecular Devices, Sunnyvale, CA, USA) at 485 nm excitation and 520 nm emission wavelengths at room temperature, every 2 min for 3 h. Data were normalized to those for bacterial controls composed of 100 mL *E. coli* HT115 (DE3) expressing dsRNA against prdx-3 or an empty vector containing M9 buffer in separate wells of the same plate. In cells, DCFDA is deacetylated by endogenous esterases to dichlorofluorescein, which reacts with ROS and generates the fluorophore DCF. Although the specific ROS responsible for DCFDA fluorescence is uncertain, DCFDA can be used in whole *C. elegans* as a marker of general ROS production [79,80]. Three independent studies were performed.

4.5. Growth Assays, Lifespan Assays and Brood Size

According to the toxicity assessment results, the synchronized L1 larvae of N2 worms were treated with 1, 10, or 100 ng/mL STX for 48 h. Worms in S-medium were used as controls.

For growth assays evaluation, worms were incubated in liquid conditions for approximately 48 h and allowed to grow to early adulthood in the absence of STX. After worms were paralyzed with tetramisole, a Nikon microscope (Eclipse E200 POL, Tokyo, Japan) equipped with a standard ruler was used to measure the body length from the posterior bulb of the pharynx to the anus, and the body area were determined by measurement of the flat surface areas of the *C. elegans* along the anterior–posterior axis in ImageJ software. At least 100 worms in each treatment were subjected to statistical analysis.

For lifespan assays evaluation, worms were transferred to fresh NGM plates with OP50 following treatment, and the time was considered day 0. On each succeeding day, worms were counted and scored as live or dead. Live *C. elegans* were picked and transferred to fresh plates every day during egg-laying and every other day after they ceased laying eggs until no live *C. elegans* remained. At least 100 worms were subjected to each treatment for statistical analysis. Survival time was determined from the survival curve data by Kaplan–Meier survival function log-rank test.

For brood size evaluation, worms were transferred to individual egg-laying plates. Animals were thereafter transferred every 24 h to fresh egg-laying plates until egg laying ceased. Eggs were kept at room temperature, and the progeny that hatched were counted after 2 days. In addition, we compared the variation in generation time from the P0 egg to the first F1 (final) egg. At least ten animals per condition per experiment were used.

4.6. Quantification of Fluorescent Protein Intensity in GFP Worms

To test the protein expression of *C. elegans* under the stress of STX, we incubated synchronized L1 larvae of NL5901 GFP worms in liquid conditions for 48 h in the presence of STX or S-medium (control group). After the exposure, all worms were washed three times with distilled water to remove *Escherichia coli*, then transferred to a new 96-well plate containing 50 μL S-medium. After measurement of the total fluorescence intensity under a fluorescence plate reader (Infinite 200; Tecan, Männedorf, Switzerland), we determined the protein expression with Image J software. Three independent studies were performed.

4.7. Adenosine Triphosphate (ATP) Assays

Several L4 worms were randomly selected and exposed to different concentrations of STX solution for 48 h. After being washed with S-medium, the treated worms were immediately placed in liquid nitrogen for ATP measurements. The ATP assays were performed as previously described [81], with some modifications. Briefly, flash-frozen *C. elegans* were resuspended in trichloroacetic acid, lysed and incubated on ice, then subjected to centrifugation. Supernatants were pooled, and aliquots of ATP standards or sample supernatant were pipetted into white 96-well plates, and this was followed by the addition of arsenite ATP buffer. ATP levels were measured with a fluorescence plate reader (Infinite 200; Tecan, Männedorf, Switzerland). Three independent studies were performed.

4.8. Statistical Analyses

The statistical analyses were performed using SPSS software version27.0 (IBM Corp., Armonk, NY, USA). One-way analysis of variance (ANOVA) was used to compare body size, lifespans, brood size, GFP expression, and ATP levels among the different test groups. When significant differences were detected by ANOVA, Tukey's honest significant difference test was performed. Statistical significance was set at $p < 0.05$.

5. Conclusions

PSTs and their causative dinoflagellate have been reported to reduce marine biomass by inducing oxidative stress. The presence of voltage-gated sodium channels in these organisms has complicated the study of the basic mechanisms of toxicity. This study used *C. elegans*, which lacks these channels, and found direct evidence of oxidative stress. Although our findings indicated no notable neurotoxic effects of STX in *C. elegans*, the lifespan, reproductivity, and body size of *C. elegans* all significantly declined in a dose-dependent manner during STX exposure. Therefore, STX decreases the biomass of marine organisms. Subsequently, a dynamic increase, as well as dose- and body size-dependent ROS levels in *C. elegans*, provided direct evidence of oxidative stress caused by STX. Finally, the ATP level and fluorescent protein expression both significantly increased in *C. elegans*, indicating that ATP and specific proteins might play important roles in worms or other organisms in reaction to oxidative stress caused by STX. Therefore, our study indicated the oxidative characteristics of STX on *C. elegans*. More studies are needed to clarify the mechanism of organism fitness in response to STX, particularly regarding the production, distribution, and utilization of ATP in *C. elegans* under STX stress.

Author Contributions: Conceptualization, Z.T. and B.P.; experiments and writing—original draft preparation, H.W. All authors have read and agreed to the published version of the manuscript.

Funding: This work was supported by the National Key R&D Program of China (No. 2022YFC3105203), the National Natural Science Foundation of China (No. 32072329, 41806138), the earmarked fund for CARS (CARS-49), and the Central Public-interest Scientific Institution Basal Research Fund, CAFS (No. 2020TD71).

Institutional Review Board Statement: Not applicable.

Data Availability Statement: The data presented in this study are available on request from the first author.

Conflicts of Interest: The authors declare that they have no known competing financial interests or personal relationships that could have appeared to influence the work reported in this paper.

References

1. European Food Safety Authority. Marine biotoxins in shellfish—Saxitoxin group. *EFSA J.* **2009**, *7*, 1019. [CrossRef]
2. Landsberg, J.H. The effects of harmful algal blooms on aquatic organisms. *Rev. Fish. Sci.* **2002**, *10*, 113–390. [CrossRef]
3. Burson, A.; Matthijs, H.; Bruijne, W.D.; Talens, R.; Huisman, J. Termination of a toxic *Alexandrium* bloom with hydrogen peroxide. *Harmful Algae* **2014**, *31*, 125–135. [CrossRef]
4. Cusick, K.D.; Minkin, S.C.; Dodani, S.C.; Chang, C.J.; Wilhelm, S.W.; Sayler, G.S. Inhibition of copper uptake in yeast reveals the copper transporter Ctr1p as a potential molecular target of saxitoxin. *Environ. Sci. Technol.* **2012**, *46*, 2959–2966. [CrossRef] [PubMed]
5. Kozlowsky-Suzuki, B.; Koski, M.; Hallberg, E.; Wallen, R.; Carlsson, P. Glutathione transferase activity and oocyte development in copepods exposed to toxic phytoplankton. *Harmful Algae* **2009**, *8*, 395–406. [CrossRef]
6. Tan, Z.J.; Yan, T.; Yu, R.C.; Zhou, M.J. Transfer of paralytic shellfish toxins via marine food chains: A simulated experiment. *Biomed. Environ. Sci.* **2007**, *20*, 235. [CrossRef]
7. Bricelj, V.M.; Ford, S.E.; Lambert, C.; Barbou, A.; Paillard, C. Effects of toxic *Alexandrium tamarense* on behavior, hemocyte responses and development of brown ring disease in *Manila clams*. *Mar. Ecol. Prog. Ser.* **2011**, *430*, 35–48. [CrossRef]
8. Navarro, J.M.; Aguila, B.L.; Machmar, F.; Chaparro, O.R.; Contreras, A.M. Dynamic of intoxication and detoxification in juveniles of *Mytilus chilensis* (Bivalvia: Mytilidae) exposed to paralytic shellfish toxins. *Aquat. Living Resour.* **2011**, *24*, 93–98. [CrossRef]
9. Goc, N.L.; Hégaret, H.; Fabioux, C.; Miner, P.; Soudant, P. Impact of the toxic dinoflagellate *Alexandrium catenella* on Pacific oyster reproductive output: Application of flow cytometry assays on spermatozoa. *Aquat. Living Resour.* **2013**, *26*, 221–228. [CrossRef]
10. Nielsen, M.V.; Strmgren, T. Shell growth response of mussels (*Mytilus edulis*) exposed to toxic microalgae. *Mar. Biol.* **1991**, *108*, 263–267. [CrossRef]
11. Tran, D.; Haberkorn, H.; Soudant, P.; Ciret, P.; Massabuau, J.C. Behavioral responses of *Crassostrea gigas* exposed to the harmful algae *Alexandrium minutum*. *Aquaculture* **2010**, *298*, 338–345. [CrossRef]
12. Yan, T.; Zhou, M.J.; Fu, M.; Wang, L. Study on impact of dinoflagellate *Alexandrium tamarense* on life activities of marine bivalves. *Acta Oceanol. Sin.* **2003**, *22*, 103–109.
13. Bif, M.B.; Yunes, J.S.; Resgalla, C., Jr. Evaluation of mysids and sea urchins exposed to saxitoxins. *Environ. Toxicol. Pharmacol.* **2013**, *36*, 819–825. [CrossRef]
14. Cembella, A.D.; Quilliam, M.A.; Lewis, N.I.; Bauder, A.G.; Dell'Aversano, C.; Thomas, K.; Jellett, J.; Cusack, R.R. The toxigenic marine dinoflagellate *Alexandrium tamarense* as the probable cause of mortality of caged salmon in Nova Scotia. *Harmful Algae* **2002**, *1*, 313–325. [CrossRef]
15. Fire, S.E.; Pruden, J.; Couture, D.; Wang, Z.; Bottein, M.Y.D.; Haynes, B.L.; Knott, T.; Bouchard, D.; Lichtenwalner, A.; Wippelhauser, G. Saxitoxin exposure in an endangered fish: Association of a shortnose sturgeon mortality event with a harmful algal bloom. *Mar. Ecol. Prog. Ser.* **2012**, *460*, 145–153. [CrossRef]
16. Zhang, D.; Hu, C.; Wang, G.; Li, D.; Liu, Y. Zebrafish neurotoxicity from aphantoxins-cyanobacterial paralytic shellfish poisons (PSPs) from aphanizomenon flos-aquae DC-1. *Environ. Toxicol.* **2013**, *28*, 239–254. [CrossRef] [PubMed]
17. Ferrão-Filho, A.D.S.; Kozlowsky-Suzuki, B. Cyanotoxins: Bioaccumulation and effects on aquatic animals. *Mar. Drugs* **2011**, *9*, 2729–2772. [CrossRef]
18. Niedzwiadek, B.; Scott, P.M.; Lau, P.Y. Monitoring of shrimp and farmed fish sold in Canada for cyanobacterial toxins. *J. Food Prot.* **2012**, *75*, 160–163. [CrossRef] [PubMed]
19. Jensen, S.K.; Lacaze, J.P.; Hermann, G.; Kershaw, J.; Brownlow, A.; Turner, A.; Hall, A. Detection and effects of harmful algal toxins in Scottish harbour seals and potential links to population decline. *Toxicon* **2015**, *97*, 1–14. [CrossRef]
20. Pearson, L.; Mihali, T.; Moffitt, M.; Kellmann, R.; Neilan, B. On the chemistry, toxicology and genetics of the cyanobacterial toxins, microcystin, nodularin, saxitoxin and cylindrospermopsin. *Mar. Drugs* **2010**, *8*, 1650–1680. [CrossRef]
21. Li, S.C.; Wang, W.X.; Hsieh, D. Effects of toxic dinoflagellate *Alexandrium tamarense* on the energy budgets and growth of two marine bivalves. *Mar. Environ. Res.* **2002**, *53*, 145–160. [CrossRef] [PubMed]
22. Tian, L.; Cheng, J.; Chen, X.; Cheng, S.H.; Mak, Y.L.; Lam, P.; Chan, L.L.; Wang, M. Early developmental toxicity of saxitoxin on medaka (*Oryzias melastigma*) embryos. *Toxicon* **2014**, *77*, 16–25. [CrossRef] [PubMed]
23. Al-Yamani, F.Y.; Polikarpov, I.; Saburova, M. Marine life mortalities and harmful algal blooms in the Northern Arabian Gulf. *Aquat. Ecosyst. Health Manag.* **2020**, *23*, 196–209. [CrossRef]
24. Burridge, L.E.; Martin, J.L.; Lyons, M.C.; Legresley, M.M. Lethality of microalgae to farmed Atlantic salmon (*Salmo salar*). *Aquaculture* **2010**, *308*, 101–105. [CrossRef]
25. Etheridge, S.M. Paralytic shellfish poisoning: Seafood safety and human health perspectives. *Toxicon* **2010**, *56*, 108–122. [CrossRef]
26. Hille, B. The receptor for tetrodotoxin and saxitoxin. A structural hypothesis. *Biophys. J.* **1975**, *15*, 615–619. [CrossRef] [PubMed]

27. Shen, H.; Li, Z.; Jiang, Y.; Pan, X.; Wu, J.; Ben, C.A.; Smith, J.J.; Chin, Y.; Lei, J.; Zhou, Q. Structural basis for the modulation of voltage-gated sodium channels by animal toxins. *Science* **2018**, *362*, eaau2596. [CrossRef]
28. César, M.; Christian, L. The voltage-gated sodium channel: A major target of marine neurotoxins. *Toxicon* **2014**, *91*, 84–85. [CrossRef]
29. Jiang, T.J.; Niu, T.; Xu, Y.X. Transfer and metabolism of paralytic shellfish poisoning from scallop (*Chlamys nobilis*) to spiny lobster (*Panulirus stimpsoni*). *Toxicon* **2006**, *48*, 988–994. [CrossRef]
30. Kwong, R.; Wang, W.X.; Lam, P.; Yu, P. The uptake, distribution and elimination of paralytic shellfish toxins in mussels and fish exposed to toxic dinoflagellates. *Aquat. Toxicol.* **2006**, *80*, 82–91. [CrossRef]
31. Turner, A.D.; Lewis, A.M.; Hatfield, R.G.; Galloway, A.W.; Higman, W.A. Transformation of paralytic shellfish poisoning toxins in *Crassostrea gigas* and Pecten maximus reference materials. *Toxicon* **2012**, *60*, 1117–1134. [CrossRef] [PubMed]
32. Cusick, K.D.; Wetzel, R.K.; Minkin, S.C.; Dodani, S.C.; Sayler, G.S. Paralytic shellfish toxins inhibit copper uptake in *Chlamydomonas reinhardtii*. *Environ. Toxicol. Chem.* **2013**, *32*, 1388–1395. [CrossRef] [PubMed]
33. Choi, N.; Yeung, L.; Siu, W.; So, I.; Jack, R.W.; Hsieh, D.; Wu, R.; Lam, P. Relationships between tissue concentrations of paralytic shellfish toxins and antioxidative responses of clams, *Ruditapes philippinarum*. *Mar. Pollut. Bull.* **2006**, *52*, 572–578. [CrossRef] [PubMed]
34. Qiu, J.; Ma, F.; Hua, F.; Li, A. Effects of feeding *Alexandrium tamarense*, a paralytic shellfish toxin producer, on antioxidant enzymes in scallops (*Patinopecten yessoensis*) and mussels (*Mytilus galloprovincialis*). *Aquaculture* **2013**, *396*, 76–81. [CrossRef]
35. Fabioux, C.; Sulistiyani, Y.; Haberkorn, H.; Hegaret, H.; Amzil, Z.; Soudant, P. Exposure to toxic *Alexandrium minutum* activates the detoxifying and antioxidant systems in gills of the oyster *Crassostrea gigas*. *Harmful Algae* **2015**, *48*, 55–62. [CrossRef]
36. Gubbins, M.J.; Eddy, F.B.; Gallacher, S.; Stagg, R.M. Paralytic shellfish poisoning toxins induce xenobiotic metabolising enzymes in Atlantic salmon (*Salmo salar*). *Mar. Environ. Res.* **2000**, *50*, 479–483. [CrossRef]
37. Costa, P.R.; Pereira, P.; Guilherme, S.; Barata, M.; Santos, M.A.; Pacheco, M.; Pousao-Ferreira, P. Hydroxybenzoate paralytic shellfish toxins induce transient GST activity depletion and chromosomal damage in white seabream (*Diplodus sargus*). *Mar. Environ. Res.* **2012**, *79*, 63–69. [CrossRef]
38. Painefilú, J.C.; Bianchi, V.A.; Krock, B.; De Anna, J.S.; Kristoff, G.; Luquet, C.M. Effects of paralytic shellfish toxins on the middle intestine of *Oncorhynchus mykiss*: Glutathione metabolism, oxidative status, lysosomal function and ATP-binding cassette class C (ABCC) proteins activity. *Ecotoxicol. Environ. Saf.* **2020**, *204*, 111069. [CrossRef]
39. Durbin, E.; Teegarden, G.; Campbell, R.; Cembella, A.; Baumgartner, M.F.; Mate, B.R. North Atlantic right whales, *Eubalaena glacialis*, exposed to paralytic shellfish poisoning (PSP) toxins via a zooplankton vector, *Calanus finmarchicus*. *Harmful Algae* **2002**, *1*, 243–251. [CrossRef]
40. Doucette, G.J.; Cembella, A.D.; Martin, J.L.; Michaud, J.; Cole, T.; Rolland, R.M. Paralytic shellfish poisoning (PSP) toxins in North Atlantic right whales *Eubalaena glacialis* and their zooplankton prey in the Bay of Fundy, Canada. *Mar. Ecol. Prog.* **2006**, *306*, 303–313. [CrossRef]
41. Cusick, K.; Gary, S. An overview on the marine neurotoxin, saxitoxin: Genetics, molecular targets, methods of detection and ecological functions. *Mar. Drugs* **2013**, *11*, 911–1018. [CrossRef] [PubMed]
42. Thottumkara, A.P.; Parsons, W.H.; Du Bois, J. Saxitoxin. *Angew. Chem. Int. Ed.* **2014**, *53*, 5760–5784. [CrossRef] [PubMed]
43. Gao, S.; Zhen, M. Action potentials drive body wall muscle contractions in *Caenorhabditis elegans*. *Proc. Natl. Acad. Sci. USA* **2011**, *108*, 2557–2562. [CrossRef] [PubMed]
44. Jiang, Q.; Li, K.; Lu, W.; Li, S.; Chen, X., Liu, X.; Yuan, J.; Ding, Q.; Lan, F.; Cai, S. Identification of small-molecule ion channel modulators in *C. elegans* channelopathy models. *Nat. Commun.* **2018**, *9*, 3912–3941. [CrossRef]
45. Wang, J.; Salata, J.J.; Bennett, P.B. Saxitoxin is a gating modifier of HERG K^+ channels. *J. Gen. Physiol.* **2003**, *121*, 583–598. [CrossRef]
46. Wang, D.Y.; Wang, Y. Phenotypic and behavioral defects caused by barium exposure in nematode *Caenorhabditis elegans*. *Arch. Environ. Contam. Toxicol.* **2008**, *54*, 447–453. [CrossRef]
47. Arantes, O.N.; Apfeld, J.; Dillin, A.; Kenyon, C. Regulation of life-span by germ-line stem cells in *Caenorhabditis elegans*. *Science* **2002**, *295*, 502–505. [CrossRef]
48. Wu, Y.C.; Wang, X.; Ding, X. Methods for studying programmed cell death in *C. elegans*. *Methods Cell Biol.* **2012**, *107*, 295–320. [CrossRef]
49. Chen, X.; Zhong, Z.; Xu, Z.; Chen, L.; Wang, Y. 2′,7′-Dichlorodihydrofluorescein as a fluorescent probe for reactive oxygen species measurement: Forty years of application and controversy. *Free Radic. Res. Commun.* **2010**, *44*, 587–604. [CrossRef]
50. Meng, Q.X.; Wang, W.Y.; Lu, Q.M.; Jin, Y.; Wei, J.F.; Zhu, S.W.; Xiong, Y.L. A novel short neurotoxin, cobrotoxin c, from monocellate cobra (*Naja kaouthia*) venom: Isolation and purification, primary and secondary structure determination, and tertiary structure modeling. *Comp. Biochem. Physiol. Part C Toxicol. Pharmacol.* **2002**, *132*, 113–121. [CrossRef]
51. Zhang, Y.; Zhao, C.; Zhang, H.; Liu, R.; Yin, L. Integrating transcriptomics and behavior tests reveals how the *C. elegans* responds to copper induced aging. *Ecotoxicol. Environ. Saf.* **2021**, *222*, 112494. [CrossRef] [PubMed]
52. Ros-Santaella, J.L. Impact of oxidative stress on male reproduction in domestic and wild animals. *Antioxidants* **2021**, *10*, 1154. [CrossRef]
53. Tian, Y.; Zhou, M.; Meng, F.; Yu, R.; Wang, Y.; Li, J. Effects of the dinoflagellate *Alexandrium tamarense* on early development of the scallop *Argopecten irradians* concentricus. *Aquaculture* **2003**, *217*, 167–178. [CrossRef]

54. Yan, T.; Zhou, M.; Fu, M.; Wang, Y.; Yu, R.; Li, J. Inhibition of egg hatching success and larvae survival of the scallop, *Chlamys farreri*, associated with exposure to cells and cell fragments of the dinoflagellate *Alexandrium tamarense*. *Toxicon* **2001**, *39*, 1239–1244. [CrossRef]
55. Yurchenko, O.V.; Radashevsky, V.I.; Hsieh, H.L.; Reunov, A.A. Ultrastructural comparison of the spermatozoa of the Pacific oyster *Crassostrea gigas* inhabiting polluted and relatively clean areas in Taiwan. *Aquat. Ecol.* **2009**, *43*, 513–519. [CrossRef]
56. Haberkorn, H.; Lambert, C.; Goiec, N.L.; Moal, J.; Suquet, M.; Gueguen, M.; Sunila, I.; Soudant, P. Effects of *Alexandrium minutum* exposure on nutrition-related processes and reproductive output in oysters *Crassostrea gigas*. *Harmful Algae* **2010**, *9*, 427–439. [CrossRef]
57. Zhijun, T.; Tian, Y.; Mingjiang, Z.; Jun, L.; Rencheng, Y.; Yunfeng, W. The effects of *Alexandrium tamarense* on survival, growth and reproduction of *Neomysis awatschensis*. *Acta Ecol. Sin.* **2002**, *22*, 1635–1639. [CrossRef]
58. Piazza, V.; Ullmann, C.V.; Aberhan, M. Temperature-related body size change of marine benthic macroinvertebrates across the Early Toarcian Anoxic Event. *Sci. Rep.* **2020**, *10*, 4675. [CrossRef]
59. Turner, R.E.; Plunket, J.S. Estuarine oiling increases a long-term decline in mussel growth. *Environ. Pollut.* **2021**, *284*, 117506. [CrossRef]
60. Grande, F.R.D.; Granado, P.; Costa, T.M. Size-at-age or structure shift: Which hypothesis explains smaller body size of the fiddler crab Leptuca uruguayensis in northern populations? *Estuar. Coast. Shelf Sci.* **2021**, *254*, 107358. [CrossRef]
61. Genner, M.J.; Sims, D.W.; Southward, A.J.; Budd, G.C.; Masterson, P.; Mchugh, M.; Rendle, P.; Southall, E.J.; Wearmouth, V.J.; Hawkins, S.J. Body size-dependent responses of a marine fish assemblage to climate change and fishing over a century-long scale. *Glob. Chang. Biol.* **2010**, *16*, 517–527. [CrossRef]
62. Luckenbach, M.W.; Sellner, K.G.; Shumway, S.E.; Greene, K. Effects of two bloom-forming dinoflagellates, *Prorocentrum* mariae-lebouriae and *Gyrodinium unctenumon* on the growth and survival of the eastern oyster, *Crassostrea virginica* (Gmelin, 1791). *J. Shellfish Res.* **1993**, *12*, 411–415.
63. Turnbull, A.; Malhi, N.; Seger, A.; Jolley, J.; Fitzgibbon, Q. Accumulation of paralytic shellfish toxins by Southern Rock Lobster *Jasus edwardsii* causes minimal impact on lobster health. *Aquat. Toxicol.* **2020**, *230*, 105704. [CrossRef]
64. Duan, J.; Liu, H.; Zhu, J.; Lu, L.; Chang, L. A dynamic energy budget model for abalone, *Haliotis discus hannai* Ino. *Ecol. Model.* **2021**, *451*, 109569. [CrossRef]
65. Pousse, E.; Flye-Sainte-Marie, J.; Alunno-Bruscia, M.; Hegaret, H.; Rannou, E.; Pecquerie, L.; Marques, G.M.; Thomas, Y.; Castrec, J.; Fabioux, C. Modelling paralytic shellfish toxins (PST) accumulation in *Crassostrea gigas* by using Dynamic Energy Budgets (DEB). *J. Sea Res.* **2019**, *143*, 152–164. [CrossRef]
66. Bruynsteen, L.; Janssens, G.P.J.; Harris, P.A.; Duchateau, L.; Valle, E.; Odetti, P.; Vandevelde, K.; Buyse, J.; Hesta, M. Changes in oxidative stress in response to different levels of energy restriction in obese ponies. *Br. J. Nutr.* **2014**, *112*, 1402–1411. [CrossRef]
67. Rolfe, D.F.; Brown, G.C. Cellular energy utilization and molecular origin of standard metabolic rate in mammals. *Physiol. Rev.* **1997**, *77*, 731–758. [CrossRef] [PubMed]
68. Stenesen, D.; Suh, J.M.; Seo, J.; Yu, K.; Lee, K.S.; Kim, J.S.; Min, K.J.; Graff, J.M. Adenosine nucleotide biosynthesis and AMPK regulate adult life span and mediate the longevity benefit of caloric restriction in flies. *Cell Metab.* **2013**, *17*, 101–112. [CrossRef]
69. Ranjan, M.; Gruber, J.; Ng, L.F.; Halliwell, B. Repression of the mitochondrial peroxiredoxin antioxidant system does not shorten life span but causes reduced fitness in *Caenorhabditis elegans*. *Free Radic. Biol. Med.* **2013**, *63*, 381–389. [CrossRef]
70. Wu, S.B.; Wei, Y.H. AMPK-mediated increase of glycolysis as an adaptive response to oxidative stress in human cells: Implication of the cell survival in mitochondrial diseases. *Biochim. Biophys. Acta-Mol. Basis Dis.* **2012**, *1822*, 233–247. [CrossRef]
71. Yanase, S.; Yasuda, K.; Ishii, N. Adaptive responses to oxidative damage in three mutants of *Caenorhabditis elegans* (age-1, mev-1 and daf-16) that affect life span. *Mech. Ageing Dev.* **2002**, *123*, 1579–1587. [CrossRef] [PubMed]
72. Acba, B.; Vp, B.; Rm, B.; Am, B.; Sg, B.; Prca, C.; Mp, B. DNA damage and oxidative stress responses of mussels *Mytilus galloprovincialis* to paralytic shellfish toxins under warming and acidification conditions—Elucidation on the organ-specificity. *Aquat. Toxicol.* **2020**, *228*, 105619. [CrossRef]
73. Almansa-Ordonez, A.; Bellido, R.; Vassena, R.; Barragan, M.; Zambelli, F. Oxidative stress in reproduction: A mitochondrial perspective. *Biology* **2020**, *9*, 269. [CrossRef]
74. Napolitano, G.; Fasciolo, G.; Venditti, P. Mitochondrial management of reactive oxygen species. *Antioxidants* **2021**, *10*, 1824. [CrossRef]
75. Wang, Y.; Sun, Y.; Zhang, Z.; Li, Z.; Cai, P. Enhancement in the ATP level and antioxidant capacity of *Caenorhabditis elegans* under continuous exposure to extremely low-frequency electromagnetic field for multiple generations. *Int. J. Radiat. Biol.* **2020**, *96*, 1633–1640. [CrossRef]
76. Kanaan, G.N.; Harper, M.E. Cellular redox dysfunction in the development of cardiovascular diseases. *Biochim. Biophys Acta Gen. Subj.* **2017**, *1861*, 2822–2829. [CrossRef] [PubMed]
77. Zhang, J.; Shi, R.; Li, H.; Xiang, Y.; Xiao, L.; Hu, M.; Ma, F.; Ma, C.W.; Huang, Z. Antioxidant and neuroprotective effects of Dictyophora indusiata polysaccharide in *Caenorhabditis elegans*. *J. Ethnopharmacol.* **2016**, *192*, 413–422. [CrossRef]
78. Wen, H.; Shi, W.; Qin, J. Multiparameter evaluation of the longevity in *C. elegans* under stress using an integrated microfluidic device. *Biomed. Microdevices* **2012**, *14*, 721–728. [CrossRef]
79. *C. elegans* Sequencing Consortium. Genome Sequence of the Nematode *C. elegans*: A Platform for Investigating Biology. *Science* **1998**, *282*, 2012–2018. [CrossRef]

80. Gruber, J.; Ng, L.F.; Poovathingal, S.K.; Halliwell, B. Deceptively simple but simply deceptive—*Caenorhabditis elegans* lifespan studies: Considerations for aging and antioxidant effects. *FEBS Lett.* **2009**, *583*, 3377–3387. [CrossRef]
81. Schulz, T.J.; Zarse, K.; Voigt, A.; Urban, N.; Birringer, M.; Ristow, M. Glucose restriction extends *Caenorhabditis elegans* life span by inducing mitochondrial respiration and increasing oxidative stress. *Cell Metab.* **2007**, *6*, 280–293. [CrossRef] [PubMed]

marine drugs

MDPI

Article

Nutrient Deficiencies Impact on the Cellular and Metabolic Responses of Saxitoxin Producing *Alexandrium minutum*: A Transcriptomic Perspective

Muhamad Afiq Akbar [1,2,3,*], Nurul Yuziana Mohd Yusof [4], Gires Usup [4], Asmat Ahmad [5], Syarul Nataqain Baharum [3] and Hamidun Bunawan [3,*]

1 Department of Microbiology, Faculty of Biotechnology and Biomolecular Sciences, Universiti Putra Malaysia, Serdang 43400, Selangor, Malaysia
2 Aquatic Animal Health and Therapeutics Laboratory, Institute of Bioscience, Universiti Putra Malaysia, Serdang 43400, Selangor, Malaysia
3 Institute of System Biology, Universiti Kebangsaan Malaysia, Bangi 43600, Selangor, Malaysia; nataqain@ukm.edu.my
4 Department of Earth Science and Environment, Faculty of Science and Technology, Universiti Kebangsaan Malaysia, Bangi 43600, Selangor, Malaysia; yuziana@ukm.edu.my (N.Y.M.Y.); gires@ukm.edu.my (G.U.)
5 Department of Biological Sciences and Biotechnology, Faculty of Science and Technology, Universiti Kebangsaan Malaysia, Bangi 43600, Selangor, Malaysia; asmat@ukm.edu.my
* Correspondence: muhamadafiq@upm.edu.my (M.A.A.); hamidun.bunawan@ukm.edu.my (H.B.)

Abstract: Dinoflagellate *Alexandrium minutum* Halim is commonly associated with harmful algal blooms (HABs) in tropical marine waters due to its saxitoxin production. However, limited information is available regarding the cellular and metabolic changes of *A. minutum* in nutrient-deficient environments. To fill this gap, our study aimed to investigate the transcriptomic responses of *A. minutum* under nitrogen and phosphorus deficiency. The induction of nitrogen and phosphorus deficiency resulted in the identification of 1049 and 763 differently expressed genes (DEGs), respectively. Further analysis using gene set enrichment analysis (GSEA) revealed 702 and 1251 enriched gene ontology (GO) terms associated with nitrogen and phosphorus deficiency, respectively. Our results indicate that in laboratory cultures, nitrogen deficiency primarily affects meiosis, carbohydrate catabolism, ammonium assimilation, ion homeostasis, and protein kinase activity. On the other hand, phosphorus deficiency primarily affects the carbon metabolic response, cellular ion transfer, actin-dependent cell movement, signalling pathways, and protein recycling. Our study provides valuable insights into biological processes and genes regulating *A. minutum*'s response to nutrient deficiencies, furthering our understanding of the ecophysiological response of HABs to environmental change.

Keywords: saxitoxin; transcriptomics; gene set enrichment analysis; harmful algae blooms

1. Introduction

Alexandrium minutum can contribute to harmful algae blooms (HABs) in marine ecosystems. In addition, this species is known to produce saxitoxin, a potent neurotoxin that can be deadly to humans. HABs are influenced by various factors, such as physical conditions (currents), chemical conditions (nutrient presence), and biological factors (dinoflagellate species). Among these, nutrients are considered key drivers in HAB formation, particularly nitrogen, and phosphorus, which also play a crucial role in regulating dinoflagellate toxin production [1–5]. For instance, field measurements showed that high nitrogen and phosphorus concentrations have been associated with blooms of the harmful dinoflagellate taxa [4,5].

Nitrogen is an essential component in saxitoxin; therefore, the biosynthesis of saxitoxin in dinoflagellates is influenced by nitrogen level [6,7]. *A. minutum* has been observed to experience decreased growth and saxitoxin biosynthesis under nitrogen stress [8]. Phosphorus deficiency, conversely, has been associated with increased saxitoxin biosynthesis in most dinoflagellates, including *A. minutum* [8,9]. Thus, studying the metabolic reactions that occur under nitrogen and phosphorus stress is critical for understanding the mechanisms of HAB formation and toxin production.

Nitrogen constitutes up to 14% of the dry weight of dinoflagellate cells. The nitrogen taken up can be assimilated into active biochemicals, which can be used intracellularly for various metabolic and physiological processes [7]. Nitrogen availability also affects cell metabolism, growth, biochemical composition, photosynthesis, and other cellular activities involving cellular physiological activity. Nitrogen deficiency inhibits cell growth, alters pigment composition, lowers photosynthetic energy uptake and photosynthetic efficiency, and can also trigger cyst formation by certain dinoflagellate species [10–12].

Phosphorus, similarly, is essential for dinoflagellate growth and is required for cellular structure, metabolism, energy storage, cell signaling, and biochemical regulation [13,14]. However, in marine ecosystems, dissolved inorganic phosphorus (DIP) is often in limited supply [13]. In response to low phosphorus availability, phytoplankton species have developed various strategies such as altering phosphorus transport protein activity, using dissolved organic phosphorus (DOP) as a source of phosphorus, altering cell membrane composition, and bypassing phosphorus uptake processes in pathways such as glycolysis and phospholipid recycling [11,15].

Our previous study described a comprehensive gene catalog for *A. minutum* that highlights various metabolic pathways and critical adaptations to the marine environment [16]. *A. minutum* might exhibit physiological adaptability in response to a variety of environmental challenges, including nutrition deprivation [8,9,17,18]. Previously, Meng et al. [18] *A. minutum* is capable of employing ecological adaptation strategies when faced with stress conditions. These strategies encompass shifts in trophic modes and related physiological responses. Despite recent advancements, our understanding of how significant nutrient deficiencies impact the cellular and metabolic functions of *A. minutum* remains limited, particularly considering that dinoflagellate metabolism is regulated through a complex interplay of transcriptional and post-translational mechanisms [17,18]. This study uses transcriptome analysis to determine the effects of nitrogen and phosphorus stress on *A. minutum* to identify biological effects resulting from changes in the gene expression of *A. minutum*. This study demonstrated gene expression profiles under nutrient-replete, nitrogen-deficient, and phosphorus-deficient conditions. In this study, we employed gene set enrichment analysis, a more sensitive approach for interpreting and analyzing transcriptional changes, in contrast to the over-representation analysis used in previous studies [18–20]. These will provide important insights into the mechanisms *A. minutum*, and other dinoflagellates might use to overcome nitrogen and phosphorus deficiencies in the marine environment.

2. Results and Discussions

2.1. Comparison of Unigene Expression Profiles A. minutum under the Induction of Nitrogen and Phosphorus Deficiency

To compare the *A. minutum* unigene expression levels in control samples and nutrient-deficient samples (nitrogen and phosphorus-deficient), readings of high-quality raw data from each sample were mapped to a total of 124,977 *A. minutum* unigenes and the expression of each unigene in different samples was estimated using the RSEM program. The expression reading value of each unigene is also expressed as a raw count value as well as a TPM value. Next, the reliability of the biological replicas used in this study was tested using the scatter plot comparison method (Figure 1A). A high correlation between the two biological repeat samples indicated a high level of repeatability, with the Pearson correlation coefficient R equal to 0.99 for all samples. This suggests that repeated biological samples produced acceptable data for further analysis. Finally, to evaluate the influence of nitrogen and phosphorus on the gene expression profile of *A. minutum*, this study adopted a multivariate analytical approach through principal component analysis (PCA). Figure 1B clearly shows the PCA plots of gene expression profiles from all six samples successfully differentiated into three groups in line with the experimental design.

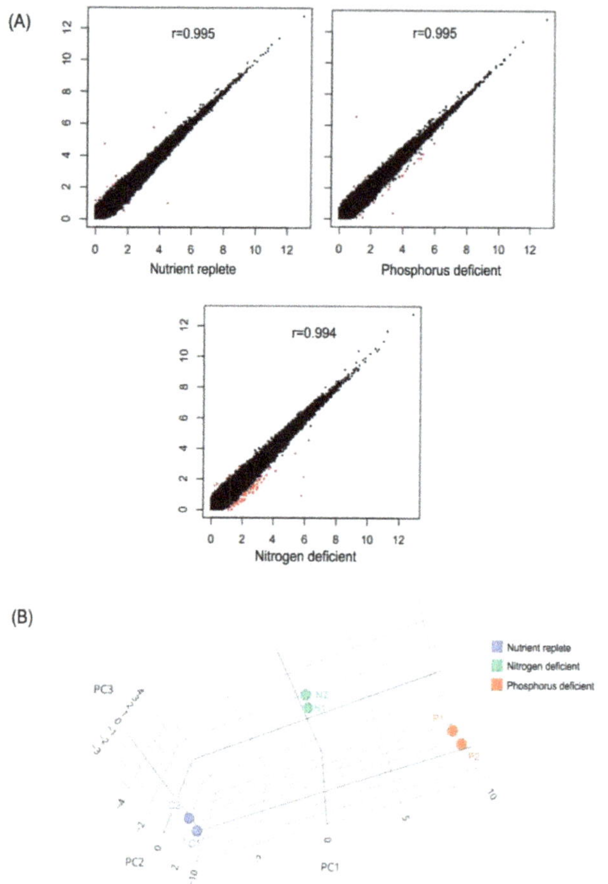

Figure 1. Comparison of unigene expression values between all samples. (**A**) Scatter plot comparing unigene expression values for biological replicates and (**B**) 3D PCA plot of unigene expression values for all samples.

To identify unigenes that showed significant differences during nitrogen and phosphorus-deficient conditions, comparisons of normalized gene expression profiles using the TMM method were performed using the edgeR program. In total, 1812 unigenes had a threshold value of FDR < 0.05 for differently expressed genes (DEGs). Of these, 0.4% of the studied unigenes (538/124 977) were identified as a gene with a higher expression during nitrogen deficiency, and 0.4% of the studied unigenes (511/124 977) had a lower expression when compared with the control group. In the case of phosphorus deficiency, 0.1% of the studied unigenes (127/124 977) had a higher expression, while 0.5% of the studied unigenes (636/124 977) had a lower expression. The set of DEGs identified in this study was subsequently plotted as MA plots and heat maps (Figure 2). However, most of these unigenes do not have a protein sequence match on the NCBI database nr. The full list of DEGs identified in this study is available in Table S1.

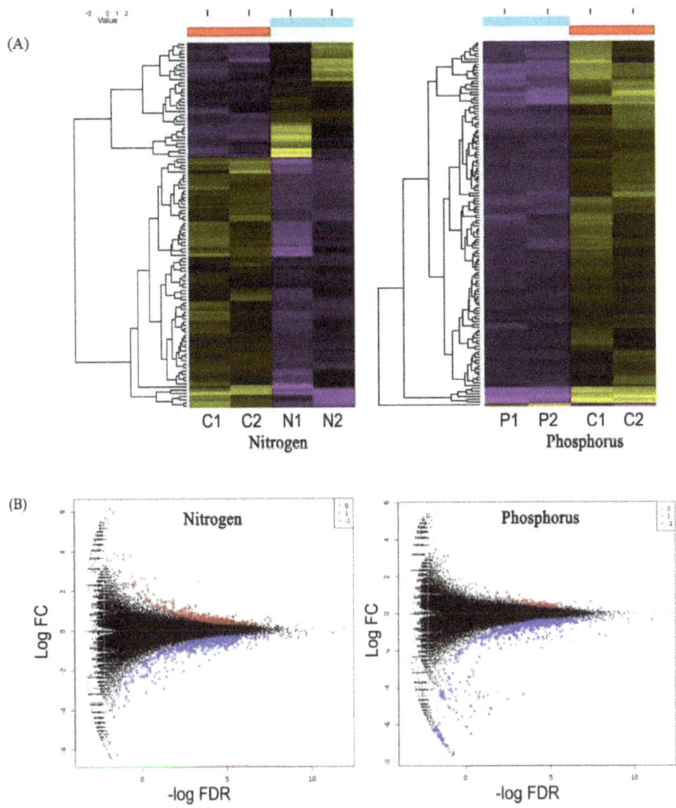

Figure 2. *A. minutum* unigenes expressed differently under the induction of nitrogen and phosphorus deficiency. (**A**) The heat map based on TPM log values (**B**) MA plot displays all DEGs. Red dots represent up-regulated DEGs, while blue dots represent down-regulated DEGs.

2.2. GSEA Analysis of A. minutum under the Induction of Nitrogen and Phosphorus Deficiency

Furthermore, an enrichment analysis of the GO term was performed to provide an overview of the biological functions and processes associated with DEGs. This study utilized the GSEA approach for enrichment analysis, which is known to be more effective and sensitive than other methods, such as the Fisher exact test and over-representative analysis (ORA). The GSEA approach analyzes the position and expression level of all unigenes in the study sample without being restricted by arbitrary threshold values such as FDR or DEG multiplier log changes, making it a more comprehensive and unbiased method for identifying enriched gene sets [19,20]. The results of GO enrichment analysis using the GSEA program successfully identified 702 GO terms which were all found to be enriched (all up-regulated) under the condition of nitrogen deficiency, while 1251 GO terms, of which 1246 were up-regulated, and 5 were down-regulated, were enriched when phosphorus was deficient with a threshold value of $FDR < 0.25$. The top three GO terms in terms of normalized enrichment score (NES) for each category are shown in Figure 3. A higher NES score indicates a gene shift in a particular GO term to one end on the gene ranking file, determining whether the GO term is up-regulated or down-regulated (via positive or negative NES values). A complete list of enriched GO terms identified in this study is in Table S2.

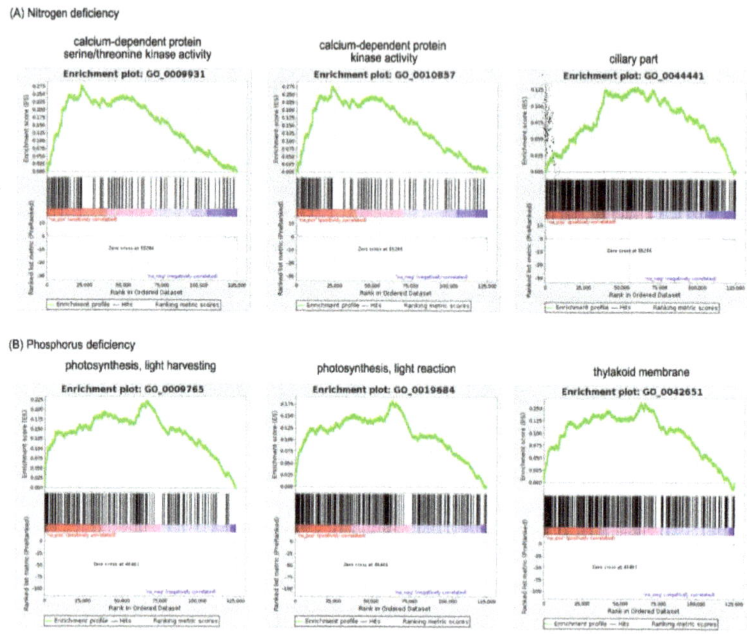

Figure 3. Top three enriched GO terms by *A. minutum* under the following conditions. (**A**) Nitrogen deficiency (**B**) Phosphorus deficiency.

GO enrichment analysis will often show the enrichment of several GO terms that are similar hence complicating the interpretation of data or study results. Concentrating or grouping GO terms according to general functions is one of the strategies used to facilitate the data interpretation process. The "enrichment map" function in Cytoscape software is a method to describe these enriched GO terms more simply. The results of the subsequent enrichment analysis can be viewed simultaneously in a single enrichment map. Only GO terms with an FDR threshold value < 0.1 were used to reduce repetition on the enrichment map. The resulting enrichment map for nitrogen-deficient samples displays 209 nodes and 1423 edges connected with an edge cut-off threshold value < 0.375. The "auto-annotate" function then classifies the mapped GO terms into 30 different clusters according to a shared general function. The largest cluster is the meiosis-related process cluster containing 47 enriched GO terms, followed by the metabolic and catabolic process cluster containing 18 enriched GO terms, and the protein kinase activity cluster containing 13 enriched GO terms. Next, for the enrichment of phosphorus-deficient samples, the resulting enrichment map displays 286 nodes and 1196 edges connected. The "auto-annotate" function then classifies the mapped GO terms into 42 different clusters according to a shared general function. The largest cluster is the metabolic process cluster containing 54 enriched GO terms, the ion transfer cluster containing 18 enriched GO terms, and the actin filament organization cluster containing 17 enriched GO terms. The enrichment map was generated to illustrate GO-enriched terms in *A. minutum* under nitrogen and phosphorus-deficient conditions (Figures 4 and 5).

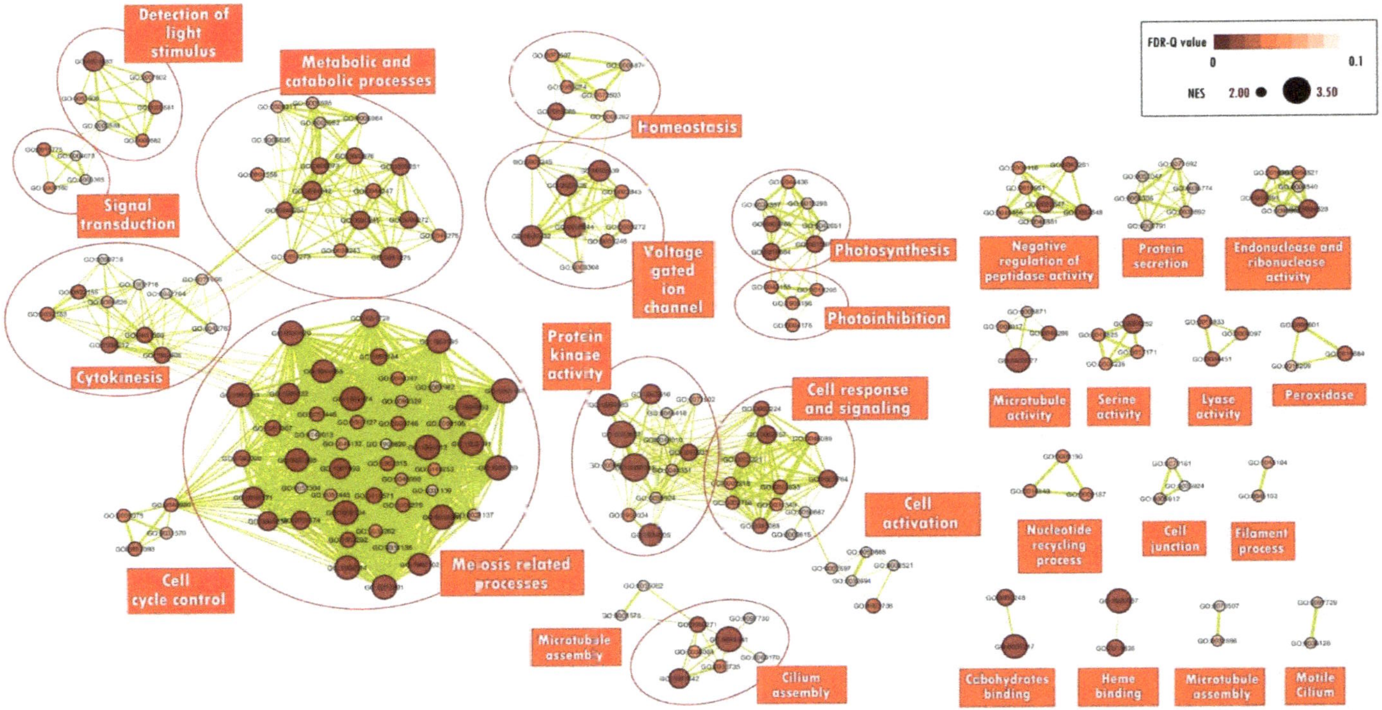

Figure 4. Enrichment map visualization for enriched GO terms by *A. minutum* under the induction of nitrogen deficiency. Each node corresponds to enriched GO terms in response to nitrogen deficiency, and the nodes' size is related to the NES value for the enriched GO term. Edges (green lines) connect GO terms to shared genes, and line thickness is associated with the number of genes shared between the GO terms.

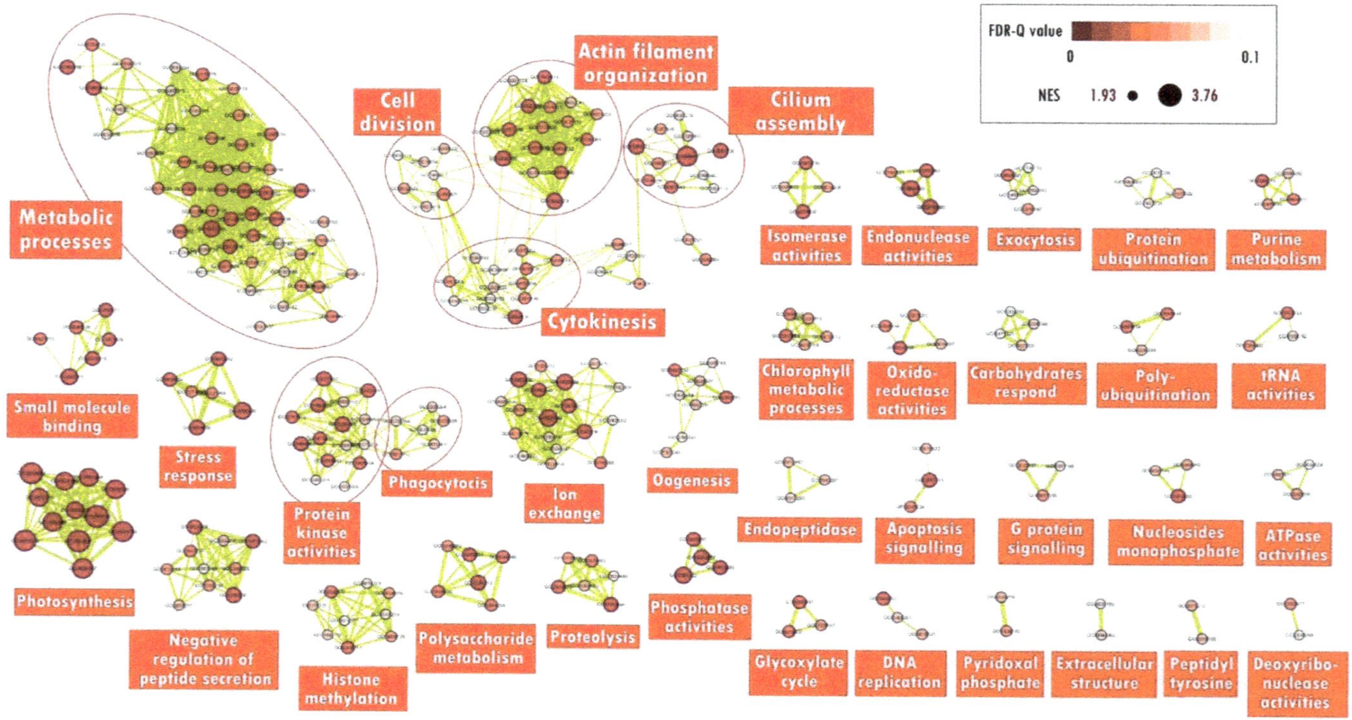

Figure 5. The Enrichment map visualizes the enriched gene ontology (GO) terms in response to phosphorus deficiency in *A. minutum*. Each node in the map represents an enriched GO term, with node size reflecting the normalized enrichment score (NES) for the term. The green edges indicate shared genes between the GO terms, with the thickness of the line indicating the number of shared genes.

2.3. Enriched GO Term by A. minutum under the Induction of Nitrogen Deficiency

2.3.1. Meiosis

A total of 47 GO terms related to the meiosis process were enriched in *A. minutum* under nitrogen deficiency (Figure 4). Some of these terms are related to the conversion of the cell cycle from mitosis to meiosis, a strategy used by dinoflagellates to form temporary cysts that can survive in unfavorable environments [7]. Homologs of meiosis protein mei2, DNA topoisomerase II, and serine/threonine kinase protein were found to contribute to the enrichment of GO:0045930. Mei2 is required for cells to migrate from mitosis to meiosis, while DNA topoisomerase II plays a key role in homologous chromosome separation during meiosis I and serine/threonine kinase proteins are involved in the termination of the mitotic cell cycle [21,22].

In addition, a previously unidentified unigene was found to contribute to the enrichment of GO:0045930. The unigene encodes a hypothetical protein of 151 amino acids and contains a forkhead-associated domain (FHA) sequence, which has been shown in past studies to be involved in DNA damage responses and cell cycle termination [23,24]. Mutations in the FHA domain of the fungus *Neurospora crassa* cause meiosis phenotypes and abnormal ascospore development [24]. These findings suggest that *A. minutum* may encode a novel protein involved in the process of cell division, consistent with the phenotypic characteristics of dinoflagellates, which have large and compact chromosomes [25].

2.3.2. Catabolic Carbohydrate Metabolism

This study also identified several GO terms related to the catabolic carbohydrate metabolism process enriched in *A. minutum* under the induction of nitrogen deficiency. These results indicate that under the induction of nitrogen deficiency, *A. minutum* is also likely to mobilize carbon sources for cellular use as opposed to storing these carbon sources in the form of carbohydrates. Although carbohydrate metabolism processes under the induction of nitrogen deficiency have not been extensively studied in dinoflagellates, similar study results have been reported in several marine diatom species such as *Thalassiosira pseudonana* and *Phaeodactylum tricornutum* [26,27]. In fact, further LEA analysis also showed that the enrichment of GO terms in metabolic and catabolic process clusters was contributed to by unigenes encoding catabolic proteins such as alpha-amylase, exoglucanase, alpha-1,4 glucan phosphorylase, glucanase, and others. The alpha-amylase protein serves to catalyze the hydrolysis of glycoside bonds α-1 $\rightarrow$ 4 polysaccharides such as starch and glycogen while reducing cellular carbohydrate accumulation [28]. Wase et al. [28] found that Alpha-amylase protein expression by the microalgae *Chlamydomonas reinhardtii* also increased under the induction of nitrogen deficiency. Furthermore, the enrichment of the term GO related to glycolysis, "glycolytic process through fructose-6-phosphate" (GO:0061615), also strengthens the hypothesis that carbohydrate metabolism in *A. minutum* is catabolic under the induction of nitrogen deficiency. An increase in the rate of glycolysis under the induction of nitrogen deficiency in *A. minutum* indicates that carbon sources, such as intracellular carbohydrate storage, are transferred to central carbon metabolism. Hockin et al. [26] suggested that in nitrogen-restricted conditions, the degradation of intracellular carbon sources through catabolic metabolism is important for the nitrogen re-assimilation processes using the GS/GOGAT pathway as well as the urea cycle.

2.3.3. Nitrogen Metabolism

Genes encoding the proteins involved in the GS/GOGAT cycle pathway contribute to the enrichment of the GO term "glutamine family amino acid metabolic process" (GO:0009064). These unigenes include genes that encode GS proteins and glutamate synthase-NADPH (GOGAT-NADPH) as well as glutamate synthase-ferredoxin (GOGAT-Fe). The GS protein catalyzes the fixation of ATP-dependent ammonium to the d-carboxyl group of glutamate to form glutamine. Next, the GOGAT protein catalyzes one molecule of glutamine and one molecule of 2-oxoglutarate to form two molecules of glutamate [29].

Thus, under the induction of nitrogen deficiency, *A. minutum* is likely to increase the activity of GS/GOGAT pathways to increase ammonium assimilation as a result of some cellular catabolic activity to compensate for nitrogen deficiency. In fact, some marine diatom species, such as *T. pseudonana* and *P. tricornutum*, also show increased expression of genes/proteins involved in the GS/GOGAT pathway under the induction of nitrogen deficiency [26,30]. In addition, the *A. minutum* unigenes encoding the GDH protein also contributed to the enrichment of the same GO term. In addition to the GS/GOGAT pathway, GDH protein activity can also catalyze the production of glutamate from ammonium and 2-oxoglutarate without the need for ATP [29].

2.3.4. Ion Homeostasis

Ion channels are important components in cell signaling and are involved in environmental adaptation, cell growth and differentiation, cell movement, exo, and endocytosis [31,32]. Dinoflagellates are also likely to utilize ion channels in various cellular processes [33]. The enrichment map shows nine GO terms related to voltage-gated ion channels enriched in *A. minutum* (Figure 4). These results suggest that the movement of ions across the cell membrane is enhanced under nitrogen deficiency. This theory is supported by several enriched GO terms related to homeostatic processes. In dinoflagellates *Prorocentrum minimum*, the expression of genes involved in cellular calcium ion homeostasis increased under the induction of copper sulphate biocides, indicating that calcium ion homeostasis is a defense mechanism against stressful environments. However, the uncontrolled and excessive entry of calcium ions into the cell can also lead to oxidative stress and cell death [34]. Increased oxidative stress due to nitrogen deficiency has also been reported in dinoflagellates *P. lima* and the microalgae *Chlorella sorokiniana* [35,36]. In summary, the homeostasis of calcium ions by *A. minutum* under nitrogen deficiency is likely to serve as a signaling system to influence cellular processes to cope with this limitation. However, continuous nitrogen deficiency in the environment can lead to oxidative stress on cells. Further studies are needed to determine the detailed role of ion homeostasis processes in dinoflagellates under certain inductions.

2.3.5. Protein Kinase Activities

In parallel with the increased entry of calcium ions into the cells of *A. minutum* discussed earlier, the activity of ion-dependent protein kinases was also enriched in *A. minutum* under nitrogen deficiency induction. LEA analysis revealed that *A. minutum* unigenes contributing to the enrichment of GO terms related to protein kinase activity are dominated by genes encoding calcium-dependent protein kinase, calmodulin-dependent protein kinase, and putative serine/threonine-protein kinase. Calcium ion binding catalyzes the activation of these proteins, which in turn react with other downstream proteins through phosphorylation activity [37]. Phosphorylation is a post-translational protein modification mechanism that activates or deactivates proteins and is a way to control signal transduction between metabolic pathways [38]. The fraction of proteins regulated using phosphorylation ranges from 10% to 60%, depending on the organism [39,40]. Protein phosphorylation activity also regulates diurnal rhythmic protein expression at the post-translational stage in dinoflagellates *L. polyedrum* [41,42]. Changes in protein phosphorylation activity under nitrogen deficiency induction have also been reported in plants and marine diatoms [39,43]. Thus, *A. minutum* is likely to regulate cellular processes in the post-translational stage to overcome nitrogen deficiency.

2.4. Enriched GO Term by A. minutum under the Induction of Phosphorus Deficiency

2.4.1. Metabolic Processes

1. Glycolysis;

LEA analysis for the enrichment of the GO term "glycolytic process" (GO:0006096) includes the unigene encoding the pyrophosphate-fructose 6-phosphate 1-phosphotransferase (PFP) protein. This protein catalyzes an alternative glycolysis pathway which allows cells to bypass enzymatic reactions that require the use of phosphorus sources [44]. Unlike classical glycolysis pathways, these alternative glycolysis pathways are considered more advantageous in terms of cellular energy savings since these alternative pathways bypass ATP utilization [45]. In addition, the increased expression of PFP proteins is also associated in part with the mechanism of cell metabolic adaptation in a phosphorus-deficient environment during the occurrence of the HAB phenomenon [46]. Increased expression of genes/proteins involved in the glycolysis pathway under the induction of phosphorus deficiency has also been reported in most marine diatom species and some dinoflagellate species such as *Karlodinium veneficum* and *P. donhaiense* [47–49]. However, increased glycolysis rates in dinoflagellates under the induction of phosphorus deficiency is not a universal response, as there are also several past studies reporting decreased gene/protein expression associated with the glycolysis pathway [15,50]. Thus, in contrast to diatoms, alternative glycolysis pathways that are less dependent on organic phosphate sources are unlikely to be conserved by all dinoflagellate species [45].

2. Nucleic Acids Metabolism;

Nucleic acids comprise approximately 9% of dry-weight phosphorus [51]. Therefore, when there is a shortage of phosphorus sources for synthesizing new nucleic acids, cells will recycle phosphorus from existing nucleic acids in the most cost-effective way. *A. minutum* in this study showed enrichment of several GO terms related to nucleic acid metabolism. Unigene-encoding proteins such as phosphodiesterase, putative 5′-nucleotidase, phosphoglycerate kinase, and enolase with nucleic acid degradation activity contribute to the enrichment of GO terms related to nucleic acid metabolism. The recycling of phosphorus sources from nucleic acids has also been reported in several past studies [47,52]. Gene expression studies of the marine diatom *P. tricornutum* have also shown increased expression of the gene encoding the 5′-nucleotidase protein as early as 72 h after the induction of phosphorus deficiency [52]. Similar results have also been reported for the dinoflagellate *P. donghaiense* [49].

3. Amino acids metabolism.

The study by Müller et al. [53] indicated that *Lupinus albus* accumulated most free amino acids under phosphorus deficiency, with tryptophan showing the highest accumulation rate. Similarly, *A. thaliana* showed a significant increase in the accumulation of aromatic amino acid groups under phosphorus deficiency [54]. *A. minutum* is also likely to experience increased accumulation of free amino acids as some GO terms related to amino acid metabolism and biosynthesis are enriched under phosphorus deficiency. Enrichment of GO terms related to the biosynthesis of aromatic amino acid groups is contributed to by proteins such as fatty acid amide hydrolase, while the enrichment of the term GO related to the glutamine amino acid group is contributed to by unigenes encoding proteins involved in GS-GOGAT pathways and the urea cycle. The increase in expression of unigenes involved in the GS-GOGAT pathway and the urea cycle by *A. minutum* under phosphorus deficiency is likely due to increased intracellular ammonia, a by-product of protein degradation, which is a universal response of most organisms under phosphorus deficiency [44,47,55].

2.4.2. Ion Transfer

The second largest cluster in the *A. minutum* enrichment map under phosphorus deficiency is the ion transfer cluster, with 18 GO terms mainly related to potassium and calcium transfer. Some plant species, including soy, have reported decreased intracellular potassium under phosphorus deficiency [56,57]. Potassium ions are known to play an important role in protein activation [58], but their function in eukaryotic photosynthetic organisms such as diatoms and dinoflagellates is still unclear. Calcium, on the other hand, is an essential macronutrient involved in cellular growth and development as well as functions as a signaling molecule, triggering cellular biochemical responses to stress [59,60]. Although the role of calcium ions in stress responses such as cold and hot temperatures and salinity has been studied, their role in signaling low nutrient stress, such as phosphorus deficiency, is not well explored, especially in dinoflagellates [60,61].

2.4.3. Organization of Actin Filaments

The *A. minutum* enrichment map revealed a cluster of 17 GO terms related to actin filament organization enriched under phosphorus deficiency induction (Figure 5). Actin filaments participate in various protein–protein interactions and have been identified to regulate transcription processes, cell division, cell motility, and shape [62]. In dinoflagellates, actin filaments are also involved in bioillumination, chloroplast translocation, cell shape maintenance, and ecdysis [63–66]. LEA analysis of the enriched terms showed that proteins involved in cell motility processes, such as Fibronectin protein, casein kinase i isoform alpha, Catenin beta-1, and Myosin II, were encoded by contributing unigenes. This suggests that *A. minutum* cell motility is enhanced under phosphorus deficiency induction, allowing it to migrate to areas with optimal nutrient concentrations. Dinoflagellates such as *Akashiwo sanguinea* and *Prorocentrum micans* have been reported to undergo vertical migration to deep waters rich in phosphorus sources, which gives an adaptive advantage to some dinoflagellate species [67,68]. Thus, the increased motility of *A. minutum* under phosphorus deficiency induction might provide an advantage for this species over other phytoplankton groups to adapt to the marine environmental change.

2.4.4. Cell Signaling Activity

GO terms related to mitogen-activated protein kinase (MAPK) signaling were enriched in response to phosphorus deficiency and concomitant with an increased influx of calcium ions acting as cell signaling molecules. MAPK is activated through the induction of cofactors, such as the presence of ions, and engages in responses to biotic and abiotic stresses [69]. MAPK signaling is also involved in the regulation of cell division processes and other reproductive activities through protein phosphorylation activities. In yeast, the organisms in which MAPK signaling is most studied, MAPK transduction signaling functions in modulating pheromone response, cell cycle control, and cell integrity [70,71]. In dinoflagellates, MAPK signaling pathways are reported to be involved in response to the induction of chemicals such as polychlorinated biphenyl (PCBs) and copper sulphate biocides [72,73].

In addition to MAPK signaling, G protein-coupled receptor (GPCR) pathways are crucial for signaling and responding to external stimuli such as hormones, nutrients, and phospholipids [74,75]. However, these pathways have not been well studied in eukaryotic photosynthetic organisms until recently. Mojib and Kubanek [76] demonstrated the presence of GPCR signaling pathways in several dinoflagellate species using a transcriptomic approach, and our study found that several GO terms related to the GPCR pathway were enriched under phosphorus deficiency. LEA analysis revealed that unigenes encoding major components of the GPCR pathway, such as G protein-coupled receptors, Ras-like, and cAMP-dependent protein kinase regulatory subunits, contributed to the enrichment of these GO terms. Previous studies have also shown that GPCR signaling plays a vital role in crosstalk activation of the MAPK signaling pathway [77,78].

2.4.5. Methylation of Histones and Proteins

In response to environmental changes, epigenetic mechanisms reprogram gene expression to prepare cells for survival [79]. *A. minutum* enriched GO terms related to histone and protein methylation, a branch of epigenetic modification, under the induction of phosphorus deficiency. The role of histones in dinoflagellates is thought to be related to gene expression regulation, not chromosome arrangement. Several previous studies have shown unchanged gene expression encoding histone proteins throughout the cell cycle in some dinoflagellate species process [25,80]. The enrichment of histone methylation-related GO terms in this study supports the hypothesis that histone proteins regulate gene expression through epigenetic modifications in dinoflagellates. Histone methylation is also a component of the response to phosphorus deficiency in most plants [81,82].

2.4.6. Ubiquitin Pathway

Our study showed that several GO terms related to the ubiquitin pathway were enriched under the induction of phosphorus deficiency and that all three major components of the pathway, including mitochondrial protein ubiquitin ligase activator, ubiquitin-conjugating enzyme E2-17 kDa, and putative E3 ubiquitin-protein ligase rbrA, contributed to this enrichment [83]. The ubiquitin pathway, which involves the covalent bonding of small ubiquitin proteins to target proteins and subsequent elimination of these pre-labeled target proteins via proteolysis by 26S proteasome proteins, is one form of post-translational regulation [84]. Proteolysis induced by the ubiquitin pathway plays an important role in basic cellular processes such as cell cycle and division regulation, stress response, and protein recycling processes [84–86]. Additionally, protein degradation through proteolysis produces free amino acids and ammonia as downstream products, which are important under phosphorus deficiency [87]. Further studies are needed to fully understand the role of ubiquitin pathways in the regulation of dinoflagellate cellular processes at the post-translational stage.

2.5. *Implications of Nitrogen and Phosphorus Deficiency on Saxitoxin Biosynthesis*

Our study found that the expression of the *A. minutum* unigene, which encodes homologs of proteins involved in saxitoxin biosynthesis, does not significantly change under conditions of phosphorus or nitrogen deficiency. These results are consistent with previous studies [3,6,88] stating that saxitoxin biosynthesis involves complex regulation at the post-transcriptional and post-translational levels. For example, Perini et al. [88] also found that the expression of homologs of two major genes in saxitoxin biosynthesis, sxtA, and sxtG, did not correlate directly with intracellular toxin changes in *A. minutum*.

The increased glycolytic activity during phosphorus deficiency in *A. minutum* may also result in the production of more acetyl-CoA and malonyl-CoA molecules, which are required to initiate saxitoxin biosynthesis. Under nitrogen deficiency, the low rate of saxitoxin biosynthesis is indirectly influenced by a decrease in cellular growth rate [89,90]. Additionally, during nitrogen deficiency, acetyl-CoA is redirected towards fatty acid synthesis, thereby reducing the pool of precursors available for saxitoxin biosynthesis in *A. minutum*.

3. Materials and Methods

3.1. *RNA-Seq Data Retrieval*

RNA-seq datasets used throughout this study were obtained from our previous research. These data can be downloaded from the NCBI Sequence Read Archive (SRA) Database under the BioProject id: PRJNA914202 [16]. In brief, *A. minutum* cells were grown in nutrient-replete ES-DK media until they reached the exponential phase before being transferred using a 15 m mesh filter to new ES-DK media lacking nitrogen/phosphorus components to induce nitrogen/phosphorus limitation in *A. minutum* cultures (Figure 6). Then, the RNA was harvested from *A. minutum* cells that were starved of nitrogen and phosphorus for 72 h. A HACH DR2800 spectrophotometer was used to confirm the growth media's lack of nitrogen/phosphorus. The complete annotation pipeline for *A. minutum* transcriptome was described in our previous study [16]. A complete annotation of *A. minutum* unigenes is available in Table S3.

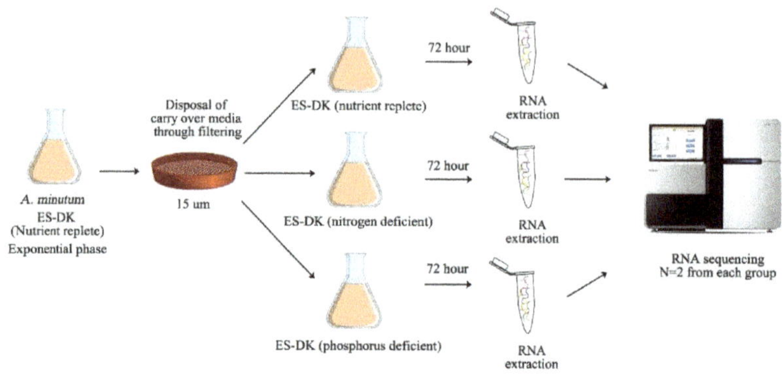

Figure 6. The experimental design adopted to induce nitrogen and phosphorus deficiency in *A. minutum*.

3.2. Estimation of Unigene Expression Levels

The expression of each *A.minutum* unigene that was assembled in our previous study was calculated using the RNA-Seq by Expectation-Maximization (RSEM) program [91] (Figure 7). The RSEM program will first re-map high-quality raw data from each sample (two control samples, two nitrogen deficiency samples, and two phosphorus deficiency samples) to assemble unigene sequences using algorithms from the bowtie2 program [92]. Next, the RSEM program will use the Expectation-Maximization algorithm to identify each unigenes' expression level. The expression reading value of each unigene is also displayed in the raw count value and the TPM value (Transcripts Per Million). TPM normalization aims to reduce the bias in unigene expression values present due to differences between unigene lengths [91].

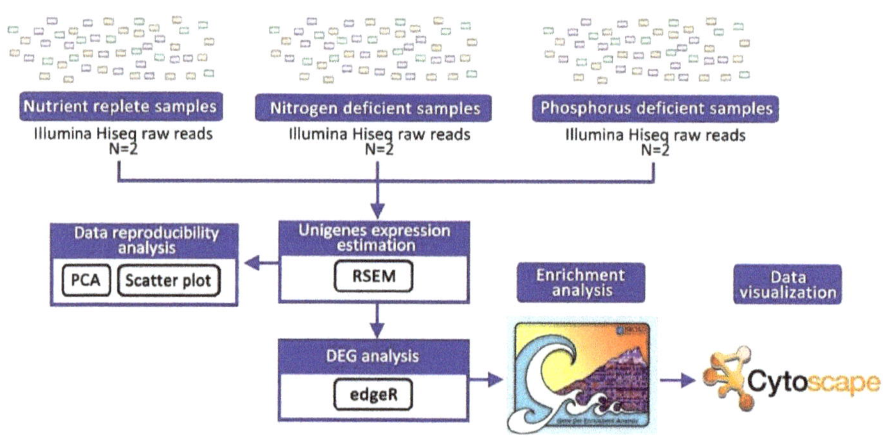

Figure 7. Workflow for transcriptomics data analysis used in this study.

Next, the raw count values for each unigenes were then used to generate scatter plots using the script "PtR compare_replicates" from the Trinity package to test the reliability of the biological replicates in this study (Haas et al. 2013). Finally, a Principal Component Analysis (PCA) plot was generated using the TCC-GUI program [93].

3.3. Differently Expressed Genes Analysis

An analysis to identify differently expressed genes (DEGs) was performed using the EdgeR program [94]. Using this program, the raw count values for each sample were normalized using the Trimmed mean of M-values (TMM) method in advance to minimize the bias resulting from the differences in the size of raw data for different samples. In this study, the EdgeR program was used with quasi-likelihood test parameter settings and glmQLFit to reduce the potential for false positive rates and increase the statistical power in detecting outlier values. The Benjamini–Hochberg method was used to calculate the false discovery rate (FDR) of unigenes expressed differently. FDR values < 0.05 were considered significant outcomes. The unigene expression levels expressed differently as per the TPM unit were then plotted as a heat map using the script "analyze_diff_expr.pl" in the Trinity program package [95].

3.4. Gene Set Enrichment Analysis

The gene set enrichment analysis (GSEA) program was used to identify gene ontology (GO) terms enriched in *A. minutum* under the induction of nitrogen and phosphorus deficiency [19]. The ranking files required as input for the GSEA program are calculated using the following equation.

$$\text{Ranking} = -\log 10\ \text{FDR} \times \sin \log \text{FC}$$

where the values of FDR and FC (fold change) are obtained from the results of the edgeR program analysis [20].

Therefore, unigenes with a high expression level will have a positive score, while unigenes with a low expression level will have a negative score on the ranking file. The rankings for all unigenes were calculated without any cut-off threshold on the FDR values. GO annotation of the *A. minutum* transcriptome from our previous study was used as the set of genes tested by the GSEA program. GO terms annotated to over 500 unigenes and fewer than 15 unigenes were removed from the set of genes tested because GO terms with large annotation sizes are typically involved in overly general biological functions, while GO terms with overly small annotation sizes would not yield results that are statistically significant [20]. GSEA analysis was then performed using a "pre-ranked" method based on 1000 times the permutation of the gene set, and FDR values below 0.25 were considered significant results.

3.5. Enrichment Map Visualisation

The "enrichment map" function in the Cytoscape version 3.8.0 program was used to visualize GO terms enriched through GSEA analysis [96,97]. To reduce the complexity of the enrichment map, only GO terms with an FDR threshold value below 0.1 are displayed on this map. Next, GO terms with nearly similar functions were clustered using the "clusterMaker" function, while the "AutoAnnotate" function was used to identify themes for the formed clusters [98]. Finally, the enrichment map is displayed in a "yFiles Organic Layout" layout with manual editing for better visualization.

4. Conclusions

In summary, this study provides new insights into the transcriptome changes of *A. minutum* in cultures under nitrogen and phosphorus deficiency, revealing distinct metabolic and cellular responses to these major nutrient deficiencies. The results highlight the significant role of various biological processes, including meiosis, carbon metabolism, ion homeostasis, and post-translational modifications, in regulating dinoflagellate cellular processes. The identification of specific unigenes that regulate these processes provides valuable information for future research in this area. Overall, these findings have implications for our understanding of how dinoflagellates acclimate to changing nutrient conditions and the potential impacts on marine ecosystems and saxitoxin biosynthesis.

Supplementary Materials: The following supporting information can be downloaded at: https://www.mdpi.com/article/10.3390/md21090497/s1, Table S1: Full list of DEGs; Table S2: Gene set enrichment analysis results; Table S3: *A. minutum* unigenes annotation.

Author Contributions: M.A.A. performed the experiment and drafted the manuscript. S.N.B. and H.B. contributed to the analysis tools. N.Y.M.Y., A.A., G.U. and H.B. edited and added valuable insights to the manuscript. All authors have read and agreed to the published version of the manuscript.

Funding: This research was funded by the Malaysia Ministry of Education under the research grant FRGS/1/2019/STG05/UKM/02/6.

Institutional Review Board Statement: Not applicable.

Data Availability Statement: The raw read sequences of transcriptome assembly obtained from Illumina Hiseq4000 (150PE) platform have been deposited in Sequence Read Archive (SRA) under the BioProject id: PRJNA914202.

Conflicts of Interest: The authors declare no conflict of interest.

References

1. Wiese, M.; Murray, S.A.; Alvin, A.; Neilan, B.A. Gene expression and molecular evolution of sxtA4 in a saxitoxin producing dinoflagellate *Alexandrium catenella*. *Toxicon* **2014**, *92*, 102–112. [CrossRef]
2. Wells, M.L.; Karlson, B.; Wulff, A.; Kudela, R.; Trick, C.; Asnaghi, V.; Dutkiewicz, S. Future HAB science: Directions and challenges in a changing climate. *Harmful Algae* **2020**, *91*, 101632. [CrossRef]
3. Wang, D.Z.; Zhang, S.F.; Zhang, Y.; Lin, L. Paralytic shellfish toxin biosynthesis in cyanobacteria and dinoflagellates: A molecular overview. *J. Proteom.* **2016**, *135*, 132–140. [CrossRef]
4. Mohd-Din, M.; Hii, K.S.; Abdul-Wahab, M.F.; Mohamad, S.E.; Gu, H.; Leaw, C.P.; Lim, P.T. Spatial-temporal variability of microphytoplankton assemblages including harmful microalgae in a tropical semi-enclosed strait (Johor Strait, Malaysia). *Mar. Environ. Res.* **2022**, *175*, 105589. [CrossRef]
5. Chai, X.; Li, X.; Hii, K.S.; Zhang, Q.; Deng, Q.; Wan, L.; Zheng, L.; Lim, P.T.; Tan, S.N.; Mohd-Din, M.; et al. Blooms of diatom and dinoflagellate associated with nutrient imbalance driven by cycling of nitrogen and phosphorus in anaerobic sediments in Johor Strait (Malaysia). *Mar. Environ. Res.* **2021**, *169*, 105398. [CrossRef]
6. Akbar, M.A.; Yusuf, N.Y.M.; Tahir, N.I.; Ahmad, A.; Usup, G.; Sahrani, F.K.; Bunawan, H. Biosynthesis of saxitoxin in marine dinoflagellates: An omics perspective. *Mar. Drugs* **2020**, *18*, 103. [CrossRef] [PubMed]
7. Dagenais-Bellefeuille, S.; Morse, D. Putting the N in dinoflagellates. *Front. Microbiol.* **2013**, *4*, 369. [CrossRef]
8. Lim, P.T.; Leaw, C.P.; Kobiyama, A.; Ogata, T. Growth and toxin production of tropical Alexandrium minutum Halim (Dinophyceae) under various nitrogen to phosphorus ratios. *J. Appl. Phycol.* **2010**, *22*, 203–210. [CrossRef]
9. Hii, K.S.; Lim, P.T.; Kon, N.F.; Takata, Y.; Usup, G.; Leaw, C.P. Physiological and transcriptional responses to inorganic nutrition in a tropical Pacific strain of Alexandrium minutum: Implications for the saxitoxin genes and toxin production. *Harmful Algae* **2016**, *56*, 9–21. [CrossRef] [PubMed]
10. Glibert, P.M.; Wilkerson, F.P.; Dugdale, R.C.; Raven, J.A.; Dupont, C.L.; Leavitt, P.R.; Kana, T.M. Pluses and minuses of ammonium and nitrate uptake and assimilation by phytoplankton and implications for productivity and community composition, with emphasis on nitrogen-enriched conditions. *Limnol. Oceanogr.* **2016**, *61*, 165–197. [CrossRef]
11. Lai, J.; Yu, Z.; Song, X.; Cao, X.; Han, X. Responses of the growth and biochemical composition of Prorocentrum donghaiense to different nitrogen and phosphorus concentrations. *J. Exp. Mar. Biol. Ecol.* **2011**, *405*, 6–17. [CrossRef]
12. Ou, L.J.; Huang, K.X.; Li, J.J.; Jing, W.Y.; Dong, H.P. Transcriptomic responses of harmful dinoflagellate Prorocentrum donghaiense to nitrogen and light. *Mar. Pollut. Bull.* **2019**, *149*, 110617. [CrossRef]
13. Karl, D.M. Microbially mediated transformations of phosphorus in the sea: New views of an old cycle. *Ann. Rev. Mar. Sci.* **2014**, *6*, 279–337. [CrossRef] [PubMed]
14. Lin, S.J.; Litaker, R.W.; Sunda, W.G. Phosphorus physiological ecology and molecular mechanisms in marine phytoplankton. *J. Phycol.* **2016**, *52*, 10–36. [CrossRef]
15. Zhang, S.F.; Yuan, C.J.; Chen, Y.; Lin, L.; Wang, D.Z. Transcriptomic response to changing ambient phosphorus in the marine dinoflagellate Prorocentrum donghaiense. *Sci. Total Environ.* **2019**, *692*, 1037–1047. [CrossRef]
16. Akbar, M.A.; Yusof, N.Y.M.; Sahrani, F.K.; Usup, G.; Ahmad, A.; Baharum, S.N.; Muhammad, N.A.N.; Bunawan, H. Insights into Alexandrium minutum Nutrient Acquisition, Metabolism and Saxitoxin Biosynthesis through Comprehensive Transcriptome Survey. *Biology* **2021**, *10*, 826. [CrossRef]
17. Yang, I.; Beszteri, S.; Tillmann, U.; Cembella, A.; John, U. Growth- and nutrient- dependent gene expression in the toxigenic marine dinoflagellate Alexandrium minutum. *Harmful Algae* **2011**, *12*, 55–69. [CrossRef]
18. Meng, F.Q.; Song, J.T.; Zhou, J.; Cai, Z.H. Transcriptomic profile and sexual reproduction-relevant genes of Alexandrium minutum in response to nutritional deficiency. *Front. Microbiol.* **2019**, *10*, 2629. [CrossRef] [PubMed]
19. Subramanian, A.; Tamayo, P.; Mootha, V.K.; Mukherjee, S.; Ebert, B.L.; Gillette, M.A.; Mesirov, J.P. Gene set enrichment analysis: A knowledge-based approach for interpreting genome-wide expression profiles. *Proc. Natl. Acad. Sci. USA* **2005**, *102*, 15545–15550. [CrossRef] [PubMed]
20. Reimand, J.; Isserlin, R.; Voisin, V.; Kucera, M.; Tannus-Lopes, C.; Rostamianfar, A.; Merico, D. Pathway enrichment analysis and visualization of omics data using g: Profiler, GSEA, Cytoscape and EnrichmentMap. *Nat. Protoc.* **2019**, *14*, 482–517. [CrossRef]

21. Jaramillo-Lambert, A.; Fabritius, A.S.; Hansen, T.J.; Smith, H.E.; Golden, A. The identification of a novel mutant allele of topoisomerase II in Caenorhabditis elegans reveals a unique role in chromosome segregation during spermatogenesis. *Genetics* **2016**, *204*, 1407–1422. [CrossRef]
22. Pereira, C.; Smolka, M.B.; Weiss, R.S.; Brieño-Enríquez, M.A. ATR signaling in mammalian meiosis: From upstream scaffolds to downstream signaling. *Environ. Mol. Mutagen.* **2020**, *61*, 752–766. [CrossRef] [PubMed]
23. Durocher, D.; Jackson, S.P. The FHA domain. *FEBS Lett.* **2002**, *513*, 58–66. [CrossRef] [PubMed]
24. Simonin, A.R.; Rasmussen, C.G.; Yang, M.; Glass, N.L. Genes encoding a striatin-like protein (ham-3) and a forkhead associated protein (ham-4) are required for hyphal fusion in Neurospora crassa. *Fungal Genet. Biol.* **2010**, *47*, 855–868. [CrossRef] [PubMed]
25. Akbar, M.A.; Ahmad, A.; Usup, G.; Bunawan, H. Current knowledge and recent advances in marine dinoflagellate transcriptomic research. *J. Mar. Sci. Eng.* **2018**, *6*, 13. [CrossRef]
26. Hockin, N.L.; Mock, T.; Mulholland, F.; Kopriva, S.; Malin, G. The response of diatom central carbon metabolism to nitrogen starvation is different from that of green algae and higher plants. *Plant Physiol.* **2012**, *158*, 299–312. [CrossRef]
27. Longworth, J.; Wu, D.; Huete-Ortega, M.; Wright, P.C.; Vaidyanathan, S. Proteome response of Phaeodactylum tricornutum, during lipid accumulation induced by nitrogen depletion. *Algal Res.* **2016**, *18*, 213–224. [CrossRef]
28. Wase, N.; Black, P.N.; Stanley, B.A.; DiRusso, C.C. Integrated quantitative analysis of nitrogen stress response in Chlamydomonas reinhardtii using metabolite and protein profiling. *J. Proteome Res.* **2014**, *13*, 1373. [CrossRef]
29. Liu, L.; Wang, J.; Han, Z.; Sun, X.; Li, H.; Zhang, J.; Lu, Y. Molecular analyses of tomato GS, GOGAT and GDH gene families and their response to abiotic stresses. *Acta Physiol. Plant* **2016**, *38*, 229. [CrossRef]
30. Alipanah, L.; Rohloff, J.; Winge, P.; Bones, A.M.; Brembu, T. Whole-cell response to nitrogen deprivation in the diatom Phaeodactylum tricornutum. *J. Exp. Bot.* **2015**, *66*, 6281–6296. [CrossRef]
31. Demidchik, V.; Straltsova, D.; Medvedev, S.S.; Pozhvanov, G.A.; Sokolik, A.; Yurin, V. Stress-induced electrolyte leakage: The role of K+-permeable channels and involvement in programmed cell death and metabolic adjustment. *J. Exp. Bot.* **2014**, *65*, 1259–1270. [CrossRef]
32. Dietrich, P.; Anschütz, U.; Kugler, A.; Becker, D. Physiology and biophysics of plant ligand-gated ion channels. *Plant Biol.* **2010**, *12*, 80–93. [CrossRef]
33. Pozdnyakov, I.A.; Skarlato, S.O. Analysis of the dinoflagellate Prorocentrum minimum transcriptome: Identifying the members of the voltage-gated cation channel superfamily. *Cell Tissue Biol.* **2015**, *9*, 483–492. [CrossRef]
34. Demidchik, V. Mechanisms of oxidative stress in plants: From classical chemistry to cell biology. *Environ. Exp. Bot.* **2015**, *109*, 212–228. [CrossRef]
35. Hou, D.Y.; Mao, X.T.; Gu, S.; Li, H.Y.; Liu, J.S.; Yang, W.D. Systems-level analysis of metabolic mechanism following nitrogen limitation in benthic dinoflagellate Prorocentrum lima. *Algal Res.* **2018**, *33*, 389–398. [CrossRef]
36. Zhang, Y.M.; Chen, H.; He, C.L.; Wang, Q. Nitrogen starvation induced oxidative stress in an oil-producing green alga Chlorella sorokiniana C3. *PLoS ONE* **2013**, *8*, e69225. [CrossRef]
37. DeFalco, T.A.; Bender, K.W.; Snedden, W.A. Breaking the code: Ca^{2+} sensors in plant signalling. *Biochem. J.* **2010**, *425*, 27–40. [CrossRef]
38. Conradi, C.; Shiu, A. Dynamics of posttranslational modification systems: Recent progress and future directions. *Biophys. J.* **2018**, *114*, 507–515. [CrossRef] [PubMed]
39. Tan, M.H.; Smith, S.R.; Hixson, K.K.; Tan, J.; McCarthy, J.K.; Kustka, A.B.; Allen, A.E. The importance of protein phosphorylation for signaling and metabolism in response to diel light cycling and nutrient availability in a marine diatom. *Biology* **2020**, *9*, 155. [CrossRef] [PubMed]
40. Vlastaridis, P.; Kyriakidou, P.; Chaliotis, A.; Van de Peer, Y.; Oliver, S.G.; Amoutzias, G.D. Estimating the total number of phosphoproteins and phosphorylation sites in eukaryotic proteomes. *Gigascience* **2017**, *6*, giw015. [CrossRef]
41. Liu, B.; Lo, S.C.L.; Matton, D.P.; Lang, B.F.; Morse, D. Daily changes in the phosphoproteome of the dinoflagellate Lingulodinium. *Protist* **2012**, *163*, 746–754. [CrossRef] [PubMed]
42. Roy, S.; Letourneau, L.; Morse, D. Cold-induced cysts of the photosynthetic dinoflagellate Lingulodinium polyedrum have an arrested circadian bioluminescence rhythm and lower levels of protein phosphorylation. *Plant Physiol.* **2014**, *164*, 966–977. [CrossRef] [PubMed]
43. Li, X.; Sanagi, M.; Lu, Y.; Nomura, Y.; Stolze, S.C.; Yasuda, S.; Lunn, J.E. Protein phosphorylation dynamics under carbon/nitrogen-nutrient stress and identification of a cell death-related receptor-like kinase in Arabidopsis. *Front. Plant Sci.* **2020**, *11*, 377. [CrossRef] [PubMed]
44. Plaxton, W.C.; Tran, H.T. Metabolic adaptations of phosphate-starved plants. *Plant Physiol.* **2011**, *156*, 1006–1015. [CrossRef]
45. Smith, S.R.; Abbriano, R.M.; Hildebrand, M. Comparative analysis of diatom genomes reveals substantial differences in the organization of carbon partitioning pathways. *Algal Res.* **2012**, *1*, 2–16. [CrossRef]
46. Xu, X.; Yu, Z.; He, L.; Cao, X.; Chen, N.; Song, X. Metabolic analyses by metatranscriptomics highlight plasticity in phosphorus acquisition during monospecific and multispecies algal blooms. *Hydrobiologia* **2020**, *847*, 1071–1085. [CrossRef]
47. Brembu, T.; Mühlroth, A.; Alipanah, L.; Bones, A.M. The effects of phosphorus limitation on carbon metabolism in diatoms. *Philos. Trans. R. Soc. B* **2017**, *372*, 20160406. [CrossRef]
48. Cui, Y.; Zhang, H.; Lin, S. Enhancement of non-photochemical quenching as an adaptive strategy under phosphorus deprivation in the dinoflagellate Karlodinium veneficum. *Front. Microbiol.* **2017**, *8*, 404. [CrossRef]

49. Shi, X.; Lin, X.; Li, L.; Li, M.; Palenik, B.; Lin, S. Transcriptomic and microRNAomic profiling reveals multi-faceted mechanisms to cope with phosphate stress in a dinoflagellate. *ISME J.* **2017**, *11*, 2209–2218. [CrossRef]
50. Lei, Q.Y.; Lu, S.H. Molecular ecological responses of the dinoflagellate Karenia mikimotoi to phosphate stress. *Harmful Algae* **2011**, *12*, 39–45. [CrossRef]
51. Feng, T.Y.; Yang, Z.K.; Zheng, J.W.; Xie, Y.; Li, D.W.; Murugan, S.B.; Yang, W.D.; Liu, J.S.; Li, H.Y. Examination of metabolic responses to phosphorus limitation via proteomic analyses in the marine diatom Phaeodactylum tricornutum. *Sci. Rep.* **2015**, *5*, 10373. [CrossRef] [PubMed]
52. Alipanah, L.; Winge, P.; Rohloff, J.; Najafi, J.; Brembu, T.; Bones, A.M. Molecular adaptations to phosphorus deprivation and comparison with nitrogen deprivation responses in the diatom Phaeodactylum tricornutum. *PLoS ONE* **2018**, *13*, e0193335. [CrossRef] [PubMed]
53. Müller, J.; Gödde, V.; Niehaus, K.; Zörb, C. Metabolic adaptations of white lupin roots and shoots under phosphorus deficiency. *Front. Plant Sci.* **2015**, *6*, 1014. [CrossRef] [PubMed]
54. Pant, B.D.; Pant, P.; Erban, A.; Huhman, D.; Kopka, J.; Scheible, W.R. Identification of primary and secondary metabolites with phosphorus status-dependent abundance in Arabidopsis, and of the transcription factor PHR 1 as a major regulator of metabolic changes during phosphorus limitation. *Plant Cell Environ.* **2015**, *38*, 172–187. [CrossRef] [PubMed]
55. Dyhrman, S.T. Nutrients and their acquisition: Phosphorus physiology in microalgae. In *The Physiology of Microalgae: Developments in Applied Phycology*; Springer: Cham, Switzerland, 2016; Volume 6, pp. 155–183.
56. Krueger, K.; Goggi, A.S.; Mallarino, A.P.; Mullen, R.E. Phosphorus and potassium fertilization effects on soybean seed quality and composition. *Crop Sci.* **2013**, *53*, 602–610. [CrossRef]
57. Singh, S.K.; Barnaby, J.Y.; Reddy, V.R.; Sicher, R.C. Varying response of the concentration and yield of soybean seed mineral elements, carbohydrates, organic acids, amino acids, protein, and oil to phosphorus starvation and CO_2 enrichment. *Front. Plant Sci.* **2016**, *7*, 1967. [CrossRef]
58. Benito, B.; Garciadeblás, B.; Fraile-Escanciano, A.; Rodríguez-Navarro, A. Potassium and sodium uptake systems in fungi. The transporter diversity of Magnaporthe oryzae. *Fungal Genet. Biol.* **2011**, *48*, 812–822. [CrossRef]
59. Liu, K.H.; Niu, Y.; Konishi, M.; Wu, Y.; Du, H.; Chung, H.S.; Ishida, T. Discovery of nitrate–CPK–NLP signalling in central nutrient–growth networks. *Nature* **2017**, *545*, 311–316. [CrossRef]
60. Thor, K. Calcium- nutrient and messenger. *Front. Plant Sci.* **2019**, *10*, 440. [CrossRef]
61. Choi, W.G.; Toyota, M.; Kim, S.H.; Hilleary, R.; Gilroy, S. Salt stress-induced Ca^{2+} waves are associated with rapid, long-distance root-to-shoot signaling in plants. *Proc. Natl. Acad. Sci. USA* **2014**, *111*, 6497–6502. [CrossRef]
62. Dominguez, R.; Holmes, K.C. Actin structure and function. *Annu. Rev. Biophys.* **2011**, *40*, 169–186. [CrossRef]
63. Berdieva, M.; Pozdnyakov, I.; Matantseva, O.; Knyazev, N.; Skarlato, S. Actin as a cytoskeletal basis for cell architecture and a protein essential for ecdysis in Prorocentrum minimum (Dinophyceae, Prorocentrales). *Phycol. Res.* **2018**, *66*, 127–136. [CrossRef]
64. Heimann, K.; Klerks, P.L.; Hasenstein, K.H. Involvement of actin and microtubules in regulation of bioluminescence and translocation of chloroplasts in the dinoflagellate Pyrocystis lunula. *Bot. Mar.* **2009**, *52*, 170–177. [CrossRef]
65. Stires, J.C.; Latz, M.I. Contribution of the cytoskeleton to mechanosensitivity reported by dinoflagellate bioluminescence. *Cytoskeleton* **2018**, *75*, 12–21. [CrossRef]
66. Vera, K.; Mariia, B.; Olga, M. The role of the cytoskeleton in the ecdysis of the dinoflagellate Prorocentrum minimum. *Protistology* **2020**, *14*, 38–44.
67. Peacock, M.B.; Kudela, R.M. Evidence for active vertical migration by two dinoflagellates experiencing iron, nitrogen, and phosphorus limitation. *Limnol. Oceanogr.* **2014**, *59*, 660–673. [CrossRef]
68. Tilney, C.L.; Hoadley, K.D.; Warner, M.E. Comparing the diel vertical migration of Karlodinium veneficum (dinophyceae) and Chattonella subsalsa (Raphidophyceae): PSII photochemistry, circadian control, and carbon assimilation. *J. Photochem. Photobiol. B Biol.* **2015**, *143*, 107–119. [CrossRef]
69. Chardin, C.; Schenk, S.T.; Hirt, H.; Colcombet, J.; Krapp, A. Mitogen-Activated protein kinases in nutritional signaling in Arabidopsis. *Plant Sci.* **2017**, *260*, 101–108. [CrossRef]
70. González-Rubio, G.; Fernández-Acero, T.; Martín, H.; Molina, M. Mitogen-activated protein kinase phosphatases (MKPs) in fungal signaling: Conservation, function, and regulation. *Int. J. Mol. Sci.* **2019**, *20*, 1709. [CrossRef] [PubMed]
71. Jiang, C.; Zhang, X.; Liu, H.; Xu, J.R. Mitogen-activated protein kinase signaling in plant pathogenic fungi. *PLoS Pathog.* **2018**, *14*, e1006875. [CrossRef] [PubMed]
72. Guo, R.; Wang, H.; Suh, Y.S.; Ki, J.S. Transcriptomic profiles reveal the genome-wide responses of the harmful dinoflagellate Cochlodinium polykrikoides when exposed to the algicide copper sulfate. *BMC Genom.* **2016**, *17*, 29. [CrossRef]
73. Wang, H.; Guo, R.; Ki, J.S. 6.0 K microarray reveals differential transcriptomic responses in the dinoflagellate Prorocentrum minimum exposed to polychlorinated biphenyl (PCB). *Chemosphere* **2018**, *195*, 398–409. [CrossRef]
74. Alhadeff, R.; Vorobyov, I.; Yoon, H.W.; Warshel, A. Exploring the free-energy landscape of GPCR activation. *Proc. Natl. Acad. Sci. USA* **2018**, *115*, 10327–10332. [CrossRef]
75. Erlandson, S.C.; McMahon, C.; Kruse, A.C. Structural basis for G protein–coupled receptor signaling. *Annu. Rev. Biophys.* **2018**, *47*, 1–18. [CrossRef] [PubMed]
76. Mojib, N.; Kubanek, J. Comparative transcriptomics supports the presence of G protein-coupled receptor-based signaling in unicellular marine eukaryotes. *Limnol. Oceanogr.* **2020**, *65*, 762–774. [CrossRef]

77. Jain, R.; Watson, U.; Vasudevan, L.; Saini, D.K. ERK activation pathways downstream of GPCRs. *Int. Rev. Cell Mol. Biol.* **2018**, *338*, 79–109.
78. Grimsey, N.J.; Lin, Y.; Narala, R.; Rada, C.C.; Mejia-Pena, H.; Trejo, J. G protein–coupled receptors activate p38 MAPK via a non-canonical TAB1–TAB2–and TAB1–TAB3–dependent pathway in endothelial cells. *J. Biol. Chem.* **2019**, *294*, 5867–5878. [CrossRef] [PubMed]
79. Rodriguez-Casariego, J.A.; Ladd, M.C.; Shantz, A.A.; Lopes, C.; Cheema, M.S.; Kim, B.; Eirin-Lopez, J.M. Coral epigenetic responses to nutrient stress: Histone H2A. X phosphorylation dynamics and DNA methylation in the staghorn coral Acropora cervicornis. *Ecol. Evol.* **2018**, *8*, 12193–12207. [CrossRef] [PubMed]
80. Riaz, S.; Niaz, Z.; Khan, S.; Liu, Y.; Sui, Z. Detection, characterization and expression dynamics of histone proteins in the dinoflagellate Alexandrium pacificum during growth regulation. *Harmful Algae* **2019**, *87*, 101630. [CrossRef]
81. Chandrika, N.N.P.; Sundaravelpandian, K.; Schmidt, W. A PHD in histone language: On the role of histone methylation in plant responses to phosphate deficiency. *Plant Signal. Behav.* **2013**, *8*, e24381. [CrossRef]
82. Secco, D.; Whelan, J.; Rouached, H.; Lister, R. Nutrient stress-induced chromatin changes in plants. *Curr. Opin. Plant Biol.* **2017**, *39*, 1–7. [CrossRef]
83. Chen, Z.; Zhou, Y.; Zhang, Z.; Song, J. Towards more accurate prediction of ubiquitination sites: A comprehensive review of current methods, tools and features. *Brief. Bioinform.* **2015**, *16*, 640–657. [CrossRef] [PubMed]
84. Xu, F.Q.; Xue, H.W. The ubiquitin-proteasome system in plant responses to environments. *Plant Cell Environ.* **2019**, *42*, 2931–2944. [CrossRef]
85. Gilberto, S.; Peter, M. Dynamic ubiquitin signaling in cell cycle regulation. *J. Cell Biol.* **2017**, *216*, 2259–2271. [CrossRef] [PubMed]
86. Stone, S.L. Role of the ubiquitin proteasome system in plant response to abiotic stress. *Int. Rev. Cell Mol. Biol.* **2019**, *34*, 65–110.
87. Holdsworth, M.J.; Vicente, J.; Sharma, G.; Abbas, M.; Zubrycka, A. The plant N-degron pathways of ubiquitin-mediated proteolysis. *J. Integr. Plant Biol.* **2020**, *62*, 70–89. [CrossRef] [PubMed]
88. Perini, F.; Galluzzi, L.; Dell'Aversano, C.; Dello Iacovo, E.; Tartaglione, L.; Ricci, F.; Forino, M.; Ciminiello, P.; Penna, A. SxtA and sxtG gene expression and toxin production in the Mediterranean Alexandrium minutum (Dinophyceae). *Mar. Drugs* **2014**, *12*, 5258–5276. [CrossRef]
89. Vanucci, S.; Pezzolesi, L.; Pistocchi, R.; Ciminiello, P.; Aversano, C.D.; Iacovo, E.D. Nitrogen and phosphorus limitation effects on cell growth, biovolume, and toxin production in Ostreopsis cf. ovata. *Harmful Algae* **2012**, *15*, 78–90. [CrossRef]
90. Vico, P.; Aubriot, L.; Martigani, F.; Rigamonti, N.; Bonilla, S.; Piccini, C. Influence of nitrogen availability on the expression of genes involved in the biosynthesis of saxitoxin and analogs in Cylindrospermopsis raciborskii. *Harmful Algae* **2016**, *56*, 37–43. [CrossRef]
91. Li, B.; Dewey, C.N. RSEM: Accurate transcript quantification from RNA-Seq data with or without a reference genome. *BMC Bioinform.* **2011**, *12*, 323. [CrossRef]
92. Langmead, B.; Salzberg, S.L. Fast gapped-read alignment with Bowtie 2. *Nat. Methods* **2012**, *9*, 357. [CrossRef]
93. Su, W.; Sun, J.; Shimizu, K.; Kadota, K. TCC-GUI: A shiny-based application for differential expression analysis of RNA-Seq count data. *BMC Res. Notes* **2019**, *12*, 133. [CrossRef]
94. Robinson, M.D.; McCarthy, D.J.; Smyth, G.K. edgeR: A Bioconductor package for differential expression analysis of digital gene expression data. *Bioinformatics* **2010**, *26*, 139–140. [CrossRef]
95. Haas, B.J.; Papanicolaou, A.; Yassour, M.; Grabherr, M.; Blood, P.D.; Bowden, J.; MacManes, M.D. De novo transcript sequence reconstruction from RNA-seq using the Trinity platform for reference generation and analysis. *Nat. Protoc.* **2013**, *8*, 1494–1512. [CrossRef] [PubMed]
96. Merico, D.; Isserlin, R.; Stueker, O.; Emili, A.; Bader, G.D. Enrichment map: A network-based method for gene-set enrichment visualization and interpretation. *PLoS ONE* **2010**, *5*, e13984. [CrossRef] [PubMed]
97. Morris, J.H.; Apeltsin, L.; Newman, A.M.; Baumbach, J.; Wittkop, T.; Su, G.; Ferrin, T.E. clusterMaker: A multi-algorithm clustering plugin for Cytoscape. *BMC Bioinform.* **2011**, *12*, 436. [CrossRef] [PubMed]
98. Kucera, M.; Isserlin, R.; Arkhangorodsky, A.; Bader, G.D. AutoAnnotate: A Cytoscape app for summarizing networks with semantic annotations. *F1000Res* **2016**, *5*, 1717. [CrossRef] [PubMed]

marine drugs

MDPI

Review

The Toxic Effects of Environmental Domoic Acid Exposure on Humans and Marine Wildlife

Ami E. Krasner [1,*], Margaret E. Martinez [2], Cara L. Field [2] and Spencer E. Fire [1]

1 Florida Institute of Technology, Department of Ocean Engineering and Marine Sciences, Melbourne, FL 32901, USA; sfire@fit.edu
2 The Marine Mammal Center, Sausalito, CA 94965, USA
* Correspondence: akrasner2021@my.fit.edu

Abstract: Biotoxins produced by harmful algal blooms (HABs) are a substantial global threat to ocean and human health. Domoic acid (DA) is one such biotoxin whose negative impacts are forecasted to increase with climate change and coastal development. This manuscript serves as a review of DA toxicosis after environmental exposure in humans and wildlife, including an introduction to HAB toxins, the history of DA toxicosis, DA production, toxicokinetic properties of DA, susceptibility, clinical signs, DA detection methods and other diagnostic tests, time course of toxicosis, treatment, prognostics, and recommendations for future research. Additionally, we highlight the utility of California sea lions (CSLs; *Zalophus californianus*) as a model and sentinel of environmental DA exposure.

Keywords: domoic acid toxicosis; harmful algal blooms; biotoxin; marine wildlife; amnesic shellfish poisoning; glutamate receptor; sentinels

1. Introduction to Harmful Algal Biotoxins

Biotoxins naturally produced by harmful algal blooms (HABs) are increasingly documented worldwide as agents of disease impacting marine wildlife and public health [1]. The apparent escalation in HAB events over the last several decades may be due to improved awareness, monitoring efforts, and detection methods [1–3]. Climatic shifts, aquatic eutrophication, overfishing, marine regime shifts, and industrialization are suspected contributors to an expanding distribution, incidence, intensity, and/or number of toxicogenic HAB species in certain regions [3–5]. Harmful algal biotoxins are a primary health threat to marine wildlife in the USA and a major cause of strandings and mass mortality events [6,7].

There are at least a dozen HAB toxins of major, minor, and/or emerging concern to human and animal health worldwide [2]. Simultaneous HAB toxin exposure may occur in humans and marine wildlife [8–12], as well as in the seafood they consume [13,14]. Exposure to multiple HAB toxins may increase individual potencies [15]. Seafood can also be contaminated with concurrent harmful agents (e.g., heavy metals, *Vibro* spp.) [16].

Routes of exposure to HAB toxins are primarily via oral ingestion of contaminated seawater or prey but can include dermal contact or respiratory inhalation of aerosolized toxins [2]. Prey can bioaccumulate HAB toxins prior to vectoring these potentially harmful agents to higher trophic-level predators [2]. Harmful algal bloom toxins typically disrupt cellular signaling pathways through binding to membrane (i.e., receptor or channel) or intracellular proteins, commonly resulting in neurotoxicity, cardiotoxicity, hepatotoxicity, gastrointestinal (GI) upset, or tumor promotion [2]. Observable health effects can be immediate (acute toxicosis) or delayed weeks to years after exposure or accumulate over multiple exposures (chronic toxicosis) [17–19]. Hydrophilic toxins are rapidly depurated within hours to days, while hydrophobic toxins may persist for weeks to months [2]. The GI tract and kidneys often act as primary routes of excretion. Biotoxin kinetics and effects are impacted by (1) exposure route, duration, dose, frequency, and history, (2) biotoxin properties (e.g., hydrophilicity and potency), (3) anatomy and physiology of the exposed taxa, and (4) underlying health status, developmental stage, and sex of the exposed individual. Data is lacking regarding toxicokinetic properties, HAB toxin interactions, subtle health impacts or those from repeated exposure(s), and diagnostic confirmation of previous exposure(s) in marine wildlife and humans [19].

This manuscript focuses on the biotoxin domoic acid (DA) and related disease in humans and marine wildlife. Domoic acid can cause acute and chronic neurobehavioral deficits, cardiac disease, reproductive failure, and mortality in numerous species [2,17–19]. Domoic acid exerts its harmful effects by mimicking the excitatory neurotransmitter L-glutamate in neural and extra-neural tissues [2,17]. Research has focused on characterizing disease in the central nervous system (CNS) of laboratory models of DA toxicosis, as well as environmentally exposed marine wildlife and humans with frequent DA consumption. The history of DA toxicosis, DA production, toxicokinetics, susceptibility, clinical signs, diagnostic tests, medical treatment options, prognosis, and future recommendations in DA-related clinical research are discussed. Additionally, we highlight the California sea lion (CSL; *Zalophus californianus*) as a model and sentinel of environmental DA exposure.

2. History of DA Toxicosis

Domoic acid was first isolated from a red macroalgae in Japan in the 1950s [20]. For several decades, DA was used to treat GI parasites in humans. Minimal to no side effects were reported, except for mild headaches, GI upset, and fatigue with doses over 0.66 mg/kg body weight [21–23]. The toxin was recognized as a public health threat in 1987 after causing acute poisoning in over 100 individuals, leading to nineteen hospitalizations and four deaths [24,25]. The outbreak of human illness, called Amnesic Shellfish Poisoning (ASP) from the distinctive anterograde memory loss described in 25% of patients, occurred after consumption of contaminated mussels in Prince Edward Island, Canada [25]. Amnesic shellfish poisoning was defined as the development of GI signs within 24 h and/or neurologic signs (e.g., amnesia, headache, confusion, hallucinations, seizures, or coma) within 48 h after DA exposure [24]. Aged individuals and those with underlying health conditions (i.e., diabetes mellitus, hypertension, or chronic kidney disease) were most severely affected [26]. Except for an unconfirmed outbreak in Washington State in 1991 [26], ASP has been minimized through monitoring DA levels in seafood for public consumption [19]. Recreational and commercial seafood closures occur in the USA, Canada, Australia, New Zealand, and the European Union if levels exceed the regulatory limit (20 ppm shellfish tissue; 30 ppm Dungeness crab viscera) [27–29]. Yet, this limit is not specific to age, sex, body weight, DA consumption rate and exposure history, pregnancy, lactation, or underlying health status; thus, it may inadequately protect sensitive or high-risk groups or those with long-term, low-level ingestion [19].

A chronic condition secondary to DA toxicosis was first identified when an elderly patient who recovered from severe ASP presented one year later with temporal lobe epilepsy (TLE), presumed to be a delayed ASP side effect [30]. Native American tribe members, participants of the Communities Advancing the Studies of Tribal Nations Across the Lifespan (CoASTAL) cohort [31], exhibited memory deficits from repeated DA exposure below the safety threshold [19]. Thus, tolerable daily intake (TDI) recommendations were modified to 0.003 mg/kg/day for adults or a maximum of 15 razor clams/month, particularly for young, aged, pregnant, or renal-compromised individuals [29,32].

In numerous marine wildlife species, DA has been associated with acute and chronic illness, death, mass stranding events, and population declines, including those that are a conservation priority. In 1991, around 150 seabirds in Monterey Bay, California, developed acute GI and neurologic illness or died from the toxin [33,34]. Subsequently, DA exposure or toxicosis has been identified in numerous seabird species spanning the Pacific coast from Alaska to Mexico, as well as Rhode Island and Spain [35–39]. Since at least 1998, CSLs have stranded from acute DA toxicosis (an immediate consequence of high-level exposure) or domoic acid-induced epileptic disease (a condition caused by repeated/chronic DA exposure or a delayed consequence of previous sublethal acute exposure), and they are the most well-documented, systemically researched, and visibly impacted marine species [28,40]. Primarily seasonal, nearly annual DA-producing blooms impact dozens to hundreds of CSLs at a time in California, with more isolated cases of DA toxicosis in Pacific Northwest populations [36]. The toxin has caused similar disease outcomes in the threatened (federally listed) keystone subspecies southern sea otters (SSOs; *Enhydra lutris nereis*) in central California [41,42]. Other marine wildlife experience less frequent, subtle, or unclear impacts from DA exposure. The first known cases of DA toxicosis in cetaceans involved several species that were stranded with marked neurologic signs during a 2002 marine mammal unusual mortality event (UME) in southern California [43]. Since then, DA exposure has been documented in other cetaceans worldwide, correlated with mass stranding events and an increased vulnerability to injury/mortality from fisheries interaction and vessel strikes, and linked to the failed recovery of several species of conservation concern, including northern (*Eubalaena glacialis*) and southern (*Eubalaena australis*) right whales [9,44–51]. The toxin was also suspected to have caused the strandings and subsequent deaths of two endangered marine reptiles: green sea turtles (*Chelonia mydas*) in Florida and leatherback sea turtles (*Dermochelys coriacea*) in California [52–54].

3. Domoic Acid Production

Domoic acid is naturally synthesized by at least half a dozen species of red macroalgae and over twenty-five species of pennate diatoms in the genus *Pseudo-nitzschia* and related *Nitzschia* species [1,55]. The toxin is produced in diverse marine waters worldwide, including the coast and open ocean, as well as arctic and equatorial environments. Domoic acid has been detected in the offshore waters of all continents but Antarctica [4,9,10,13,45,56–69] (Figure 1). Domoic acid production may be occurring with increasing frequency, potency, duration, and geographic extent, as well as by a greater number of algal species in some regions of North America [1,3]. Detecting the toxin can be challenging in offshore environments and regions with bloom patchiness. Monitoring DA production can also be challenging in communities or nations with limited access to the necessary resources.

Triggers of bloom formation and DA synthesis are multifactorial and complex, as there are regional, seasonal, anthropogenic, environmental, and hydrographic influences [70–72]. While DA is detectable year-round in many regions, seasonal blooms occur in California (late spring to fall), the northeast USA (late fall to winter), and the Gulf of Mexico (spring) [1,66,73–75]. Triggers may be linked with climatic cycles or extreme ocean events, such as the "Warm Water Anomaly" in 2015–2016 along the USA Pacific coast, which was associated with the most toxic, extensive, and persistent DA-producing bloom ever recorded [36]. As ocean warming, acidification, and nutrient levels are major contributors to HAB blooms and toxin synthesis, the global threat of exposure is expected to magnify with climate change projections and urban expansion.

Domoic acid can persist in pelagic and benthic water and food webs [44,76], including marine snow and sediments [77,78], representing exposure sources in lieu of toxigenic blooms. The toxin is stable under ambient environmental conditions but may degrade upon exposure to ultraviolet radiation, oxygen, or extreme pH or temperature fluctuations [79]. Conventional cooking of DA-contaminated seafood only minimally degrades the toxin [80].

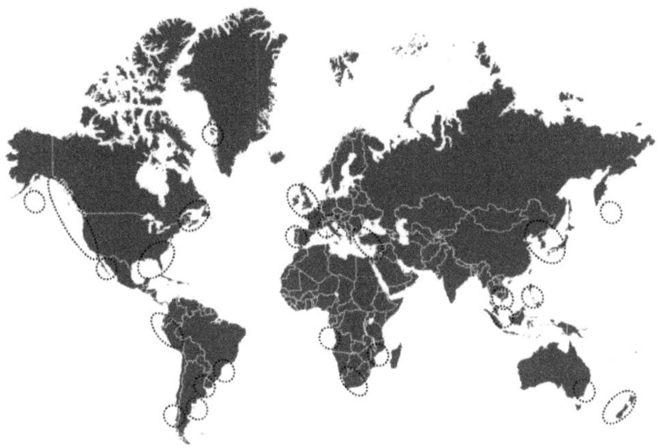

Figure 1. Worldwide distribution of domoic acid-producing blooms (dashed circles) (Adapted from [2]).

4. Toxicokinetic Properties of DA

Domoic acid is a polar, hydrophilic, excitatory amino acid (EAA) that is similar in biochemical structure and function to kainic acid (KA), L-aspartate, and L-glutamate [1,81] (Figure 2). Kainic acid is a biotoxin naturally produced by a macroalgae [82]. L-aspartate and L-glutamate are excitatory neurotransmitters in vertebrates, but L-glutamate is dominant. Both DA and KA can bind to glutamate receptors (GluRs) and mimic glutamate's biological activity [83,84]. However, DA is three times more potent than KA and up to 100× more potent than L-glutamate [85]. Though there are about a dozen known DA isomers with variable potency, isomers are not considered a substantial threat to humans and marine wildlife due to their lower GluR affinity and environmental concentrations [19,25]

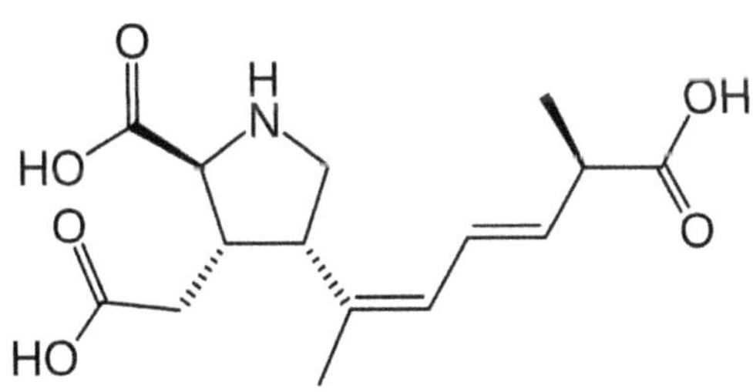

Figure 2. The structure of DA, an excitatory amino acid and harmful algal bloom (HAB) toxin.

Toxicokinetic data is primarily from healthy laboratory animals (murine, finfish, and nonhuman primates) administered parental DA. Extrapolation to environmental exposure is complicated by differential sensitivity based on species, exposure route and amount, exposure duration, and individual health history. Rodents and finfish, two of the most common models for DA toxicosis, are more resistant to the toxic effects of DA than humans or marine wildlife [86,87]. There is variable uptake, storage, and elimination between taxa [88]. Laboratory studies utilize pure DA, but seafood may also contain isomers or other contaminants with varying impacts on toxin cycling and health. Dose-effect and toxin cycling elucidation is challenging in environmentally exposed wildlife due to opportunistic testing, rapid DA elimination, and unknown consumption dose, duration, frequency, history, and time elapse between DA ingestion and work-up. Thus, the toxicokinetic properties of DA after environmental exposure are presumptive.

Humans and marine wildlife may be exposed to DA via the ingestion of contaminated zooplankton, shellfish, finfish, or seawater, or through maternal toxin transfer [89]. Prey that bioaccumulate DA demonstrate markedly different elimination rates from days in finfish to years in some shellfish [82–84]. In nonhuman primate and murine models, DA is slowly and poorly absorbed from the GI tract and diffuses poorly through the placenta and blood-brain barrier (BBB) [90–92]. The toxin can bioaccumulate in amniotic fluid and breast milk for weeks, magnifying perinatal exposure [18,93–96]. Access to the CNS may be augmented by an increased BBB permeability from DA-induced seizures or via the less protected circumventricular organs [26,97]. Due to DA's high potency, small quantities may be capable of extreme toxicity. Active transporters for DA distribution into body compartments may be present in some species [92,98].

Domoic acid primarily exerts its toxic effects through excessive stimulation of GluRs, resulting in cell death and organ dysfunction [99]. Mammalian GluRs are categorized into two families: (1) ionotropic (iGluRs; fast-acting ligand-gated cation channels responsible for excitatory neurotransmission) and (2) metabotropic (mGluRs; slower-acting G-protein-coupled receptors that modulate excitatory neurotransmission) [81]. Since DA primarily interacts with iGluRs, toxicity is most substantial in cells and body regions (e.g., the hippocampus) where they are highly concentrated. Neural and extra-neural tissues (e.g., heart, kidneys, liver, lungs, reproductive organs, adrenal and pituitary glands, and GI tract) contain iGluRs and may experience toxicity [39,94–98]. Domoic acid may also interact to a lesser degree with mGluRs [100,101]. In addition to direct activation of local GluRs, toxicity in peripheral tissues (e.g., heart and GI tract) may be due to central activation [97,100,102–105].

Domoic acid induces a complicated cascade of cellular events depending on the exposure scenario. In acute high-dose exposures, DA stimulates all CNS iGluR subtypes, causing elevated intracellular calcium levels, reactive oxygen species (ROS) production, and subsequent necrosis, whereas low-level exposures likely largely involve apoptosis without all iGluR subtypes or ROS [99,106,107]. Acute toxicosis, followed by an asymptomatic latency period with structural and functional damage, culminating in progressive disease, also occurs [30,108]. Other mechanisms of CNS or peripheral injury include the following: (1) osmotic tissue damage, (2) mitochondrial disruption, (3) inadequate neuroprotection provided by astrocytes, (4) inflammation, (5) dysregulation of gamma-aminobutyric acid (GABA, the primary inhibitory CNS neurotransmitter), (6) immune imbalance, (7) gene expression changes, (8) altered endocrine homeostasis, (9) vascular damage, and (10) cardiac conducting system dysfunction [41,106,108–117].

Domoic acid is eliminated from the body unchanged, suggesting limited metabolism [26]. The kidneys serve as the primary systemic elimination route [88,98,118,119]. Preferential renal accumulation with equal levels in the medulla and cortex, glomerular filtration with active tubular resorption, tubular secretion using organic anion transport, biliary secretion, and cerebral spinal fluid (CSF) elimination have been observed [25,88,90,92,95,118–121]. While depuration rates vary among impacted species, from hours in CSLs to over a week in humans, clearance is generally considered rapid [122,123]. Elimination may be delayed in dehydrated individuals [124].

A dose-effect continuum occurs where lower DA doses cause acute GI upset, higher doses lead to acute neurotoxicity or death [102], and long-term, low-level exposure can result in subtle neurobehavioral deficits [125]. A DA dose-response estimate for acute toxicosis and death in a common minke whale (*Balaenoptera acutorostrata*; 1.1–6.8 mg/kg) was consistent with the estimated lethal dose in humans (4 mg/kg) [24,27,89]. Exposure dose estimates based on prey consumption suggest that doses below 1.1 mg/kg are not overtly neurotoxic in mysticetes [44,126]. Pre-conditioning permitted a neuroprotective tolerance to DA in the hippocampus of young, but not aged, rats in situ [127], whereas enhanced neurologic sensitivity was observed after intracoelomic DA administration in finfish models of repeated exposure [128]. A potential adaptive response was observed in the CNS of nonhuman primate models of long-term oral DA exposure [129]. These studies highlight the potential contribution of exposure history, exposure route, and species to dose-response relationships.

5. Susceptibility

Biological and environmental factors affect susceptibility to DA toxicosis, including individual (i.e., species, age class, sex, underlying health status, and exposure history) and population-level (i.e., home range, prey availability, and conservation status) influences.

5.1. Species

Humans are one of the most DA-sensitive taxa due to physiologic, geographic, and socioeconomic factors. Groups at an increased risk based on seafood consumption rates include (1) coastal residents [62], (2) commercial or recreational anglers [130], (3) individuals who perceive health benefits from, and can afford, omega-3-rich seafood, and (4) indigenous communities who rely on shellfish for culture and nutrition [131]. Domoic acid susceptibility may be mitigated through seafood monitoring programs in humans and piscivorous wildlife under human-managed care. Long-lived taxa, such as humans and marine mammals, are more susceptible to repetitive exposure or likely to experience delayed or progressive impacts [132]. Nonhuman primate models may be most translatable to human DA sensitivity [133].

Though DA has been detected in a range of marine fauna, from zooplankton to large whales [44], linking exposure to disease susceptibility is often challenging due to inaccessible open-ocean-dwelling species, unknown mechanisms of action, and subtle or nonspecific lesions. Marine mammals have enhanced susceptibility to DA-induced systemic impacts due to their anatomy and physiology, diet, and foraging characteristics. During diving, blood is shunted to the most vital, yet DA-sensitive organs (i.e., CNS and heart) and restricted to those responsible for elimination (i.e., GI tract and kidneys) [134]. Marine mammals are exposed to a larger DA dose per body weight due to the hydrophilic toxins' poor distribution to blubber [89]. Southern sea otters are uniquely susceptible because of their high daily dietary intake of primarily marine invertebrates, which biomagnify and slowly depurate DA [41]. Avians may be susceptible across taxa since seabirds and landbirds have demonstrated environmental and experimental DA toxicosis, respectively [135]. However, marine birds have an enhanced risk of environmental exposure. Marine megafauna with DA exposure and/or individual or population-level impacts are summarized (Table 1).

Table 1. Marine megafauna with suspected (S) or confirmed (C) environmental DA exposure, population-level impacts, or individual-level toxicosis due to exposure. Dashed line (-) represents no data available.

Species	Exposure	Population-Level	Individual-Level	Citation
Pinnipeds				
California sea lion (*Zalophus californianus*)	C	C	C	[40,93]

Table 1. *Cont.*

Species	Exposure	Population-Level	Individual-Level	Citation
Stellar sea lion (*Eumetopias jubatus*)	C	-	S	[9,136]
Guadalupe fur seal (*Arctocephalus townsendi*)	C	S	C	[137,138]
South American sea lion (*Otaria byronia*)	C	-	-	[10]
Peruvian fur seal (*Arctocephalus australis*)	C	-	-	[10]
Northern fur seal (*Callorhinus ursinus*)	C	C	C	[9,139]
Harbor seal – Pacific, Scottish (*Phoca vitulina*)	C, C	-, S	C, S	[9,95,140–142]
Ringed seal (*Phoca hispida*)	C	-	-	[9,143]
Bearded seal (*Erignathus barbatus*)	C	-	-	[9,143]
Spotted seal (*Phoca largha*)	C	-	-	[9,143]
Ribbon seal (*Histriophoca fasciata*)	C	-	-	[9,143]
Pacific walrus (*Odobenus rosmarus*)	C	-	-	[9]
Fissipeds				
Sea otter— Southern, Northern (*Enhydra lutra*)	C, C	C, -	C, -	[9,41,42]
Cetaceans				
Bowhead whale (*Balaena mysticetus*)	C	-	-	[9,144]
Right whale- Northern, Southern (*Eubalaena glacialis, australis*)	C, C	S, S	-, -	[46,47,50]
Blue whale (*Balaenoptera musculus*)	C	-	-	[9,44]
Pygmy sperm whale (*Kogia breviceps*)	C	-	-	[45]
Dwarf sperm whale (*Kogia sima*)	C	-	-	[45]
Long-beaked common dolphin (*Delphinus capensis*)	C	-	C	[11,43]
Short-beaked common dolphin (*Delphinus delphis*)	C	-	C	[11,43]

Table 1. *Cont.*

Species	Exposure	Population-Level	Individual-Level	Citation
Bottlenose dolphin (*Tursiops truncatus*)	C	-	C	[8,11,43,67,145–147]
Risso's dolphin (*Grampus griseus*)	C	-	-	[11,43]
Harbor porpoise (*Phocoena phocoena*)	C	-	C	[9,11,95]
Dall's porpoise (*Phocoenoides dalli*)	C	-	-	[11]
Minke whale (*Balaenoptera acutorostrata*)	C	-	C	[11,89]
Humpback whale (*Megaptera novaeangliae*)	C	-	-	[9,11,43,44]
Cuvier's beaked whale (*Ziphius cavirostris*)	C	-	-	[11,43]
Gray whale (*Eschrichtius robustus*)	C	-	S	[11,43]
Fin whale (*Balaenoptera physalus*)	C	-	-	[11]
Northern right whale dolphin (*Lissodelphis borealis*)	C	-	-	[11]
Pacific white sided dolphin (*Lagenorhyncus obliquidens*)	C	-	-	[11]
Beluga whale (*Delphinapterus leucas*)	C	-	-	[9,148]
Long-finned pilot whale (*Globicephala macrocephalus*)	S	S	S	[49]
Seabirds				
Brandt's cormorant (*Phalacrocorax penicillatus*)	C	S	C	[34,39]
Brown pelican (*Pelecanus occidentalis*)	C	S	C	[34,35,39]
Clark's grebe (*Aechmophorus clarkii*)	C	-	C	[39]
Pacific loon (*Gavia pacifica*)	C	-	C	[39]
Red-throated loon (*Gavia stellata*)	C	-	C	[39]
Surf scoter (*Melanitta perspicillata*)	C	-	S	[39]
Common murre (*Uria aalge*)	C	-	C	[39,149]
White-winged scoter (*Melanitta deglandi*)	C	-	S	[39]
Double-crested cormorant (*Phalacrocorax auratus*)	C	-	C	[39]

Table 1. *Cont.*

Species	Exposure	Population-Level	Individual-Level	Citation
Ring-billed gull (*Larus delawarensis*)	C	-	S	[39]
Cassin's auklet (*Ptychoramphus aleuticus*)	C	-	C	[39,149]
Northern fulmar (*Fulmarus glacialis*)	C	-	S	[39,149]
Sooty shearwater (*Puffinus griseus*)	S	-	S	[150]
Marbled murrelet (*Brachyramphus marmoratus*)	C	S	C	[151]
Marine Reptiles				
Green sea turtle (*Chelonia mydas*)	C	-	S	[52,54]
Leatherback sea turtle (*Dermochelys coriacea*)	C	-	S	[53]

Domoic acid toxicosis from environmental exposure has not been definitively diagnosed in finfish, elasmobranchs, or invertebrates. However, mortality concurrent with DA-producing blooms, as well as DA toxicosis after experimental administration, have been observed in these species [2,35,152–156]. Moreover, DA can accumulate in their neural and extra-neural tissues, which also possess GluRs [87,152–156]. These taxa may avoid DA toxicosis from environmental exposure due to molecular mechanisms [157] or rapid toxin depuration [88], or they may experience subclinical disease [152]. Experimental DA administration caused developmental anomalies and death in finfish [155,158], as well as genotoxicity, immunomodulation, oxidative stress, respiratory dysfunction, and developmental and neurobehavioral disease in marine invertebrates [121,126,154,159–172]. Along with the toxins' historical use as an anthelmintic, these studies underscore the potential for sensitivity and subsequent impacts on the fitness and survival of free-ranging marine finfish, elasmobranchs, and invertebrates.

5.2. Age

Age-related DA sensitivity and health effects vary depending on taxa and exposure scenarios. Age may influence DA resistance, as young, but not aged, murine developed a preconditioned tolerance in the CNS in situ [127]. Developing and aged individuals are generally the most sensitive. Susceptibility and health impacts due to environmental DA exposure should be evaluated over the lifespan [173,174], particularly in long-lived species.

The increased DA susceptibility in developing and aged individuals is primarily due to less effective physiological protections. Renal clearance and BBB integrity are reduced [106,175]. Toxin recirculation via the amniotic fluid can further amplify prenatal exposure dose and duration [93]. Windows of increased susceptibility and impacts of exposure are based on species-specific milestones, such as the developmental stage (e.g., neurogenesis, neuromigration, or synaptogenesis) during which exposure occurs [125,173,175,176]. Most species have an enhanced susceptibility during early gestation due to the intensity/stage of neurodevelopment. Perinatal CSLs are more vulnerable in certain regions due to the spatiotemporal overlap of blooms, the pupping season, and maternal foraging areas [175]. Decreased levels of neuroprotective hormones and antioxidants may contribute to the increased sensitivity in aged individuals [177–179]. Health impacts associated with an enhanced DA sensitivity in these age groups include (1) a higher incidence of seizures and hospitalization, greater death rates, and prolonged neurologic deficits in aged ASP patients [24]; (2) delayed and permanent neurobehavioral disease, socioemotional deficits, and cardiac and spinal cord defects, as well as altered concentrations of CNS neurotransmitters and their receptors after developmental exposure [125,132,158,174,180,181]; (3) an increased incidence of domoic acid-induced epileptic disease in younger CSLs [108]; and (4) decreased survival rates in CSLs exposed in utero [17,182].

Adults may experience enhanced DA susceptibility under some conditions. The spatiotemporal overlap of foraging sites with toxigenic blooms results in an increased risk to some mature cetaceans and pinnipeds [11,93,139]. For this reason, adult female CSLs and northern fur seals (NFSs, *Callorhinus ursinus*) in California are at the highest risk of developing acute toxicosis [108,139]. Physiology may also contribute to prime-age DA sensitivity, as evidenced in some model species [183,184], as well as by increased severity of brain damage and risk of cardiomyopathy in marine mammals [185,186].

5.3. Sex

Marine wildlife exhibit population-specific sex predilections where males or females are more at risk for acute toxicosis based on DA loads in preferred prey and foraging locations, as well as unknown factors. Regional sex predilections to acute DA toxicosis include mature male CSLs in Baja, Mexico [187], mature female CSLs in central California (predilection of 24:1) [108,188], female NFSs in California [139], and male common dolphins in California [11,43]. Sex-related feeding niche segregation, as well as a wider isotopic niche, may have contributed to a higher risk of male CSLs in Baja, Mexico [187]. Conversely, male CSLs in central California are fasting while breeding during their temporary occupancy in areas highly impacted by DA-producing blooms prior to migrating to less impacted regions [108]. The higher risk to female CSLs and NFSs in California is attributed to foraging near breeding rookeries with high DA loads [108,139]. Male common dolphins often forage in bachelor pods that may target highly impacted areas or prey items [11,43]. For example, DA load was associated with higher consumption of an efficient DA vector, the northern anchovy (*Engraulis mordax*), in male common dolphins compared to females [11]. Male brown pelicans (*Pelicanus occidentalis*) may also have an increased susceptibility to acute toxicosis, but the potential cause is unknown [34]. No sex predilection is apparent with domoic acid-induced epileptic disease or chronic toxicosis in wildlife [108].

Sex-based sensitivity varies in humans and laboratory models based on organ systems, exposure scenarios, and hormonal fluctuations during the lifespan [177,179]. Males tend to be more sensitive to DA-induced neurobehavioral impacts [189,190], while females may have an enhanced susceptibility to acute renal damage [191]. The most seriously ill ASP patients were male, possibly due to an increased exposure dose, more compromising underlying health conditions, or socio-emotional or biological influences [26]. Male rodents are most sensitive to long-term, low-level DA exposure [192], and there are sex-specific differences in socioemotional impacts in developing individuals [190,193–196].

5.4. Underlying Health Status

Co-morbidities, especially those affecting vital organs or tissues with high concentrations of iGluRs, may increase an individual's DA sensitivity. Pre-existing conditions, including insulin-dependent diabetes, chronic renal disease, hypertension, autoimmune disease, chronic liver dysfunction, pituitary adenoma, and Parkinson's disease (PD), were associated with an increased ASP severity [197,198], partly due to less effective toxin neutralization or clearance. Humans and marine wildlife commonly experience kidney or cardiac dysfunction [199–202], thus amplifying their vulnerability to DA exposure. Gastrointestinal lesions were hypothesized as a predisposing factor to DA toxicosis related to gut absorption kinetics [97]. Though GI ulcers are common in CSLs and other marine wildlife [201], there are no published studies to date to support this hypothesis. Metabolic, hormonal, cardiovascular, and immune changes during pregnancy may increase maternal DA sensitivity [93,203].

5.5. Exposure Scenario

Extrinsic factors influence individual or population-level DA susceptibility. An individual's exposure history, including DA dose, frequency, and duration, may influence sensitivity [128,129]. Previous DA exposure may lead to a decreased or enhanced neurologic sensitivity [127,128], or an adapted CNS response [129]. Concurrent exposure to other biotoxins or contaminants may increase vulnerability to DA [15]. Factors that may affect population-level DA susceptibility include toxin bioaccumulation kinetics in preferred prey, prey availability, and the spatiotemporal overlap between toxigenic blooms, vector prey, and the impacted species' home range [204,205]. For example, anchovies are abundant planktivores that are efficient DA vectors and dominate over sardines during certain climate cycles off coastal California [2]. Domoic acid exposure may have greater impacts on the stability of endangered (e.g., northern right whales, green sea turtles, and leatherback sea turtles), threatened (e.g., SSOs, Guadalupe fur seals (*Arctocephalus Townsendi*), ringed seals (*Phoca hispida*), bearded seals (*Erignathus barbatus*)), and keystone (e.g., SSOs) populations, as well as those in decline (e.g., CSLs in the Gulf of California and Scottish harbor seals (*Phoca vitulina*)) [44,45,191–193].

6. Clinical Signs and Symptoms

Due to the broad tissue distribution of GluRs, DA may impact numerous organ systems. Clinical signs are commensurate with the organ affected, range from subtle to conspicuous and non-specific to specific, are dependent on exposure dose and time lapse since exposure, and may go unobserved in marine wildlife due to a delayed presentation to health managers.

6.1. Gastrointestinal

Domoic acid-induced GI symptoms in marine wildlife and humans include abdominal cramps, hypersalivation, gagging, retching, lip licking/smacking, regurgitation, nausea, vomiting, diarrhea, inappetence, gastric bleeding, and fecal impaction [24,34,41]. The most common symptoms in ASP patients were GI-related [24]. Domoic acid may induce vomiting and regurgitation acutely in seabirds [34], fecal impaction, inappetence, and nausea acutely to chronically in SSOs [41], and vomiting and inappetence chronically in CSLs [106].

6.2. Central Nervous System

The most common DA-induced CNS signs across species and exposure scenarios are seizures and/or cognitive deficits. Hallmark CNS signs include long-term amnesia in ASP patients, as well as hindlimb scratching and atypical aggression in wildlife [24,41,108]. Most CNS signs are from limbic system injury, but damage to other areas can also manifest in symptoms [204].

In ASP patients, CNS signs primarily occurred within 48 h and included mutism, purposeless chewing and grimacing, emotional lability, severe headache, visual disturbance, dizziness, confusion, anterograde and/or retrograde amnesia, hallucinations, cranial nerve deficits, lack of response to painful stimuli, seizures, posturing, myoclonus, generalized weakness, unsteadiness, and coma [24,197]. While these were typically reversible, permanent anterograde amnesia occurred in some patients, and another developed TLE one year after recovery [24,30]. Long-term, low-level DA ingestion caused transient or permanent memory loss depending on the exposure dose in some members of Indigenous communities in the Pacific Northwest, USA [18,31,32,206,207].

Only acute CNS signs have been reported in seabirds, and these include abnormal posturing, inability to take flight or retract legs during flight, fine motor tremors, head weaving, scratching, disorientation, loss of awareness or righting reflex, ataxia, toe clenching, paddling, weakness, paralysis, and anomalous behaviors (e.g., unusually agitated or docile and asocial behavior) [34,37,135]. California sea lions with acute toxicosis often strand in clusters, while those with the chronic disease following toxicosis are more likely to strand individually and in atypical locations [17]. Central nervous system symptoms most commonly associated with acute toxicosis in CSLs include ataxia, head weaving, and scratching, seizures, and coma [40], whereas those more commonly associated with chronic disease include spontaneous, intermittent, and progressive seizures, muscle twitching, periodic lethargy, anomalous behaviors (e.g., conspecific or human-directed aggression, frantic pacing, rocking, circular swimming, backward circular walking, and chewing), central blindness, blepharospasm, spatial memory deficits, delayed habituation, enhanced dishabituation, and foraging difficulty in the wild due to behavioral inflexibility [17,108,132]. Similar neurobehavioral signs are observed in other pinnipeds and fissipeds with DA toxicosis [137,139,140], but these diminish in favor of cardiac signs in SSOs with chronic disease [41]. Other CNS signs in SSOs include hyphema, trouble with ambulation or food prehension acutely, paresthesia, hindlimb paraparesis, fecal impaction, and urine retention chronically [41].

Laboratory models of DA toxicosis can demonstrate subtle or long-term CNS impacts that are difficult to ascertain after environmental exposure. Examples include reversible memory and learning deficits after a wash-out period from long-term exposure [181], a reduced seizure threshold in adulthood after low-dose, perinatal exposure [208], and intention tremors from long-term, low-level exposure during pregnancy [125]. These findings may have implications for the well being of environmentally exposed species.

6.3. Cardiovascular/Respiratory

Domoic acid-induced cardiorespiratory signs include (1) heart palpations and dyspnea in ASP patients and (2) dyspnea, cyanosis, oronasal froth, heart murmur, and pulse deficits in SSOs and CSLs [24,26,41,104,209]. Cardiac signs typically predominate in SSOs with subacute to chronic DA toxicosis.

6.4. Urogenital

The most prominent DA-induced urogenital symptom is reproductive failure in CSLs and SSOs with acute toxicosis, manifesting as abortion, stillbirth, premature parturition, or death of pregnant dams [41,93]. Male SSOs with subacute to chronic toxicosis may have testicular atrophy [41], which could impact reproductive success through reduced sperm production and quality, as well as lower testosterone levels.

6.5. Integumentary/Musculoskeletal

While marine mammals with acute DA toxicosis typically demonstrate normal body condition, those with chronic toxicosis may be emaciated due to reduced foraging success [40,41,47,89]. Southern sea otters with acute toxicosis usually have healthy pelages (except for peri-oronasal staining), whereas those with subacute to chronic disease may have poorly groomed, dry, and/or traumatized (fight- or self-induced) pelages [41]. Piloerection was observed in some ASP patients [197].

6.6. Other/Indirect

Clinical signs may be indirectly triggered by DA blooms. For example, the "Warm Water Anomaly" caused marine mammal habitat and prey shifts and fisheries closures, inadvertently causing an unprecedented number of large whale entanglements [210]. Similarly, DA-producing blooms were associated with injury and death from fishing gear entanglement and vessel strikes in large whales [51].

7. Diagnostic Tests

This section will review pre- and post-mortem diagnostic tests for confirming DA exposure and related disease. Few tests have been assessed for specificity or sensitivity or are routinely utilized after environmental exposure [209]. Appropriate test selection is determined by available sample(s), resources, species, and exposure scenario. No single test confirms DA toxicosis; thus, methods must be used in tandem. A combination of environmental and patient data is often required, along with a "weight of evidence" approach [211]. Definitive confirmation is complicated by the rapid clearance rate of the toxin in many species, delayed manifestations of disease, subtle or non-specific clinical signs and pathologies, inadequate resources for sample collection or testing, and unknown clinical or exposure histories.

7.1. Diagnosis of DA Exposure

To diagnose acute DA exposure, the toxin and/or toxigenic diatom should be detected in water samples in spatiotemporal proximity to a marine wildlife stranding event, as well as within vector prey and the impacted individual [204]. This may be infeasible in cases of subacute to chronic toxicosis with delayed disease manifestations [211].

7.1.1. *Pseudo-nitzschia* spp. Detection

Seawater or prey samples in known habitats of intoxicated individuals, as well as GI contents, should be assessed for *Pseudo-nitzschia* spp. frustules. A false negative result may occur if the diatom or vector prey are no longer present in the environment or GI tract at the time of clinical presentation, with cryptic blooms or limited monitoring capabilities. The presence of environmental diatoms does not guarantee exposure, and diatoms may not be toxigenic [140]. Light microscopy is typically insufficient for species identification. Scanning electron microscopy, transition electron microscopy, or molecular techniques should be used for taxonomic delineation [1,204]. Molecular techniques can distinguish between diatom species that are morphologically comparable and may also be useful for species quantification [1,212].

7.1.2. Domoic Acid Antigen Detection

Detection of DA is vital, when possible, to help confirm disease caused by acute exposure but can also support the diagnosis of subacute to chronic toxicosis if data is available regarding previous DA exposure in an individual. Matrices for DA antigen detection include diatom, seawater, or prey samples and/or body fluids and tissues (i.e., GI/cloacal contents, feces, urine, serum, bile, breast milk, kidney, liver, brain, aqueous humor, pericardial fluid, CSF, amniotic fluid, fetal meconium, and allantoic fluid) of impacted individuals [25,39,96,139]. Serum, feces, and urine are usually the most readily available and accurate pre-mortem matrices [123,213]. Measured DA antigen levels may not accurately represent the exposure dose due to toxin depuration and unknown baseline body burdens [128,214,215]. Enzyme immunoassay (ELISA) and/or liquid chromatography-tandem mass spectrometry (LC-MS/MS) are commonly used alone or jointly [123,216]. While ELISA is rapid, sensitive, and easily accessible, some yield an indirect measurement that can be skewed by isomers, DA antibodies, or matrix interference [213,214]. Alternatively, LC-MS/MS provides a direct measurement that is highly sensitive and specific but requires a substantially greater time and resource investment [213]. These tests may require toxin extraction to help reduce matrix effects despite an increased DA recovery variability. An appropriate internal standard can mitigate test limitations [213].

7.1.3. Domoic Acid Antibody Detection

Serum DA antibodies have been detected in CSLs with acute toxicosis immediately and up to several weeks later via ELISA, as well as in humans with previous or long-term exposure via ELISA and surface plasmon resonance (SPR) biosensor [123,128]. Though SPR had high sensitivity and specificity [123], the development of clinical DA antibody tests was halted due to logistical constraints (K. Lefebvre, personal communication, October 2021).

7.1.4. Vector Prey Detection

Vector prey are specific to the region and species impacted. Their identification and quantification (e.g., via otolith examination) within the GI tract of impacted individuals can explain the exposure pathway and aid exposure dose calculations [44,89,126,152].

7.2. *Diagnosis of DA Toxicosis*

There are many potential tests for diagnosing DA toxicosis, including generalized, organ-specific, and emerging methods. Findings vary based on the exposure scenario and species. Unlike the detection of the DA antigen, very few current diagnostic tests are specific for DA toxicosis.

7.2.1. Blood Parameters

Serial complete blood count (CBC) and serum biochemistry tests can evaluate for DA-induced systemic impacts, assess longitudinal health trends, and detect delayed effects. Anomalies after environmental exposure can be inconsistent and are nonspecific but may indicate immunomodulation, dehydration, and kidney or muscle damage (Table 2). Examples of DA-induced immunomodulation are leukocytosis in ASP patients and dolphins with subclinical exposure [26,67,145], lymphocytopenia and monocytosis in seals with subclinical exposure [141], and eosinophilia in CSLs and dolphins with acute or chronic toxicosis [67,145,217]. Elevated levels of creatine phosphokinase (CPK)/creatine kinase (CK) are observed in several species with acute toxicosis due to seizure-induced muscle damage [26,34,40]. Abnormal serum creatinine (sCr), blood urea nitrogen (BUN), uric acid (UA), and/or hematocrit (HCT) may be due to dehydration or kidney damage with altered water reabsorption and red blood cell (RBC) production [26,34,40,137,197,209]. Immune suppression or modulation, as well as altered hemolymph biochemical parameters (i.e., pH, bicarbonate, alkaline phosphatase (ALP), lactate dehydrogenase (LDH), and total protein (TP)), have been observed in DA-exposed marine invertebrates [159,160,162,170].

Table 2. Abnormal (↑ or ↓) routine blood work results reported after environmental DA exposure (BND = Bottlenose dolphin; CSL = California sea lion; GFS = Guadalupe fur seal; HS = Harbor seal; SB = Seabird; H = Human).

Blood Variable	Anomaly	Species	Citation
Complete Blood Count (CBC)			
Red Blood Cell (RBC)	↑	CSL	[218]
White Blood Cell (WBC)	↑	H, BND, CSL, GFS	[26,67,137,145,197,218]
Hemoglobin (HGB)	↑	CSL	[218]
Mean Corpuscular Volume (MCV)	↓	CSL	[218]
Platelet (PLT)	↑	CSL	[218]
Neutrophils	↑	CSL	[141,218]
Lymphocytes	↓	HS	[141]
Monocytes	↑	HS	[141]
Eosinophils	↑	CSL, BND	[67,145,209,217]
Hematocrit (HCT)	↑	CSL, GFS	[40,137,209]
Serum Biochemistry			
Serum Creatinine (sCr)	↑ or ↓	H, GFS, CSL	[26,137,197,218]
Blood Urea Nitrogen (BUN)	↑ or ↓	H, SB, CSL, GFS, CSL	[26,34,40,137,197,218]
Uric Acid (UA)	↑	SB	[34]
Creatine Kinase (CK)/ Creatine Phosphokinase (CPK)	↑	H, SB, CSL	[26,34,40,197]
Gamma-glutamyl transpeptidase (GGT)	↑	CSL	[218]
Alanine transaminase (ALT)	↑	CSL	[218]
Cholesterol (CHOL)	↓	CSL	[218]
Glucose (GLU)	↓	CSL	[218]
Total Bilirubin (TBili)	↑	CSL	[218]
Phosphorous (PHOS)	↓	CSL	[218]
Total Iron (TIRON)	↓	CSL	[218]
Calium (Ca^{2+})	↓	CSL	[218]
Sodium (Na^{+})	↓	CSL	[218]
Albumin (ALB)	↓	CSL	[218]
Total Protein (TP)	↑	GFS	[137]

7.2.2. Hormone Testing

Cortisol and thyroid hormone levels may be altered by DA exposure. Serum cortisol levels may be low in CSLs with acute or chronic toxicosis [217]; fecal glucocorticoid metabolites (fGCm) negatively correlate with exposure in southern right whales [219], and fecal cortisol positively associates with exposure in northern right whales [47]. Though serum thyroid hormone (i.e., T3, T4) and thyroid-stimulating hormone (TSH) alterations were observed in models of DA toxicosis [220,221], as well as the decreased transcription of a thyroid hormone receptor in northern sea otters (*Enhydra lutris kenyoni*) with suspect repetitive, low-level DA exposure [222], thyroid hormone impacts from environmental exposure require investigation. Adrenal and thyroid gland ultrasonography can evaluate for DA-induced disease in the tissues responsible for hormone production and/or release.

7.2.3. Kidney Injury Biomarkers

Serum kidney function markers (e.g., BUN and sCr) may have limited use in diagnosing DA-induced acute kidney injury [119] but may aid in monitoring the development and progression of dysfunction over time. Elevated levels of urinary markers of acute tubular injury, neutrophil gelatinase-associated lipocalin (uNGAL) and kidney injury molecule-1 (uKIM-1), occurred 72 h after DA exposure in murine models [119], but their use, along with renal ultrasonography, after environmental exposure requires investigation.

7.2.4. Cardiovascular Injury Biomarkers

Tests for DA-induced cardiac injury include serum biomarkers of heart damage, chest imaging, electrocardiogram (ECG), and blood pressure measurement. Modalities may require expertise for sedation/general anesthesia and/or interpretation in wildlife, which could impact results.

Serum troponin-1 (cTn1) and N-terminal pro-brain natriuretic peptide (NT-proBNP) were validated as cardiac injury biomarkers in several marine mammal species [223]. Though CSLs with DA toxicosis may have considerable cTnI elevations, single measurements are not predictive of cardiomyopathy and cannot differentiate acute vs. chronic disease, possibly due to cTn1's short plasma half-life [104]. Serial cTn1 levels can help diagnose and monitor DA-induced cardiac disease in SSOs [224] and may be useful in other species. While skeletal and cardiac muscle L-carnitine are not predictive of DA-induced cardiac disease in CSLs, skeletal levels may help differentiate acute vs. chronic disease [104].

Imaging of the chest cavity may allow diagnosis and monitoring of DA-induced cardiomyopathy and subsequent congestive heart failure. Echocardiographic anomalies in CSLs with DA toxicosis include reductions in cardiac output, fractional shortening, and ejection fraction, as well as valvular insufficiency and abnormal septal wall movement [209]. Transthoracic echocardiography can be challenging in sea otters due to their dense and vital pelage [225]. Serial chest radiographs can help assess DA-induced cardiac disease in SSOs and may be valuable in other species [186].

Domoic acid-induced ECG anomalies include a decreased P wave amplitude, atrial and ventricular ectopy, first or second-degree atrioventricular (AV) block, and a sinoventricular rhythm in SSOs with subacute to chronic toxicosis [224], as well as 1st or 3rd degree AV block and premature atrial and ventricular complexes in CSLs [209]. Tracings do not correlate with disease in CSLs and, thus, may not accurately detect cardiac injury [201]. Continuous ECG and blood pressure measurements are recommended in ASP patients to monitor for arrhythmias and hypotension [24,198].

7.2.5. Neurobehavioral Injury Markers

Electrography, advanced imaging, and neurobehavioral tests can support the diagnosis of DA-induced CNS disease, as well as assess the type, localization, and severity of damage.

Electroencephalograms (EEGs) and electromyograms (EMGs) can evaluate CNS and muscle electrical activity, respectively. Nonspecific EEG anomalies in ASP patients included generalized slowing of background activity and epileptogenic activity in the frontotemporal region in the acute phase, as well as generalized disturbance of background activity in the temporal lobes and periodic epileptiform discharges in the chronic phase [24,25,30]. In CSLs, EEG findings include epileptiform discharges characterized by spikes, sharp waves, slow waves, and/or spike waves of varying severity and localization [226], as well as intermittent rhythmic delta activity often followed by background attenuation localized to the posterior cerebral hemispheres with chronic toxicosis [17,108]. However, these findings cannot differentiate between acute and chronic cases and may be observed with other etiologies [17,108,226]. In ASP patients, EMG anomalies a few months after exposure included spontaneous activity, neurogenic recruitment, and unstable motor unit potentials [25,26]. Anomalous EEGs and EMGs in models of DA toxicosis may have relevance to environmental exposure [227,228].

Magnetic resonance imaging (MRI) and positron emission tomography (PET) scans can help diagnose DA-induced CNS disease. Anomalies are primarily localized to the limbic system, specifically the hippocampus, amygdala, and parahippocampal gyrus. Neuron loss and decreased glucose metabolism were observed in ASP cases [197,229] vs. bilateral atrophy and glucose metabolic deficits in the TLE patient [30]. Anomalies in CSLs with DA toxicosis can be unilateral to bilateral and progressive and include structural loss/atrophy, mossy fiber sprouting, altered connectivity, pathologic T2 hyperintensity, gliosis, decreased glucose metabolism, neuropil neocapillarization, and temporal horn enlargement [17,106,132,185,230–233]. Regional encephalitis and swelling that progressed to atrophy were observed in a Guadalupe fur seal with chronic toxicosis [137]. Imaging anomalies may be absent in the acute phase even with severe symptoms and tissue damage [101]. Long-term DA administration in nonhuman primates caused decreased white matter integrity in hippocampal motor tracts and increased brain lactate [133], but findings may not extrapolate to environmental exposure.

Neurobehavioral tests may aid in diagnosing DA toxicosis and related deficits. Patients with ASP and individuals who developed TLE exhibited memory deficits, whereas language function, concept formation, and verbal comprehension were normal [30,197,234]. Decreased cognition and memory in CoASTAL cohort participants was likely due to repetitive, low-dose DA exposure [18,32,206,207]. Tests in CSLs with DA toxicosis include (1) behavioral ethograms, in which head weaving, muscle fasciculations, hind flipper dragging, and swift scanning are diagnostic for acute or chronic disease [124], (2) auditory responsive behavioral tests, in which delayed habituation can help diagnose acute or chronic disease [132,235], (3) spatial memory assays, in which performance may positively correlate with the dorsal right hippocampal volume on MRI [230], (4) "clinical assessment scores" and "neuroscores/(NS)", which negatively correlate with prognosis [40,209], and (5) olfactory function testing, which may identify impairment in chronic disease [236]. Studies in models of acute, long-term, or developmental DA exposure demonstrated impairment of learning, memory, cognition, passive avoidance, startle response, motor coordination, conditioned avoidance response, circadian activity, gait, habituation, and exploratory behavior [125,176,181,190,237–241], which may extrapolate to environmental exposure.

7.2.6. Reproductive Failure Confirmation

Abdominal ultrasound and/or radiographs can assess for pregnancy, fetus viability, and DA-induced reproductive disease (i.e., uterine torsion/rupture or testicular atrophy) [209].

7.2.7. Gastrointestinal Lesion Identification

Abdominal ultrasound, endoscopy, and/or radiographs can evaluate for DA-induced GI ulceration, torsion, and intussusception.

7.3. *Gross and Microscopic Histopathology*

Gross and microscopic histopathology are performed post-mortem due to the invasiveness of pre-mortem collection of the primary organs impacted. Microscopic histopathology is often the gold standard for diagnosing DA toxicosis. Gross pathology can be distinctive but is frequently absent. Limited pathology may be present even with substantial clinical signs, particularly in seabirds and cetaceans [34,146]. Most histopathologic scrutiny focuses on the CNS, but peripheral organs can also suffer damage. Specific immunostains (e.g., neurodegeneration biomarkers) can support histopathologic findings [242].

7.3.1. Gastrointestinal Tract

On gross exam, GI tracts may be empty due to nausea, inappetence, or vomiting, and erosions, ulcers, or bleeding may be present [24,41,101]. Findings in SSOs include full GI tracts acutely, atrophic empty tracts chronically, GI torsion, ileus, or intussusception acutely to subacutely, and congestion [41].

7.3.2. Central Nervous System

There are minor differences in the type, distribution, and severity of DA-induced CNS lesions based on species, exposure scenario, and time lapse since exposure. Damage to limbic structures is most common and severe, particularly the hippocampal pyramidal neurons in CA1, CA3, and CA4 sectors and the dentate gyrus granule cells. Other impacted regions may include the amygdala, thalamus, hypothalamus, subfrontal cortex, medulla oblongata, pons, cerebellum, piriform and entorhinal cortices, olfactory bulbs and tracts, septum, nucleus accumbens, arcuate nucleus, circumventricular organs, ependyma, cingulate gyrus, occipital lobe, choroid plexus, ventricles, pituitary, retina, and spinal cord [25,30,41,108,137,222]. Lesion symmetry is typically bilateral in ASP cases and unilateral or bilateral in pinnipeds and fissipeds [17,41,243].

The most prominent CNS lesions in ASP cases were neuronal necrosis and sclerosis in the amygdala, hippocampal pyramidal neurons, and dentate gyrus [25], which progressed to hippocampal atrophy, ventricular dilation, and neuronal loss and gliosis of all CA sectors, the amygdala, and secondary olfactory areas in the TLE patient [30]. While CSLs with acute toxicosis lack gross CNS lesions, parahippocampal and hippocampal atrophy and lateral ventricle dilation may be observed in chronic cases [101]. Microscopic lesions include (1) per-acute microvesicular hydropic degeneration in the neuropil of limbic structures and laminar vacuolization in CA3 pyramidal cells, (2) acute ischemic neural necrosis in the hippocampal formation, and (3) chronic neuronal loss, atrophy, gliosis, non-suppurative inflammation, laminar disorganization, sclerosis, and mossy fiber sprouting predominately in the hippocampus and parahippocampal gyrus [17,101,108,244]. Focal meningeal hemorrhage and brain edema in parahippocampal areas are also observed in immature CSLs with chronic toxicosis [17]. Brain edema may be present in pinniped pups exposed in utero [140,182]. Compared to humans, CSLs exhibit more frequent dentate gyrus and asymmetric injury, as well as less damage in sector CA1 [17]. Other pinnipeds have similar CNS lesions as CSLs, though time-dependent patterns are not established. Differences include less common perivascular cuffing with lymphocytes in NFSs [139] and CA2 involvement in Pacific harbor seals [139,140], as well as prominent cerebellar folia lesions in a Guadalupe fur seal [137].

Southern sea otters with acute toxicosis primarily demonstrate gross congestion and hemorrhage of the meninges, neuropil, and ventricles, as well as pink discoloration of the neuropil, whereas hippocampal atrophy, ventricular enlargement, and pale tan neuropil predominate in chronic cases [41]. Instead of the nerve and glial cell necrosis and degeneration observed acutely in other species, SSOs demonstrate congestion and microhemorrhages in the brain, spinal cord, and meninges. Pituitary gland lesions are a distinguishing feature of acute disease in SSOs. Those with chronic toxicosis typically demonstrate neuronal and glial cell loss with necrosis, apoptosis, gliosis, spongiosis, scarring, and "moth-eaten" cells, and SSOs more frequently experience severe damage in the CA2 sector and spinal cord involvement than other species [41].

Seabirds are the only other wildlife with documented DA-induced CNS damage, which includes acute neuronal necrosis with capillary endothelial cell hyperplasia [34].

7.3.3. Ocular

Domoic acid-induced retinal damage is in the ganglion cell layer in CSLs vs. the inner nuclear and outer plexiform layer in model species [26,101,102,245]. Fibrinous ophthalmitis may be present in CSLs with acute disease [101]. Congestion and microhemorrhages in the choroid and ciliary body, as well as hyphema, may aid the diagnosis of acute toxicosis in SSOs [41].

7.3.4. Cardiovascular/Respiratory

The following lesions are observed in SSOs with DA-induced degenerative cardiomyopathy: (1) gross pulmonary hypersecretion and evidence of congestive heart failure, such as discolored myocardium, hemorrhage, venodilation, congestion, venous shunts, cardiomegaly, and pericardial, pleural, peritoneal, or pulmonary edema and fibrosis, and (2) microscopic acute myocardial and vascular congestion, microhemorrhages, necrosis, apoptosis, edema, and vacuolation [41]. Subacute to chronic lesions are more severe, advanced, likely to involve non-suppurative inflammation, and predominated by cardiomyocyte loss, fatty replacement, fibrosis, and coronary arteriosclerosis [41]. Pinnipeds with DA toxicosis may have similar cardiac lesions as SSOs, as well as bronchopneumonia [104,139,140], but lesion progression is not described. Domoic acid-induced cardiac lesions are presumptive in cetaceans [246], have not been described in ASP patients, and are predominated by focal myofiber necrosis in seabirds [34].

7.3.5. Urogenital

Urinary tract anomalies from environmental DA exposure include kidney edema/congestion in SSOs and CSLs with acute to chronic disease, renal tubular regeneration in CSLs with acute disease, and bladder distention in SSOs with acute to chronic disease [40,41,187]. Proximal tubular necrosis with vascular and tubular damage, cytoplasmic eosinophilia, cell desquamation, mitochondrial swelling, and vacuolization were observed in the kidneys of rodent models of acute toxicosis [119].

Gross genital tract lesions in pinnipeds and SSOs include abortion, fetal resorption, stillborn or preterm fetus, congestion/edema of the uterus or placenta, uterine torsion, prolapse, or rupture, and signs of forced copulation outside of estrus [41,93]. Microscopic lesions include hyper-eosinophilia, edema, hemorrhage, apoptosis, or necrosis within or around the myometrium, myometrial vasculature, and placenta, as well as testicular atrophy in SSOs with subacute to chronic disease. Fetal pathology resembling adult lesions can lead to abnormal growth, development, and survival [164,165].

7.3.6. Integumentary/Musculoskeletal

Gross anomalies include focal hemorrhages in the forelimb and hindlimb musculature of seabirds with acute toxicosis [34], as well as a dry unhealthy pelage, decreased fat/blubber stores, cachexia, and skeletal muscle or serous atrophy in marine mammals with chronic toxicosis [41,47,101]. Seabirds may demonstrate focal to multifocal skeletal muscle necrosis microscopically [34].

7.3.7. Other

Damage to other organs occurs with less consistency or severity, such as vascular congestion in the adrenal glands and progressive congestion, hemosiderosis, and necrosis in the liver [41]. It is unclear whether these lesions correlate with clinical signs.

7.4. Molecular Diagnostics

Molecular diagnostics can elucidate (1) mechanisms of DA-induced tissue injury [119,247], (2) biomarkers of exposure [248–251], and (3) differential diagnoses (e.g., acute vs. chronic DA toxicosis vs. other) [248,252]. These tests require specialized equipment and expertise and, thus, may not be feasible in the clinical setting. Some methods only permit analysis of a limited number of proteins or genes, though others allow global evaluation. While many studies used molecular diagnostics in models of DA toxicosis [121,128,129,158,169,170,172,241,253–255], Table 3 summarizes findings after environmental exposure.

Table 3. Summary of findings from studies utilizing molecular diagnostics after environmental DA exposure in California sea lions (CSL) and northern sea otters (NSO); CSF = cerebral spinal fluid; GluR = glutamate receptor.

Method	Species	Matrix	Findings	Citation
Gene microarray	CSL	Whole blood	Discriminated CSLs with DA toxicosis vs. other illnesses; Identified upregulation of inflammatory mediators and TNFAIP6 as a candidate biomarker of DA toxicosis	[248]
Peptidomics (MALDI-TOF with artificial neural networks)	CSL	Serum	This method can be a sensitive or specific diagnostic test for acute DA toxicosis	[252]
Proteomics (2D-GE/MS)	CSL	Plasma	Apolipoprotein E is a sensitive, but not specific biomarker of chronic DA toxicosis	[250]
Shotgun proteomics (LC-MS/MS)	CSL	CSF	Identified candidate biomarkers of DA toxicosis, as well as molecular mechanisms for neurodegeneration	[247]
Proteomics (MS)	CSL	CSF	Identified candidate biomarkers to diagnose and differentiate acute vs. chronic DA toxicosis	[256]
Polymerase Chain Reaction (PCR)	NSO	Whole blood	Evidence of persistent, low-level DA exposure to Kachemak Bay, Alaska sea otters; Noted differences in neurologic, cardiac, immune, and detoxification function gene expression in DA exposed vs. reference population	[222]
Immuno-histochemistry (IHC) + PCR	CSL	Heart	Pathway of DA-induced cardiac damage may involve direct activation of local GluRs and apoptosis	[100,104]

Table 3. *Cont.*

Method	Species	Matrix	Findings	Citation
Immuno-fluorescence (IF)	CSL	Brain	Fluoro-jade staining did not identify ischemic neuronal degeneration per-acutely before standard HE staining	[101]
Immuno-fluorescence (IF)	CSL	Hippocampus	Correlated increased oxidative stress and glial activation with disease severity and glial activation and nitric oxide with the development of chronic toxicosis; Gliosis and alterations in glutamine synthetase may be part of mechanism for DA-induced seizures	[113]
Immuno-histochemistry (IHC)	CSL	Hippocampus	Oxidative stress is involved in acute and chronic DA toxicosis, whereas glutamine synthetase redistribution is only involved in chronic toxicosis	[117]
Immuno-cytochemistry (ICC)	CSL	Hippocampus	Supported similarity between human temporal lobe epilepsy (TLE) and chronic toxicosis	[244]

8. Time Course of DA Toxicosis

A range of time-dependent processes are possible after DA exposure, including per-acute, acute, subacute, chronic, and acute-on-chronic disease [209]. These can be difficult to differentiate due to a disease spectrum based on DA exposure dose and history. Criteria for delineating these processes are incomplete and only available for SSOs and CSLs. In SSOs, the spectrum of disease from acute to subacute to chronic is characterized through clinical, biochemical, gross, and microscopic pathological findings [41]. Similar criteria (Table 4) may be used to differentiate acute and chronic toxicosis in CSLs [17,108,209]. Per-acute and acute-on-chronic descriptions are not available for any species.

Table 4. Summary of criteria used to differentiate acute vs. chronic DA toxicosis in CSLs. (Condensed from [17,108,209]). BCS = body condition score.

	Acute DA Toxicosis	Chronic DA Toxicosis
Brain histopathology	None-hippocampal necrosis +/− involvement of other limbic system regions	Hippocampal atrophy +/− gliosis +/− involvement of other limbic system regions
Clinical signs	Ataxia, head weaving, seizures, tremors, coma, decreased responsiveness to stimuli, scratching behavior +/− good BCS	Intermittent seizures, episodic lethargy and inappetence, vomiting, central blindness, abnormal behaviors (stereotypic scratching, conspecific or human-directed aggression) +/− poor BCS
Case history	Strand in clusters (≥5 individuals within 48 h and 80 km) concurrent with a DA-producing bloom	Strand individually, possibly in an atypical location without a concurrent DA-producing bloom +/− previous treatment for acute toxicosis
DA levels	Not detectable-detectable	Not detectable

9. Medical Treatment for DA Toxicosis

Treatment of individuals with acute DA toxicosis is focused on symptomatic and palliative care to manage seizures, maintain hydration and electrolyte balance, minimize patient discomfort, reduce inflammation, and prevent tissue damage. There are few therapies for toxin neutralization or elimination; thus, treatment is primarily supportive. The goal of treatment is to release recovered wildlife and discharge human ASP patients with a favorable prognosis by preventing long-term sequelae if possible. Wildlife with chronic toxicosis are not considered candidates for release. Except for CSLs and NFSs, treatment recommendations for marine wildlife with acute DA toxicosis are anecdotal or lacking. Successful treatment may be limited by DA's potency, delayed presentation of individuals for care, and the challenge of antemortem diagnosis.

Seizure management in humans and pinnipeds with DA toxicosis involves a combination of anti-epileptic drugs (AEDs), typically phenobarbital and benzodiazepines [40,139,198]. Intoxicated marine mammals are usually anorexic; thus, parenteral treatment is commonly needed. The current recommendation for pinnipeds with acute DA toxicosis is a loading dose of phenobarbital (4 mg/kg BID IM for 2 d), followed by a maintenance dose (2 mg/kg BID IM or PO for 5 d) with the addition of midazolam, lorazepam, and/or diazepam (0.2–0.6 mg/kg PRN IM) for breakthrough seizures [209]. Lorazepam is more effective than diazepam for breakthrough seizures in CSLs [40], but a low dose of diazepam may assist with behavioral modification [257]. Anti-epileptic drug therapy is typically more successful in CSLs with acute toxicosis, as those with chronic disease often progress to a refractory state [17]. This was the case in a subadult CSL with suspected domoic acid-induced epileptic disease treated with phenobarbital (~1.5 mg/kg SID PO or IM) and lorazepam (0.1–0.2 mg/kg PRN IM) in managed care [257]. However, a Guadalupe fur seal with suspected chronic toxicosis was managed for two decades with phenobarbital alone (0.57–1.5 mg/kg BID PO or 0.57–0.8 mg/kg PRN IM) [137]. Dose adjustments are based on seasonal weight fluctuations, clinical status (i.e., abnormal behavior, reduced appetite, or breakthrough seizures), and trough phenobarbital levels [137,257]. Serum phenobarbital should be monitored at least annually, as well as when initiating or adjusting therapy or potential side effects are observed. A target serum level of 20–30 μg/mL is recommended since levels below 18 μg/mL often yield inadequate seizure control, while ataxia and sedation have been observed above 35 μg/mL [137,257]. Levetiracetam has also been used successfully for long-term epilepsy management in a few pinnipeds (C. Field, personal communication, December 2024). A CSL in managed care with progressive domoic acid-induced epileptic disease surgically transplanted with GABAergic interneuron porcine progenitor cells intrahippocampally demonstrated improved appetite, body weight, behavior, and seizure control [257,258]. Though xenotransplantation may be an effective therapy in pinnipeds with chronic DA toxicosis refractory to AEDs, the substantial expertise and resources required greatly limit its use, and xenotransplantation from domestic species may be inappropriate for wildlife intended for release. There are no AED dose recommendations for DA toxicosis treatment in other species, but phenobarbital and benzodiazepines may be appropriate [259,260]. Seizure control is critical for aquatic mammals that would be at risk of drowning; however, over-sedation can also occur, thus close monitoring and/or dry docking may be required for marine wildlife with unmanaged seizures and when medication is adjusted.

Individuals with DA toxicosis should receive palliative therapies. The purpose of fluid supplementation is to help maintain hydration and electrolyte balance, as well as support DA elimination from the body. Colloid fluids can be administered intravenously in hospitalized ASP patients and/or subcutaneously in rehabilitating or managing pinnipeds or fissipeds [40,137,198]. The ideal fluid dose is dependent on hydration and electrolyte status, but 20–25 mL/kg/day may be appropriate for anorexic pinnipeds and fissipeds [40,137,209]. Further correction of electrolyte imbalances, assisted feeding, and vitamin supplementation may be necessary [198,211]. Prokinetics (e.g., metoclopramide) can be administered to ASP patients known to have DA persisting in the GI tract to increase the speed of evacuation [198]. Endotracheal intubation may be required in patients with excessive bronchial secretions [197]. Antiemetics (e.g., maropitant) and antacids (e.g., famotidine) may be administered to treat GI upset [137]. Analgesics may be indicated for ASP patients with abdominal pain or headaches [198]. Induction of abortion is recommended in pregnant CSLs with acute DA toxicosis as the prognosis for the fetus is poor and the prognosis for the dam will improve [93,209]. Administration of dexamethasone sodium phosphate (40 mg or 0.25 mg/kg SID IM for 3 d) is usually an effective abortifacient in CSLs and should be followed by prostaglandin F2alpha (500 μg) or oxytocin if needed and possibly systemic antibiotics (e.g., ceftiofur crystalline free acid) to prevent pyometra.

Anti-inflammatory, antioxidant, and/or neuroprotective agents may aid in preventing or neutralizing DA-induced toxic impacts. The potential benefits of these agents have primarily been evaluated in model species and are as follows: (1) vitamin B6 (10 mg/kg) may reduce seizure activity [261], (2) troxuretin (150–225 mg/kg/d PO for 3 wk) may reduce memory deficits [262], (3) ursolic acid (100 mg/kg/d PO for 3 wk) may reverse memory deficits [263], (4) melatonin (10 mg/kg IP) may diminish neuronal damage, glial activation, and nitric oxide synthase induction [264], (5) purple sweet potato color (200 mg/kg/d PO for 4 wk) may attenuate cognitive deficits [265], (6) exogenous glutathione may inhibit apoptosis of cerebellar granular neurons [266], (7) ascorbate acid (10 mM for 24 h) may alleviate CNS deficits [267], (8) naringin (40–80 mg/kg IP) and sesamin extract (30 mg/kg SID PO for 3 d) may ameliorate seizures, cognitive dysfunction, CNS oxidative stress, and mortality rates [268,269], and (9) kynurenic acid (300–600 mg/kg IP after exposure) may reduce excitotoxic and convulsant stimulation [103,270–273]. Kynurenic acid, a generalized excitatory amino acid receptor antagonist, may also protect against DA-induced GI lesions. Neuroprotective strategies against DA may include (1) downregulation or knockdown of protein kinase C zeta (PKC-ζ) to reduce cognitive deficits and (2) upregulation or activation of CNS serotonergic 5-HT_{1A} receptors to attenuate seizures [274,275]. While these treatments were effective in models of DA toxicosis, they must be used with caution in marine wildlife and humans pending evaluation for safety and efficacy. The CNS antioxidant alpha-lipoic acid (ALA; 10 mg/kg SID SQ) and an anti-inflammatory dose of dexamethasone or prednisone (if no contraindications are present) are often administered to pinnipeds with acute toxicosis [209,276]. Though increased survival rates and decreased hippocampal damage were observed from a non-steroidal anti-inflammatory drug (NSAID) in models of acute DA toxicosis [194], NSAIDs may be unsafe in patients with renal compromise [277]. Moreover, three cases of acute renal failure after administration of an NSAID in CSLs undergoing rehabilitation for acute DA toxicosis implied a reduced renal capacity in these patients and underscored the importance of cautious NSAID use in DA toxicosis patients [277]. Treatment of acute DA toxicosis with exogenous glutamatergic antagonists is undergoing preclinical trials [198].

There are no published recommendations for the treatment of DA-induced cardiac disease or other severe extra-neural impacts (e.g., uterine torsion). Euthanasia should be considered when marine wildlife with DA toxicosis develop refractory seizures and/or substantial cardiomyopathy due to a poor prognosis for survival [209,211].

10. Prognosis

The prognosis related to DA exposure depends upon exposure dose, duration, history, underlying health status, time lapse since exposure, and stability of the population or species. Damage to GluR-containing organs may have enduring consequences on memory, navigation, social interactions, fecundity, cardiac output, and kidney function with cascading impacts on reproductive success, longevity, fitness, and survival [17,182].

There are no reports of stranded cetaceans with DA toxicosis surviving, or prognostic recommendations specific to fissipeds, seabirds, and sea turtles [211]. The individual prognosis from DA toxicosis has been best studied in CSLs, where major contributing factors include the type, severity, and timing of clinical signs, response to treatment, and underlying health status. Those with acute toxicosis who develop progressive seizures (i.e., domoic acid-induced epileptic disease) have a poorer prognosis than those with a gradual reduction in frequency [40]. Cognitive deficits manifest as anomalous diving and foraging patterns, which also negatively impact survivorship in CSLs with chronic toxicosis [233]. Scoring systems were developed and honed at The Marine Mammal Center (TMMC) in Sausalito, CA, USA to link clinical signs with prognosis. These "clinical assessment scores" or "NS" are performed following a week of AED treatment and cessation of AED treatment and aid clinicians with assessing patient progress, where higher scores are associated with a poorer outcome [40,209]. The best predictors of releasability are the third NS, change in NS over time, nutritional status on admittance, and if the patient begins eating during treatment [209]. In general, CSLs released after DA toxicosis treatment have a higher re-stranding rate (6% acute/chronic; 71% chronic) than those admitted for other reasons (0.5%), and more than 50% of CSLs with DA toxicosis are euthanized or die naturally [17,209]. Right dorsal hippocampal lesions correlate with poor navigational memory, and, thus, lesion presence and severity may be used as a prognostic marker [230]. There are no studies differentiating prognosis with long-term, low-level exposure vs. long-term sequelae to previously sublethal acute exposure. Since cognitive deficits were reversible after discontinuing long-term, low-dose exposure in model species [181], recovery may also be possible in humans and marine wildlife after suspending low-level DA ingestion. Marine wildlife deemed non-releasable are often euthanized due to a poor prognosis, but clinically stable, non-releasable pinnipeds may be transferred to public display or research facilities as ambassadors for their species and for long-term monitoring following environmental exposure [137,278]. The potential for gestational DA exposure must be considered in asymptomatic pinnipeds prior to placement in permanent care or release back into the wild since delayed but progressive neurobehavioral disease can occur [278]. The prognosis may be impacted by the increasing DA exposure dose, frequency, and duration expected with climate change and urbanization projections [71,72]. Threatened, endangered, and keystone populations may be the most vulnerable, particularly since DA often impacts prime-age adults whose reproductive success is important for population stability [186]. Prognosticating is limited by the challenge of diagnosis confirmation, as well as a lack of dose-effect and exposure history data.

11. California Sea Lions as Models of DA Toxicosis

Extrapolation of findings from studies using laboratory models of DA toxicosis is often limited by the use of (1) taxa and exposure routes or doses that are not ecologically relevant, (2) healthy individuals that do not adequately represent sensitive subgroups, and (3) short study periods that do not characterize health impacts that may develop in long-lived species. California sea lions are effective mammalian sentinels of DA exposure due to their high trophic level status, susceptibility to toxin bioaccumulation and subsequent health effects, well-understood biology, capability of being researched under human-managed care, longevity, abundance, accessibility, similar DA sensitivity as humans, comparable physiology to humans, and unavoidable contact with well-documented seasonal blooms nearly annually [86,246,279–282]. California sea lions are commonly rescued and transferred to marine mammal rehabilitation facilities for DA-related care (Figure 3). Since CSLs move throughout marine environments, they can enhance stationary monitoring programs across spatiotemporal scales [283]. For example, CSLs may strand with acute DA toxicosis prior to increases of DA at collection sites (e.g., particulate DA from piers or shellfish), highlighting their utility as early warning tools [284]. However, DA-producing blooms do not always correspond to increased CSL stranding events [205]. For this reason, as well as the variability in DA uptake and depuration in commercially important seafood and the spatial and temporal heterogeneity in blooms and CSL movement patterns, such data is best used in tandem [205,284]. Finally, CSLs with DA toxicosis may serve as models of neurodegenerative and excitotoxic disorders also linked to overactivation of iGluRs in humans [285], including TLE, schizophrenia, autism spectrum disorder (ASD), Alzheimer's disease, amyotrophic lateral sclerosis, MSG toxicosis, and β-Methylamino-L-alanine (BMAA) exposure [47,108,195–197,208,244,248,286–291].

Figure 3. Two California sea lions (CSLs, *Zalophus californianus*) undergoing treatment for DA toxicosis at The Marine Mammal Center (TMMC, Sausalito, CA, USA), under NOAA permit #24359. This species is commonly rescued and transferred to a marine mammal rehabilitation center for DA-related care.

12. Future Research

The primary focus of DA-related health research over the last three decades regarded the characterization and treatment of neurotoxicity in model species. Contemporary aims include better describing disease in extra-neural tissues, exposure risks in sensitive and high-risk groups (i.e., developing, aged, pregnant, or compromised individuals or those with long-term, low-level exposure), DA interactions with other biotoxins and contaminants, and the spectrum of disease along a temporal continuum and across the lifespan in humans and marine wildlife. Such studies can help provide an improved understanding of DA-induced clinical signs and survival, as well as inform diagnostic tests and treatments, thus improving the management of impacted individuals and conservation of affected populations.

Author Contributions: Conceptualization, A.E.K., S.E.F. and M.E.M.; investigation, A.E.K.; writing—original draft preparation, A.E.K.; writing—review and editing, S.E.F., C.L.F., A.E.K. and M.E.M.; visualization, A.E.K.; funding acquisition, A.E.K., S.E.F., C.L.F. and M.E.M. All authors have read and agreed to the published version of the manuscript.

Funding: This paper was supported in part by a fellowship award under contract FA9550-21-F-0003 through the National Defense Science and Engineering Graduate (NDSEG) Fellowship Program, sponsored by the Air Force Research Laboratory (AFRL), the Office of Naval Research (ONR), and the Army Research Office (ARO), as well as the NOAA John H. Prescott Marine Mammal Rescue Assistance Grant Program.

Institutional Review Board Statement: Not applicable.

Data Availability Statement: No new data was created or analyzed in this study. Data sharing is not applicable to this article.

Acknowledgments: The authors would like to acknowledge the staff, volunteers, and donors of The Marine Mammal Center, as well as our NOAA permit 24359, for making this research possible. We also appreciate graduate committee members at the Florida Institute of Technology, as well as Michael Janech from the College of Charleston and Kathi Lefebvre from NOAA, for providing helpful insight during this study.

Conflicts of Interest: The authors declare no conflict of interest. The funders had no role in the design of the study; in the collection, analyses, or interpretation of data; in the writing of the manuscript; or in the decision to publish the results.

References

1. Bates, S.S.; Hubbard, K.A.; Lundholm, N.; Montresor, M.; Leaw, C.P. *Pseudo-Nitzschia*, *Nitzschia*, and Domoic Acid: New Research since 2011. *Harmful Algae* **2018**, *79*, 3–43. [CrossRef] [PubMed]
2. Fire, S.E.; Van Dolah, F.M. Marine Biotoxins. In *New Directions in Conservation Medicine: Applied Cases of Ecological Health*, Oxford University Press: New York, NY, USA, 2012; Volume 374.
3. Hallegraeff, G.M.; Anderson, D.M.; Belin, C.; Bottein, M.-Y.D.; Bresnan, E.; Chinain, M.; Enevoldsen, H.; Iwataki, M.; Karlson, B.; McKenzie, C.H. Perceived Global Increase in Algal Blooms Is Attributable to Intensified Monitoring and Emerging Bloom Impacts. *Commun. Earth Environ.* **2021**, *2*, 117. [CrossRef]
4. Trainer, V.L.; Bates, S.S.; Lundholm, N.; Thessen, A.E.; Cochlan, W.P.; Adams, N.G.; Trick, C.G. *Pseudo-nitzschia* Physiological Ecology, Phylogeny, Toxicity, Monitoring and Impacts on Ecosystem Health. *Harmful Algae* **2012**, *14*, 271–300. [CrossRef]
5. Trainer, V.L.; Moore, S.K.; Hallegraeff, G.; Kudela, R.M.; Clement, A.; Mardones, J.I.; Cochlan, W.P. Pelagic Harmful Algal Blooms and Climate Change: Lessons from Nature's Experiments with Extremes. *Harmful Algae* **2020**, *91*, 101591. [CrossRef] [PubMed]
6. Fey, S.B.; Siepielski, A.M.; Nusslé, S.; Cervantes-Yoshida, K.; Hwan, J.L.; Huber, E.R.; Fey, M.J.; Catenazzi, A.; Carlson, S.M. Recent Shifts in the Occurrence, Cause, and Magnitude of Animal Mass Mortality Events. *Proc. Natl. Acad. Sci. USA* **2015**, *112*, 1083–1088. [CrossRef]
7. Onens, P.; Wilkin, S.; Fauquier, D.; Spradlin, T.; Manley, S.; Greig, D.; Brill, K.; Garron, M.; Smith, A.; Fougeres, E. 2019 Report of Marine Mammal Strandings in the United States: National Overview. *Natl. Mar. Fish. Serv. Off. Prot. Resour.* **2023**, *21*, 1–20.
8. Fire, S.E.; Wang, Z.; Byrd, M.; Whitehead, H.R.; Paternoster, J.; Morton, S.L. Co-Occurrence of Multiple Classes of Harmful Algal Toxins in Bottlenose Dolphins (*Tursiops truncatus*) Stranding during an Unusual Mortality Event in Texas, USA. *Harmful Algae* **2011**, *10*, 330–336. [CrossRef]
9. Lefebvre, K.A.; Quakenbush, L.; Frame, E.; Huntington, K.B.; Sheffield, G.; Stimmelmayr, R.; Bryan, A.; Kendrick, P.; Ziel, H.; Goldstein, T. Prevalence of Algal Toxins in Alaskan Marine Mammals Foraging in a Changing Arctic and Subarctic Environment. *Harmful Algae* **2016**, *55*, 13–24. [CrossRef]
10. Fire, S.E.; Adkesson, M.J.; Wang, Z.; Jankowski, G.; Cárdenas-Alayza, S.; Broadwater, M. Peruvian Fur Seals (*Arctocephalus australis* ssp.) and South American Sea Lions (*Otaria byronia*) in Peru Are Exposed to the Harmful Algal Toxins Domoic Acid and Okadaic Acid. *Mar. Mammal Sci.* **2017**, *33*, 630–644. [CrossRef]
11. Danil, K.; Berman, M.; Frame, E.; Preti, A.; Fire, S.E.; Leighfield, T.; Carretta, J.; Carter, M.L.; Lefebvre, K. Marine Algal Toxins and Their Vectors in Southern California Cetaceans. *Harmful Algae* **2021**, *103*, 102000. [CrossRef] [PubMed]
12. Edwards, M.L.; Schaefer, A.M.; McFarland, M.; Fire, S.; Perkins, C.R.; Ajemian, M.J. Detection of Numerous Phycotoxins in Young Bull Sharks (*Carcharhinus leucas*) Collected from an Estuary of National Significance. *Sci. Total Environ.* **2023**, *857*, 159602. [CrossRef] [PubMed]

13. Stobo, L.; Lacaze, J.-P.; Scott, A.; Petrie, J.; Turrell, E. Surveillance of Algal Toxins in Shellfish from Scottish Waters. *Toxicon* **2008**, *51*, 635–648. [CrossRef]
14. Kershaw, J.L.; Jensen, S.-K.; McConnell, B.; Fraser, S.; Cummings, C.; Lacaze, J.-P.; Hermann, G.; Bresnan, E.; Dean, K.J.; Turner, A.D. Toxins from Harmful Algae in Fish from Scottish Coastal Waters. *Harmful Algae* **2021**, *105*, 102068. [CrossRef]
15. Dragunow, M.; Trzoss, M.; Brimble, M.A.; Cameron, R.; Beuzenberg, V.; Holland, P.; Mountfort, D. Investigations into the Cellular Actions of the Shellfish Toxin Gymnodimine and Analogues. *Environ. Toxicol. Pharmacol.* **2005**, *20*, 305–312. [CrossRef]
16. Xu, L.; Cai, J.; Gao, T.; Ma, A. Shellfish Consumption and Health: A Comprehensive Review of Human Studies and Recommendations for Enhanced Public Policy. *Crit. Rev. Food Sci. Nutr.* **2022**, *62*, 4656–4668. [CrossRef] [PubMed]
17. Goldstein, T.; Mazet, J.; Zabka, T.; Langlois, G.; Colegrove, K.; Silver, M.; Bargu, S.; Van Dolah, F.; Leighfield, T.; Conrad, P. Novel Symptomatology and Changing Epidemiology of Domoic Acid Toxicosis in California Sea Lions (*Zalophus californianus*): An Increasing Risk to Marine Mammal Health. *Proc. R. Soc. B Biol. Sci.* **2008**, *275*, 267–276. [CrossRef] [PubMed]
18. Grattan, L.M.; Boushey, C.J.; Liang, Y.; Lefebvre, K.A.; Castellon, L.J.; Roberts, K.A.; Toben, A.C.; Morris Jr, J. Repeated Dietary Exposure to Low Levels of Domoic Acid and Problems with Everyday Memory: Research to Public Health Outreach. *Toxins* **2018**, *10*, 103. [CrossRef]
19. Lefebvre, K.A.; Robertson, A. Domoic Acid and Human Exposure Risks: A Review. *Toxicon* **2010**, *56*, 218–230. [CrossRef] [PubMed]
20. Takemoto, T.; Daigo, K. Constituents of *Chondria armata*. *Chem. Pharm. Bull.* **1958**, *6*, 578b–5580. [CrossRef]
21. Daigo, K. Constituents of *Chondria armata*. II. Isolation of an Antihelminthic Constituent. *Yakugaku Zasshi* **1959**, *79*, 353–356. [CrossRef]
22. Komiya, Y.; Kojima, K.; Kumada, M.; Ogawa, H. The Anthelmintic Effect of Domoic Acid for Ascariasis. *Jpn. J. Parasitol.* **1960**, *9*, 85–87.
23. Seki, T. The Anthelmintic Effect of Domoic Acid against *Ascaris lumbricoides*. *Kiseichugaku Zasshi* **1961**, *10*, 614–616.
24. Perl, T.M.; Bédard, L.; Kosatsky, T.; Hockin, J.C.; Todd, E.C.; Remis, R.S. An Outbreak of Toxic Encephalopathy Caused by Eating Mussels Contaminated with Domoic Acid. *N. Engl. J. Med.* **1990**, *322*, 1775–1780. [CrossRef]
25. Jeffery, B.; Barlow, T.; Moizer, K.; Paul, S.; Boyle, C. Amnesic Shellfish Poison. *Food Chem. Toxicol.* **2004**, *42*, 545–557. [CrossRef]
26. Todd, E.C. Domoic Acid and Amnesic Shellfish Poisoning-A Review. *J. Food Prot.* **1993**, *56*, 69–83. [CrossRef] [PubMed]
27. Wekell, J.C.; Hurst, J.; Lefebvre, K.A. The Origin of the Regulatory Limits for PSP and ASP Toxins in Shellfish. *J. Shellfish Res.* **2004**, *23*, 927–930.
28. Anderson, D.M.; Fensin, E.; Gobler, C.J.; Hoeglund, A.E.; Hubbard, K.A.; Kulis, D.M.; Landsberg, J.H.; Lefebvre, K.A.; Provoost, P.; Richlen, M.L. Marine Harmful Algal Blooms (HABs) in the United States: History, Current Status and Future Trends. *Harmful Algae* **2021**, *102*, 101975. [CrossRef] [PubMed]
29. Domoic Acid in Razor Clams | Washington State Department of Health. Available online: https://doh.wa.gov/community-and-environment/shellfish/recreational-shellfish/illnesses/biotoxins/domoic-acid-razor-clams (accessed on 5 April 2024).
30. Cendes, F.; Andermann, F.; Carpenter, S.; Zatorre, R.J.; Cashman, N.R. Temporal Lobe Epilepsy Caused by Domoic Acid Intoxication: Evidence for Glutamate Receptor–Mediated Excitotoxicity in Humans. *Ann. Neurol. Off. J. Am. Neurol. Assoc. Child Neurol. Soc.* **1995**, *37*, 123–126. [CrossRef]
31. Tracy, K.; Boushey, C.J.; Roberts, S.M.; Morris Jr, J.G.; Grattan, L.M. Communities Advancing the Studies of Tribal Nations across Their Lifespan: Design, Methods, and Baseline of the CoASTAL Cohort. *Harmful Algae* **2016**, *57*, 9–19. [CrossRef]
32. Stuchal, L.D.; Grattan, L.M.; Portier, K.M.; Kilmon, K.A.; Manahan, L.M.; Roberts, S.M.; Morris, J.G., Jr. Dose-Response Assessment for Impaired Memory from Chronic Exposure to Domoic Acid among Native American Consumers of Razor Clams. *Regul. Toxicol. Pharmacol.* **2020**, *117*, 104759. [CrossRef]
33. Fritz, L.; Quilliam, M.A.; Wright, J.L.; Beale, A.M.; Work, T.M. An Outbreak of Domoic Acid Poisoning Attributed to the Pennate Diatom *Pseudonitzschia australis*. *J. Phycol.* **1992**, *28*, 439–442. [CrossRef]
34. Work, T.M.; Barr, B.; Beale, A.M.; Fritz, L.; Quilliam, M.A.; Wright, J.L. Epidemiology of Domoic Acid Poisoning in Brown Pelicans (*Pelecanus occidentalis*) and Brandt's Cormorants (*Phalacrocorax penicillatus*) in California. *J. Zoo Wildl. Med.* **1993**, *24*, 54–62.
35. Beltrán, A.S.; Palafox-Uribe, M.; Grajales-Montiel, J.; Cruz-Villacorta, A.; Ochoa, J. Sea Bird Mortality at Cabo San Lucas, Mexico: Evidence That Toxic Diatom Blooms Are Spreading. *Toxicon* **1997**, *35*, 447–453. [CrossRef] [PubMed]
36. McCabe, R.M.; Hickey, B.M.; Kudela, R.M.; Lefebvre, K.A.; Adams, N.G.; Bill, B.D.; Gulland, F.M.; Thomson, R.E.; Cochlan, W.P.; Trainer, V.L. An Unprecedented Coastwide Toxic Algal Bloom Linked to Anomalous Ocean Conditions. *Geophys. Res. Lett.* **2016**, *43*, 10–366. [CrossRef]
37. Solino, L.; Ferrer-Obiol, J.; Navarro-Herrero, L.; Gonzalez-Solis, J.; Costa, P.R. Are Pelagic Seabirds Exposed to Amnesic Shellfish Poisoning Toxins? *Harmful Algae* **2019**, *84*, 172–180. [CrossRef] [PubMed]
38. Van Hemert, C.; Schoen, S.K.; Litaker, R.W.; Smith, M.M.; Arimitsu, M.L.; Piatt, J.F.; Holland, W.C.; Hardison, D.R.; Pearce, J.M. Algal Toxins in Alaskan Seabirds: Evaluating the Role of Saxitoxin and Domoic Acid in a Large-Scale Die-off of Common Murres. *Harmful Algae* **2020**, *92*, 101730. [CrossRef] [PubMed]

39. Gibble, C.M.; Kudela, R.M.; Knowles, S.; Bodenstein, B.; Lefebvre, K.A. Domoic Acid and Saxitoxin in Seabirds in the United States between 2007 and 2018. *Harmful Algae* **2021**, *103*, 101981. [CrossRef]
40. Gulland, E.M.; Haulena, M.; Fauquier, D.; Lander, M.; Zabka, T.; Duerr, R.; Langlois, G. Domoic Acid Toxicity in Californian Sea Lions (*Zalophus californianus*): Clinical Signs, Treatment and Survival. *Vet. Rec.* **2002**, *150*, 475–480. [CrossRef]
41. Miller, M.A.; Moriarty, M.E.; Duignan, P.J.; Zabka, T.S.; Dodd, E.; Batac, F.I.; Young, C.; Reed, A.; Harris, M.D.; Greenwald, K. Clinical Signs and Pathology Associated with Domoic Acid Toxicosis in Southern Sea Otters (*Enhydra lutris nereis*). *Front. Mar. Sci.* **2021**, *8*, 585501. [CrossRef]
42. Kreuder, C.; Miller, M.A.; Jessup, D.A.; Lowenstine, L.J.; Harris, M.; Ames, J.A.; Carpenter, T.E.; Conrad, P.A.; Mazet, J. Patterns of Mortality in Southern Sea Otters (*Enhydra lutris nereis*) from 1998–2001. *J. Wildl. Dis.* **2003**, *39*, 495–509. [CrossRef] [PubMed]
43. De La Riva, G.T.; Johnson, C.K.; Gulland, F.M.; Langlois, G.W.; Heyning, J.E.; Rowles, T.K.; Mazet, J.A. Association of an Unusual Marine Mammal Mortality Event with *Pseudo-nitzschia* spp. Blooms along the Southern California Coastline. *J. Wildl. Dis.* **2009**, *45*, 109–121. [CrossRef]
44. Lefebvre, K.A.; Bargu, S.; Kieckhefer, T.; Silver, M.W. From Sanddabs to Blue Whales: The Pervasiveness of Domoic Acid. *Toxicon* **2002**, *40*, 971–977. [CrossRef]
45. Fire, S.E.; Wang, Z.; Leighfield, T.A.; Morton, S.L.; McFee, W.E.; McLellan, W.A.; Litaker, R.W.; Tester, P.A.; Hohn, A.A.; Lovewell, G. Domoic Acid Exposure in Pygmy and Dwarf Sperm Whales (*Kogia* spp.) from Southeastern and Mid-Atlantic US Waters. *Harmful Algae* **2009**, *8*, 658–664. [CrossRef]
46. Leandro, L.F.; Rolland, R.M.; Roth, P.B.; Lundholm, N.; Wang, Z.; Doucette, G.J. Exposure of the North Atlantic Right Whale *Eubalaena glacialis* to the Marine Algal Biotoxin, Domoic Acid. *Mar. Ecol. Prog. Ser.* **2010**, *398*, 287–303. [CrossRef]
47. Doucette, G.J.; Mikulski, C.M.; King, K.L.; Roth, P.B.; Wang, Z.; Leandro, L.F.; DeGrasse, S.L.; White, K.D.; De Biase, D.; Gillett, R.M. Endangered North Atlantic Right Whales (*Eubalaena glacialis*) Experience Repeated, Concurrent Exposure to Multiple Environmental Neurotoxins Produced by Marine Algae. *Environ. Res.* **2012**, *112*, 67–76. [CrossRef] [PubMed]
48. Wilson, C.; Sastre, A.V.; Hoffmeyer, M.; Rowntree, V.J.; Fire, S.E.; Santinelli, N.H.; Ovejero, S.D.; D'Agostino, V.; Marón, C.F.; Doucette, G.J. Southern Right Whale (*Eubalaena australis*) Calf Mortality at Península Valdés, Argentina: Are Harmful Algal Blooms to Blame? *Mar. Mammal Sci.* **2016**, *32*, 423–451. [CrossRef]
49. Bengtson Nash, S.; Baddock, M.; Takahashi, E.; Dawson, A.; Cropp, R. Domoic Acid Poisoning as a Possible Cause of Seasonal Cetacean Mass Stranding Events in Tasmania, Australia. *Bull. Environ. Contam. Toxicol.* **2017**, *98*, 8–13. [CrossRef]
50. D'Agostino, V.C.; Degrati, M.; Sastre, V.; Santinelli, N.; Krock, B.; Krohn, T.; Dans, S.L.; Hoffmeyer, M.S. Domoic Acid in a Marine Pelagic Food Web: Exposure of Southern Right Whales (*Eubalaena australis*) to Domoic Acid on the Península Valdés Calving Ground, Argentina. *Harmful Algae* **2017**, *68*, 248–257. [CrossRef]
51. Silber, G.K.; Silber, K.M. Co-Occurrence of Harmful Algal Blooms and Whale Deaths. *Front. Mar. Sci.* **2024**, *11*, 1454656. [CrossRef]
52. Perrault, J.R.; Perkins, C.R.; Ajemian, M.J.; Bresette, M.J.; Mott, C.R.; Page-Karjian, A. Harmful Algal and Cyanobacterial Toxins in Foraging Green Turtles (*Chelonia mydas*) in Florida's Big Bend. *Toxicon X* **2020**, *5*, 100020. [CrossRef]
53. Harris, H.; Benson, S.; Miller, M.; Fire, S.; Stacy, B.; Kudela, R.; Fahy, C.; Seminoff, J. Domoic Acid in the Leatherback Turtle Food Web on Critical Foraging Grounds in Central California. In Proceedings of the 44th Annual Conference of the International Association of Aquatic Animal Medicine, Sausalito, CA, USA, 21–26 April 2013.
54. Brown, H. Exposure to Algal Biotoxins: Exploring Health Effects in Green Sea Turtles (*Chelonia mydas*). Master's Thesis, Florida Atlantic University, Boca Raton, FL, USA, 2023.
55. Bates, S.S. Domoic-acid-producing Diatoms: Another Genus Added! *J. Phycol.* **2000**, *36*, 978–983. [CrossRef]
56. Trainer, V.L.; Adams, N.G.; Bill, B.D.; Stehr, C.M.; Wekell, J.C.; Moeller, P.; Busman, M.; Woodruff, D. Domoic Acid Production near California Coastal Upwelling Zones, June 1998. *Limnol. Oceanogr.* **2000**, *45*, 1818–1833. [CrossRef]
57. James, K.J.; Gillman, M.; Amandi, M.F.; López-Rivera, A.; Puente, P.F.; Lehane, M.; Mitrovic, S.; Furey, A. Amnesic Shellfish Poisoning Toxins in Bivalve Molluscs in Ireland. *Toxicon* **2005**, *46*, 852–858. [CrossRef]
58. Romero, M.; Lirdwitayaprasit, T.; Kotaki, Y.; Lundholm, N.; Relox, J.; Furio, E.; Terada, R.; Yokoyama, T.; Kodama, M.; Fukuyo, Y. Isolation of ASP Toxin-Producing Nitzschia from Thailand. *Mar. Res. Indones.* **2008**, *33*, 225–228. [CrossRef]
59. Ujević, I.; Ninčević-Gladan, Ž.; Roje, R.; Skejić, S.; Arapov, J.; Marasović, I. Domoic Acid-a New Toxin in the Croatian Adriatic Shellfish Toxin Profile. *Molecules* **2010**, *15*, 6835–6849. [CrossRef]
60. Silver, M.W.; Bargu, S.; Coale, S.L.; Benitez-Nelson, C.R.; Garcia, A.C.; Roberts, K.J.; Sekula-Wood, E.; Bruland, K.W.; Coale, K.H. Toxic Diatoms and Domoic Acid in Natural and Iron Enriched Waters of the Oceanic Pacific. *Proc. Natl. Acad. Sci. USA* **2010**, *107*, 20762–20767. [CrossRef]
61. Hinder, S.L.; Hays, G.C.; Brooks, C.J.; Davies, A.P.; Edwards, M.; Walne, A.W.; Gravenor, M.B. Toxic Marine Microalgae and Shellfish Poisoning in the British Isles: History, Review of Epidemiology, and Future Implications. *Environ. Health* **2011**, *10*, 1–12. [CrossRef]
62. Andjelkovic, M.; Vandevijvere, S.; Van Klaveren, J.; Van Oyen, H.; Van Loco, J. Exposure to Domoic Acid through Shellfish Consumption in Belgium. *Environ. Int.* **2012**, *49*, 115–119. [CrossRef]

63. Pitcher, G.C.; Cembella, A.D.; Krock, B.; Macey, B.; Mansfield, L.; Probyn, T. Identification of the Marine Diatom *Pseudo-nitzschia multiseries* (Bacillariophyceae) as a Source of the Toxin Domoic Acid in Algoa Bay, South Africa. *Afr. J. Mar. Sci.* **2014**, *36*, 523–528. [CrossRef]
64. Fernandes, L.F.; Hubbard, K.A.; Richlen, M.L.; Smith, J.; Bates, S.S.; Ehrman, J.; Léger, C.; Mafra, L.L., Jr.; Kulis, D.; Quilliam, M. Diversity and Toxicity of the Diatom *Pseudo-nitzschia peragallo* in the Gulf of Maine, Northwestern Atlantic Ocean. *Deep Sea Res. Part II Top. Stud. Oceanogr.* **2014**, *103*, 139–162. [CrossRef]
65. Tenorio, C.; Álvarez, G.; Quijano-Scheggia, S.; Perez-Alania, M.; Arakaki, N.; Araya, M.; Álvarez, F.; Blanco, J.; Uribe, E. First Report of Domoic Acid Production from *Pseudo-nitzschia multistriata* in Paracas Bay (Peru). *Toxins* **2021**, *13*, 408. [CrossRef]
66. Kelchner, H.; Reeve-Arnold, K.E.; Schreiner, K.M.; Bargu, S.; Roques, K.G.; Errera, R.M. Domoic Acid and *Pseudo-nitzschia* spp. Connected to Coastal Upwelling along Coastal Inhambane Province, Mozambique: A New Area of Concern. *Toxins* **2021**, *13*, 903. [CrossRef]
67. Twiner, M.J.; Fire, S.; Schwacke, L.; Davidson, L.; Wang, Z.; Morton, S.; Roth, S.; Balmer, B.; Rowles, T.K.; Wells, R.S. Concurrent Exposure of Bottlenose Dolphins (*Tursiops truncatus*) to Multiple Algal Toxins in Sarasota Bay, Florida, USA. *PLoS ONE* **2011**, *6*, e17394. [CrossRef] [PubMed]
68. Malhi, N.; Turnbull, A.; Tan, J.; Kiermeier, A.; Nimmagadda, R.; McLeod, C. A National Survey of Marine Biotoxins in Wild-Caught Abalone in Australia. *J. Food Prot.* **2014**, *77*, 1960–1967. [CrossRef]
69. Rhodes, L.; Scholin, C.; Garthwaite, I. *Pseudo-nitzschia* in New Zealand and the Role of DNA Probes and Immunoassays in Refining Marine Biotoxin Monitoring Programmes. *Nat. Toxins* **1998**, *6*, 105–111. [CrossRef]
70. McKibben, S.M.; Peterson, W.; Wood, A.M.; Trainer, V.L.; Hunter, M.; White, A.E. Climatic Regulation of the Neurotoxin Domoic Acid. *Proc. Natl. Acad. Sci. USA* **2017**, *114*, 239–244. [CrossRef] [PubMed]
71. Zhu, Z.; Qu, P.; Fu, F.; Tennenbaum, N.; Tatters, A.O.; Hutchins, D.A. Understanding the Blob Bloom: Warming Increases Toxicity and Abundance of the Harmful Bloom Diatom *Pseudo-nitzschia* in California Coastal Waters. *Harmful Algae* **2017**, *67*, 36–43. [CrossRef]
72. Wohlrab, S.; John, U.; Klemm, K.; Eberlein, T.; Grivogiannis, A.M.F.; Krock, B.; Frickenhaus, S.; Bach, L.T.; Rost, B.; Riebesell, U. Ocean Acidification Increases Domoic Acid Contents during a Spring to Summer Succession of Coastal Phytoplankton. *Harmful Algae* **2020**, *92*, 101697. [CrossRef] [PubMed]
73. Bargu, S.; Goldstein, T.; Roberts, K.; Li, C.; Gulland, F. *Pseudo-nitzschia* Blooms, Domoic Acid, and Related California Sea Lion Strandings in Monterey Bay, California. *Mar. Mammal Sci.* **2012**, *28*, 237–253. [CrossRef]
74. Liefer, J.D.; Robertson, A.; MacIntyre, H.L.; Smith, W.L.; Dorsey, C.P. Characterization of a Toxic *Pseudo-nitzschia* spp. Bloom in the Northern Gulf of Mexico Associated with Domoic Acid Accumulation in Fish. *Harmful Algae* **2013**, *26*, 20–32. [CrossRef]
75. Parsons, M.L.; Dortch, Q.; Doucette, G.J. An Assessment of *Pseudo-nitzschia* Population Dynamics and Domoic Acid Production in Coastal Louisiana. *Harmful Algae* **2013**, *30*, 65–77. [CrossRef]
76. Lefebvre, K.; Silver, M.; Coale, S.; Tjeerdema, R. Domoic Acid in Planktivorous Fish in Relation to Toxic *Pseudo-nitzschia* Cell Densities. *Mar. Biol.* **2002**, *140*, 625–631.
77. Sekula-Wood, E.; Schnetzer, A.; Benitez-Nelson, C.R.; Anderson, C.; Berelson, W.M.; Brzezinski, M.A.; Burns, J.M.; Caron, D.A.; Cetinic, I.; Ferry, J.L. Rapid Downward Transport of the Neurotoxin Domoic Acid in Coastal Waters. *Nat. Geosci.* **2009**, *2*, 272–275. [CrossRef]
78. Schnetzer, A.; Lampe, R.H.; Benitez-Nelson, C.R.; Marchetti, A.; Osburn, C.L.; Tatters, A.O. Marine Snow Formation by the Toxin-Producing Diatom, *Pseudo-nitzschia australis*. *Harmful Algae* **2017**, *61*, 23–30. [CrossRef]
79. Quilliam, M. Chemical Methods for Domoic Acid, the Amnesic Shellfish Poisoning (ASP) Toxin. *Man. Harmful Mar. Microalgae Monogr. Oceanogr. Methodol.* **2003**, *11*, 247–266.
80. McCarron, P.; Hess, P. Tissue Distribution and Effects of Heat Treatments on the Content of Domoic Acid in Blue Mussels, *Mytilus edulis*. *Toxicon* **2006**, *47*, 473–479. [CrossRef] [PubMed]
81. Gill, S.S.; Pulido, O.M. Glutamate Receptors in Peripheral Tissues: Current Knowledge, Future Research, and Implications for Toxicology. *Toxicol. Pathol.* **2001**, *29*, 208–223. [CrossRef]
82. Shimizu, Y. Microalgal Metabolites. *Chem. Rev.* **1993**, *93*, 1685–1698. [CrossRef]
83. Coyle, J.T. Kainic Acid: Insights into Excitatory Mechanisms Causing Selective Neuronal Degeneration. In *Ciba Foundation Symposium 126—Selective Neuronal Death*; Wiley Online Library: Hoboken, NJ, USA, 2007; pp. 186–203.
84. Biscoe, T.; Evans, R.; Headley, P.; Martin, M.; Watkins, J. Domoic and Quisqualic Acids as Potent Amino Acid Excitants of Frog and Rat Spinal Neurones. *Nature* **1975**, *255*, 166–167. [CrossRef]
85. Carcache, L.M.; Rodriguez, J.; Rein, K.S. The Structural Basis for Kainoid Selectivity at AMPA Receptors Revealed by Low-Mode Docking Calculations. *Bioorganic Med. Chem.* **2003**, *11*, 551–559. [CrossRef] [PubMed]
86. Costa, L.G.; Giordano, G.; Faustman, E.M. Domoic Acid as a Developmental Neurotoxin. *Neurotoxicology* **2010**, *31*, 409–423. [CrossRef] [PubMed]
87. Lefebvre, K.A.; Frame, E.R.; Kendrick, P.S. Domoic Acid and Fish Behavior: A Review. *Harmful Algae* **2012**, *13*, 126–130. [CrossRef]

88. Lefebvre, K.A.; Noren, D.P.; Schultz, I.R.; Bogard, S.M.; Wilson, J.; Eberhart, B.-T.L. Uptake, Tissue Distribution and Excretion of Domoic Acid after Oral Exposure in Coho Salmon (*Oncorhynchus kisutch*). *Aquat. Toxicol.* **2007**, *81*, 266–274. [CrossRef]
89. Fire, S.E.; Wang, Z.; Berman, M.; Langlois, G.W.; Morton, S.L.; Sekula-Wood, E.; Benitez-Nelson, C.R. Trophic Transfer of the Harmful Algal Toxin Domoic Acid as a Cause of Death in a Minke Whale (*Balaenoptera acutorostrata*) Stranding in Southern California. *Aquat. Mamm.* **2010**, *36*, 342–350. [CrossRef]
90. Iverson, F.; Truelove, J.; Nera, E.; Tryphonas, L.; Campbell, J.; Lok, E. Domoic Acid Poisoning and Mussel-Associated Intoxication: Preliminary Investigations into the Response of Mice and Rats to Toxic Mussel Extract. *Food Chem. Toxicol.* **1989**, *27*, 377–384. [CrossRef] [PubMed]
91. Iverson, F.; Truelove, J.; Tryphonas, L.; Nera, E. The Toxicology of Domoic Acid Administered Systemically to Rodents and Primates. *Can. Dis. Wkly. Rep.=Rapp. Hebd. Des Mal. Au Can.* **1990**, *16*, 15–18.
92. Jing, J.; Petroff, R.; Shum, S.; Crouthamel, B.; Topletz, A.R.; Grant, K.S.; Burbacher, T.M.; Isoherranen, N. Toxicokinetics and Physiologically Based Pharmacokinetic Modeling of the Shellfish Toxin Domoic Acid in Nonhuman Primates. *Drug Metab. Dispos.* **2018**, *46*, 155–165. [CrossRef] [PubMed]
93. Brodie, E.C.; Gulland, F.M.; Greig, D.J.; Hunter, M.; Jaakola, J.; Leger, J.S.; Leighfield, T.A.; Van Dolah, F.M. Domoic Acid Causes Reproductive Failure in California Sea Lions (*Zalophus californianus*). *Mar. Mammal Sci.* **2006**, *22*, 700–707. [CrossRef]
94. Maucher, J.M.; Ramsdell, J.S. Maternal–Fetal Transfer of Domoic Acid in Rats at Two Gestational Time Points. *Environ. Health Perspect.* **2007**, *115*, 1743–1746. [CrossRef] [PubMed]
95. Rust, L.; Gulland, F.; Frame, E.; Lefebvre, K. Domoic Acid in Milk of Free Living California Marine Mammals Indicates Lactational Exposure Occurs. *Mar. Mammal Sci.* **2014**, *30*, 1272–1278. [CrossRef]
96. Lefebvre, K.A.; Hendrix, A.; Halaska, B.; Duignan, P.; Shum, S.; Isoherranen, N.; Marcinek, D.J.; Gulland, F.M. Domoic Acid in California Sea Lion Fetal Fluids Indicates Continuous Exposure to a Neuroteratogen Poses Risks to Mammals. *Harmful Algae* **2018**, *79*, 53–57. [CrossRef] [PubMed]
97. Ramsdell, J.S. The Molecular and Integrative Basis to Domoic Acid Toxicity. In *Phycotoxins: Chemistry and Biochemistry*; Wiley Online Library: Hoboken, NJ, USA, 2007; pp. 223–250.
98. Schultz, I.R.; Skillman, A.; Sloan-Evans, S.; Woodruff, D. Domoic Acid Toxicokinetics in Dungeness Crabs: New Insights into Mechanisms That Regulate Bioaccumulation. *Aquat. Toxicol.* **2013**, *140*, 77–88. [CrossRef] [PubMed]
99. Tasker, R.A. Domoic Acid. In *Advances in Neurotoxicology*; Elsevier: Amsterdam, The Netherlands, 2021; Volume 6, pp. 49–88. ISBN 2468-7480.
100. Gill, S.; Goldstein, T.; Situ, D.; Zabka, T.S.; Gulland, F.M.; Mueller, R.W. Cloning and Characterization of Glutamate Receptors in Californian Sea Lions (*Zalophus californianus*). *Mar. Drugs* **2010**, *8*, 1637–1649. [CrossRef]
101. Silvagni, P.; Lowenstine, L.; Spraker, T.; Lipscomb, T.; Gulland, F. Pathology of Domoic Acid Toxicity in California Sea Lions (*Zalophus californianus*). *Vet. Pathol.* **2005**, *42*, 184–191. [CrossRef]
102. Tryphonas, L.; Truelove, J.; Nera, E.; Iverson, F. Acute Neurotoxicity of Domoic Acid in the Rat. *Toxicol. Pathol.* **1990**, *18*, 1–9. [CrossRef]
103. Glavin, G.; Pinsky, C.; Bose, R. Gastrointestinal Effects of Contaminated Mussels and Putative Antidotes Thereof. *Can. Dis. Wkly. Rep.=Rapp. Hebd. Des Mal. Au Can.* **1990**, *16*, 111–115.
104. Zabka, T.; Goldstein, T.; Cross, C.; Mueller, R.; Kreuder-Johnson, C.; Gill, S.; Gulland, F. Characterization of a Degenerative Cardiomyopathy Associated with Domoic Acid Toxicity in California Sea Lions (*Zalophus californianus*). *Vet. Pathol.* **2009**, *46*, 105–119. [CrossRef]
105. Vranyac-Tramoundanas, A.; Harrison, J.C.; Sawant, P.M.; Kerr, D.S.; Sammut, I.A. Ischemic Cardiomyopathy Following Seizure Induction by Domoic Acid. *Am. J. Pathol.* **2011**, *179*, 141–154. [CrossRef]
106. Pulido, O.M. Domoic Acid Toxicologic Pathology: A Review. *Mar. Drugs* **2008**, *6*, 180–219. [CrossRef]
107. Petroff, R.; Hendrix, A.; Shum, S.; Grant, K.S.; Lefebvre, K.A.; Burbacher, T.M. Public Health Risks Associated with Chronic, Low-Level Domoic Acid Exposure: A Review of the Evidence. *Pharmacol. Ther.* **2021**, *227*, 107865. [CrossRef]
108. Ramsdell, J.S.; Gulland, F.M. Domoic Acid Epileptic Disease. *Mar. Drugs* **2014**, *12*, 1185–1207. [CrossRef] [PubMed]
109. Ross, I.; Johnson, W.; Sapienza, P.; Kim, C. Effects of the Seafood Toxin Domoic Acid on Glutamate Uptake by Rat Astrocytes. *Food Chem. Toxicol.* **2000**, *38*, 1005–1011. [CrossRef]
110. Gill, S.S.; Hou, Y.; Ghane, T.; Pulido, O.M. Regional Susceptibility to Domoic Acid in Primary Astrocyte Cells Cultured from the Brain Stem and Hippocampus. *Mar. Drugs* **2008**, *6*, 25–38. [CrossRef] [PubMed]
111. Levin, M.; Leibrecht, H.; Ryan, J.; Van Dolah, F.; De Guise, S. Immunomodulatory Effects of Domoic Acid Differ between in Vivo and in Vitro Exposure in Mice. *Mar. Drugs* **2008**, *6*, 636–659. [CrossRef] [PubMed]
112. Levin, M.; Joshi, D.; Draghi, A.; Gulland, F.M.; Jessup, D.; De Guise, S. Immunomodulatory Effects upon in Vitro Exposure of California Sea Lion and Southern Sea Otter Peripheral Blood Leukocytes to Domoic Acid. *J. Wildl. Dis.* **2010**, *46*, 541–550. [CrossRef] [PubMed]

113. Kirkley, K.S.; Madl, J.E.; Duncan, C.; Gulland, F.M.; Tjalkens, R.B. Domoic Acid-Induced Seizures in California Sea Lions (*Zalophus californianus*) Are Associated with Neuroinflammatory Brain Injury. *Aquat. Toxicol.* **2014**, *156*, 259–268. [CrossRef] [PubMed]
114. Hiolski, E.; Ito, S.; Beggs, J.; Lefebvre, K.; Litke, A.; Smith, D. Domoic Acid Disrupts the Activity and Connectivity of Neuronal Networks in Organotypic Brain Slice Cultures. *Neurotoxicology* **2016**, *56*, 215–224. [CrossRef]
115. Vieira, A.C.; Cifuentes, J.M.; Bermúdez, R.; Ferreiro, S.F.; Castro, A.R.; Botana, L.M. Heart Alterations after Domoic Acid Administration in Rats. *Toxins* **2016**, *8*, 68. [CrossRef] [PubMed]
116. Moyer, C.E.; Hiolski, E.M.; Marcinek, D.J.; Lefebvre, K.A.; Smith, D.R.; Zuo, Y. Repeated Low Level Domoic Acid Exposure Increases CA1 VGluT1 Levels, but Not Bouton Density, VGluT2 or VGAT Levels in the Hippocampus of Adult Mice. *Harmful Algae* **2018**, *79*, 74–86. [CrossRef]
117. Madl, J.; Duncan, C.; Stanhill, J.; Tai, P.-Y.; Spraker, T.; Gulland, F. Oxidative Stress and Redistribution of Glutamine Synthetase in California Sea Lions (*Zalophus californianus*) with Domoic Acid Toxicosis. *J. Comp. Pathol.* **2014**, *150*, 306–315. [CrossRef] [PubMed]
118. Suzuki, C.; Hierlihy, S. Renal Clearance of Domoic Acid in the Rat. *Food Chem. Toxicol.* **1993**, *31*, 701–706. [CrossRef]
119. Funk, J.A.; Janech, M.G.; Dillon, J.C.; Bissler, J.J.; Siroky, B.J.; Bell, P.D. Characterization of Renal Toxicity in Mice Administered the Marine Biotoxin Domoic Acid. *J. Am. Soc. Nephrol.* **2014**, *25*, 1187–1197. [CrossRef] [PubMed]
120. Maucher Fuquay, J.; Muha, N.; Wang, Z.; Ramsdell, J.S. Elimination Kinetics of Domoic Acid from the Brain and Cerebrospinal Fluid of the Pregnant Rat. *Chem. Res. Toxicol.* **2012**, *25*, 2805–2809. [CrossRef]
121. Ventoso, P.; Pazos, A.J.; Blanco, J.; Pérez-Parallé, M.L.; Triviño, J.C.; Sánchez, J.L. Transcriptional Response in the Digestive Gland of the King Scallop (*Pecten maximus*) after the Injection of Domoic Acid. *Toxins* **2021**, *13*, 339. [CrossRef] [PubMed]
122. Gulland, F.M. *Domoic Acid Toxicity in California Sea Lions (Zalophus californianus) Stranded along the Central California Coast, May-October 1998: Report to the National Marine Fisheries Service Working Group on Unusual Marine Mammal Mortality Events*; US Department of Commerce, National Oceanic and Atmospheric Administration: Washington, DC, USA, 2000; Volume 2, pp. 1–299.
123. Lefebvre, K.A.; Yakes, B.J.; Frame, E.; Kendrick, P.; Shum, S.; Isoherranen, N.; Ferriss, B.E.; Robertson, A.; Hendrix, A.; Marcinek, D.J. Discovery of a Potential Human Serum Biomarker for Chronic Seafood Toxin Exposure Using an SPR Biosensor. *Toxins* **2019**, *11*, 293. [CrossRef]
124. Wittmaack, C.; Lahvis, G.P.; Keith, E.O.; Self-Sullivan, C. Diagnosing Domoic Acid Toxicosis in the California Sea Lion (*Zalophus californianus*) Using Behavioral Criteria: A Novel Approach. *Zoo Biol.* **2015**, *34*, 314–320. [CrossRef]
125. Burbacher, T.M.; Grant, K.S.; Petroff, R.; Shum, S.; Crouthamel, B.; Stanley, C.; McKain, N.; Jing, J.; Isoherranen, N. Effects of Oral Domoic Acid Exposure on Maternal Reproduction and Infant Birth Characteristics in a Preclinical Nonhuman Primate Model. *Neurotoxicol. Teratol.* **2019**, *72*, 10–21. [CrossRef] [PubMed]
126. Bargu, S.; Lefebvre, K.; Silver, M.W. Effect of Dissolved Domoic Acid on the Grazing Rate of Krill *Euphausia pacifica*. *Mar. Ecol. Prog. Ser.* **2006**, *312*, 169–175. [CrossRef]
127. Kerr, D.S.; Razak, A.; Crawford, N. Age-Related Changes in Tolerance to the Marine Algal Excitotoxin Domoic Acid. *Neuropharmacology* **2002**, *43*, 357–366. [CrossRef]
128. Lefebvre, K.A.; Frame, E.R.; Gulland, F.; Hansen, J.D.; Kendrick, P.S.; Beyer, R.P.; Bammler, T.K.; Farin, F.M.; Hiolski, E.M.; Smith, D.R. A Novel Antibody-Based Biomarker for Chronic Algal Toxin Exposure and Sub-Acute Neurotoxicity. *PLoS ONE* **2012**, *7*, e36213. [CrossRef]
129. Petroff, R.L.; Williams, C.; Li, J.-L.; MacDonald, J.W.; Bammler, T.K.; Richards, T.; English, C.N.; Baldessari, A.; Shum, S.; Jing, J. Prolonged, Low-Level Exposure to the Marine Toxin, Domoic Acid, and Measures of Neurotoxicity in Nonhuman Primates. *Environ. Health Perspect.* **2022**, *130*, 097003. [CrossRef]
130. Mazzillo, F.F.; Pomeroy, C.; Kuo, J.; Ramondi, P.T.; Prado, R.; Silver, M.W. Angler Exposure to Domoic Acid via Consumption of Contaminated Fishes. *Aquat. Biol.* **2010**, *9*, 1–12. [CrossRef]
131. Cisneros-Montemayor, A.M.; Pauly, D.; Weatherdon, L.V.; Ota, Y. A Global Estimate of Seafood Consumption by Coastal Indigenous Peoples. *PLoS ONE* **2016**, *11*, e0166681. [CrossRef]
132. Cook, P.F.; Reichmuth, C.; Rouse, A.; Dennison, S.; Van Bonn, B.; Gulland, F. Natural Exposure to Domoic Acid Causes Behavioral Perseveration in Wild Sea Lions: Neural Underpinnings and Diagnostic Application. *Neurotoxicol. Teratol.* **2016**, *57*, 95–105. [CrossRef]
133. Petroff, R.; Richards, T.; Crouthamel, B.; McKain, N.; Stanley, C.; Grant, K.S.; Shum, S.; Jing, J.; Isoherranen, N.; Burbacher, T.M. Chronic, Low-Level Oral Exposure to Marine Toxin, Domoic Acid, Alters Whole Brain Morphometry in Nonhuman Primates. *Neurotoxicology* **2019**, *72*, 114–124. [CrossRef]
134. Bejarano, A.C.; VanDola, F.M.; Gulland, F.M.; Rowles, T.K.; Schwacke, L.H. Production and Toxicity of the Marine Biotoxin Domoic Acid and Its Effects on Wildlife: A Review. *Hum. Ecol. Risk Assess.* **2008**, *14*, 544–567. [CrossRef]
135. Silvagni, P.A. Comparative Pathology and Diagnosis of Domoic Acid Toxicity. Ph.D. Thesis, University of California, Davis, CA, USA, 2003; 139p. ISBN 0-496-55169-8.
136. Akmajian, A.M.; Scordino, J.J.; Acevedo-Gutiérrez, A. Year-Round Algal Toxin Exposure in Free-Ranging Sea Lions. *Mar. Ecol. Prog. Ser.* **2017**, *583*, 243–258. [CrossRef]

137. Schmitt, T.L.; St. Leger, J.; Inglis, B.A.; Michal, I.; Stedman, N.; Nollens, H.H.; Dennison-Gibby, S.; Herrick, K.; Clarke, E.O.; Mena, A. Twenty Years of Managed Epilepsy for a Stranded Male Guadalupe Fur Seal (*Arctocephalus townsendi*) Secondary to Suspect Domoic Acid Toxicosis. *J. Zool. Bot. Gard.* **2023**, *4*, 665–679. [CrossRef]
138. D'Agnese, E.; Lambourn, D.; Rice, J.; Duffield, D.; Huggins, J.; Spraker, T.; Raverty, S.; Kuzmina, T.; Grigg, M.E.; Wilkinson, K. Reemergence of Guadalupe Fur Seals in the US Pacific Northwest: The Epidemiology of Stranding Events during 2005–2016. *Mar. Mammal Sci.* **2020**, *36*, 828–845. [CrossRef]
139. Lefebvre, K.A.; Robertson, A.; Frame, E.R.; Colegrove, K.M.; Nance, S.; Baugh, K.A.; Wiedenhoft, H.; Gulland, F.M. Clinical Signs and Histopathology Associated with Domoic Acid Poisoning in Northern Fur Seals (*Callorhinus ursinus*) and Comparison of Toxin Detection Methods. *Harmful Algae* **2010**, *9*, 374–383. [CrossRef]
140. McHuron, E.A.; Greig, D.J.; Colegrove, K.M.; Fleetwood, M.; Spraker, T.R.; Gulland, F.M.; Harvey, J.T.; Lefebvre, K.A.; Frame, E.R. Domoic Acid Exposure and Associated Clinical Signs and Histopathology in Pacific Harbor Seals (*Phoca vitulina richardii*). *Harmful Algae* **2013**, *23*, 28–33. [CrossRef]
141. Jensen, S.-K.; Lacaze, J.-P.; Hermann, G.; Kershaw, J.; Brownlow, A.; Turner, A.; Hall, A. Detection and Effects of Harmful Algal Toxins in Scottish Harbour Seals and Potential Links to Population Decline. *Toxicon* **2015**, *97*, 1–14. [CrossRef] [PubMed]
142. Hall, A.J.; Frame, E. Evidence of Domoic Acid Exposure in Harbour Seals from Scotland: A Potential Factor in the Decline in Abundance? *Harmful Algae* **2010**, *9*, 489–493. [CrossRef]
143. Hendrix, A.M.; Lefebvre, K.A.; Quakenbush, L.; Bryan, A.; Stimmelmayr, R.; Sheffield, G.; Wisswaesser, G.; Willis, M.L.; Bowers, E.K.; Kendrick, P. Ice Seals as Sentinels for Algal Toxin Presence in the Pacific Arctic and Subarctic Marine Ecosystems. *Mar. Mammal Sci.* **2021**, *37*, 1292–1308. [CrossRef]
144. Bowers, E.K.; Stimmelmayr, R.; Lefebvre, K.A. Stability of Domoic Acid in 50% Methanol Extracts and Raw Fecal Material from Bowhead Whales (*Balaena mysticetus*). *Mar. Drugs* **2021**, *19*, 423. [CrossRef]
145. Schwacke, L.H.; Twiner, M.J.; De Guise, S.; Balmer, B.C.; Wells, R.S.; Townsend, F.I.; Rotstein, D.C.; Varela, R.A.; Hansen, L.J.; Zolman, E.S. Eosinophilia and Biotoxin Exposure in Bottlenose Dolphins (*Tursiops truncatus*) from a Coastal Area Impacted by Repeated Mortality Events. *Environ. Res.* **2010**, *110*, 548–555. [CrossRef] [PubMed]
146. Twiner, M.J.; Flewelling, L.J.; Fire, S.E.; Bowen-Stevens, S.R.; Gaydos, J.K.; Johnson, C.K.; Landsberg, J.H.; Leighfield, T.A.; Mase-Guthrie, B.; Schwacke, L. Comparative Analysis of Three Brevetoxin-Associated Bottlenose Dolphin (*Tursiops truncatus*) Mortality Events in the Florida Panhandle Region (USA). *PLoS ONE* **2012**, *7*, e42974. [CrossRef] [PubMed]
147. Venn-Watson, S.; Colegrove, K.M.; Litz, J.; Kinsel, M.; Terio, K.; Saliki, J.; Fire, S.; Carmichael, R.; Chevis, C.; Hatchett, W. Adrenal Gland and Lung Lesions in Gulf of Mexico Common Bottlenose Dolphins (*Tursiops truncatus*) Found Dead Following the Deepwater Horizon Oil Spill. *PLoS ONE* **2015**, *10*, e0126538. [CrossRef]
148. Burek-Huntington, K.A.; Dushane, J.L.; Goertz, C.E.; Romero, C.H.; Raverty, S.A. Morbidity and Mortality in Stranded Cook Inlet Beluga Whales (*Delphinapterus leucas*). *Dis. Aquat. Org.* **2015**, *114*, 45–60. [CrossRef] [PubMed]
149. Greenwald, K.M.; Gibble, C.M.; Miller, M.A.; Donnelly-Greenan, E.; Kudela, R.M. Investigation of a Mass Stranding Event Reveals a Novel Pattern of Cascading Comorbidities in Northern Fulmars (*Fulmarus glacialis*). *J. Wildl. Dis.* **2024**, *60*, 171–178. [CrossRef]
150. Bargu, S.; Silver, M.W.; Ohman, M.D.; Benitez-Nelson, C.R.; Garrison, D.L. Mystery behind Hitchcock's Birds. *Nat. Geosci.* **2012**, *5*, 2–3. [CrossRef]
151. Peery, M.Z.; Beissinger, S.R.; Burkett, E.; Newman, S.H. Local Survival of Marbled Murrelets in Central California: Roles of Oceanographic Processes, Sex, and Radiotagging. *J. Wildl. Manag.* **2006**, *70*, 78–88. [CrossRef]
152. Lefebvre, K.A.; Dovel, S.L.; Silver, M.W. Tissue Distribution and Neurotoxic Effects of Domoic Acid in a Prominent Vector Species, the Northern Anchovy *Engraulis mordax*. *Mar. Biol.* **2001**, *138*, 693–700. [CrossRef]
153. Reizopoulou, S.; Strogyloudi, E.; Giannakourou, A.; Granéli, E.; Pagou, K. Toxin Accumulation in Benthic Populations under Blooms of *Dinophysis acuminata* and *Pseudo-nitzschia multiseries*. *Mar. Drugs* **2012**, *121*, 156–167.
154. Cook, K.B.; Lacaze, J.-P.; Machairopoulou, M.; Bresnan, E. Investigations into the Relationship between Domoic Acid and Copepods in Scottish Waters. *ICES J. Mar. Sci.* **2022**, *79*, 963–973. [CrossRef]
155. Beltrán-Solís, K.; García-Mendoza, E.; Sánchez-Serrano, S.; López, L.M. Domoic Acid Affects Brain Morphology and Causes Behavioral Alterations in Two Fish Species. *Sci. Rep.* **2023**, *13*, 21729. [CrossRef]
156. Panlilio, J.M.; Hammar, K.M.; Aluru, N.; Hahn, M.E. Developmental Exposure to Domoic Acid Targets Reticulospinal Neurons and Leads to Aberrant Myelination in the Spinal Cord. *Sci. Rep.* **2023**, *13*, 2587. [CrossRef]
157. Schaffer, P.; Reeves, C.; Casper, D.; Davis, C. Absence of Neurotoxic Effects in Leopard Sharks, *Triakis semifasciata*, Following Domoic Acid Exposure. *Toxicon* **2006**, *47*, 747–752. [CrossRef]
158. Hong, Z.; Zhang, Y.; Zuo, Z.; Zhu, R.; Gao, Y. Influences of Domoic Acid Exposure on Cardiac Development and the Expression of Cardiovascular Relative Genes in Zebrafish (*Danio rerio*) Embryos. *J. Biochem. Mol. Toxicol.* **2015**, *29*, 254–260. [CrossRef]
159. Jones, T.O.; Whyte, J.N.; Ginther, N.G.; Townsend, L.D.; Iwama, G.K. Haemocyte Changes in the Pacific Oyster, Crassostrea Gigas, Caused by Exposure to Domoic Acid in the Diatom *Pseudonitzschia pungens* F. *multiseries*. *Toxicon* **1995**, *33*, 347–353. [CrossRef]

160. Jones, T.; Whyte, J.; Townsend, L.; Ginther, N.; Iwama, G. Effects of Domoic Acid on Haemolymph pH, PCO2 and PO2 in the Pacific Oyster, *Crassostrea gigas* and the California Mussel, *Mytilus californianus*. *Aquat. Toxicol.* **1995**, *31*, 43–55. [CrossRef]
161. Shaw, B.; Andersen, R.; Harrison, P. Feeding Deterrent and Toxicity Effects of Apo-Fucoxanthinoids and Phycotoxins on a Marine Copepod (*Tigriopus californicus*). *Mar. Biol.* **1997**, *128*, 273–280. [CrossRef]
162. Dizer, H.; Fischer, B.; Harabawy, A.; Hennion, M.-C.; Hansen, P.-D. Toxicity of Domoic Acid in the Marine Mussel *Mytilus edulis*. *Aquat. Toxicol.* **2001**, *55*, 149–156. [CrossRef] [PubMed]
163. Bargu, S.; Marinovic, B.; Mansergh, S.; Silver, M.W. Feeding Responses of Krill to the Toxin-Producing Diatom *Pseudo-nitzschia*. *J. Exp. Mar. Biol. Ecol.* **2003**, *284*, 87–104. [CrossRef]
164. Liu, H.; Kelly, M.S.; Campbell, D.A.; Dong, S.L.; Zhu, J.X.; Wang, S.F. Exposure to Domoic Acid Affects Larval Development of King Scallop *Pecten maximus* (Linnaeus, 1758). *Aquat. Toxicol.* **2007**, *81*, 152–158. [CrossRef]
165. Liu, H.; Kelly, M.S.; Campbell, D.A.; Fang, J.; Zhu, J. Accumulation of Domoic Acid and Its Effect on Juvenile King Scallop *Pecten maximus* (Linnaeus, 1758). *Aquaculture* **2008**, *284*, 224–230. [CrossRef]
166. Tammilehto, A.; Nielsen, T.G.; Krock, B.; Møller, E.F.; Lundholm, N. Calanus spp.—Vectors for the Biotoxin, Domoic Acid, in the Arctic Marine Ecosystem? *Harmful Algae* **2012**, *20*, 165–174. [CrossRef]
167. Harðardóttir, S.; Pančić, M.; Tammilehto, A.; Krock, B.; Møller, E.F.; Nielsen, T.G.; Lundholm, N. Dangerous Relations in the Arctic Marine Food Web: Interactions between Toxin Producing *Pseudo-nitzschia* Diatoms and *Calanus* Copepodites. *Mar. Drugs* **2015**, *13*, 3809–3835. [CrossRef]
168. Harðardóttir, S.; Krock, B.; Wohlrab, S.; John, U.; Nielsen, T.G.; Lundholm, N. Can Domoic Acid Affect Escape Response in Copepods? *Harmful Algae* **2018**, *79*, 50–52. [CrossRef] [PubMed]
169. Pazos, A.J.; Ventoso, P.; Martínez-Escauriaza, R.; Pérez-Parallé, M.L.; Blanco, J.; Triviño, J.C.; Sánchez, J.L. Transcriptional Response after Exposure to Domoic Acid-Producing *Pseudo-nitzschia* in the Digestive Gland of the Mussel *Mytilus galloprovincialis*. *Toxicon* **2017**, *140*, 60–71. [CrossRef] [PubMed]
170. Chi, C.; Zhang, C.; Liu, J.; Zheng, X. Effects of Marine Toxin Domoic Acid on Innate Immune Responses in Bay Scallop *Argopecten irradians*. *J. Mar. Sci. Eng.* **2019**, *7*, 407. [CrossRef]
171. Song, J.A.; Choi, C.Y.; Park, H.-S. Exposure to Domoic Acid Causes Oxidative Stress in Bay Scallops *Argopecten irradians*. *Fish. Sci.* **2020**, *86*, 701–709. [CrossRef]
172. Ventoso, P.; Pazos, A.J.; Pérez-Parallé, M.L.; Blanco, J.; Triviño, J.C.; Sánchez, J.L. RNA-Seq Transcriptome Profiling of the Queen Scallop (*Aequipecten opercularis*) Digestive Gland after Exposure to Domoic Acid-Producing *Pseudo-nitzschia*. *Toxins* **2019**, *11*, 97. [CrossRef]
173. Grant, K.S.; Burbacher, T.M.; Faustman, E.M.; Grattan, L. Domoic Acid: Neurobehavioral Consequences of Exposure to a Prevalent Marine Biotoxin. *Neurotoxicol. Teratol.* **2010**, *32*, 132–141. [CrossRef] [PubMed]
174. Dakshinamurti, K.; Sharma, S.; Sundaram, M.; Watanabe, T. Hippocampal Changes in Developing Postnatal Mice Following Intrauterine Exposure to Domoic Acid. *J. Neurosci.* **1993**, *13*, 4486–4495. [CrossRef] [PubMed]
175. Ramsdell, J.S.; Zabka, T.S. In Utero Domoic Acid Toxicity: A Fetal Basis to Adult Disease in the California Sea Lion (*Zalophus californianus*). *Mar. Drugs* **2008**, *6*, 262–290. [CrossRef] [PubMed]
176. Grant, K.S.; Crouthamel, B.; Kenney, C.; McKain, N.; Petroff, R.; Shum, S.; Jing, J.; Isoherranen, N.; Burbacher, T.M. Preclinical Modeling of Exposure to a Global Marine Bio-Contaminant: Effects of in Utero Domoic Acid Exposure on Neonatal Behavior and Infant Memory. *Neurotoxicol. Teratol.* **2019**, *73*, 1–8. [CrossRef]
177. Azcoitia, I.; Sierra, A.; Veiga, S.; Honda, S.; Harada, N.; Garcia-Segura, L.M. Brain Aromatase Is Neuroprotective. *J. Neurobiol.* **2001**, *47*, 318–329. [CrossRef] [PubMed]
178. Sack, R.L.; Lewy, A.J.; Erb, D.L.; Vollmer, W.M.; Singer, C.M. Human Melatonin Production Decreases with Age. *J. Pineal Res.* **1986**, *3*, 379–388. [CrossRef]
179. Hendrix, A.M.; Lefebvre, K.A.; Bowers, E.K.; Stuppard, R.; Burbacher, T.; Marcinek, D.J. Age and Sex as Determinants of Acute Domoic Acid Toxicity in a Mouse Model. *Toxins* **2023**, *15*, 259. [CrossRef] [PubMed]
180. Tiedeken, J.A.; Ramsdell, J.S.; Ramsdell, A.F. Developmental Toxicity of Domoic Acid in Zebrafish (*Danio rerio*). *Neurotoxicol. Teratol.* **2005**, *27*, 711–717. [CrossRef] [PubMed]
181. Lefebvre, K.A.; Kendrick, P.S.; Ladiges, W.; Hiolski, E.M.; Ferriss, B.E.; Smith, D.R.; Marcinek, D.J. Chronic Low-Level Exposure to the Common Seafood Toxin Domoic Acid Causes Cognitive Deficits in Mice. *Harmful Algae* **2017**, *64*, 20–29. [CrossRef] [PubMed]
182. Goldstein, T.; Zabka, T.S.; DeLong, R.L.; Wheeler, E.A.; Ylitalo, G.; Bargu, S.; Silver, M.; Leighfield, T.; Dolah, F.V.; Langlois, G. The Role of Domoic Acid in Abortion and Premature Parturition of California Sea Lions (*Zalophus californianus*) on San Miguel Island, California. *J. Wildl. Dis.* **2009**, *45*, 91–108. [CrossRef] [PubMed]
183. Scallet, A.; Binienda, Z.; Caputo, F.; Hall, S.; Paule, M.; Rountree, R.; Schmued, L.; Sobotka, T.; Slikker Jr, W. Domoic Acid-Treated Cynomolgus Monkeys (*M. fascicularis*): Effects of Dose on Hippocampal Neuronal and Terminal Degeneration. *Brain Res.* **1993**, *627*, 307–313. [CrossRef] [PubMed]

184. Truelove, J.; Mueller, R.; Pulido, O.; Iverson, F. Subchronic Toxicity Study of Domoic Acid in the Rat. *Food Chem. Toxicol.* **1996**, *34*, 525–529. [CrossRef]
185. Montie, E.W.; Wheeler, E.; Pussini, N.; Battey, T.W.; Van Bonn, W.; Gulland, F. Magnetic Resonance Imaging Reveals That Brain Atrophy Is More Severe in Older California Sea Lions with Domoic Acid Toxicosis. *Harmful Algae* **2012**, *20*, 19–29. [CrossRef]
186. Moriarty, M.E.; Tinker, M.T.; Miller, M.A.; Tomoleoni, J.A.; Staedler, M.M.; Fujii, J.A.; Batac, F.I.; Dodd, E.M.; Kudela, R.M.; Zubkousky-White, V. Exposure to Domoic Acid Is an Ecological Driver of Cardiac Disease in Southern Sea Otters. *Harmful Algae* **2021**, *101*, 101973. [CrossRef] [PubMed]
187. Elorriaga-Verplancken, F.R.; Hernandez-Camacho, C.J.; Álvarez-Santamaría, L.; Paniagua-Mendoza, A.; Robles-Hernandez, R.; Rebolledo-Villa, F.; Rosales-Nanduca, H.; Ramos-Rodriguez, A.; Acevedo-Whitehouse, K. Largest Mortality Event to Date of California Sea Lions in Mexico Might Be Linked to a Harmful Algal Bloom. *Aquat. Mamm.* **2022**, *48*, 59–67. [CrossRef]
188. Greig, D.J.; Gulland, F.M.; Kreuder, C. A Decade of Live California Sea Lion (*Zalophus californianus*) Strandings along the Central California Coast: Causes and Trends, 1991–2000. *Aquat. Mamm.* **2005**, *31*, 11. [CrossRef]
189. Marriott, A.L.; Ryan, C.L.; Doucette, T.A. Neonatal Domoic Acid Treatment Produces Alterations to Prepulse Inhibition and Latent Inhibition in Adult Rats. *Pharmacol. Biochem. Behav.* **2012**, *103*, 338–344. [CrossRef]
190. Zuloaga, D.; Lahvis, G.; Mills, B.; Pearce, H.; Turner, J.; Raber, J. Fetal Domoic Acid Exposure Affects Lateral Amygdala Neurons, Diminishes Social Investigation and Alters Sensory-Motor Gating. *Neurotoxicology* **2016**, *53*, 132–140. [CrossRef] [PubMed]
191. Thompson, R.G. Gender Differences in the Renal Toxicity of the Neurotoxin Domoic Acid Griffin Thompson, Hernan Grenett, Lan He, Noor Alsirafi, Amanda Hanninen, and P. Darwin Bell University of Alabama Birmingham, and Birmingham VA, Birmingham, AL. *FASEB J.* **2017**, *31*, 1030.23. [CrossRef]
192. Baron, A.W.; Rushton, S.P.; Rens, N.; Morris, C.M.; Blain, P.G.; Judge, S.J. Sex Differences in Effects of Low Level Domoic Acid Exposure. *Neurotoxicology* **2013**, *34*, 1–8. [CrossRef] [PubMed]
193. Doucette, T.A.; Ryan, C.L.; Tasker, R. Gender-Based Changes in Cognition and Emotionality in a New Rat Model of Epilepsy. *Amino Acids* **2007**, *32*, 317–322. [CrossRef] [PubMed]
194. Ryan, C.L.; Robbins, M.A.; Smith, M.T.; Gallant, I.C.; Adams-Marriott, A.L.; Doucette, T.A. Altered Social Interaction in Adult Rats Following Neonatal Treatment with Domoic Acid. *Physiol. Behav.* **2011**, *102*, 291–295. [CrossRef] [PubMed]
195. Gill, D.A.; Perry, M.A.; McGuire, E.P.; Pérez-Gómez, A.; Tasker, R.A. Low-Dose Neonatal Domoic Acid Causes Persistent Changes in Behavioural and Molecular Indicators of Stress Response in Rats. *Behav. Brain Res.* **2012**, *230*, 409–417. [CrossRef] [PubMed]
196. Mills, B.D.; Pearce, H.L.; Khan, O.; Jarrett, B.R.; Fair, D.A.; Lahvis, G.P. Prenatal Domoic Acid Exposure Disrupts Mouse Pro-Social Behavior and Functional Connectivity MRI. *Behav. Brain Res.* **2016**, *308*, 14–23. [CrossRef]
197. Teitelbaum, J.S.; Zatorre, R.J.; Carpenter, S.; Gendron, D.; Evans, A.C.; Gjedde, A.; Cashman, N.R. Neurologic Sequelae of Domoic Acid Intoxication Due to the Ingestion of Contaminated Mussels. *N. Engl. J. Med.* **1990**, *322*, 1781–1787. [CrossRef] [PubMed]
198. Schroeder, G.; Bates, S.; Spallino, J. Amnesic Shellfish Poisoning. Emergency Medical Management. *J. Mar. Sci. Res. Dev.* **2015**, *6*, 2. [CrossRef]
199. Kreuder, C.; Miller, M.A.; Lowenstine, L.J.; Conrad, P.A.; Carpenter, T.E.; Jessup, D.A.; Mazet, J.A. Evaluation of Cardiac Lesions and Risk Factors Associated with Myocarditis and Dilated Cardiomyopathy in Southern Sea Otters (*Enhydra lutris nereis*). *Am. J. Vet. Res.* **2005**, *66*, 289–299. [CrossRef]
200. Colegrove, K.; Gulland, F.; Harr, K.; Naydan, D.; Lowenstine, L. Pathological Features of Amyloidosis in Stranded California Sea Lions (*Zalophus californianus*). *J. Comp. Pathol.* **2009**, *140*, 105–112. [CrossRef] [PubMed]
201. Field, C.L.; Gulland, F.M.; Johnson, S.P.; Simeone, C.A.; Whoriskey, S.T. Seal and Sea Lion Medicine. In *CRC Handbook of Marine Mammal Medicine*; CRC Press: Boca Raton, FL, USA, 2018; pp. 909–934.
202. Knowles, S.; Lynch, D.; Thomas, N. Leptospirosis in Northern Sea Otters (*Enhydra lutris kenyoni*) from Washington, USA. *J. Wildl. Dis.* **2020**, *56*, 466–471. [CrossRef] [PubMed]
203. Varshavsky, J.; Smith, A.; Wang, A.; Hom, E.; Izano, M.; Huang, H.; Padula, A.; Woodruff, T.J. Heightened Susceptibility: A Review of How Pregnancy and Chemical Exposures Influence Maternal Health. *Reprod. Toxicol.* **2020**, *92*, 14–56. [CrossRef] [PubMed]
204. Broadwater, M.H.; Van Dolah, F.M.; Fire, S.E. Vulnerabilities of Marine Mammals to Harmful Algal Blooms. In *Harmful Algal Blooms: A Compendium Desk Reference*; Wiley Online Library: Hoboken, NJ, USA, 2018; pp. 191–222.
205. Bargu, S.; Silver, M.; Goldstein, T.; Roberts, K.; Gulland, F. Complexity of Domoic Acid-Related Sea Lion Strandings in Monterey Bay, California: Foraging Patterns, Climate Events, and Toxic Blooms. *Mar. Ecol. Prog. Ser.* **2010**, *418*, 213–222. [CrossRef]
206. Grattan, L.M.; Holobaugh, S.; Morris, J.G., Jr. Harmful Algal Blooms and Public Health. *Harmful Algae* **2016**, *57*, 2–8. [CrossRef]
207. Grattan, L.M.; Kaddis, L.; Tracy, J.K.; Morris, J.G. Long Term Memory Outcome of Repetitive, Low-Level Dietary Exposure to Domoic Acid in Native Americans. *Int. J. Environ. Res. Public Health* **2021**, *18*, 3955. [CrossRef]
208. Gill, D.; Bastlund, J.; Watson, W.; Ryan, C.; Reynolds, D.; Tasker, R. Neonatal Exposure to Low-Dose Domoic Acid Lowers Seizure Threshold in Adult Rats. *Neuroscience* **2010**, *169*, 1789–1799. [CrossRef]

209. McClain, A.M.; Field, C.L.; Norris, T.A.; Borremans, B.; Duignan, P.J.; Johnson, S.P.; Whoriskey, S.T.; Thompson-Barbosa, L.; Gulland, F.M. The Symptomatology and Diagnosis of Domoic Acid Toxicosis in Stranded California Sea Lions (*Zalophus californianus*): A Review and Evaluation of 20 Years of Cases to Guide Prognosis. *Front. Vet. Sci.* **2023**, *10*, 1245864. [CrossRef]
210. Santora, J.A.; Mantua, N.J.; Schroeder, I.D.; Field, J.C.; Hazen, E.L.; Bograd, S.J.; Sydeman, W.J.; Wells, B.K.; Calambokidis, J.; Saez, L. Habitat Compression and Ecosystem Shifts as Potential Links between Marine Heatwave and Record Whale Entanglements. *Nat. Commun.* **2020**, *11*, 536. [CrossRef] [PubMed]
211. Fauquier, D.; Landsberg, J. Harmful Algae and Biotoxins. In *CRC handbook of Marine Mammal Medicine*; CRC Press: Boca Raton, FL, USA, 2018; pp. 319–328.
212. Hubbard, K.A.; Villac, M.C.; Chadwick, C.; DeSmidt, A.A.; Flewelling, L.; Granholm, A.; Joseph, M.; Wood, T.; Fachon, E.; Brosnahan, M.L. Spatiotemporal Transitions in *Pseudo-nitzschia* Species Assemblages and Domoic Acid along the Alaska Coast. *PLoS ONE* **2023**, *18*, e0282794. [CrossRef] [PubMed]
213. Shum, S.; Kirkwood, J.S.; Jing, J.; Petroff, R.; Crouthamel, B.; Grant, K.S.; Burbacher, T.M.; Nelson, W.L.; Isoherranen, N. Validated HPLC-MS/MS Method to Quantify Low Levels of Domoic Acid in Plasma and Urine after Subacute Exposure. *ACS Omega* **2018**, *3*, 12079–12088. [CrossRef] [PubMed]
214. Ben-Gigirey, B.; Soliño, L.; Bravo, I.; Rodríguez, F.; Casero, M.V. Paralytic and Amnesic Shellfish Toxins Impacts on Seabirds, Analyses and Management. *Toxins* **2021**, *13*, 454. [CrossRef] [PubMed]
215. Frame, E.R.; Lefebvre, K.A. *ELISA Methods for Domoic Acid Quantification in Multiple Marine Mammal Species and Sample Matrices*; NOAA Technical Memorandum NMFS-NWFSC-122; National Oceanic and Atmospheric Administration: Washington, DC, USA, 2013; 20p.
216. Seubert, E.L.; Howard, M.D.; Kudela, R.M.; Stewart, T.N.; Litaker, R.W.; Evans, R.; Caron, D.A. Development, Comparison, and Validation Using ELISAs for the Determination of Domoic Acid in California Sea Lion Body Fluids. *J. AOAC Int.* **2014**, *97*, 345–355. [CrossRef]
217. Gulland, F.M.; Hall, A.J.; Greig, D.J.; Frame, E.R.; Colegrove, K.M.; Booth, R.K.; Wasser, S.K.; Scott-Moncrieff, J.C.R. Evaluation of Circulating Eosinophil Count and Adrenal Gland Function in California Sea Lions Naturally Exposed to Domoic Acid. *J. Am. Vet. Med. Assoc.* **2012**, *241*, 943–949. [CrossRef]
218. Williams, K.M. Clinical Values of Blood Variables in Wild and Stranded California Sea Lions (*Zalophus californianus*) and Blood Sample Storage Stability. Master's Thesis, California State University Monterey Bay and Moss Landing Marine Laboratories, Moss Landing, CA, USA, 2013.
219. D'Agostino, V.C.; Fernández Ajó, A.; Degrati, M.; Krock, B.; Hunt, K.E.; Uhart, M.M.; Buck, C.L. Potential Endocrine Correlation with Exposure to Domoic Acid in Southern Right Whale (*Eubalaena australis*) at the Península Valdés Breeding Ground. *Oecologia* **2022**, *198*, 21–34. [CrossRef]
220. Arufe, M.; Arias, B.; Duran, R.; Alfonso, M. Effects of Domoic Acid on Serum Levels of TSH and Thyroid Hormones. *Endocr. Res.* **1995**, *21*, 671–680. [CrossRef] [PubMed]
221. Alfonso, M.; Duran, R.; Arufe, M. Effect of Excitatory Amino Acids on Serum TSH and Thyroid Hormone Levels in Freely Moving Rats. *Horm. Res.* **2000**, *54*, 78–83. [CrossRef]
222. Bowen, L.; Knowles, S.; Lefebvre, K.; St. Martin, M.; Murray, M.; Kloecker, K.; Monson, D.; Weitzman, B.; Ballachey, B.; Coletti, H. Divergent Gene Expression Profiles in Alaskan Sea Otters: An Indicator of Chronic Domoic Acid Exposure? *Oceans* **2022**, *3*, 401–418. [CrossRef]
223. Joblon, M.J.; Flower, J.E.; Thompson, L.A.; Biddle, K.E.; Burt, D.A.; Zabka, T.S.; Adkesson, M.J.; Halaska, B.; Goertz, C.E.; Rouse, N. Investigation of the Use of Serum Biomarkers for the Detection of Cardiac Disease in Marine Mammals. *J. Zoo Wildl. Med.* **2022**, *53*, 373–382. [CrossRef]
224. Moriarty, M.E.; Miller, M.A.; Murray, M.J.; Duignan, P.J.; Gunther-Harrington, C.T.; Field, C.L.; Adams, L.M.; Schmitt, T.L.; Johnson, C.K. Exploration of Serum Cardiac Troponin I as a Biomarker of Cardiomyopathy in Southern Sea Otters (*Enhydra lutris nereis*). *Am. J. Vet. Res.* **2021**, *82*, 529–537. [CrossRef]
225. Gunther-Harrington, C.T.; Moriarty, M.E.; Field, C.L.; Adams, L.M.; Johnson, C.K.; Murray, M.J. Transthoracic Echocardiographic Evaluation and Serum Cardiac Troponin Values in Anesthetized Healthy Female Southern Sea Otters (*Enhydra lutris nereis*). *J. Zoo Wildl. Med.* **2021**, *52*, 490–498. [CrossRef]
226. Williams, D.C.; Haulena, M.; Dennison, S.; Waugh, L.; Goldstein, T.; Nutter, F.; Bonn, B.V.; Hoard, V.; Laxer, K.D.; Buckmaster, P.S. Pinniped Electroencephalography: Methodology and Findings in California Sea Lions (*Zalophus californianus*). *Front. Vet. Sci.* **2023**, *10*, 1040125. [CrossRef] [PubMed]
227. Fujita, T.; Tanaka, T.; Yonemasu, Y.; Cendes, F.; Cashman, N.R.; Andermann, F. Electroclinical and Pathological Studies after Parenteral Administration of Domoic Acid in Freely Moving Nonanesthetized Rats: An Animal Model of Excitotoxicity. *J. Epilepsy* **1996**, *9*, 87–93. [CrossRef]

228. Petroff, R.; Murias, M.; Grant, K.; Crouthamel, B.; McKain, N.; Shum, S.; Jing, J.; Isoherranen, N.; Burbacher, T. Power Spectrum Analysis of EEG in a Translational Nonhuman Primate Model after Chronic Exposure to Low Levels of the Common Marine Neurotoxin, Domoic Acid. *Neurotoxicology* **2020**, *80*, 124–129. [CrossRef]
229. Gjedde, A.; Evans, A. PET Studies of Domoic Acid Poisoning in Humans: Excitotoxic Destruction of Brain Glutamatergic Pathways, Revealed in Measurements of Glucose Metabolism by Positron Emission Tomography. *Can. Dis. Wkly. Rep.=Rapp. Hebd. Des Mal. Au Can.* **1990**, *16*, 105–109.
230. Cook, P.F.; Reichmuth, C.; Rouse, A.A.; Libby, L.A.; Dennison, S.E.; Carmichael, O.T.; Kruse-Elliott, K.T.; Bloom, J.; Singh, B.; Fravel, V.A. Algal Toxin Impairs Sea Lion Memory and Hippocampal Connectivity, with Implications for Strandings. *Science* **2015**, *350*, 1545–1547. [CrossRef] [PubMed]
231. Krucik, D.D.; Cook, P.; Cathey, M.; Meegan, J.M.; Gomez, F.M.; Van Bonn, W.; Le-Bert, C. Adult-Onset Epilepsy and Hippocampal Pathology in a California Sea Lion (*Zalophus californianus*): A Case Study of Suspected in Utero Exposure to Domoic Acid. *Neurotoxicology* **2023**, *96*, 13–18. [CrossRef] [PubMed]
232. Cook, P.F.; Berns, G.S.; Colegrove, K.; Johnson, S.; Gulland, F. Postmortem DTI Reveals Altered Hippocampal Connectivity in Wild Sea Lions Diagnosed with Chronic Toxicosis from Algal Exposure. *J. Comp. Neurol.* **2018**, *526*, 216–228. [CrossRef] [PubMed]
233. Thomas, K.; Harvey, J.T.; Goldstein, T.; Barakos, J.; Gulland, F. Movement, Dive Behavior, and Survival of California Sea Lions (*Zalophus californianus*) Posttreatment for Domoic Acid Toxicosis. *Mar. Mammal Sci.* **2010**, *26*, 36–52. [CrossRef]
234. Zatorre, R. Memory Loss Following Domoic Acid Intoxication from Ingestion of Toxic Mussels. *Can. Dis. Wkly. Rep.=Rapp. Hebd. Des Mal. Au Can.* **1990**, *16*, 101–104.
235. Cook, P.; Reichmuth, C.; Gulland, F. Rapid Behavioural Diagnosis of Domoic Acid Toxicosis in California Sea Lions. *Biol. Lett.* **2011**, *7*, 536–538. [CrossRef]
236. De Maio, L.M.; Cook, P.F.; Reichmuth, C.; Gulland, F.M. The Evaluation of Olfaction in Stranded California Sea Lions (*Zalophus californianus*) and Its Relevance to Domoic Acid Toxicosis. *Aquat. Mamm.* **2018**, *44*, 231–238. [CrossRef]
237. Nakajima, S.; Potvin, J.L. Neural and Behavioural Effects of Domoic Acid, an Amnesic Shellfish Toxin, in the Rat. *Can. J. Psychol./Rev. Can. De Psychol.* **1992**, *46*, 569. [CrossRef]
238. Petrie, B.; Pinsky, C.; Standish, N.; Bose, R.; Glavin, G. Parenteral Domoic Acid Impairs Spatial Learning in Mice. *Pharmacol. Biochem. Behav.* **1992**, *41*, 211–214. [CrossRef] [PubMed]
239. Sobotka, T.; Brown, R.; Quander, D.; Jackson, R.; Smith, M.; Long, S.; Barton, C.; Rountree, R.; Hall, S.; Eilers, P. Domoic Acid: Neurobehavioral and Neurohistological Effects of Low-Dose Exposure in Adult Rats. *Neurotoxicol. Teratol.* **1996**, *18*, 659–670. [CrossRef]
240. Shiotani, M.; Cole, T.B.; Hong, S.; Park, J.J.Y.; Griffith, W.C.; Burbacher, T.M.; Workman, T.; Costa, L.G.; Faustman, E.M. Neurobehavioral Assessment of Mice Following Repeated Oral Exposures to Domoic Acid during Prenatal Development. *Neurotoxicol. Teratol.* **2017**, *64*, 8–19. [CrossRef] [PubMed]
241. Panlilio, J.M.; Aluru, N.; Hahn, M.E. Developmental Neurotoxicity of the Harmful Algal Bloom Toxin Domoic Acid: Cellular and Molecular Mechanisms Underlying Altered Behavior in the Zebrafish Model. *Environ. Health Perspect.* **2020**, *128*, 117002. [CrossRef]
242. Scallet, A.C.; Schmued, L.C.; Johannessen, J.N. Neurohistochemical Biomarkers of the Marine Neurotoxicant, Domoic Acid. *Neurotoxicol. Teratol.* **2005**, *27*, 745–752. [CrossRef] [PubMed]
243. Sutherland, R.; Hoesing, J.; Whishaw, I. Domoic Acid, an Environmental Toxin, Produces Hippocampal Damage and Severe Memory Impairment. *Neurosci. Lett.* **1990**, *120*, 221–223. [CrossRef] [PubMed]
244. Buckmaster, P.S.; Wen, X.; Toyoda, I.; Gulland, F.M.; Van Bonn, W. Hippocampal Neuropathology of Domoic Acid–Induced Epilepsy in California Sea Lions (*Zalophus californianus*). *J. Comp. Neurol.* **2014**, *522*, 1691–1706. [CrossRef]
245. Tryphonas, L.; Truelove, J.; Iverson, F. Acute Parenteral Neurotoxicity of Domoic Acid in Cynomolgus Monkeys (*M. fascicularis*). *Toxicol. Pathol.* **1990**, *18*, 297–303. [CrossRef] [PubMed]
246. Bossart, G.D. Marine Mammals as Sentinel Species for Oceans and Human Health. *Vet. Pathol.* **2011**, *48*, 676–690. [CrossRef]
247. Neely, B.A.; Soper, J.L.; Gulland, F.M.; Bell, P.D.; Kindy, M.; Arthur, J.M.; Janech, M.G. Proteomic Analysis of Cerebrospinal Fluid in California Sea Lions (*Zalophus californianus*) with Domoic Acid Toxicosis Identifies Proteins Associated with Neurodegeneration. *Proteomics* **2015**, *15*, 4051–4063. [CrossRef] [PubMed]
248. Mancia, A.; Ryan, J.C.; Chapman, R.W.; Wu, Q.; Warr, G.W.; Gulland, F.M.; Van Dolah, F.M. Health Status, Infection and Disease in California Sea Lions (*Zalophus californianus*) Studied Using a Canine Microarray Platform and Machine-Learning Approaches. *Dev. Comp. Immunol.* **2012**, *36*, 629–637. [CrossRef]
249. Hiolski, E.M.; Kendrick, P.S.; Frame, E.R.; Myers, M.S.; Bammler, T.K.; Beyer, R.P.; Farin, F.M.; Wilkerson, H.; Smith, D.R.; Marcinek, D.J. Chronic Low-Level Domoic Acid Exposure Alters Gene Transcription and Impairs Mitochondrial Function in the CNS. *Aquat. Toxicol.* **2014**, *155*, 151–159. [CrossRef] [PubMed]

250. Neely, B.A.; Ferrante, J.A.; Chaves, J.M.; Soper, J.L.; Almeida, J.S.; Arthur, J.M.; Gulland, F.M.; Janech, M.G. Proteomic Analysis of Plasma from California Sea Lions (*Zalophus californianus*) Reveals Apolipoprotein E as a Candidate Biomarker of Chronic Domoic Acid Toxicosis. *PLoS ONE* **2015**, *10*, e0123295. [CrossRef]
251. Lefebvre, K.A.; Tilton, S.C.; Bammler, T.K.; Beyer, R.P.; Srinouanprachan, S.; Stapleton, P.L.; Farin, F.M.; Gallagher, E.P. Gene Expression Profiles in Zebrafish Brain after Acute Exposure to Domoic Acid at Symptomatic and Asymptomatic Doses. *Toxicol. Sci.* **2009**, *107*, 65–77. [CrossRef] [PubMed]
252. Neely, B.A.; Soper, J.L.; Greig, D.J.; Carlin, K.P.; Favre, E.G.; Gulland, F.M.; Almeida, J.S.; Janech, M.G. Serum Profiling by MALDI-TOF Mass Spectrometry as a Diagnostic Tool for Domoic Acid Toxicosis in California Sea Lions. *Proteome Sci.* **2012**, *10*, 1–12. [CrossRef]
253. Ryan, J.; Morey, J.; Ramsdell, J.; Van Dolah, F. Acute Phase Gene Expression in Mice Exposed to the Marine Neurotoxin Domoic Acid. *Neuroscience* **2005**, *136*, 1121–1132. [CrossRef]
254. Guillotin, S.; Delcourt, N. Marine Neurotoxins' Effects on Environmental and Human Health: An OMICS Overview. *Mar. Drugs* **2021**, *20*, 18. [CrossRef]
255. Hidayat, A.S.; Lefebvre, K.A.; MacDonald, J.; Bammler, T.; Aluru, N. Symptomatic and Asymptomatic Domoic Acid Exposure in Zebrafish (*Danio rerio*) Revealed Distinct Non-Overlapping Gene Expression Patterns in the Brain. *Aquat. Toxicol.* **2022**, *252*, 106310. [CrossRef]
256. Ghosh, G. Identification of Candidate Protein Biomarkers Associated with Domoic Acid Toxicosis in Cerebrospinal Fluid of California Sea Lions (*Zalophus californianus*). *J. Proteome Res.* **2024**, *23*, 2419–2430. [CrossRef]
257. Simeone, C.A.; Scott, G.; Navarro, R.A.; Procter, D. Clinical Observations Associated with Phenobarbital Serum Monitoring to Manage Epilepsy in a California Sea Lion with Domoic Acid Toxicosis. *Oceans* **2022**, *3*, 331–339. [CrossRef]
258. Simeone, C.A.; Andrews, J.P.; Johnson, S.P.; Casalia, M.; Kochanski, R.; Chang, E.F.; Cameron, D.; Dennison, S.; Inglis, B.; Scott, G. Xenotransplantation of Porcine Progenitor Cells in an Epileptic California Sea Lion (*Zalophus californianus*): Illustrative Case. *J. Neurosurg. Case Lessons* **2022**, 3. [CrossRef]
259. Simeone, C.A.; Stoskopf, M.K. Pharmaceuticals and Formularies. In *CRC Handbook of Marine Mammal Medicine*; CRC Press: Boca Raton, FL, USA, 2018; pp. 607–674.
260. Hedley, J. *BSAVA Small Animal Formulary. Part B: Exotic Pets*; John Wiley & Sons, Inc.: Hoboken, NJ, USA, 2020; ISBN 1-910443-71-9.
261. Dakshinamurti, K.; Sharma, S.; Geiger, J.D. Neuroprotective Actions of Pyridoxine. *Biochim. Biophys. Acta (BBA)-Proteins Proteom.* **2003**, *1647*, 225–229. [CrossRef]
262. Lu, J.; Wu, D.; Zheng, Y.; Hu, B.; Cheng, W.; Zhang, Z.; Li, M. Troxerutin Counteracts Domoic Acid–Induced Memory Deficits in Mice by Inhibiting CCAAT/Enhancer Binding Protein β–Mediated Inflammatory Response and Oxidative Stress. *J. Immunol.* **2013**, *190*, 3466–3479. [CrossRef] [PubMed]
263. Wu, D.; Lu, J.; Zhang, Y.; Zheng, Y.; Hu, B.; Cheng, W.; Zhang, Z.; Li, M. Ursolic Acid Improves Domoic Acid-Induced Cognitive Deficits in Mice. *Toxicol. Appl. Pharmacol.* **2013**, *271*, 127–136. [CrossRef] [PubMed]
264. Ananth, C.; Dheen, S.T.; Gopalakrishnakone, P.; Kaur, C. Distribution of NADPH-diaphorase and Expression of nNOS, N-methyl-D-aspartate Receptor (NMDAR1) and non-NMDA Glutamate Receptor (GlutR2) Genes in the Neurons of the Hippocampus after Domoic Acid-induced Lesions in Adult Rats. *Hippocampus* **2003**, *13*, 260–272. [CrossRef] [PubMed]
265. Lu, J.; Wu, D.; Zheng, Y.; Hu, B.; Cheng, W.; Zhang, Z. Purple Sweet Potato Color Attenuates Domoic Acid-Induced Cognitive Deficits by Promoting Estrogen Receptor-α-Mediated Mitochondrial Biogenesis Signaling in Mice. *Free Radic. Biol. Med.* **2012**, *52*, 646–659. [CrossRef] [PubMed]
266. Giordano, G.; White, C.C.; Mohar, I.; Kavanagh, T.J.; Costa, L.G. Glutathione Levels Modulate Domoic Acid–Induced Apoptosis in Mouse Cerebellar Granule Cells. *Toxicol. Sci.* **2007**, *100*, 433–444. [CrossRef]
267. Tian, D.; Zhang, G. Toxic Effects of Domoic Acid on *Caenorhabditis elegans* and the Underlying Mechanism. *Int. J. Biol.* **2019**, *11*, 1–9. [CrossRef]
268. Golechha, M.; Chaudhry, U.; Bhatia, J.; Saluja, D.; Arya, D.S. Naringin Protects against Kainic Acid-Induced Status Epilepticus in Rats: Evidence for an Antioxidant, Anti-Inflammatory and Neuroprotective Intervention. *Biol. Pharm. Bull.* **2011**, *34*, 360–365. [CrossRef]
269. Hsieh, P.F.; Hou, C.-W.; Yao, P.-W.; Wu, S.-P.; Peng, Y.-F.; Shen, M.-L.; Lin, C.-H.; Chao, Y.-Y.; Chang, M.-H.; Jeng, K.-C. Sesamin Ameliorates Oxidative Stress and Mortality in Kainic Acid-Induced Status Epilepticus by Inhibition of MAPK and COX-2 Activation. *J. Neuroinflamm.* **2011**, *8*, 1–10. [CrossRef] [PubMed]
270. Pinsky, C.; Giavim, G.B.; Bose, R. Kynurenic Acid Protects against Neurotoxicity and Lethality of Toxic Extracts from Contaminated Atlantic Coast Mussels. *Prog. Neuro-Psychopharmacol. Biol. Psychiatry* **1989**, *13*, 595–598. [CrossRef] [PubMed]
271. Glavin, G.B.; Pinsky, C. Kynurenic Acid Attenuates Experimental Ulcer Formation and Basal Gastric Acid Secretion in Rats. *Res. Commun. Chem. Pathol. Pharmacol.* **1989**, *64*, 111–119. [PubMed]
272. Glavin, G.B.; Bose, R.; Pinsky, C. Kynurenic Acid Protects against Gastroduodenal Ulceration in Mice Injected with Extracts from Poisonous Atlantic Shellfish. *Prog. Neuro-Psychopharmacol. Biol. Psychiatry* **1989**, *13*, 569–572. [CrossRef]

273. Glavin, G.B.; Pinsky, C.; Bose, R. Domoic Acid-Induced Neuro Visceral Toxic Syndrome: Characterization of an Animal Model and Putative Antidotes. *Brain Res. Bull.* **1990**, *24*, 701–703. [CrossRef]
274. Wu, D.; Lu, J.; Zheng, Y.; Zhang, Y.; Hu, B.; Cheng, W.; Zhang, Z.; Li, M. Small Interfering RNA–Mediated Knockdown of Protein Kinase C Zeta Attenuates Domoic Acid–Induced Cognitive Deficits in Mice. *Toxicol. Sci.* **2012**, *128*, 209–222. [CrossRef] [PubMed]
275. Sharma, S.; Dakshinamurti, K. Suppression of Domoic Acid Induced Seizures by 8-(OH)-DPAT. *J. Neural Transm./Gen. Sect. JNT* **1993**, *93*, 87–98. [CrossRef]
276. Field, C.L.; Whoriskey, S.T.; Zhao, X.; Papich, M.G. Pharmacokinetics of Subcutaneous Alpha Lipoic Acid, a Proposed Therapeutic Aid for Domoic Acid Intoxication in California Sea Lions (*Zalophus californianus*). *J. Zoo Wildl. Med.* **2021**, *52*, 872–879. [CrossRef]
277. Peters, M.; Duignan, P.; Martinez, M.; Papich, M.G.; Field, C. Meloxicam-Associated Nephrotoxicity in Three Adult Female California Sea Lions (*Zalophus californianus*) in Treatment for Domoic Acid Toxicosis. In Proceedings of the 54th Annual Conference of the International Association of Aquatic Animal Medicine, Salt Lake City, UT, USA, 20–24 May 2023.
278. Simeone, C.; Fauquier, D.; Skidmore, J.; Cook, P.; Colegrove, K.; Gulland, F.; Dennison, S.; Rowles, T.K. Clinical Signs and Mortality of Non-released Stranded California Sea Lions Housed in Display Facilities: The Suspected Role of Prior Exposure to Algal Toxins. *Vet. Rec.* **2019**, *185*, 304. [CrossRef] [PubMed]
279. Aguirre, A.A.; Tabor, G.M. Introduction: Marine Vertebrates as Sentinels of Marine Ecosystem Health. *EcoHealth* **2004**, *1*, 236–238. [CrossRef]
280. Wells, R.S.; Rhinehart, H.L.; Hansen, L.J.; Sweeney, J.C.; Townsend, F.I.; Stone, R.; Casper, D.R.; Scott, M.D.; Hohn, A.A.; Rowles, T.K. Bottlenose Dolphins as Marine Ecosystem Sentinels: Developing a Health Monitoring System. *EcoHealth* **2004**, *1*, 246–254. [CrossRef]
281. Basu, N.; Scheuhammer, A.M.; Bursian, S.J.; Elliott, J.; Rouvinen-Watt, K.; Chan, H.M. Mink as a Sentinel Species in Environmental Health. *Environ. Res.* **2007**, *103*, 130–144. [CrossRef] [PubMed]
282. Moore, S.E. Marine Mammals as Ecosystem Sentinels. *J. Mammal.* **2008**, *89*, 534–540. [CrossRef]
283. Hazen, E.L.; Abrahms, B.; Brodie, S.; Carroll, G.; Jacox, M.G.; Savoca, M.S.; Scales, K.L.; Sydeman, W.J.; Bograd, S.J. Marine Top Predators as Climate and Ecosystem Sentinels. *Front. Ecol. Environ.* **2019**, *17*, 565–574. [CrossRef]
284. Smith, J.; Cram, J.A.; Berndt, M.P.; Hoard, V.; Shultz, D.; Deming, A.C. Quantifying the linkages between California sea lion (*Zalophus californianus*) strandings and particulate domoic acid concentrations at piers across Southern California. *Front. Mar. Sci.* **2023**, *10*, 1278293. [CrossRef]
285. Mehta, A.; Prabhakar, M.; Kumar, P.; Deshmukh, R.; Sharma, P. Excitotoxicity: Bridge to Various Triggers in Neurodegenerative Disorders. *Eur. J. Pharmacol.* **2013**, *698*, 6–18. [CrossRef]
286. Stewart, I. Environmental Risk Factors for Temporal Lobe Epilepsy–Is Prenatal Exposure to the Marine Algal Neurotoxin Domoic Acid a Potentially Preventable Cause? *Med. Hypotheses* **2010**, *74*, 466–481. [CrossRef] [PubMed]
287. Sharma, A. Monosodium Glutamate-Induced Oxidative Kidney Damage and Possible Mechanisms: A Mini-Review. *J. Biomed. Sci.* **2015**, *22*, 1–6. [CrossRef]
288. Lahvis, G.P. What California Sea Lions Exposed to Domoic Acid Might Teach Us about Autism: Lessons for Predictive and Preventive Medicine. *EPMA J.* **2017**, *8*, 229–235. [CrossRef]
289. Syed, M.; Shangloo, P.; Gupte, B.; Gupta, S. Effect Of AJI-NO-MOTO On Kidney Of Adult Albino Rat: A Histopathological Evaluation. *Perspect. Med. Res.* **2022**, *10*, 20–30. [CrossRef]
290. Onwubiko, G.N.; Nwankwo, N.E.; Egbuonu, A.C.C.; Soribe, B.A.; Okeke, E.S.; Eze, C.J. Amelioration of High Levels of Serum Kidney Function Biomarkers by Vernonia Amygdalina in Monosodium Glutamate Induced Rats. *Indian J. Nat. Prod. Resour. (IJNPR) [Former. Nat. Prod. Radiance (NPR)]* **2022**, *13*, 197–205.
291. Gill, D.A.; Ramsay, S.L.; Tasker, R.A. Selective Reductions in Subpopulations of GABAergic Neurons in a Developmental Rat Model of Epilepsy. *Brain Res.* **2010**, *1331*, 114–123. [CrossRef] [PubMed]

marine drugs

MDPI

Review

Marine Algal Toxins and Public Health: Insights from Shellfish and Fish, the Main Biological Vectors

Kuan-Kuan Yuan, Hong-Ye Li and Wei-Dong Yang *

Key Laboratory of Aquatic Eutrophication and Control of Harmful Algal Blooms of Guangdong Higher Education Institute, College of Life Science and Technology, Jinan University, Guangzhou 510632, China; yk9723@stu2019.jnu.edu.cn (K.-K.Y.); thyli@jnu.edu.cn (H.-Y.L.)

* Correspondence: tywd@jnu.edu.cn

Abstract: Exposure to toxigenic harmful algal blooms (HABs) can result in widely recognized acute poisoning in humans. The five most commonly recognized HAB-related illnesses are diarrhetic shellfish poisoning (DSP), paralytic shellfish poisoning (PSP), amnesic shellfish poisoning (ASP), neurotoxic shellfish poisoning (NSP), and ciguatera poisoning (CP). Despite being caused by exposure to various toxins or toxin analogs, these clinical syndromes share numerous similarities. Humans are exposed to these toxins mainly through the consumption of fish and shellfish, which serve as the main biological vectors. However, the risk of human diseases linked to toxigenic HABs is on the rise, corresponding to a dramatic increase in the occurrence, frequency, and intensity of toxigenic HABs in coastal regions worldwide. Although a growing body of studies have focused on the toxicological assessment of HAB-related species and their toxins on aquatic organisms, the organization of this information is lacking. Consequently, a comprehensive review of the adverse effects of HAB-associated species and their toxins on those organisms could deepen our understanding of the mechanisms behind their toxic effects, which is crucial to minimizing the risks of toxigenic HABs to human and public health. To this end, this paper summarizes the effects of the five most common HAB toxins on fish, shellfish, and humans and discusses the possible mechanisms.

Keywords: harmful algal blooms; diarrhetic shellfish poisoning; paralytic shellfish poisoning; neurotoxic shellfish poisoning; domoic acid; ciguatera; public health

1. Introduction

Harmful algal blooms (HABs) generally refer to the proliferation of toxic and non-toxic algae that pose a threat to humans, aquatic organisms, and/or aquatic environments [1,2]. At high cell densities, non-toxic algae may deplete dissolved oxygen, thereby causing harm to the environment and aquatic organisms [3,4]. In this article, the term "HABs" specifically relates toxigenic HABs only to those algal species that produce toxins and pose a threat to human health. The serious impact of HABs on the safety of aquatic organisms and public health is increasingly appreciated with the dramatic rise in the number, frequency, and intensity of HABs in coastal areas around the world [5]. To date, the public Harmful Algae Events Database (HAEDAT, http://haedat.iode.org/, accessed on 5 November 2024) has more than 10,000 global data records.

In recent years, many toxic algal species and phycotoxins have been studied, and more geographic areas are affected than ever before. The proliferation and geographic dispersal of these HAB-related algae are generally attributed to natural environmental factors, anthropogenic activities, and climatic change [6–8]. Outbreaks of HABs may result in the death of many marine organisms, including fish and shellfish, and cause illness and even death in humans through the food chain [9–11]. As the demand for seafood rises, coastal urbanization accelerates, and more people engage in marine recreational activities, the risk of human illness associated with HABs is increasing [12–14]. As a result, a growing number of studies have focused on the toxicological assessment of HAB-related species and their toxins to aquatic organisms. However, due to the HAB toxins diversity and the complexity of toxic mechanisms, the toxicity to different aquatic organisms varies widely [10,15–18], which makes it challenging to monitor the adverse effects of HABs on these organisms and complicates the assessment of the potential threat of HABs to human health.

The consumption of seafood has become one of the main causes of food-borne diseases with a known etiology, including poisoning, allergies, and infections [19]. Additionally, HAB-related toxins are a significant source of seafood contamination globally, posing a serious threat to marine organisms, consumer health, and ecosystems [19,20]. The primary human health concerns associated with HABs stem from the consumption of fish and shellfish contaminated with these toxins [21,22]. A comprehensive review of the adverse effects of HABs on aquatic organisms can, therefore, advance our understanding of the mechanisms underlying their toxic effects and inform the development of effective public health measures to mitigate the threat to human health.

Based on the nature of the toxins associated with HABs and the symptoms of human intoxication, these toxins can be categorized into the following groups: diarrhetic shellfish poisoning (DSP) toxins, paralytic shellfish poisoning (PSP) toxins, amnesic shellfish poisoning (ASP) toxins, neurotoxic shellfish poisoning (NSP) toxins, and ciguatera poisoning (CP) toxins. This article outlines the effects of the five most common HAB toxins on aquatic organisms and humans. The aim is to summarize the current research status and suggest directions for future studies that seek to mitigate the impacts of HABs on human health.

2. Phycotoxins

2.1. DSP Toxins

2.1.1. DSP Toxins: Physicochemical Properties and Structure

DSP toxins are mainly produced by two marine dinoflagellate genera, *Dinophysis* and *Prorocentrum* [23]. Initially, the DSP toxin group included okadaic acid (OA), dinophysistoxins (DTXs), and their derivatives, as well as yessotoxins (YTXs) and pectenotoxin (PTXs) [24]. However, since 2002, YTXs and PTXs have been excluded from the DSP toxin classification, as they have been demonstrated to cause liver necrosis and myocardial damage but not diarrhea [25]. Currently, only OA, DTX-1, DTX-2, and its C-7 acyl derivative (DTX-3), which are found in shellfish [26–28], have been confirmed to cause diarrheal effects [29,30]. The LD_{50} values for the main DSP toxins in mice (intraperitoneally) are 185.6 μg kg^{-1} for OA, 150.4 μg kg^{-1} for DTX-1, and 352 μg kg^{-1} for DTX-2 [31]. These toxins exhibit lipophilic properties, accumulating in adipose tissues and being metabolized and eliminated with difficulty [32]. The principal action mechanism of DSP toxins is the inhibition of serine/threonine protein phosphatases 1 and 2A (PP1 and PP2A), resulting in the deregulation of numerous intracellular processes [33]. Structurally, OA and DTXs share the same carbon backbone, the differences among various analogs are only in the number or position of methyl groups (see Figure 1) [34]. Moreover, as polyether compounds, DSP toxins are fat-soluble and thermally stable [35,36]. Given its lipophilicity, DSP toxins easily accumulate in the tissues of filter-feeding marine organisms that ingest toxic algae, which may result in direct or indirect transmission to humans through the food chain [37]. Thermal stability allows them to remain active even after exposure to cooking temperatures [38]. Since the first DSP incident was reported in 1961, DSP incidents have been continuously occurring worldwide [39].

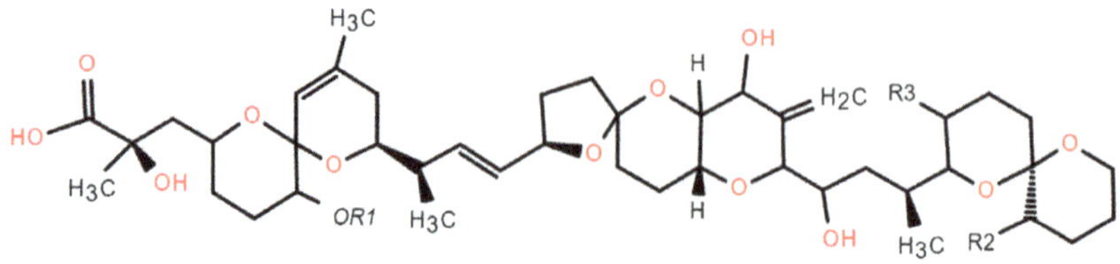

OA: R1=H; R2=H; R3=CH_3

DTX-1: R1=H; R2=CH_3; R3=CH_3

DTX2: R1=H; R2=CH_3; R3=H

Figure 1. General chemical structure of DSP toxins.

2.1.2. Main Biological Vectors of DSP Toxins

Filter-Feeding Bivalves

DSP is characterized by the onset of acute gastrointestinal symptoms such as nausea, abdominal pain, and diarrhea after consumption of seafood contaminated with OA and DTXs [40,41]. Filter-feeding bivalves, including mussels, scallops, and oysters, are recognized as significant vectors of DSP toxins. DSP toxins and related HAB species are known to induce a number of physiological changes in these bivalves, including valve closure response, pseudo-feces production, and clearance rate reduction [42–45]. In addition, DSP toxins have been shown to cause histopathological changes as well as immunotoxic, cytotoxic, and genotoxic effects [46–48]. Despite all these biological effects, no mortality has been documented in bivalves naturally exposed to HAB species associated with DSP toxins. Many studies have shown that bivalves are tolerant to DSP toxins compared to human susceptibility, as evidenced by the fact that the toxicity of DSP toxins to bivalves decreases with exposure time [46,48–50].

Studies have demonstrated that after exposure to the DSP toxin-producing dinoflagellate *D. acuminata* or *D. acuta*, cytochrome 450 (CYP450) enzyme activity in the digestive gland tissues of *Mytilus galloprovincialis* significantly increased. This increase was negatively correlated with the accumulated OA content in the tissue, suggesting the potential involvement of the CYP450 enzyme system in the metabolism of DSP toxins in mussels [51]. In addition, after exposure to *P. lima*, differential expression of *CYP3A4*, *CYP2D14*, *CYP3L3*, and *CYP2C8* was observed in the digestive gland and gill tissues of *Perna viridis*. Inhibition of CYP3A4 activity significantly altered the metabolic processing of DSP toxins in the mussel, indicating a pivotal role for CYP3A4 in the metabolism of DSP toxins [52]. After exposure of *Crassostrea ariakensis* to *P. lima*, glutathione S-transferase-ω (*GST-ω*) in gill tissues and *GST-α* in digestive gland tissues were significantly up-regulated, suggesting that different isoforms of GST in different tissues are involved in the metabolic process of DSP toxins [53]. Exposure to *P. lima* can significantly induce the expression of ATP-binding cassette (ABC) transporters in mussels, with different expression levels among different isoforms and tissues [54,55]. The P-glycoprotein (P-gp) specific inhibitor significantly decreased the activity of P-gp in the gill tissues of *P. viridis* but did not affect the accumulation of DSP toxins in gill tissue. In contrast, a broad-spectrum inhibitor of ABC transporters significantly increased the accumulation of DSP toxins in gill tissues [54–56]. These findings indicate that different isoforms of ABC transporters, including P-gp and multidrug resistance-associated protein (MRP), may collaborate to form a complex defense network during the metabolism of DSP toxins in bivalve species [55–57].

In bivalves, the digestive gland is the main organ for the accumulation of DSP toxins, where DSP toxins undergo a transformation during the digestive process [58]. Studies have shown that free DSP toxins (OA, DTX1, and DTX2) can be esterified with fatty acids of varying carbon chain lengths to form acyl derivatives such as DTX3 [26,27,59], which has lower affinity to target proteins, resulting in reduced toxicity [60]. This may also contribute to the higher tolerance of bivalves to DSP toxins. In addition, exposure to *P. lima* can activate the Nrf2 signaling pathway in the gill and digestive gland tissues of bivalves, thereby mitigating the impact of DSP toxins [49,61]. The addition of DHA can significantly enhance the expression of genes and enzyme activity related to the Nrf2 signaling pathway and reduce the damage caused by DSP toxins to the digestive gland of *P. viridis*. This further confirms the crucial role of the Nrf2 signaling pathway in the response and detoxification process of bivalves to DSP toxins [62]. However, the molecular mechanisms regulating the Nrf2 signaling pathway during the detoxification of DSP toxins in bivalves remain unclear. Taken together, the current findings suggest that bivalves may have developed cytoprotective mechanisms to mitigate the deleterious effects of DSP toxins, including the esterification of DSP toxins, the up-regulation of the three-phase metabolic detoxification, and the antioxidant system.

Fish

In addition to bivalves being the main vectors of DSP toxins, fish, as an important marine biological group, have been shown to be exposed to DSP toxins through their diet [16]. However, unlike bivalve species, fish appear to be susceptible to DSP toxins. Studies have shown that exposure of larvae of *Minidia beryllina* and *Cyprinodon variegatus* to DSP toxin-producing dinoflagellates results in adverse effects on growth, feeding rates, and swimming activity [63,64]. The juveniles and adult fish exposed to DSP toxins experience oxidative stress, multiorgan histopathology, behavioral changes, and even death [65,66]. In terms of the toxicological effects of OA and DTX1 in medaka and zebrafish models, medaka fish embryo exposure to concentrations of 0.75 $\mu g\ mL^{-1}$ of OA and toxic extract containing DSP toxins, with a concentration of 0.22 μg OA eq mL^{-1}, resulted in 100% mortality [67]. Conversely, juvenile zebrafish exposed to sublethal concentrations of OA or DTX-1 exhibited developmental delays, significant pericardial edema, monocular malformations, reduced anterior–posterior axes, and oxidative stress [68]. In addition, Neves et al. (2020) assessed the sublethal toxicity of DSP toxins in fish through a food chain transfer experiment and revealed that brine shrimp were effectively used as a link for the transfer of DSP toxins to higher trophic levels, leading to the accumulation of DSP toxins in wild ringneck blenny (*Parablennius pilicornis*: Blenniidae) [66]. Although current data suggest that exposure to DSP toxins induces a variety of adverse effects in fish, including behavioral and morphological changes and increased mortality, information regarding the lethal and sublethal impacts, as well as chronic effects of HAB species associated with DSP toxins on fish in the aquatic environment is still very scarce [63,66,69], which should be of concern.

2.1.3. The Mechanism and Possible Pathway of Toxicity to Humans

DSP toxins can accumulate in marine organisms and subsequently be transferred through the food chain to higher trophic levels, including humans [70]. The clinical symptoms of human poisoning by DSP toxins are mainly incapacitating diarrhea, along with gastrointestinal discomfort, nausea, vomiting, abdominal pain, and sometimes headache, chills, and fever [70]. Symptoms of DSP are very similar to those of gastrointestinal illness and can occur a few hours after consumption of contaminated seafood. Acute symptoms usually disappear within 2 to 3 days, which may lead to DSP incidents often being overlooked, so the actual DSP incidence may be much higher than that reported in the data [71]. The chronic toxicity of DSP toxins is of greater concern due to their carcinogenicity. Several studies have associated chronic exposure to DSP toxins with an increased risk of cancer, including gastrointestinal, pancreatic, and liver cancers [72–74]. Notably, most studies attribute the toxic mechanism of DSP toxins to the inhibition of serine/threonine protein phosphatases, specifically PP1 and PP2A; however, this inhibition does not fully account for all toxic effects [75]. Recently, studies have shown that OA exposure can cause the dysregulation of the intestinal microenvironment in rats by altering gut microbiota composition and disrupting the epithelial barrier. This suggests that OA-degrading bacteria and opportunistic pathogens play a non-negligible role in OA toxicity [76,77], resulting in the complex toxicity profile of DSP toxins.

2.2. PSP Toxins

2.2.1. PSP Toxins: Physicochemical Properties and Structure

PSP toxins are a group of neurotoxic alkaloids, more than 60 of which have been identified to date and are widely distributed in marine environments [78,79]. These analogs are generally classified into three major groups based on their distinct chemical structures at the R4 site (see Figure 2): carbamate toxins (e.g., GTX1&4, GTX2&3, STX, neoSTX), N-sulfocarbamoyl toxins (e.g., C1&2, C3&4, GTX5, GTX6), and decarbamoyl toxins (e.g., dcGTX1&4, dcGTX2&3, dcSTX, dcneoSTX) [80]. The toxicity of PSP toxins is ranked from highest to lowest as follows: carbamate toxins, decarbamoyl toxins, and N-sulfocarbamoyl toxins [81]. Within these PSP toxins, STX is the most potent neurotoxin, with an LD_{50} value of 10 $\mu g\ kg^{-1}$ in mice (intraperitoneally) [82]. The minimum lethal dose (MLD) of PSP toxins for humans is reported to be 3000 mouse units (MUs), which is primarily determined based on death cases caused by PSP toxins. A single MU of PSP toxins corresponds to the quantity of toxin capable of killing a 20 g male ddY strain mouse within 15 min following intraperitoneal injection [83]. According to the above data, the PSP toxin-related toxicity characterized by MUs is presented in Table 1. Currently, several marine dinoflagellates, such as *Pyrodinium bahamense*, *Gymnodinium catenatum*, and various species of *Alexandrium*, along with some freshwater cyanobacterial species, have been identified as producers of PSP toxins [84–86]. PSP toxins in those marine dinoflagellates exist almost exclusively as β-epimers (GTX3&4, C2&4), while α-epimers (GTX1&2, C1&3) predominate in the shellfish profile, with an epimer α:β ratio close to 3:1 at equilibrium [82]. In addition to common PSP toxin derivatives, many new C-11 hydroxyl metabolites (so-called M-toxins) of PSP toxins have been discovered in contaminated shellfish. Trace amounts of such substances have also been detected in some strains of microalgae that produce PSP toxins [87]. M-toxins are regarded as low-toxic derivatives of typical PSP toxins based on their structure–activity relationships [88]. Considering that M-toxins quickly accumulate in high concentrations within shellfish following the ingestion of PSP toxin-producing microalgae, the M-toxins are regarded as metabolic products of PSP toxins during shellfish detoxification [89]. The toxins M1&M3 and M7&M9 can be transformed from C1&2 and C3&4, respectively, while M2&M4&M6 and M8&M10 can be transformed from GTX2&3 and GTX1&4, respectively. The transformation rates of C&2 and GTX2&3 are significantly higher than those of C3&4 and GTX1&4, and the metabolic transformation of C1&2 to M1&3 occurs most rapidly, which is also faster than the production rate of other M-toxins in scallops and mussels [90]. Furthermore, a new class of PSP toxins, referred to as GC toxins, has been found in *G. catenatum*, which contain hydroxybenzoate R4 substituents [91]. Unlike the hydrophilic nature of the three main classes of PSP toxins, the hydroxybenzoate side chains in GC toxins introduce a degree of lipophilicity [92]; however, the exact toxicity of GC toxins has not yet been determined.

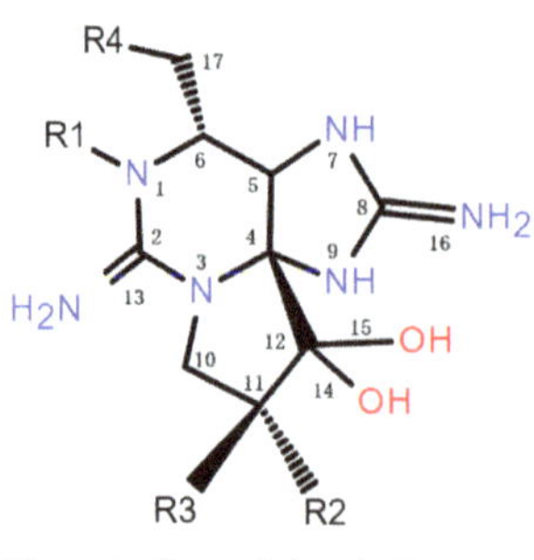

Figure 2. General chemical structure of PSP toxins.

Table 1. PSP toxin components and their specific toxicities (MU/μmol) adapted from [82].

PSP Toxins		R1	R2	R3	R4	Toxicity
Carbamate toxins	STX	H	H	H	$CONH_2$	2483
	neoSTX	OH	H	H	$CONH_2$	2295
	GTX1	OH	OSO_3^-	H	$CONH_2$	2468
	GXT2	H	OSO_3^-	H	$CONH_2$	892
	GXT3	H	H	OSO_3^-	$CONH_2$	1584
	GXT4	OH	H	OSO_3^-	$CONH_2$	1803
N-sulfocarbamoyl toxins	GTX5	H	H	H	$CONHSO_3^-$	160
	GXT6	OH	H	H	$CONHSO_3^-$	180
	C3	OH	OSO_3^-	H	$CONHSO_3^-$	33
	C1	H	OSO_3^-	H	$CONHSO_3^-$	15
	C2	H	H	OSO_3^-	$CONHSO_3^-$	239
	C4	OH	H	OSO_3^-	$CONHSO_3^-$	143
Decarbamoyl toxins	dcSTX	H	H	H	H	1274
	dcneoSTX	OH	H	H	H	33
	dcGTX2	H	OSO_3^-	H	H	1617
	dcGTX3	H	H	OSO_3^-	H	1872

2.2.2. Main Biological Vectors of PSP Toxins

Bivalves

Although PSP toxins transfer through the food web and accumulate from zooplankton to whales [93,94], this review only focuses on the vectors known to pose a potential public health risk to human consumers. The traditional route of PSP toxins through the food web involves filter-feeding mollusks, particularly bivalves, which ingest and concentrate PSP toxins and associated HAB species [79,95–97]. PSP toxins exert detrimental impacts on various physiological functions of bivalves, including reduced clearance rates, adductor paralysis, mantle retraction, decreased valve closure response, and even death [98–101]. In addition, PSP toxins show obvious immunotoxicity to bivalves, mainly manifested as infiltration of hemocytes (especially in the connective tissues surrounding the digestive gland), suppression of hemocyte phagocytic activity, and induction of apoptosis [100–105].

Despite these adverse biological effects, bivalves were initially observed to survive and thrive during the periods of PSP toxin-producing dinoflagellate blooms and exhibited an ability to accumulate and tolerate higher concentrations of PSP toxins compared to humans [106]. Similar to tetrodotoxin (TTX), PSP toxins attack the nervous system by blocking sodium channels on the nerve membrane and inhibiting action potential conduction [107]. Voltage-gated sodium channel 1 (Nav1) is the major sodium channel in the animal nervous system and the target of PSP toxins [106,108]. Li et al. (2017) analyzed the scallop genome and identified two sodium channel genes, Nav1 and Nav2. Further analysis revealed the presence of a potential antitoxin T mutation at 1425 of Nav1 in scallops [106], which, in rats, corresponds to a 15-fold increase in resistance to saxitoxin (STX) (the most toxic PSP toxin) and a similar 15-fold increase in resistance to TTX [107,109]. Furthermore, analysis of Nav1 in *C. gigas* and the Atlantic clam *Solemya velum* revealed the presence of a Q mutation at site 945, which has been demonstrated to confer a 19,880-fold increase in resistance to STX in rats [110]. These findings suggest that the two novel mutations on Nav1 can enhance toxin resistance in other organisms and that their presence in bivalves may account for the ability to tolerate PSP toxins.

It is found that the hepatopancreas and kidneys are the organs with the highest PSP toxins concentrations in *C. farferri* after exposure to PSP toxin-producing microalga *A. minutum*. Interestingly, there is little expression of sodium channel genes in either tissue [106], implying that bivalves may possess alternative mechanisms to tolerate high concentrations of PSP toxins. *A. minutum* showed greater toxicity to the kidney than the hepatopancreas, possibly due to the conversion of GTX into highly toxic STX in the kidney [106]. The elevated expression of the cytoplasmic sulfotransferase (*Sult*) gene, which likely mediates the transfer of sulfate groups from donor molecules (such as GTX) to various receptor molecules, endogenous metabolites, and xenobiotics [111], corresponds with the transformation of PSP toxins in the kidney. Furthermore, the *Sult* gene family is notably expanded in the *C. farreri* genome, with 83 genes compared to 26 in oysters, 31 in pearl oysters, 13 in humans, and 8 in flies [106]. All these findings suggest that the expanded *Sult* genes may be responsible for converting GTX into the more potent STX, potentially endowing scallops with a robust deterrent mechanism against predation. In addition, the expression of detoxification- and antioxidant-related genes such as superoxide dismutase (*SOD*) [112], *GST* [113], ABC transporters [114], and heat shock proteins [115] was markedly induced in bivalves as a response to counter the adverse impacts of PSP toxins.

Unfortunately, the mass mortality of bivalves has been increasing globally in recent years. PSP toxins have been linked to the mass mortality of many bivalve species [101]. While there is no definitive evidence that HABs directly cause mortality events in bivalves [116–118], recent studies have found that anthropogenic global factors such as global warming, ocean acidification, and widespread microplastic pollution may exacerbate bivalves' susceptibility to toxigenic HABs (including PSP toxins and DSP toxins) [119–122]. These factors could potentially explain the mass mortality of bivalves caused by HABs in recent years.

Fish

Recently, the rising global demand for protein has spurred the expansion of aquaculture worldwide, with fish being one of the most widely farmed species in this sector [123]. Given the impact of HABs on human health, the action of shellfish has been thoroughly documented, prompting the establishment of marine toxin monitoring programs in most coastal nations [3]. However, since fish are not the primary focus of these monitoring efforts, data on the bioaccumulation of toxins in fish is comparatively scarce. PSP toxins can accumulate in fish that prey on zooplankton and subsequently exert detrimental effects on these fish [96,124]. Early field observations indicate that poisoning symptoms in fish affected by PSP toxins include loss of balance, lateral or inverted swimming, and swimming with open mouths [125,126]. Upon exposure to extracts from PSP toxin-producing algae via gavage, fish engaged in rapid circular swimming in confined spaces for a brief duration before ultimately becoming motionless on the bottom, with some individuals displaying significantly reduced respiratory rates [127]. Salierno et al. (2006) found that killifish (*Fundulus heteroclitus*) subjected to high doses of STX showed similar toxic symptoms, possibly due to the inhibited expression of c-FOS protein in the anterior and posterior lobes, which govern swimming and other motor functions [128]. In addition, PSP toxins have immunotoxic and hepatotoxic effects on fish and can induce oxidative stress that damages the gills, thereby increasing the oxygen consumption and demand of fish [129–131]. The similarity in poisoning symptoms observed among different wild fish species and those in laboratory conditions suggests that the physiological response to PSP toxins may be a common occurrence across the marine fish population. Moreover, Bianchi et al. (2021) found that other undefined soluble extracellular compounds produced by *Alexandrium* spp. were highly ichthyotoxic to fish [132]. This suggests that future studies of PSP-associated species should broaden their scope to include these extracellular substances.

2.2.3. Mechanism of PSP Toxin Toxicity to Humans

PSP toxins pose a serious threat to public health primarily through contamination of the human food supply chain [133]. Symptoms of PSP begin with numbness and tingling in the lips and mouth and may progress to muscle weakness, ataxia, neurological symptoms, and, in severe cases, muscle paralysis and possibly death from respiratory paralysis [70,134,135]. Voltage-gated sodium channels serve as the molecular targets for PSP toxins, which share a common structural motif consisting of a primary α subunit with a molecular size of 220–260 kDa and one or two auxiliary β subunits with a molecular size of 33–36 kDa [136–138]. PSP toxins act by reversibly binding to sodium ion channels in the neural membrane, thus interrupting neuronal transmission and inducing systemic neurological dysfunction [135,139,140]. As mentioned above, unlike the susceptibility in humans, natural selection pressure has led to mutations in sodium channels within bivalves [106,107], resulting in tolerance to PSP toxins.

2.3. ASP Toxins

2.3.1. ASP Toxins: Physicochemical Properties and Structure

ASP toxins include domoic acid (DA) and its isomers, such as isodomoic acid. DA is a naturally occurring crystalline, water-soluble, acidic amino acid with a secondary amino group, and the LD_{50} of DA in mice (intraperitoneally) was assessed as 2.4 mg kg^{-1} [17,141]. ASP toxins are synthesized exclusively by specific species of diatoms, mainly the genus *Pseudo-nitzschia*, as well as certain members of the *Nitzschia* and *Amphora* genera, red alga *Chondria armata*, and other related species [17,142]. DA consists of a proline ring, three carboxyl groups, and one amino group. Its high polarity is due to the presence of three carboxyl groups and one amino group, with pK_a values of 1.85, 4.47, 4.75, and 10.60, respectively, at 20 °C. Therefore, depending on pH, DA may exist in five different states of charge, ranging from −3 to +1, which can affect its retention in different chromatographic systems [143]. The side chain of the DA molecule encompasses two conjugated double bonds, giving DA the ability to absorb ultraviolet light. At neutral pH, the maximum emission wavelength is 242 nm, which can be utilized for the detection of DA through liquid chromatography [144]. These structural features of DA are directly associated with the interaction with glutamate receptors and the toxicity of the molecule [145]. Despite its relative stability at room temperature [143], DA can undergo substantial decomposition under high temperatures (>50 °C) and in highly acidic (pH $\leq$ 2) or alkaline (pH $\geq$ 12) conditions [144]. Within shellfish tissue, however, the situation is slightly different. McCarron and Hess (2006) indicated that conventional cooking and pressure sterilization at 121 °C in mussel tissue only decreases the total DA concentration by approximately 3%, suggesting that cooking does not significantly improve the safety of DA-contaminated shellfish [146]. Moreover, the chemistry properties of DA and its analogs and geometrical isomers of iso-domoic acid A, B, and C were investigated in *C. armata* [147]. Three other iso-domoic acids, D, E, and F were isolated from plankton cells and shellfish tissues [148]. In addition, two new isomers of DA from *C. armata*, namely, iso-domoic acid G and H, were isolated in 1997 (see Figure 3) [147]. Iso-domoic acids are secondary components relative to DA and are not always found in bivalves contaminated with ASP toxins [149]. When exposed to ultraviolet light or heat, DA can be converted to iso-domoic acid [150,151]. Therefore, it has been proposed that domoic acid may be a biosynthetic product that is subsequently converted to iso-domoic acid under suitable environmental conditions [150].

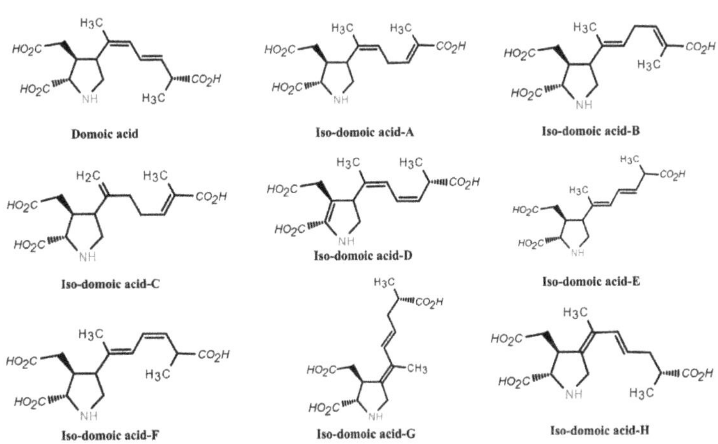

Figure 3. General chemical structure for DA and its isomers.

2.3.2. Main Biological Vectors of ASP Toxins

Bivalves

Bivalves are the main carriers of DA, and most DA in these organisms is not metabolized and remains unchanged when excreted [152]. The accumulation kinetics and depuration rates of DA vary significantly among different shellfish species [153,154]. Based on this, bivalves can be categorized into two groups: rapid and slow detoxifiers of DA. Rapid detoxifiers can eliminate toxins within days to weeks, including some mussel species, such as the mussels *M. galloprovincialis* [155], *M. edulis* [156], *M. californianus* [157], and *P. canalicus* [158], clams, such as *Mesodesma donacium* [159] and *Ruditapes philippinarum* [160], oysters, such as *C. virginica* [161] and *C. gigas* [162], and pectinids, like *Argopecten purpuratus* [153]. The slow detoxifiers, which may take months or even years to depurate DA, mainly include some commercially important bivalves such as *Pecten maximus* [163], *Placopecten magellanicus* [164], *Siliqua patula* [165], and *Spondylus cruentus* [166]. However, the mechanisms underlying the extensive variability in DA accumulation and detoxification kinetics in bivalves remain unclear.

Trainer and Bill (2004) suggested that the expression of low-affinity glutamate receptors across all tissues and the selective expression of high DA capacity sites in tissues such as the siphon might be responsible for the tissue-specific retention of ASP toxins in species like *S. patula* [167]. Mauriz and Blanco (2010) proposed that the lack of effective membrane transport proteins for toxin excretion could account for the high accumulation and/or slow detoxification rate of DA in *P. maximus* [168]. In addition, by using a novel immunohistochemical technique based on an anti-DA antibody, García-Corona et al. (2022; 2024a) observed that in the digestive glands of species such as *P. maximus*, *Aequipecten opercularis*, and *Crepidula fornicate*, most DA staining was confined to small autophagic vesicles within the cytoplasm of digestive cells during early autophagy and in the post-digestive residues remaining in the distal cytoplasmic region of digestive cells during late autophagy [169,170]. García-Corona et al. (2024b) further analyzed the dynamics of DA subcellular localization in the digestive glands of *P. maximus*, suggesting that autophagy may be a contributing factor to the long-term retention of DA in *P. maximus* [171].

Several studies have shown that exposure to ASP toxin-producing microalgae or DA can trigger a range of detrimental physiological alterations in bivalves. For instance, *M. edulis* exhibited transient DNA damage after DA injection [172]. *M. californianus* developed mild respiratory alkalosis upon exposure to ASP toxin-producing microalgae *P. nitzschia* [173]. Similarly, *C. gigas* displayed transient respiratory acidosis, hypoxia, and an increase in hemocyte activity and count following exposure to *P. nitzschia* [162]. Dissolved DA can negatively affect the growth and survival of *P. maximus* larvae and impair immune function and induce oxidative stress in *A. irradians* [174–176]. Furthermore, transcriptomic analysis of the digestive glands in *M. galloprovincialis* and *A. opercularis* exposed to *P. nitzschia* revealed differential expression of genes involved in antioxidant, metabolic detoxification, transmembrane transport, and immune processes [177,178]. Oxidative stress is the main effect observed in other invertebrates exposed to DA [179,180]. Song et al. (2020) reported that exposure to pure DA induces oxidative stress in bay scallops, disrupts their metabolic processes, and negatively affects their defense response [181]. These findings suggest that oxidative stress may contribute significantly to the toxicity of DA in bivalves. Despite being a neurotoxic amino acid, available data suggest that DA has no significant neurotoxic effect on bivalves [172].

Fish

Although planktivorous fish can accumulate considerable amounts of DA during toxic *P. nitzschia* blooms [182], there is no evidence that fish are affected under natural exposure conditions. Fish exhibit neurotoxic symptoms when DA is administered intraperitoneally, whereas no such effects are observed when the toxins are orally ingested, which is the most ecologically relevant exposure route [183]. Even though oral DA builds up in the brain, liver, and kidneys of species such as salmon (*Oncorhynchus kisutch*) [184], rainbow trout (*Oncorhynchus mykiss*) [185], and anchovy (*Engraulis mordax*) [186], they do not exhibit excitotoxic behaviors. Consequently, the prevailing dogma is that fish are relatively insensitive to DA acquired through ecological exposure routes [183,187], but the reasons for this are unclear.

2.3.3. Mechanism of ASP Toxin Toxicity to Humans

The first case of human DA poisoning occurred in Canada in 1987, with clinical symptoms of acute poisoning such as confusion, disorientation, seizures, permanent short-term memory loss, and, in the most severe cases, even death, so it is known as ASP [188]. The structure of DA is similar to neurotoxin kainic acid (KA) and the endogenous neurotransmitters glutamate (Glu) and aspartate (Asp) [183,189]. Consequently, it is not surprising that DA interacts with glutamate receptors (GluR) in nerve terminals. Glu is the main excitatory neurotransmitter in the brain; however, excessive Glu can lead to neurodegeneration [190], seizures [191], and apoptosis [192,193]. For example, overactivation of endogenous Glu or Asp on GluR triggers a cascade of biochemical events that can lead to neuronal damage or cell death [194,195]. Additionally, the rigid structure of DA significantly reduces its mobility at glutamate receptor binding sites, thereby improving the binding efficiency [196]. DA-induced excitotoxicity acts in a manner similar to Glu and KA, with a cohesive effect on both sides of the synapse, resulting in depolarization and the release of Glu into the synaptic gap [197]. The stimulation by DA results in a rapid influx of Ca^{2+}, transporting Glu-containing vesicles to the membrane surface and releasing them into the synaptic space via exocytosis. Excessive Ca^{2+} influx disrupts Ca^{2+}-dependent cascades [189], ultimately leading to neuronal death, including membrane, cytoplasmic, and nuclear events that lead to neurotoxicity.

2.4. NSP Toxins

2.4.1. NSP Toxins: Physicochemical Properties and Structure

NSP toxins are mainly composed of brevetoxins (BTXs), which are produced mainly by the marine dinoflagellate *Karenia brevis* [198,199]. In addition, some strains of *K. papilionacea* are very low producers of BTXs [200]. BTXs are a class of lipid-soluble cyclic polyether compounds, which can be divided into A-type and B-type based on their backbone structure [201,202]. A-type BTXs have a backbone structure consisting of ten trans-fused ether rings, while that of B-type BTXs is composed of eleven ether rings (see Figure 4) [203]. Through molecular modeling, Rein et al. (1994a, 1994b), as well as Gawley et al. (1995), found that the structures of PbTx-1 and PbTx-2 are relatively linear, with a bend approximately in the middle of the molecule, a lactone function in the A ring, a strict rigid region in the four terminal rings, a side chain at the molecular end that allows moderate modification, and a spacer region that separates the rigid region from the A ring lactone [204–206]. All natural BTXs and derivatives with full activity have all these characteristics [202]. BTXs can also be distinguished by side chains located in the tail region. Both A-type brevetoxin-1 (PbTx-1) and B-type brevetoxin-2 (PbTx-2) possess the same side chain, with a highly active α, β-unsaturated aldehyde group at the end of the side chain [203]. Of the BTXs group, PbTx-1 is considered to be the most potent, while PbTx-2 is the most abundant toxin produced by *K. brevis* observed to date [207].

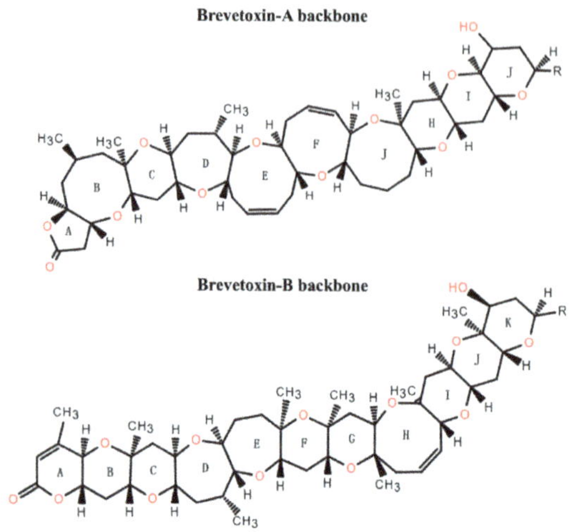

Figure 4. BTXs are based on two different structural backbones, which are perceived to be the two parent molecules, PbTx-1 (brevetoxin-A backbone) and PbTx-2 (brevetoxin-B backbone).

2.4.2. Main Biological Vectors of NSP Toxins

Bivalves

Several studies have shown that various bivalve species can accumulate PbTxs [208–210]. In general, bivalves do not appear to suffer adverse effects from the toxins [199,201], although some studies suggest that exposure to NSP toxins may affect the development and clearance of bivalve larvae [210,211]. Despite the limited studies on the molecular mechanisms underlying toxic responses in bivalves to NSP toxins, the metabolism of NSP toxins across bivalve species may partially explain the toxicological responses of NSP toxins in bivalves. Ishida et al. (1995) identified a conjugate of oxidized PbTx-2 with C-42 N-taurine in cockle *Austrovenus stutchburyi*, naming it PbTx-B1 [212]. Morohashi et al. (1995) determined that PbTx-B3 is formed from PbTx-2 through fatty acid esterification at the open D-ring and oxidation at the aldehyde terminus [213]. Subsequently, Murata et al. (1998) discovered a conjugate with S-cysteine (sulfoxide) at the C-50 S-cysteine in *P. canaliculus*, termed PbTx-B2 [214]. Morohashi et al. (1999) further identified conjugates of the PbTx-B2 cysteine sulfoxide moiety with *N*-myristoyl and *N*-palmitoyl groups in *P. canaliculus* and named it PbTx -B4 [215]. Furthermore, the main metabolites of PbTx-2 identified in Easter oysters include the reduction derivative PbTx-3, the cysteine sulfoxide conjugate PbTx-B2, and the newly identified S-deoxy-PbTx-B2, which is easily oxidized to PbTx-B2 [216]. Ishida et al. (2004) characterized the oxidized derivative of PbTx-2, specifically, the terminal aldehyde in cockle, and named it PbTx-B5 [217]. The discovery of a large number of PbTx metabolites in shellfish increases the difficulty of detecting these toxins. However, the current studies suggest that PbTxs ingested by humans through shellfish are metabolites of PbTxs, rather than the parent NSP toxins from algae [202,218], which undoubtedly emphasizes the importance of detecting these metabolites.

The depuration time of PbTxs in bivalves varies but is usually within 2 to 8 weeks, although there are records of PbTxs being retained for much longer (nearly a year) [216,219,220]. Given that PbTxs can be metabolized in various molluscan shellfish, as mentioned above, the metabolism of PbTxs in shellfish may be partially responsible for their low impact on shellfish. The limited acute animal toxicity data of PbTxs and their metabolites is summarized in Table 2. The 24 h LD_{50} of PbTx-2 administered intraperitoneally in mice was 200 $\mu g\ kg^{-1}$ body weight, whereas the oral toxicity of PbTx-2 was significantly lower in comparison. With a few exceptions, the metabolism of PbTx-2 mitigates its acute toxicity under the same experimental conditions. Specifically, the 24 h LD_{50} of PbTx-3, S-deoxy-PbTx-B2, and PbTx-B2 administered intraperitoneal (i.p.) in mice was 875, 211, and 400 μg kg^{-1} body weight, respectively. However, the current understanding of the mechanisms associated with the toxic response to PbTxs in bivalves remains limited.

Table 2. Acute toxicity of BTXs in mice.

BTXs	Route	Toxicity, μg kg^{-1} Body wt.	Reference
PbTx-2	LD_{50}, 24 h, i.p.	200	[221]
	LD_{50}, 24 h, oral	6600	[222]
PbTx-3	LD_{50}, 24 h, i.p.	875	[222]
	LD_{50}, 24 h, oral	520	[221]
S-deoxy-PbTx-B2	LD_{50}, 24 h, i.p.	211	[223]
PbTx-B1	MLD [a], <2 h, i.p.	50	[212]
PbTx-B2	LD_{50}, 24 h, i.p.	400	[223]
	MLD, <1 h, i.p.	306	[214]
PbTx-B3	MLD, 24 h, i.p.	>300 [b]	[213]
PbTx-B4	MLD, 6–24 h, i.p.	100	[215]
PbTx-B5	MLD [c], i.p.	300–500	[224]

a. Minimum lethal dose. b. No deaths recorded at 300 μg kg^{-1}. c. Time to death not reported.

Fish

Additionally, PbTxs are ichthyotoxins that typically kill millions of fish and affect hundreds of species [78,225,226]. Fish mortality typically results from the rupture of algal cells and the subsequent release of PbTxs, which are absorbed directly through the gill membranes [207]. However, some studies indicate that fish can endure HABs and act as a significant vector for the transfer of PbTxs to higher trophic levels [227,228]. Naar et al. (2007) found that PbTxs can accumulate in live fish via dietary exposure, and fish fed with PbTx-contaminated shellfish or *K. brevis* cultures remain healthy while accumulating high levels of PbTxs in their tissues [228]. Moreover, Griffin et al. (2023) demonstrated that Atlantic tarpon (*Megalops atlanticus*) is tolerant to *K. brevis* blooms [229]. Although, it has been suggested that species-specific susceptibility to both lethal and sublethal *K. brevis* effects is largely dependent on the intersection of innate tolerance, life history stage, dispersal ability, and behavioral ecology/environmental conditions, as well as bloom toxicity dynamics and distribution [230], and even the synergistic interaction between ROS (superoxide anion) and polyunsaturated fatty acids (PUFAs) [231]. To date, these contradictory findings have not been adequately explained. Therefore, this emphasizes the need to better assess the role of fish in the transfer of PbTxs, including their ability to metabolize PbTxs. However, compared to the advances in PbTx metabolism in bivalves, the metabolic fate of PbTxs in fish remains largely unexplored.

2.4.3. Mechanism of NSP Toxin Toxicity to Humans

NSP usually occurs after consumption of bivalve mollusks contaminated with BTXs [20]. The characteristic clinical features of NSP mainly involve the gastrointestinal tract and the nervous systems, and the nervous system manifestations last longer than the gastrointestinal disturbances [232,233]. Specifically, these symptoms include nausea, vomiting, abdominal pain, and diarrhea, as well as partial limb paralysis, slurred speech, dizziness, ataxia, and a generalized loss of coordination [199,234]. The neurotoxic effects of BTXs have been widely substantiated, mainly due to their high affinity with the voltage-gated sodium channel (VGSC) receptor site 5 in nerve, skeletal, and cardiac cells [37], resulting in prolonged channel opening, particularly the inability of open channels to be inactivated, thereby leading to sustained depolarization of neurons [235,236]. This depolarization, in turn, triggers a cascade of downstream neurochemical effects, such as enhanced release of glutamate, overstimulation of N-methyl-d-aspartic acid (NMDA) receptors, and unregulated influx of Ca^{2+} into postsynaptic cells [237]. Ultimately, these events lead to the obstruction of ion conduction and action potential generation, resulting in the loss of neuromuscular function and muscle paralysis [238]. Given the similarity between the targets of BTXs and PSP toxins, the insensitivity of bivalves to BTXs may also be attributed to mutations in sodium channel sites in these organisms. However, this hypothesis requires further investigation to be substantiated.

2.5. CP Toxins

2.5.1. CP Toxins: Physicochemical Properties and Structure

CP toxins, including ciguatoxins (CTXs), maitotoxins (MTXs), gambierol, and related compounds, are associated with foodborne ciguatera fish poisoning [239]. These toxins are produced by the dinoflagellates *Gambierdiscus* and *Fukuyoa* (more than 18 species produce MTX, but only 3 or 4 produce CTX) [240]. They are bioactive molecules consisting of ether rings, with significant differences in carbon number, structure, and mechanism of action. Among them, CTXs are polyether neurotoxins that can directly cause CP and are currently one of the most studied CP toxins [239]. They consist of a ladder-like structure composed of 13 or 14 ether rings and can be classified into Pacific (P-), Caribbean (C-), and Indian Ocean (I-) groups based on the number of rings, backbone structure, and region (see Figure 5) [241]. Although the geographic classification of CTXs is still in use, it does not adequately characterize the diversity of CTXs, and classification and nomenclature based on structural features have evolved. Structurally, CTXs can be categorized into three types of molecules: the oxopene (CTX1B type), oxocene (CTX3C type), and Caribbean/Indian CTXs structures [239,242,243]. The first two are commonly associated with fish and microalgae in the Pacific, and the main CTX toxins include P-CTX-1, P-CTX-2, and P-CTX-3. The derivative found in Pacific fish, CTX1B, is one of the most toxic analogs; hence, the Toxicity Equivalence Factor (TEF) recommended by the European Food Safety Authority (EFSA) is assigned with CTX1B as the reference (TEF = 1). CTX-2 and CTX-3 both have a TEF of 0.3, while CTX3C has a TEF of 0.2, and C-CTX-1 has a TEF of 0.1. These values were determined by acute toxicity testing via i.p. in mice [60,244]. However, a recent study by Raposo-García et al. (2023) showed that the effects of the analogs CTX3C and C-CTX1 were only 0.3 times lower than that of the Pacific analog CTX1B, both in terms of cytotoxicity and reduced sodium current intensity [245]. These data contradict the statements of the EFSA and the U.S. Food and Drug Administration (FDA) that the relative potency of C-CTX1 is 10 times lower than that of the Pacific analogs. Therefore, it is necessary to re-evaluate the relative toxicity and potency of ciguatoxins to determine the safety limits for these emerging toxins in commercialized fish.

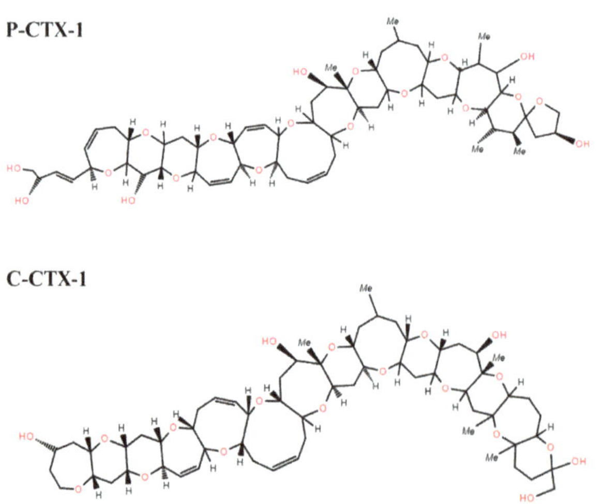

Figure 5. General chemical structure of Pacific ciguatoxin-1 (P-CTX-1) and Caribbean ciguatoxin-1 (C-CTX-1). The major CTXs from the Indian region (I-CTXs) have a similar structure to C-CTX-1.

2.5.2. Main Biological Vectors of CP Toxins

CTXs originate from the benthic dinoflagellate genus *Gambierdiscus*, which is mainly found in tropical and subtropical seas. When toxin-producing algae are consumed by herbivorous fish, which are consumed by carnivorous fish, CTXs are transferred through the food web, ultimately resulting in poisoning incidents when humans consume fish contaminated with CTXs [244–248]. A survey of CTXs in a coral reef system showed that P-CTX was detected in 54% of herbivorous fish (total P-CTXs < 0.500–1670 $pg \cdot g^{-1}$ wet weight (ww)), in 72% of omnivorous fish (<0.500–1810 $pg \cdot g^{-1}$ ww), and in 76% of predatory fish (<0.500–69,500 $pg \cdot g^{-1}$ ww) [249].

Numerous studies have confirmed that P-CTX-1 can induce developmental, reproductive, and cardiovascular toxicity in fish [250–252]. Furthermore, juvenile fish exposed to P-CTX-1 exhibit reduced, delayed, or paralyzed responses to external stimuli, indicating that P-CTX-1 may impair the physiological behaviors of fish, such as foraging and evading predators [251]. Yan et al. (2017) found that injecting P-CTX-1 into *Oryzias melastigma* embryos resulted in malformations, visceral damage, immune dysfunction, and muscle physiological changes. Subsequent 21-day exposure reduced spawning and hatching rates in the progeny [253]. Jiang et al. (2012) conducted proteomic analysis on fish in CTX-contaminated fields and found that CTXs disrupted the influx/efflux of Na^+ and Ca^{2+} in fish livers and interfered with various biological functions, such as signal transduction, metabolic processes, detoxification, anti-apoptosis, and immune defense [250]. Li et al. (2022) showed that direct or indirect exposure to CTX-producing microalga *G. caribaeus* affected the swimming behavior of *O. melastigma*, characterized by reduced swimming ability and abnormal swimming patterns. Further transcriptomic analysis revealed that *G. caribaeus* ingestion could significantly affect energy metabolism, induce immune stress, and up-regulate the expression of pro-inflammatory factors, causing liver tissue damage [254]. These findings indicate a range of adverse effects of CTXs on fish. However, Xu et al. (2021) found that *G. caribaeus* from Weizhou Island, China, has relatively low toxicity, yet fish contaminated with CTXs exceeded regulatory thresholds, suggesting that the long-term presence of low-toxicity *G. caribaeus* may lead to the accumulation of CTXs in fish [255]. The ability of fish to accumulate CTXs implies that they may have some resistance. Consistent with the findings of Xu et al., juvenile *Naso brevirostris* exposed to a gel food embedded with CTX-producing microalga *G. polynesiensis* for four months exhibited no signs of intoxication [256], which seems to contradict the toxic effects of CTX on fish mentioned above [250–254]. In addition to species-specific tolerance to CTXs, the discrepancy in outcomes may be related to the microalgae species/strains used in the experimental test. It has been demonstrated that the toxicity of specific *Gambierdiscus* strains is 100 times higher than that of other strains within the same species [257,258]. This variability seems to partially account for the conflicting results.

Currently, CP is an emerging threat in Europe. The first outbreak was recorded in 2004 in the Canary Islands of Spain [259]. Since then, Portugal reported its first case in 2008, when 11 of 16 crew members on a fishing vessel developed CP symptoms after consuming 30 kg amberjack (*Seriola* sp.) caught in the Selvagens Islands [260]. In the same year, CP symptoms were also observed in 20–30 individuals who consumed a smaller amberjack (*Seriola rivoliana*) that was purchased in the Canary Islands but caught in the Selvagens Islands [261]. Recently, red porgy (*Pagrus pagrus*) from the Selvagens Islands was found to have toxicity levels as high as 0.76 μg C-CTX1 kg^{-1} [262]. In addition, other mid-trophic level marine fish species, including zebra bream (*Diplodus cervinus*) and barred hogfish (*Bodianus scrofa*), also showed CTX-like toxicity up to 0.75 μg C-CTX1 kg^{-1}. Furthermore, several fish species have the potential to act as vectors for the transmission of ciguatoxin to humans, including the parrotfish (*Sparisoma cretense*), the grey triggerfish (*Balistes capriscus*), and the blacktail comber (*Serranus atricauda*), highlighting that the risk of CP is not limited to top predators [263]. More notably, CTX-associated mass spectrum signals and ion fragmentation patterns were detected in random samples of benthic organisms, including the starfish *Ophidiaster ophidianus* off the southeast coast of Madeira Island [264]. Moreover, invertebrates, such as shellfish, could potentially be responsible for CP [265]. Ciguatera shellfish poisoning events involving giant clams (*Tridacna maxima*) are reported occasionally in Pacific island communities, and even after purification, the CP toxin in *T. maxima* will not be eliminated [266]. These findings highlight the necessity for further investigation of the toxic response and transfer kinetics of CTX in benthic organisms.

2.5.3. Mechanism of CP Toxin Toxicity to Humans

CP is well-known in tropical regions, but with global warming, the risk of CP is increasing with the geographical spread of CP toxin-producing algae [267,268]. CP can cause neurological, gastrointestinal, and cardiovascular symptoms, depending on the amount and type of toxin ingested [247,269,270], and in severe cases, CP can be fatal [271]. P-CTX-1 as low as 0.08–0.1 $\mu g\ kg^{-1}$ in fish meat can pose a risk to human health [248]. CP symptoms usually appear within 12 h of consuming toxic fish and begin with severe gastrointestinal issues (nausea, vomiting, diarrhea, abdominal pain) and typically subside within 24 h [272]. Cardiovascular issues (usually a combination of bradycardia and hypotension) and/or neurological symptoms (generally confusion, memory impairment, difficulty concentrating, depression or irritability, and anxiety) may also accompany this acute episode [246,273–275].

CTXs activate VGSC in cell membranes, thereby increasing sodium permeability and depolarizing nerve cells. This nerve cell depolarization is thought to cause a range of neurological symptoms associated with CP [276]. Despite detailed case reports, there is a lack of accurate identification and comprehensive understanding of the characteristics of CP, and no definitive biomarkers have been developed to diagnose CTX exposure in humans. Currently, the "gold standard" for CP diagnosis includes the detection of CTXs in relevant fish species through appropriate testing methods, such as animal mortality tests, biological methods (cytotoxicity assays, receptor binding assays, and immunoassays), and chemical methods (HPLC with fluorescence detection, LC-MS/MS) [277], along with symptoms consistent with CP.

3. Conclusions

Collectively, shellfish and fish are the main vectors of HAB toxins. Unlike the susceptibility of humans to HAB toxins, shellfish and fish may have developed some tolerance due to natural selection pressures, though the underlying mechanisms of this tolerance remain poorly understood. Since there are currently no antidotes for HAB toxins, symptom management and supportive care are the only treatment options available once a case is confirmed. In view of this, public health surveillance and intervention for HABs are critical to managing risks related to human health. Further studies are necessary to elucidate the underlying molecular mechanisms of fish and shellfish tolerance to HABs and related toxins, identify key biomarkers for HAB occurrence, develop rapid monitoring methods for toxins, especially real-time online monitoring methods, and develop effective antidotes or scavengers to mitigate public health concerns arising from HABs. Effective public health interventions include but are not limited to enhancing environmental monitoring for HAB toxins, identifying biomarkers for human diseases, further describing HAB-associated syndrome characteristics, revising disease reporting and monitoring schemes, and promoting policy development. The current approach for monitoring HAB toxin levels in contaminated aquatic organisms is the most direct, mainly relying on LC-MS/MS, but the high time and financial costs associated, as well as the limited availability of the necessary equipment, pose a challenge to detection. The future direction is to initially assess the contamination of HAB toxins by rapidly detecting changes in molecular markers or target molecules sensitive to specific HAB toxins in fish or shellfish, followed by a detailed analysis of the accumulation of HAB toxins using LC-MS/MS to formulate targeted public health intervention policies. Additionally, there are currently no effective antidotes or toxin removal agents, but the results from some in vivo studies in aquatic organisms suggest that adding natural active substances such as curcumin and cinnamaldehyde can significantly reduce the accumulation of DSP toxins in mussels and alleviate damage by affecting CYP3A4 in mussels, which has been proven to be involved in the metabolism of DSP toxins in shellfish [278,279]. Therefore, identifying the key biomarkers for the occurrence of HABs can not only effectively reduce algal toxins in aquatic organisms but also reduce the occurrence of human poisoning.

Author Contributions: K.-K.Y.: methodology, visualization, and writing—original draft. H.-Y.L.: supervision. W.-D.Y.: conceptualization, supervision, validation, and writing—review and editing. All authors have read and agreed to the published version of the manuscript.

Funding: This work was funded by the National Natural Science Foundation of China (42076143, 42476125).

Institutional Review Board Statement: Not applicable.

Data Availability Statement: Not applicable.

Conflicts of Interest: The authors declare no competing financial interests.

References

1. Wells, M.L.; Karlson, B.; Wulff, A.; Kudela, R.; Trick, C.; Asnaghi, V.; Berdalet, E.; Cochlan, W.; Davidson, K.; De Rijcke, M. Future HAB science: Directions and challenges in a changing climate. *Harmful Algae* **2020**, *91*, 101632. [CrossRef]
2. Pal, M.; Yesankar, P.J.; Dwivedi, A.; Qureshi, A. Biotic control of harmful algal blooms (HABs): A brief review. *J. Environ. Manag.* **2020**, *268*, 110687. [CrossRef]
3. Anderson, D.M. Approaches to monitoring, control and management of harmful algal blooms (HABs). *Ocean Coast. Manag.* **2009**, *52*, 342–347. [CrossRef]
4. Zohdi, E.; Abbaspour, M. Harmful algal blooms (red tide): A review of causes, impacts and approaches to monitoring and prediction. *Int. J. Environ. Sci. Technol.* **2019**, *16*, 1789–1806. [CrossRef]
5. Hallegraeff, G.; Enevoldsen, H.; Zingone, A. Global harmful algal bloom status reporting. *Harmful Algae* **2021**, *102*, 101992. [CrossRef]
6. Glibert, P.M. Harmful algae at the complex nexus of eutrophication and climate change. *Harmful Algae* **2020**, *91*, 101583. [CrossRef]
7. Yan, Z.; Kamanmalek, S.; Alamdari, N. Predicting coastal harmful algal blooms using integrated data-driven analysis of environmental factors. *Sci. Total Environ.* **2024**, *912*, 169253. [CrossRef]
8. Paerl, H.W.; Gardner, W.S.; Havens, K.E.; Joyner, A.R.; McCarthy, M.J.; Newell, S.E.; Qin, B.; Scott, J.T. Mitigating cyanobacterial harmful algal blooms in aquatic ecosystems impacted by climate change and anthropogenic nutrients. *Harmful Algae* **2016**, *54*, 213–222. [CrossRef]
9. Oh, J.W.; Pushparaj, S.S.C.; Muthu, M.; Gopal, J. Review of harmful algal blooms (HABs) causing marine fish kills: Toxicity and mitigation. *Plants* **2023**, *12*, 3936. [CrossRef]
10. Tan, K.; Sun, Y.; Zhang, H.; Zheng, H. Effects of harmful algal blooms on the physiological, immunity and resistance to environmental stress of bivalves: Special focus on paralytic shellfish poisoning and diarrhetic shellfish poisoning. *Aquaculture* **2023**, *563*, 739000. [CrossRef]
11. Turner, A.D.; Lewis, A.M.; Bradley, K.; Maskrey, B.H. Marine invertebrate interactions with harmful algal blooms—Implications for one health. *J. Invertebr. Pathol.* **2021**, *186*, 107555. [CrossRef]
12. Lenzen, M.; Li, M.; Murray, S.A. Impacts of harmful algal blooms on marine aquaculture in a low-carbon future. *Harmful Algae* **2021**, *110*, 102143. [CrossRef]
13. Trottet, A.; George, C.; Drillet, G.; Lauro, F.M. Aquaculture in coastal urbanized areas: A comparative review of the challenges posed by Harmful Algal Blooms. *Crit. Rev. Environ. Sci. Technol.* **2021**, *52*, 2888–2929. [CrossRef]
14. Beet, A.; Kite-Powell, H.; Ralston, E.P. An estimate of the cost of acute health effects from food- and water-borne marine pathogens and toxins in the USA. *J. Water Health* **2011**, *9*, 680–694.
15. Soliño, L.; Costa, P.R. Global impact of ciguatoxins and ciguatera fish poisoning on fish, fisheries and consumers. *Environ. Res.* **2020**, *182*, 109111. [CrossRef]
16. Corriere, M.; Soliño, L.; Costa, P.R. Effects of the marine biotoxins okadaic acid and dinophysistoxins on fish. *J. Mar. Sci. Eng.* **2021**, *9*, 293. [CrossRef]
17. Saeed, A.F.; Awan, S.A.; Ling, S.; Wang, R.; Wang, S. Domoic acid: Attributes, exposure risks, innovative detection techniques and therapeutics. *Algal Res.* **2017**, *24*, 97–110. [CrossRef]
18. Devillier, V.M.; Hall, E.R.; Lovko, V.; Pierce, R.; Anderson, D.M.; Lewis, K.A. Mesocosm study of PAC-modified clay effects on *Karenia brevis* cells and toxins, chemical dynamics, and benthic invertebrate physiology. *Harmful Algae* **2024**, *134*, 102609. [CrossRef]
19. Pulido, O.M. Marine algal toxins: Seafood safety, human health and beyond. *Oceanography* **2014**, *02*, e111. [CrossRef]
20. Grattan, L.M.; Holobaugh, S.; Morris, J.G. Harmful algal blooms and public health. *Harmful Algae* **2016**, *57*, 2–8. [CrossRef]

21. Young, N.; Sharpe, R.A.; Barciela, R.; Nichols, G.; Davidson, K.; Berdalet, E.; Fleming, L.E. Marine harmful algal blooms and human health: A systematic scoping review. *Harmful Algae* **2020**, *98*, 101901. [CrossRef]
22. Heil, C.A.; Muni-Morgan, A.L. Florida's Harmful algal bloom (HAB) problem: Escalating risks to human, environmental and economic health with climate change. *Front. Ecol. Evol.* **2021**, *9*, 646080. [CrossRef]
23. Reguera, B.; Riobo, P.; Rodriguez, F.; Diaz, P.A.; Pizarro, G.; Paz, B.; Franco, J.M.; Blanco, J. Dinophysis toxins: Causative organisms, distribution and fate in shellfish. *Mar. Drugs* **2014**, *12*, 394–461. [CrossRef]
24. Lee, T.C.; Fong, F.L.; Ho, K.C.; Lee, F.W. The mechanism of diarrhetic shellfish poisoning toxin production in *Prorocentrum* spp.: Physiological and molecular perspectives. *Toxins* **2016**, *8*, 272. [CrossRef]
25. Dominguez, H.J.; Paz, B.; Daranas, A.H.; Norte, M.; Franco, J.M.; Fernández, J.J. Dinoflagellate polyether within the yessotoxin, pectenotoxin and okadaic acid toxin groups: Characterization, analysis and human health implications. *Toxicon* **2010**, *56*, 191–217. [CrossRef]
26. Yanagi, T.; Murata, M.; Torigoe, K.; Yasumoto, T. Biological activities of semisynthetic analogs of Dinophysistoxin-3, the major diarrhetic shellfish toxin. *Agric. Biol. Chem.* **1989**, *53*, 525–529.
27. Suzuki, T.; Ota, H.; Yamasaki, M. Direct evidence of transformation of dinophysistoxin-1 to 7-O-acyl-dinophysistoxin-1 (dinophysistoxin-3) in the scallop *Patinopecten yessoensis*. *Toxicon* **1999**, *37*, 187–198. [CrossRef]
28. Rossignoli, A.E.; Fernandez, D.; Regueiro, J.; Marino, C.; Blanco, J. Esterification of okadaic acid in the mussel *Mytilus galloprovincialis*. *Toxicon* **2011**, *57*, 712–720. [CrossRef]
29. Yasumoto, T.; Murata, M.; Oshima, Y.; Sano, M.; Matsumoto, G.K.; Clardy, J. Diarrhetic shellfish toxins. *Tetrahedron* **1985**, *41*, 1019–1025. [CrossRef]
30. García, C.; Truan, D.; Lagos, M.; Santelices, J.P.; Diaz, J.C.; Lagos, N. Metabolic transformation of dinophysistoxin-3 into dinophysistoxin-1 causes human intoxication by consumption of o-acyl-derivates dinophysistoxins contaminated shellfish. *J. Toxicol. Sci.* **2005**, *30*, 287–296. [CrossRef]
31. Suzuki, H.; Okada, Y. Comparative toxicity of dinophysistoxin-1 and okadaic acid in mice. *J. Vet. Med. Sci.* **2018**, *80*, 616–619. [CrossRef]
32. Bauder, A.; Cembella, A.; Bricelj, V.; Quilliam, M. Uptake and fate of diarrhetic shellfish poisoning toxins from the dinoflagellate *Prorocentrum lima* in the bay scallop *Argopecten irradians*. *Mar. Ecol. Prog. Ser.* **2001**, *213*, 39–52. [CrossRef]
33. Suganuma, M.; Fujiki, H.; Okabe, S.; Nishiwaki, S.; Brautigan, D.; Ingebritsen, T.S.; Rosner, M.R. Structurally different members of the okadaic acid class selectively inhibit protein serine/threonine but not tyrosine phosphatase activity. *Toxicon* **1992**, *30*, 873–878. [CrossRef]
34. Garibo, D.; de la Iglesia, P.; Diogène, J.; Campàs, M. Inhibition equivalency factors for dinophysistoxin-1 and dinophysistoxin-2 in protein phosphatase assays: Applicability to the analysis of shellfish samples and comparison with LC-MS/MS. *J. Agric. Food Chem.* **2013**, *61*, 2572–2579. [CrossRef]
35. McCarthy, M.; O'Halloran, J.; O'Brien, N.M.; van Pelt, F.F. Does the marine biotoxin okadaic acid cause DNA fragmentation in the blue mussel and the pacific oyster? *Mar. Environ. Res.* **2014**, *101*, 153–160. [CrossRef]
36. Wu, H.; Chen, J.; Peng, J.; Zhong, Y.; Zheng, G.; Guo, M.; Tan, Z.; Zhai, Y.; Lu, S. Nontarget screening and toxicity evaluation of diol esters of okadaic acid and dinophysistoxins reveal intraspecies difference of *Prorocentrum lima*. *Environ. Sci. Technol.* **2020**, *54*, 12366–12375. [CrossRef]
37. Vilariño, N.; Louzao, M.C.; Abal, P.; Cagide, E.; Carrera, C.; Vieytes, M.R.; Botana, L.M. Human poisoning from marine toxins: Unknowns for optimal consumer protection. *Toxins* **2018**, *10*, 324. [CrossRef]
38. González, J.C.; Fontal, O.I.; Vieytes, M.R.; Vieites, J.M.; Botana, L.M. Basis for a new procedure to eliminate diarrheic shellfish toxins from a contaminated matrix. *J. Agric. Food Chem.* **2002**, *50*, 400–405. [CrossRef]
39. Karlson, B.; Andersen, P.; Arneborg, L.; Cembella, A.; Eikrem, W.; John, U.; West, J.J.; Klemm, K.; Kobos, J.; Lehtinen, S. Harmful algal blooms and their effects in coastal seas of Northern Europe. *Harmful Algae* **2021**, *102*, 101989. [CrossRef]
40. Louzao, M.C.; Abal, P.; Costas, C.; Suzuki, T.; Watanabe, R.; Vilariño, N.; Botana, A.M.; Vieytes, M.R.; Botana, L.M. DSP toxin distribution across organs in mice after acute oral administration. *Mar. Drugs* **2021**, *19*, 23. [CrossRef]
41. Park, S.Y.; Kang, J.H.; Jung, H.J.; Hwang, J.H.; Chun, H.S.; Yoon, Y.S.; Oh, S.H. Okadaic acid is at least as toxic as dinophysistoxin-1 after repeated administration to mice by gavage. *Toxins* **2023**, *15*, 587. [CrossRef]
42. Vaughan, L.; de Jesús Romero-Geraldo, R.; García-Lagunas, N.; Hernández-Saavedra, N.Y. Effects of in vitro exposure to diarrheic toxin producer *Prorocentrum lima* on gene expressions related to cell cycle regulation and immune response in *Crassostrea gigas*. *PLoS ONE* **2014**, *9*, e97181.
43. Romero-Geraldo Rde, J.; Garcia-Lagunas, N.; Hernandez-Saavedra, N.Y. *Crassostrea gigas* exposure to the dinoflagellate *Prorocentrum lima*: Histological and gene expression effects on the digestive gland. *Mar. Environ. Res.* **2016**, *120*, 93–102. [CrossRef]
44. Neves, R.A.F.; Santiago, T.C.; Carvalho, W.F.; Silva, E.D.S.; da Silva, P.M.; Nascimento, S.M. Impacts of the toxic benthic dinoflagellate *Prorocentrum lima* on the brown mussel *Perna perna*: Shell-valve closure response, immunology, and histopathology. *Mar. Environ. Res.* **2019**, *146*, 35–45. [CrossRef]
45. Nielsen, P.; Krock, B.; Hansen, P.J.; Vismann, B. Effects of the DSP-toxic dinoflagellate *Dinophysis acuta* on clearance and respiration rate of the blue mussel, *Mytilus edulis*. *PLoS ONE* **2020**, *15*, e0230176. [CrossRef]
46. Prado-Alvarez, M.; Florez-Barros, F.; Sexto-Iglesias, A.; Mendez, J.; Fernandez-Tajes, J. Effects of okadaic acid on haemocytes from *Mytilus galloprovincialis*: A comparison between field and laboratory studies. *Mar. Environ. Res.* **2012**, *81*, 90–93. [CrossRef]

47. Prado-Alvarez, M.; Florez-Barros, F.; Mendez, J.; Fernandez-Tajes, J. Effect of okadaic acid on carpet shell clam (*Ruditapes decussatus*) haemocytes by in vitro exposure and harmful algal bloom simulation assays. *Cell Biol. Toxicol.* **2013**, *29*, 189–197. [CrossRef]
48. Prego-Faraldo, M.V.; Valdiglesias, V.; Laffon, B.; Mendez, J.; Eirin-Lopez, J.M. Early genotoxic and cytotoxic effects of the toxic dinoflagellate *Prorocentrum lima* in the mussel *Mytilus galloprovincialis*. *Toxins* **2016**, *8*, 159. [CrossRef]
49. He, Z.B.; Duan, G.F.; Liang, C.Y.; Li, H.Y.; Liu, J.S.; Yang, W.D. Up-regulation of Nrf2-dependent antioxidant defenses in *Perna viridis* after exposed to *Prorocentrum lima*. *Fish Shellfish Immunol.* **2019**, *90*, 173–179. [CrossRef]
50. Svensson, S.; Särngren, A.; Förlin, L. Mussel blood cells, resistant to the cytotoxic effects of okadaic acid, do not express cell membrane p-glycoprotein activity (multixenobiotic resistance). *Aquat. Toxicol.* **2003**, *65*, 27–37. [CrossRef]
51. Vidal, A.; Ruiz, Y.; Suárez, P.; Martinez, A.A.; Rossignoli, A.; Blanco, J.; Garcia, O.; San Juan, F. Accumulation of okadaic acid and detoxifying enzymes in the digestive gland of *Mytilus galloprovincialis* during exposure to DSP. In *Molluscan Shellfish Safety*; Springer: Berlin/Heidelberg, Germany, 2014; pp. 217–225.
52. Wei, X.M.; Lu, M.Y.; Duan, G.F.; Li, H.Y.; Liu, J.S.; Yang, W.D. Responses of CYP450 in the mussel *Perna viridis* after short-term exposure to the DSP toxins-producing dinoflagellate *Prorocentrum lima*. *Ecotoxicol. Environ. Saf.* **2019**, *176*, 178–185. [CrossRef]
53. Zou, Y.; Wei, X.M.; Weng, H.W.; Li, H.Y.; Liu, J.S.; Yang, W.D. Expression profile of eight glutathione S-transferase genes in *Crassostrea ariakensis* after exposure to DSP toxins producing dinoflagellate *Prorocentrum lima*. *Toxicon* **2015**, *105*, 45–55. [CrossRef]
54. Martinez-Escauriaza, R.; Lozano, V.; Perez-Paralle, M.L.; Blanco, J.; Sanchez, J.L.; Pazos, A.J. Expression analyses of genes related to multixenobiotic resistance in *Mytilus galloprovincialis* after exposure to okadaic acid-producing *Dinophysis acuminata*. *Toxins* **2021**, *13*, 614. [CrossRef]
55. Lv, J.J.; Yuan, K.K.; Lu, G.X.; Li, H.Y.; Kwok, H.F.; Yang, W.D. Responses of ABCB and ABCC transporters to the toxic dinoflagellate *Prorocentrum lima* in the mussel *Perna viridis*. *Aquat. Toxicol.* **2022**, *254*, 106368. [CrossRef]
56. Huang, L.; Wang, J.; Chen, W.C.; Li, H.Y.; Liu, J.S.; Tao, J.; Yang, W.D. P-glycoprotein expression in *Perna viridis* after exposure to *Prorocentrum lima*, a dinoflagellate producing DSP toxins. *Fish Shellfish Immunol.* **2014**, *39*, 254–262. [CrossRef]
57. Lozano, V.; Martínez-Escauriaza, R.; Pérez-Parallé, M.L.; Pazos, A.J.; Sánchez, J.L. Two novel multidrug resistance associated protein (MRP/ABCC) from the Mediterranean mussel (*Mytilus galloprovincialis*): Characterization and expression patterns in detoxifying tissues. *Can. J. Zool.* **2015**, *93*, 567–578. [CrossRef]
58. Blanco, J. Accumulation of dinophysis toxins in bivalve molluscs. *Toxins* **2018**, *10*, 453. [CrossRef]
59. Qiu, J.; Ji, Y.; Fang, Y.; Zhao, M.; Wang, S.; Ai, Q.; Li, A. Response of fatty acids and lipid metabolism enzymes during accumulation, depuration and esterification of diarrhetic shellfish toxins in mussels (*Mytilus galloprovincialis*). *Ecotoxicol. Environ. Saf.* **2020**, *206*, 111223. [CrossRef]
60. European Food Safety Authority (EFSA). Opinion of the scientific panel on contaminants in the food chain on a request from the European Commission on marine biotoxins in shellfish –okadaic acid and analoguesn. *EFSA* **2008**, *589*, 1–62.
61. Dou, M.; Jiao, Y.H.; Zheng, J.W.; Zhang, G.; Li, H.Y.; Liu, J.S.; Yang, W.D. De novo transcriptome analysis of the mussel *Perna viridis* after exposure to the toxic dinoflagellate *Prorocentrum lima*. *Ecotoxicol. Environ. Saf.* **2020**, *192*, 110265. [CrossRef]
62. Yuan, K.K.; Chen, Z.M.; Liu, Y.X.; Li, H.Y.; Yang, W.D. Possible role of docosahexaenoic acid in response to diarrhetic shellfish toxins in the mussel *Perna viridis*. *Mar. Drugs* **2023**, *21*, 155. [CrossRef]
63. Rountos, K.J.; Kim, J.J.; Hattenrath-Lehmann, T.K.; Gobler, C.J. Effects of the harmful algae, *Alexandrium catenella* and *Dinophysis acuminata*, on the survival, growth, and swimming activity of early life stages of forage fish. *Mar. Environ. Res.* **2019**, *148*, 46–56. [CrossRef]
64. Ajuzie, C.C. Toxic *Prorocentrum lima* induces abnormal behaviour in juvenile sea bass. *J. Appl. Phycol.* **2007**, *20*, 19–27. [CrossRef]
65. Souid, G.; Souayed, N.; Haouas, Z.; Maaroufi, K. Does the phycotoxin okadaic acid cause oxidative stress damages and histological alterations to seabream (*Sparus aurata*)? *Toxicon* **2018**, *144*, 55–60. [CrossRef]
66. Neves, R.A.F.; Nascimento, S.M.; Santos, L.N. Sublethal fish responses to short-term food chain transfer of DSP toxins: The role of somatic condition. *J. Exp. Mar. Biol. Ecol.* **2020**, *524*, 151317. [CrossRef]
67. Escoffier, N.; Gaudin, J.; Mezhoud, K.; Huet, H.; Chateau-Joubert, S.; Turquet, J.; Crespeau, F.; Edery, M. Toxicity to medaka fish embryo development of okadaic acid and crude extracts of *Prorocentrum* dinoflagellates. *Toxicon* **2007**, *49*, 1182–1192. [CrossRef]
68. Figueroa, D.; Signore, A.; Araneda, O.; Contreras, H.R.; Concha, M.; Garcia, C. Toxicity and differential oxidative stress effects on zebrafish larvae following exposure to toxins from the okadaic acid group. *J. Toxicol. Environ. Health A* **2020**, *83*, 573–588. [CrossRef]
69. Rolton, A.; Rhodes, L.; Hutson, K.S.; Biessy, L.; Bui, T.; MacKenzie, L.; Symonds, J.E.; Smith, K.F. Effects of harmful algal blooms on fish and shellfish species: A case study of New Zealand in a changing environment. *Toxins* **2022**, *14*, 341. [CrossRef]
70. James, K.J.; Carey, B.; O'Halloran, J.; van Pelt, F.N.A.M.; ŠKrabÁKovÁ, Z. Shellfish toxicity: Human health implications of marine algal toxins. *Epidemiol. Infect.* **2010**, *138*, 927–940. [CrossRef]
71. Munday, R. Is protein phosphatase inhibition responsible for the toxic effects of okadaic acid in animals? *Toxins* **2013**, *5*, 267–285. [CrossRef]
72. Manerio, E.; Rodas, V.L.; Costas, E.; Hernandez, J.M. Shellfish consumption: A major risk factor for colorectal cancer. *Med. Hypotheses* **2008**, *70*, 409–412. [CrossRef]
73. Cordier, S.; Monfort, C.; Miossec, L.; Richardson, S.; Belin, C. Ecological analysis of digestive cancer mortality related to contamination by diarrhetic shellfish poisoning toxins along the coasts of France. *Environ. Res.* **2000**, *84*, 145–150. [CrossRef]

74. López-Rodas, V.; Maneiro, E.; Martínez, J.; Navarro, M.; Costas, E. Harmful algal blooms, red tides and human health: Diarrhetic shellfish poisoning and colorectal cancer. *An. Real Acad. Nac. Farm.* **2006**, *72*, 391–408.
75. Kiguchi, K.; Glesne, D.; Chubb, C.H.; Fujiki, H.; Huberman, E. Differential induction of apoptosis in human breast tumor cells by okadaic acid and related inhibitors of protein phosphatases 1 and 2A. *Cell Growth Differ.* **1994**, *5*, 995–1004. [PubMed]
76. Liu, Y.; Yuan, T.Q.; Zheng, J.W.; Li, D.W.; Jiao, Y.H.; Li, H.Y.; Li, R.M.; Yang, W.D. Exposure to okadaic acid could disrupt the colonic microenvironment in rats. *Ecotoxicol. Environ.Saf.* **2023**, *263*, 115376. [CrossRef] [PubMed]
77. Liu, Y.; Zheng, J.W.; Peng, X.C.; Li, H.Y.; Huang, L.; Li, D.W.; Liu, J.S.; Yang, W.D. Changes in colonic microbiotas in rat after long-term exposure to low dose of okadaic acid. *Chemosphere* **2020**, *254*, 126874. [CrossRef]
78. Wiese, M.; D'Agostino, P.M.; Mihali, T.K.; Moffitt, M.C.; Neilan, B.A. Neurotoxic alkaloids: Saxitoxin and its analogs. *Mar. Drugs* **2010**, *8*, 2185–2211. [CrossRef]
79. Lin, Z.R.; Geng, H.X.; Yu, R.C. Potential roles of hydroxybenzoate paralytic shellfish toxins in modulating toxin biokinetics in scallops. *J. Hazard. Mater.* **2024**, *469*, 133896. [CrossRef]
80. Leal, J.F.; Cristiano, M.L.S. Marine paralytic shellfish toxins: Chemical properties, mode of action, newer analogues, and structure–toxicity relationship. *Nat. Prod. Rep.* **2022**, *39*, 33–57. [CrossRef] [PubMed]
81. Qiu, J.B.; Zhang, L.; Cao, Y.D.; Zhao, P.; Zhang, J.R.; Li, A.F. Bioaccessibility of paralytic shellfish toxins in different cooked shellfish using the simulated digestive model. *Food Chem.* **2022**, *390*, 133094. [CrossRef]
82. Asakawa, M.; Gomez-Delan, G.; Barte-Quilantang, M.; Ito, K. Paralytic shellfish poison (PSP)–producing dinoflagellate and PSP-infested organisms. In *Marine Protists: Diversity and Dynamics*; Springer: Tokyo, Japan, 2015; pp. 567–596.
83. Hashimoto, K.; Noguchi, T. Recent studies on paralytic shellfish poison in Japan. *Pure Appl. Chem.* **1989**, *61*, 7–18. [CrossRef]
84. Landsberg, J.H. The effects of harmful algal blooms on aquatic organisms. *Rev. Fish. Sci.* **2010**, *10*, 113–390. [CrossRef]
85. Núñez-Vázquez, E.J.; Poot-Delgado, C.A.; Turner, A.D.; Hernández-Sandoval, F.E.; Okolodkov, Y.B.; Fernández-Herrera, L.J.; Bustillos-Guzmán, J.J. Paralytic shellfish toxins of *Pyrodinium bahamense* (Dinophyceae) in the southeastern Gulf of Mexico. *Toxins* **2022**, *14*, 760. [CrossRef]
86. Kwon, H.K.; Kim, G.; Lim, W.A.; Park, J.W.; Park, T.G. Conditions of nutrients and dissolved organic matter for the outbreaks of Paralytic Shellfish Poisoning (PSP) in Jinhae Bay, Korea. *Mar. Pollut. Bull.* **2020**, *158*, 111381. [CrossRef]
87. Che, Y.; Ding, L.; Qiu, J.; Ji, Y.; Li, A. Conversion and stability of new metabolites of paralytic shellfish toxins under different temperature and pH conditions. *J. Agric. Food Chem.* **2020**, *68*, 1427–1435. [CrossRef]
88. Dell'Aversano, C.; Walter, J.A.; Burton, I.W.; Stirling, D.J.; Fattorusso, E.; Quilliam, M.A. Isolation and structure elucidation of new and unusual saxitoxin analogues from mussels. *J. Nat. Prod.* **2008**, *71*, 1518–1523. [CrossRef]
89. Ding, L.; Qiu, J.; Li, A. Proposed biotransformation pathways for new metabolites of paralytic shellfish toxins based on field and experimental mussel samples. *J. Agric. Food Chem.* **2017**, *65*, 5494–5502. [CrossRef]
90. Qiu, J.; Meng, F.; Ding, L.; Che, Y.; McCarron, P.; Beach, D.G.; Li, A. Dynamics of paralytic shellfish toxins and their metabolites during timecourse exposure of scallops *Chlamys farreri* and mussels *Mytilus galloprovincialis* to *Alexandrium pacificum*. *Aquat. Toxicol.* **2018**, *200*, 233–240. [CrossRef]
91. Negri, A.; Stirling, D.; Quilliam, M.; Blackburn, S.; Bolch, C.; Burton, I.; Eaglesham, G.; Thomas, K.; Walter, J.; Willis, R. Three novel hydroxybenzoate saxitoxin analogues isolated from the dinoflagellate *Gymnodinium catenatum*. *Chem. Res. Toxicol.* **2003**, *16*, 1029–1033. [CrossRef]
92. Llewellyn, L.; Negri, A.; Quilliam, M. High affinity for the rat brain sodium channel of newly discovered hydroxybenzoate saxitoxin analogues from the dinoflagellate *Gymnodinium catenatum*. *Toxicon* **2004**, *43*, 101–104. [CrossRef]
93. Geraci, J.R.; Anderson, D.M.; Timperi, R.J.; St. Aubin, D.J.; Early, G.A.; Prescott, J.H.; Mayo, C.A. Humpback whales (*Megaptera novaeangliae*) fatally poisoned by dinoflagellate toxin. *Can. J. Fish. Aquat. Sci.* **1989**, *46*, 1895–1898. [CrossRef]
94. Doucette, G.J.; Turner, J.T.; Powell, C.L.; Keafer, B.A.; Anderson, D.M. Trophic accumulation of PSP toxins in zooplankton during *Alexandrium fundyense* blooms in Casco Bay, Gulf of Maine, April–June 1998. I. Toxin levels in *A. fundyense* and zooplankton size fractions. *Deep Sea Res. II Top. Stud. Oceanogr.* **2005**, *52*, 2764–2783. [CrossRef]
95. Santana-Viera, S.; Lara-Martin, P.A. Analysis of paralytic shellfish toxins in seafood by liquid chromatography: A critical review of the methodologies developed. *Trends Environ. Anal. Chem.* **2023**, *37*, e00190. [CrossRef]
96. Silva, M.; Rey, V.; Barreiro, A.; Kaufmann, M.; Neto, A.I.; Hassouani, M.; Sabour, B.; Botana, A.; Botana, L.M.; Vasconcelos, V. Paralytic shellfish toxins occurrence in non-traditional invertebrate vectors from north Atlantic waters (Azores, Madeira, and Morocco). *Toxins* **2018**, *10*, 362. [CrossRef]
97. Goya, A.B.; Tarnovius, S.; Hatfield, R.G.; Coates, L.; Lewis, A.M.; Turner, A.D. Paralytic shellfish toxins and associated toxin profiles in bivalve mollusc shellfish from Argentina. *Harmful Algae* **2020**, *99*, 101910. [CrossRef]
98. Castrec, J.; Soudant, P.; Payton, L.; Tran, D.; Miner, P.; Lambert, C.; Le Goïc, N.; Huvet, A.; Quillien, V.; Boullot, F. Bioactive extracellular compounds produced by the dinoflagellate *Alexandrium minutum* are highly detrimental for oysters. *Aquat. Toxicol.* **2018**, *199*, 188–198. [CrossRef]
99. Tran, D.; Haberkorn, H.; Soudant, P.; Ciret, P.; Massabuau, J.C. Behavioral responses of *Crassostrea gigas* exposed to the harmful algae *Alexandrium minutum*. *Aquaculture* **2010**, *298*, 338–345. [CrossRef]
100. Lassudrie, M.; Wikfors, G.H.; Sunila, I.; Alix, J.H.; Dixon, M.S.; Combot, D.; Soudant, P.; Fabioux, C.; Hégaret, H. Physiological and pathological changes in the eastern oyster *Crassostrea virginica* infested with the trematode *Bucephalus* sp. and exposed to the toxic dinoflagellate *Alexandrium fundyense*. *J. Invertebr. Pathol.* **2015**, *126*, 51–63. [CrossRef]

101. Álvarez, G.; Díaz, P.A.; Godoy, M.; Araya, M.; Ganuza, I.; Pino, R.; Álvarez, F.; Rengel, J.; Hernández, C.; Uribe, E.; et al. Paralytic shellfish toxins in surf clams *Mesodesma donacium* during a large bloom of *Alexandrium catenella* dinoflagellates associated to an intense shellfish mass mortality. *Toxins* **2019**, *11*, 188. [CrossRef]
102. Astuya, A.; Carrera, C.; Ulloa, V.; Aballay, A.; Núñez-Acuña, G.; Hégaret, H.; Gallardo-Escárate, C. Saxitoxin modulates immunological parameters and gene transcription in *Mytilus chilensis* hemocytes. *Int. J. Mol. Sci.* **2015**, *16*, 15235–15250. [CrossRef]
103. Mello, D.F.; Silva, P.M.d.; Barracco, M.A.; Soudant, P.; Hégaret, H. Effects of the dinoflagellate *Alexandrium minutum* and its toxin (saxitoxin) on the functional activity and gene expression of *Crassostrea gigas* hemocytes. *Harmful Algae* **2013**, *26*, 45–51. [CrossRef]
104. Cao, R.; Wang, D.; Wei, Q.; Wang, Q.; Yang, D.; Liu, H.; Dong, Z.; Zhang, X.; Zhang, Q.; Zhao, J. Integrative biomarker assessment of the influence of saxitoxin on marine bivalves: A comparative study of the two bivalve species oysters, *Crassostrea gigas*, and Scallops, *Chlamys farreri*. *Front. Physiol.* **2018**, *9*, 1173. [CrossRef] [PubMed]
105. Estrada, N.; Ascencio, F.; Shoshani, L.; Contreras, R.G. Apoptosis of hemocytes from lions-paw scallop *Nodipecten subnodosus* induced with paralyzing shellfish poison from *Gymnodinium catenatum*. *Immunobiology* **2014**, *219*, 964–974. [CrossRef]
106. Li, Y.; Sun, X.; Hu, X.; Xun, X.; Zhang, J.; Guo, X.; Jiao, W.; Zhang, L.; Liu, W.; Wang, J. Scallop genome reveals molecular adaptations to semi-sessile life and neurotoxins. *Nat. Commun.* **2017**, *8*, 1721. [CrossRef]
107. Bricelj, V.M.; Connell, L.; Konoki, K.; MacQuarrie, S.P.; Scheuer, T.; Catterall, W.A.; Trainer, V.L. Sodium channel mutation leading to saxitoxin resistance in clams increases risk of PSP. *Nature* **2005**, *434*, 763–767. [CrossRef] [PubMed]
108. Zakon, H.H. Adaptive evolution of voltage-gated sodium channels: The first 800 million years. *Proc. Natl. Acad. Sci. USA* **2012**, *109*, 10619–10625. [CrossRef]
109. Jost, M.C.; Hillis, D.M.; Lu, Y.; Kyle, J.W.; Fozzard, H.A.; Zakon, H.H. Toxin-resistant sodium channels: Parallel adaptive evolution across a complete gene family. *Mol. Biol. Evol.* **2008**, *25*, 1016–1024. [CrossRef]
110. Kontis, K.J.; Goldin, A.L. Site-directed mutagenesis of the putative pore region of the rat IIA sodium channel. *Mol. Pharmacol.* **1993**, *43*, 635–644.
111. Bojarová, P.; Williams, S.J. Sulfotransferases, sulfatases and formylglycine-generating enzymes: A sulfation fascination. *Curr. Opin. Chem. Biol.* **2008**, *12*, 573–581. [CrossRef]
112. Lian, S.; Zhao, L.; Xun, X.; Lou, J.; Li, M.; Li, X.; Wang, S.; Zhang, L.; Hu, X.; Bao, Z. Genome-wide identification and characterization of SODs in Zhikong scallop reveals gene expansion and regulation divergence after toxic dinoflagellate exposure. *Mar. Drugs* **2019**, *17*, 700. [CrossRef]
113. Fabioux, C.; Sulistiyani, Y.; Haberkorn, H.; Hegaret, H.; Amzil, Z.; Soudant, P. Exposure to toxic *Alexandrium minutum* activates the detoxifying and antioxidant systems in gills of the oyster *Crassostrea gigas*. *Harmful Algae* **2015**, *48*, 55–62. [CrossRef]
114. Wang, H.; Liu, S.; Xun, X.; Li, M.; Lou, J.; Zhang, Y.; Shi, J.; Hu, J.; Bao, Z.; Hu, X. Toxin- and species-dependent regulation of ATP-binding cassette (ABC) transporters in scallops after exposure to paralytic shellfish toxin-producing dinoflagellates. *Aquat. Toxicol.* **2021**, *230*, 105697. [CrossRef]
115. Hu, B.; Li, M.; Yu, X.; Xun, X.; Lu, W.; Li, X.; Li, Y.; Lou, J.; Wang, S.; Zhang, L. Diverse expression regulation of Hsp70 genes in scallops after exposure to toxic *Alexandrium* dinoflagellates. *Chemosphere* **2019**, *234*, 62–69. [CrossRef]
116. Soon, T.K.; Zheng, H. Climate change and bivalve mass mortality in temperate regions. *Rev. Environ. Contam. Toxicol.* **2020**, *251*, 109–129.
117. Matsuyama, Y. Impacts of the harmful dinoflagellate *Heterocapsa circularisquama* bloom on shellfish aquaculture in Japan and some experimental studies on invertebrates. *Harmful Algae* **2012**, *14*, 144–155. [CrossRef]
118. Griffith, A.W.; Shumway, S.E.; Gobler, C.J. Differential mortality of north atlantic bivalve molluscs during harmful algal blooms caused by the dinoflagellate, *Cochlodinium* (aka *Margalefidinium*) polykrikoides. *Estuaries Coasts* **2019**, *42*, 190–203. [CrossRef]
119. Mellado, C.; Chaparro, O.R.; Duarte, C.; Villanueva, P.A.; Ortiz, A.; Valdivia, N.; Torres, R.; Navarro, J.M. Ocean acidification exacerbates the effects of paralytic shellfish toxins on the fitness of the edible mussel *Mytilus chilensis*. *Sci. Total Environ.* **2019**, *653*, 455–464. [CrossRef]
120. Funesto, E.G.M.; Lewis, A.M.; Turner, A.D.; Cameron, T.C.; Steinke, M. Immediate and delayed effects of a heatwave and *Prorocentrum lima* ((Ehrenberg) Stein 1878) bloom on the toxin accumulation, physiology, and survival of the oyster *Magallana gigas* (Thunberg, 1793). *Sci. Total Environ.* **2023**, *892*, 164485. [CrossRef]
121. Tang, X.W.; Yu, R.; Fan, M.H.; Yang, Z.; Liao, Z.; Yang, Z.X.; Xie, C.Y.; Xuan, Y.K.; Wang, J.X.; Zhang, X.L.; et al. Physiological and transcriptome analysis of *Mytilus coruscus* in response to *Prorocentrum lima* and microplastics. *Front. Mar. Sci.* **2022**, *9*, 1087667. [CrossRef]
122. Yuan, K.K.; Yu, Y.Y.; Mo, Y.H.; Liu, Y.J.; Zhang, W.X.; Lv, J.J.; Shi, W.; Liu, G.X.; Li, H.Y.; Yang, W.D. Exposure to microplastics renders immunity of the thick-shell mussel more vulnerable to diarrhetic shellfish toxin-producing harmful algae. *Sci. Total Environ.* **2024**, *926*, 172125. [CrossRef] [PubMed]
123. Arechavala-Lopez, P.; Cabrera-Álvarez, M.J.; Maia, C.M.; Saraiva, J.L. Environmental enrichment in fish aquaculture: A review of fundamental and practical aspects. *Rev. Aquacult.* **2021**, *14*, 704–728. [CrossRef]
124. Reis Costa, P. Impact and effects of paralytic shellfish poisoning toxins derived from harmful algal blooms to marine fish. *Fish Fish.* **2014**, *17*, 226–248. [CrossRef]
125. White, A.W. Recurrence of kills of Atlantic herring (*Clupea harengus harengus*) caused by dinoflagellate toxins transferred through herbivorous zooplankton. *Can. J. Fish. Aquat. Sci.* **1980**, *37*, 2262–2265. [CrossRef]

126. Montoya, N.; Akselman, R.; Franco, J.; Carreto, J. Paralytic shellfish toxins and mackerel (*Scomber japonicus*) mortality in the Argentine Sea. *Harmful Toxic Algal Bloom.* **1996**, *1*, 417–420.
127. White, A.W. Dinoflagellate toxins as probable cause of an Atlantic herring (*Clupea harengus harengus*) kill, and pteropods as apparent vector. *J. Fish. Board Can.* **1977**, *34*, 2421–2424. [CrossRef]
128. Salierno, J.; Snyder, N.; Murphy, A.; Poli, M.; Hall, S.; Baden, D.; Kane, A. Harmful algal bloom toxins alter c-Fos protein expression in the brain of killifish, *Fundulus heteroclitus. Aquat. Toxicol.* **2006**, *78*, 350–357. [CrossRef]
129. Chen, C.Y.; Chou, H.N. Ichthyotoxicity studies of milkfish *Chanos chanos* fingerlings exposed to a harmful dinoflagellate *Alexandrium minutum. J. Exp. Mar. Biol. Ecol.* **2001**, *262*, 211–219. [CrossRef]
130. Haque, M.N.; Nam, S.E.; Shin, Y.K.; Rhee, J.S. The dinoflagellate *Alexandrium affine* acutely induces significant modulations on innate immunity, hepatic function, and antioxidant defense system in the gill and liver tissues of red seabream. *Aquat. Toxicol.* **2021**, *240*, 105985. [CrossRef]
131. Möller, K.; Tillmann, U.; Pöchhacker, M.; Varga, E.; Krock, B.; Porreca, F.; Koch, F.; Harris, T.M.; Meunier, C.L. Toxic effects of the emerging *Alexandrium pseudogonyaulax* (Dinophyceae) on multiple trophic levels of the pelagic food web. *Harmful Algae* **2024**, *138*, 102705. [CrossRef]
132. Bianchi, V.A.; Bickmeyer, U.; Tillmann, U.; Krock, B.; Müller, A.; Abele, D. In vitro effects of paralytic shellfish toxins and lytic extracellular compounds produced by *Alexandrium* Strains on hemocyte integrity and function in *Mytilus edulis. Toxins* **2021**, *13*, 544. [CrossRef]
133. Berdalet, E.; Fleming, L.E.; Gowen, R.; Davidson, K.; Hess, P.; Backer, L.C.; Moore, S.K.; Hoagland, P.; Enevoldsen, H. Marine harmful algal blooms, human health and wellbeing: Challenges and opportunities in the 21st century. *J. Mar. Biol. Assoc. UK* **2015**, *96*, 61–91. [CrossRef]
134. García, C.; Lagos, M.; Truan, D.; Lattes, K.; Véjar, O.; Chamorro, B.; Iglesias, V.; Andrinolo, D.; Lagos, N. Human intoxication with paralytic shellfish toxins: Clinical parameters and toxin analysis in plasma and urine. *Biol. Res.* **2005**, *38*, 197–205. [CrossRef]
135. Batoréu, M.C.C.; Dias, E.; Pereira, P.; Franca, S. Risk of human exposure to paralytic toxins of algal origin. *Environ. Toxicol. Pharmacol.* **2005**, *19*, 401–406. [CrossRef]
136. Catterall, W.A. The molecular basis of neuronal excitability. *Science* **1984**, *223*, 653–661. [CrossRef]
137. Catterall, W.A. From ionic currents to molecular mechanisms: The structure and function of voltage-gated sodium channels. *Neuron* **2000**, *26*, 13–25. [CrossRef]
138. Catterall, W.A.; Cestèle, S.; Yarov-Yarovoy, V.; Yu, F.H.; Konoki, K.; Scheuer, T. Voltage-gated ion channels and gating modifier toxins. *Toxicon* **2007**, *49*, 124–141. [CrossRef]
139. Catterall, W.A.; Morrow, C.S.; Hartshorne, R.P. Neurotoxin binding to receptor sites associated with voltage-sensitive sodium channels in intact, lysed, and detergent-solubilized brain membranes. *J. Biol. Chem.* **1979**, *254*, 11379–11387. [CrossRef]
140. Catterall, W.A. Neurotoxins that act on voltage-sensitive sodium channels in excitable membranes. *Annu. Rev. Pharmacol. Toxicol.* **1980**, *20*, 15–43. [CrossRef]
141. Jeffery, B.; Barlow, T.; Moizer, K.; Paul, S.; Boyle, C. Amnesic shellfish poison. *Food Chem. Toxicol.* **2004**, *42*, 545–557. [CrossRef]
142. Botana, L.M. *Seafood and Freshwater Toxins*, 3rd ed.; CRC Press: Boca Raton, FL, USA, 2014.
143. La Barre, S.; Bates, S.S.; Quilliam, M.A. Domoic acid. In *Outstanding Marine Molecules*; Wiley Blackwell: Hoboken, NJ, USA, 2014; pp. 189–216.
144. Quilliam, M. Chemical methods for domoic acid, the amnesic shellfish poisoning (ASP) toxin. *Man. Harmful Mar. Microalgae Monogr. Oceanogr. Methodol.* **2003**, *11*, 247–266.
145. Hampson, D.R.; Huang, X.P.; Wells, J.W.; Walter, J.A.; Wright, J.L. Interaction of domoic acid and several derivatives with kainic acid and AMPA binding sites in rat brain. *Eur. J. Pharmacol.* **1992**, *218*, 1–8. [CrossRef]
146. McCarron, P.; Hess, P. Tissue distribution and effects of heat treatments on the content of domoic acid in blue mussels, *Mytilus edulis. Toxicon* **2006**, *47*, 473–479. [CrossRef]
147. Zaman, L.; Arakawa, O.; Shimosu, A.; Onoue, Y.; Nishio, S.; Shida, Y.; Noguchi, T. Two new isomers of domoic acid from a red alga, *Chondria armata. Toxicon* **1997**, *35*, 205–212. [CrossRef]
148. Sawant, P.M.; Tyndall, J.D.A.; Holland, P.T.; Peake, B.M.; Mountfort, D.O.; Kerr, D.S. In vivo seizure induction and affinity studies of domoic acid and isodomoic acids-D, -E and -F. *Neuropharmacology* **2010**, *59*, 129–138. [CrossRef]
149. Zhao, J.Y.; Thibault, P.; Quilliam, M.A. Analysis of domoic acid isomers in seafood by capillary electrophoresis. *Electrophoresis* **1997**, *18*, 268–276. [CrossRef]
150. Wright, J.; Bird, C.; De Freitas, A.; Hampson, D.; McDonald, J.; Quilliam, M. Chemistry, biology, and toxicology of domoic acid and its isomers. *Can. Dis. Wkly. Rep.* **1990**, *16*, 21–26.
151. Wright, J.L.; Falk, M.; McInnes, A.G.; Walter, J.A. Identification of isodomoic acid D and two new geometrical isomers of domoic acid in toxic mussels. *Can. J. Chem.* **1990**, *68*, 22–25. [CrossRef]
152. Novaczek, I.; Madhyastha, M.; Ablett, R.; Johnson, G.; Nijjar, M.; Sims, D.E. Uptake, disposition and depuration of domoic acid by blue mussels (*Mytilus edulis*). *Aquat. Toxicol.* **1991**, *21*, 103–118. [CrossRef]
153. Álvarez, G.; Rengel, J.; Araya, M.; Álvarez, F.; Pino, R.; Uribe, E.; Díaz, P.A.; Rossignoli, A.E.; López-Rivera, A.; Blanco, J. Rapid domoic acid depuration in the scallop *Argopecten purpuratus* and its transfer from the digestive gland to other organs. *Toxins* **2020**, *12*, 698. [CrossRef]

154. Blanco, J.; Mauríz, A.; Álvarez, G. Distribution of domoic acid in the digestive gland of the king scallop *Pecten maximus*. *Toxins* **2020**, *12*, 371. [CrossRef]
155. Blanco, J.; Acosta, C.; De La Puente, M.B.; Salgado, C. Depuration and anatomical distribution of the amnesic shellfish poisoning (ASP) toxin domoic acid in the king scallop *Pecten maximus*. *Aquat. Toxicol.* **2002**, *60*, 111–121. [CrossRef] [PubMed]
156. Novaczek, I.; Madhyastha, M.; Ablett, R.; Donald, A.; Johnson, G.; Nijjar, M.; Sims, D.E. Depuration of domoic acid from live blue mussels (*Mytilus edulis*). *Can. J. Fish. Aquat.Sci.* **1992**, *49*, 312–318. [CrossRef]
157. Whyte, J.; Ginther, N.; Townsend, T. Accumulation and depuration of domoic acid by the mussel, *Mytilus californianus*. In *Harmful Marine Algal Blooms*; Lassus, P., Arzul, G., Erard, E., Gentien, P., Marcaillou, C., Eds.; Technique et Documentation-Lavoisier, Intercept Ltd.: París, France, 1995; pp. 531–537.
158. Mackenzie, A.; White, D.; Sim, P.; Holland, A. Domoic acid and the New Zealand Greenshell mussel (Perna canaliculus). In *Toxic Phytoplankton Blooms in the Sea, Proceedings of the Fifth International Conference on Toxic Marine Phytoplankton, Newport, RI, USA, 28 October–1 November 1991*; Smayda, T.J., Shimizu, Y., Eds.; Elsevier: Amsterdam, NY, USA, 1993; pp. 607–612.
159. Álvarez, G.; Uribe, E.; Regueiro, J.; Martin, H.; Gajardo, T.; Jara, L.; Blanco, J. Depuration and anatomical distribution of domoic acid in the surf clam *Mesodesma donacium*. *Toxicon* **2015**, *102*, 1–7. [CrossRef]
160. Dusek Jennings, E.; Parker, M.S.; Simenstad, C.A. Domoic acid depuration by intertidal bivalves fed on toxin-producing *Pseudo-nitzschia multiseries*. *Toxicon X* **2020**, *6*, 100027. [CrossRef]
161. Mafra, L.L.; Bricelj, V.M.; Fennel, K. Domoic acid uptake and elimination kinetics in oysters and mussels in relation to body size and anatomical distribution of toxin. *Aquat. Toxicol.* **2010**, *100*, 17–29. [CrossRef]
162. Jones, T.O.; Whyte, J.N.; Ginther, N.G.; Townsend, L.D.; Iwama, G.K. Haemocyte changes in the Pacific oyster, *Crassostrea gigas*, caused by exposure to domoic acid in the diatom *Pseudonitzschia pungens* f. *multiseries*. *Toxicon* **1995**, *33*, 347–353. [CrossRef] [PubMed]
163. Bresnan, E.; Fryer, R.J.; Fraser, S.; Smith, N.; Stobo, L.; Brown, N.; Turrell, E. The relationship between *Pseudo-nitzschia* (Peragallo) and domoic acid in Scottish shellfish. *Harmful Algae* **2017**, *63*, 193–202. [CrossRef]
164. Douglas, D.J.; Kenchington, E.R.; Bird, C.J.; Pocklington, R.; Bradford, B.; Silvert, W. Accumulation of domoic acid by the sea scallop (*Placopecten magellanicus*) fed cultured cells of toxic *Pseudo-nitzschia multiseries*. *Can. J. Fish. Aquat. Sci.* **1997**, *54*, 907–913. [CrossRef]
165. Horner, R.; Kusske, M.; Moynihan, B.; Skinner, R.; Wekell, J. Retention of domoic acid by Pacific razor clams, *Siliqua patula*(Dixon, 1789): Preliminary study. *J. Shellfish Res.* **1993**, *12*, 451–456.
166. Dao, V.H.; Takata, Y.; Sato, S.; Fukuzo, Y.; Kodama, M. Domoic acid in a bivalve *Sepondylus cruentus* in Nha Trang Bay, Khanh Hoa Province, Vietnam. *Coast. Mar. Sci.* **2006**, *30*, 130–132.
167. Trainer, V.L.; Bill, B.D. Characterization of a domoic acid binding site from Pacific razor clam. *Aquat. Toxicol.* **2004**, *69*, 125–132. [CrossRef] [PubMed]
168. Mauriz, A.; Blanco, J. Distribution and linkage of domoic acid (amnesic shellfish poisoning toxins) in subcellular fractions of the digestive gland of the scallop *Pecten maximus*. *Toxicon* **2010**, *55*, 606–611. [CrossRef] [PubMed]
169. García-Corona, J.L.; Hégaret, H.; Deléglise, M.; Marzari, A.; Rodríguez-Jaramillo, C.; Foulon, V.; Fabioux, C. First subcellular localization of the amnesic shellfish toxin, domoic acid, in bivalve tissues: Deciphering the physiological mechanisms involved in its long-retention in the king scallop *Pecten maximus*. *Harmful Algae* **2022**, *116*, 102251. [CrossRef]
170. García-Corona, J.L.; Hegaret, H.; Lassudrie, M.; Derrien, A.; Terre-Terrillon, A.; Delaire, T.; Fabioux, C. Comparative study of domoic acid accumulation, isomer content and associated digestive subcellular processes in five marine invertebrate species. *Aquat. Toxicol.* **2024**, *266*, 106793. [CrossRef]
171. García-Corona, J.L.; Fabioux, C.; Vanmaldergem, J.; Petek, S.; Derrien, A.; Terre-Terrillon, A.; Bressolier, L.; Breton, F.; Hegaret, H. The amnesic shellfish poisoning toxin, domoic acid: The tattoo of the king scallop *Pecten maximus*. *Harmful Algae* **2024**, *133*, 102607. [CrossRef]
172. Dizer, H.; Fischer, B.; Harabawy, A.; Hennion, M.C.; Hansen, P.D. Toxicity of domoic acid in the marine mussel *Mytilus edulis*. *Aquat. Toxicol.* **2001**, *55*, 149–156. [CrossRef] [PubMed]
173. Jones, T.; Whyte, J.; Townsend, L.; Ginther, N.; Iwama, G. Effects of domoic acid on haemolymph pH, PCO_2 and PO_2 in the Pacific oyster, *Crassostrea gigas* and the California mussel, *Mytilus californianus*. *Aquat. Toxicol.* **1995**, *31*, 43–55. [CrossRef]
174. Liu, H.; Kelly, M.S.; Campbell, D.A.; Fang, J.; Zhu, J. Accumulation of domoic acid and its effect on juvenile king scallop *Pecten maximus* (Linnaeus, 1758). *Aquaculture* **2008**, *284*, 224–230. [CrossRef]
175. Liu, H.; Kelly, M.S.; Campbell, D.A.; Dong, S.L.; Zhu, J.X.; Wang, S.F. Exposure to domoic acid affects larval development of king scallop *Pecten maximus* (Linnaeus, 1758). *Aquat. Toxicol.* **2007**, *81*, 152–158. [CrossRef]
176. Chi, C.; Zhang, C.; Liu, J.; Zheng, X. Effects of marine toxin domoic acid on innate immune responses in bay scallop *Argopecten irradians*. *J. Mar. Sci. Eng.* **2019**, *7*, 407. [CrossRef]
177. Pazos, A.J.; Ventoso, P.; Martínez-Escauriaza, R.; Pérez-Parallé, M.L.; Blanco, J.; Triviño, J.C.; Sánchez, J.L. Transcriptional response after exposure to domoic acid-producing *Pseudo-nitzschia* in the digestive gland of the mussel *Mytilus galloprovincialis*. *Toxicon* **2017**, *140*, 60–71. [CrossRef] [PubMed]
178. Ventoso, P.; Pazos, A.J.; Pérez-Parallé, M.L.; Blanco, J.; Triviño, J.C.; Sánchez, J.L. RNA-Seq transcriptome profiling of the queen scallop (*Aequipecten opercularis*) digestive gland after exposure to domoic acid-producing *Pseudo-nitzschia*. *Toxins* **2019**, *11*, 97. [CrossRef]

179. Cabrera, J.; González, P.M.; Puntarulo, S. The phycotoxin domoic acid as a potential factor for oxidative alterations enhanced by climate change. *Front. Plant Sci.* **2020**, *11*, 576971. [CrossRef] [PubMed]
180. Tian, D.; Zhang, G. Toxic effects of domoic acid on caenorhabditis elegans and the underlying mechanism. *Int. J. Biol.* **2019**, *11*, 1–9. [CrossRef]
181. Song, J.A.; Choi, C.Y.; Park, H.-S. Exposure to domoic acid causes oxidative stress in bay scallops *Argopecten irradians*. *Fish. Sci.* **2020**, *86*, 701–709. [CrossRef]
182. Lefebvre KA, M.W.; Silver, M.; Coale, S.; Tjeerdema, R. Domoic acid in planktivorous fish in relation to toxic *Pseudo-nitzschia* cell densities. *Mar. Biol.* **2002**, *140*, 625–631.
183. Lefebvre, K.A.; Frame, E.R.; Kendrick, P.S. Domoic acid and fish behavior: A review. *Harmful Algae* **2012**, *13*, 126–130. [CrossRef]
184. Lefebvre, K.; Noren, D.; Schultz, I.; Bogard, S.; Wilson, J.; Eberhart, B. Uptake, tissue distribution and excretion of domoic acid after oral exposure in coho salmon (*Oncorhynchus kisutch*). *Aquat. Toxicol.* **2007**, *81*, 266–274. [CrossRef]
185. Hardy, R.W.; Scott, T.M.; Hatfield, C.L.; Barnett, H.J.; Gauglitz Jr, E.J.; Wekell, J.C.; Eklund, M.W. Domoic acid in rainbow trout (*Oncorhynchus mykiss*) feeds. *Aquaculture* **1995**, *131*, 253–260. [CrossRef]
186. Lefebvre, K.A.; Dovel, S.L.; Silver, M.W. Tissue distribution and neurotoxic effects of domoic acid in a prominent vector species, the northern anchovy *Engraulis mordax*. *Mar. Biol.* **2001**, *138*, 693–700. [CrossRef]
187. Beltrán-Solís, K.; García-Mendoza, E.; Sánchez-Serrano, S.; López, L.M. Domoic acid affects brain morphology and causes behavioral alterations in two fish species. *Sci. Rep.* **2023**, *13*, 21729. [CrossRef] [PubMed]
188. Bates, S.; Bird, C.J.; Freitas, A.d.; Foxall, R.; Gilgan, M.; Hanic, L.A.; Johnson, G.R.; McCulloch, A.; Odense, P.; Pocklington, R. Pennate diatom *Nitzschia pungens* as the primary source of domoic acid, a toxin in shellfish from eastern Prince Edward Island, Canada. *Can. J. Fish. Aquat. Sci.* **1989**, *46*, 1203–1215. [CrossRef]
189. Lefebvre, K.A.; Robertson, A. Domoic acid and human exposure risks: A review. *Toxicon* **2010**, *56*, 218–230. [CrossRef] [PubMed]
190. Nijjar, M.; Nijjar, S.S. Domoic acid-induced neurodegeneration resulting in memory loss is mediated by Ca^{2+} overload and inhibition of Ca^{2++} calmodulin-stimulated adenylate cyclase in rat brain. *Int. J. Mol. Med.* **2000**, *6*, 377–466. [CrossRef]
191. Cendes, F.; Andermann, F.; Carpenter, S.; Zatorre, R.J.; Cashman, N.R. Temporal lobe epilepsy caused by domoic acid intoxication: Evidence for glutamate receptor–mediated excitotoxicity in humans. *Annu. Neurol.* **1995**, *37*, 123–126. [CrossRef]
192. Giordano, G.; Klintworth, H.; Kavanagh, T.; Costa, L.G. Apoptosis induced by domoic acid in mouse cerebellar granule neurons involves activation of p38 and JNK MAP kinases. *Neurochem. Int.* **2008**, *52*, 1100–1105. [CrossRef]
193. Giordano, G.; White, C.C.; Mohar, I.; Kavanagh, T.J.; Costa, L.G. Glutathione levels modulate domoic acid–induced apoptosis in mouse cerebellar granule cells. *Toxicol. Sci.* **2007**, *100*, 433–444. [CrossRef]
194. Lipton, S.A.; Rosenberg, P.A. Excitatory amino acids as a final common pathway for neurologic disorders. *N. Engl. J. Med.* **1994**, *330*, 613–622.
195. Adla, S.K.; Virtanen, H.; Thongsodsaeng, T.; Huttunen, K.M. Amino acid transporters in neurological disorders and neuroprotective effects of cysteine derivatives. *Neurochem. Int.* **2024**, *177*, 105771. [CrossRef]
196. Carcache, L.M.; Rodriguez, J.; Rein, K.S. The structural basis for kainoid selectivity at AMPA receptors revealed by low-mode docking calculations. *Bioorg. Med. Chem.* **2003**, *11*, 551–559. [CrossRef]
197. Hampson, D.R.; Manalo, J.L. The activation of glutamate receptors by kainic acid and domoic acid. *Nat. Toxins* **1998**, *6*, 153–158. [CrossRef]
198. Wang, D.Z. Neurotoxins from marine dinoflagellates: A brief review. *Mar. Drugs* **2008**, *6*, 349–371. [CrossRef] [PubMed]
199. Watkins, S.M. Neurotoxic shellfish poisoning. *Mar. Drugs* **2008**, *6*, 430–455. [CrossRef]
200. McNabb, P.; Rhodes, L.; Adamson, J.; Holland, P. Brevetoxin—An elusive toxin in New Zealand waters. *Afr. J. Mar. Sci.* **2006**, *28*, 375–377. [CrossRef]
201. Plakas, S.M.; Dickey, R.W. Advances in monitoring and toxicity assessment of brevetoxins in molluscan shellfish. *Toxicon* **2010**, *56*, 137–149. [CrossRef]
202. Baden, D.G.; Bourdelais, A.J.; Jacocks, H.; Michelliza, S.; Naar, J. Natural and derivative brevetoxins: Historical background, multiplicity, and effects. *Environ. Health Perspect.* **2005**, *113*, 621–625. [CrossRef]
203. Abraham, A.; Flewelling, L.J.; El Said, K.R.; Odom, W.; Geiger, S.P.; Granholm, A.A.; Jackson, J.T.; Bodager, D. An occurrence of neurotoxic shellfish poisoning by consumption of gastropods contaminated with brevetoxins. *Toxicon* **2021**, *191*, 9–17. [CrossRef]
204. Rein, K.S.; Baden, D.G.; Gawley, R.E. Conformational analysis of the sodium channel modulator brevetoxin a, comparison with brevetoxin B conformations, and an hypothesis about the common pharmacophore of the "site 5" toxins. *J. Org. Chem.* **1994**, *59*, 2101–2106. [CrossRef]
205. Rein, K.S.; Lynn, B.; Gawley, R.E.; Baden, D.G. Brevetoxin B: Chemical modifications, synaptosome binding, toxicity, and an unexpected conformational effect. *J. Org. Chem.* **1994**, *59*, 2107–2113. [CrossRef]
206. Gawley, R.E.; Rein, K.S.; Jeglitsch, G.; Adams, D.J.; Theodorakis, E.A.; Tiebes, J.; Nicolaou, K.; Baden, D.G. The relationship of brevetoxins "length" and A-ring functionality to binding and activity in neuronal sodium channels. *Chem. Biol.* **1995**, *2*, 533–541. [CrossRef]
207. Hort, V.; Abadie, E.; Arnich, N.; Dechraoui Bottein, M.Y.; Amzil, Z. Chemodiversity of brevetoxins and other potentially toxic metabolites produced by *Karenia* spp. and their metabolic products in marine organisms. *Mar. Drugs* **2021**, *19*, 656. [CrossRef] [PubMed]

208. Plakas, S.M.; Jester, E.L.E.; El Said, K.R.; Granade, H.R.; Abraham, A.; Dickey, R.W.; Scott, P.S.; Flewelling, L.J.; Henry, M.; Blum, P.; et al. Monitoring of brevetoxins in the *Karenia brevis* bloom-exposed Eastern oyster (*Crassostrea virginica*). *Toxicon* **2008**, *52*, 32–38. [CrossRef]
209. Abraham, A.; Wang, Y.; El Said, K.R.; Plakas, S.M. Characterization of brevetoxin metabolism in *Karenia brevis* bloom-exposed clams (*Mercenaria* sp.) by LC-MS/MS. *Toxicon* **2012**, *60*, 1030–1040. [CrossRef]
210. Leverone, J.R.; Shumway, S.E.; Blake, N.J. Comparative effects of the toxic dinoflagellate *Karenia brevis* on clearance rates in juveniles of four bivalve molluscs from Florida, USA. *Toxicon* **2007**, *49*, 634–645. [CrossRef]
211. Leverone, J.R.; Blake, N.J.; Pierce, R.H.; Shumway, S.E. Effects of the dinoflagellate *Karenia brevis* on larval development in three species of bivalve mollusc from Florida. *Toxicon* **2006**, *48*, 75–84. [CrossRef]
212. Ishida, H.; Nozawa, A.; Totoribe, K.; Muramatsu, N.; Nukaya, H.; Tsuji, K.; Yamaguchi, K.; Yasumoto, T.; Kaspar, H.; Berkett, N. Brevetoxin B1, a new polyether marine toxin from the New Zealand shellfish, *Austrovenus stutchburyi*. *Tetrahedron Lett.* **1995**, *36*, 725–728. [CrossRef]
213. Morohashi, A.; Satake, M.; Murata, K.; Naoki, H.; Kaspar, H.F.; Yasumoto, T. Brevetoxin B3, a new brevetoxin analog isolated from the greenshell mussel *Perna canaliculus* involved in neurotoxic shellfish poisoning in New Zealand. *Tetrahedron Lett.* **1995**, *36*, 8995–8998. [CrossRef]
214. Murata, K.; Satake, M.; Naoki, H.; Kaspar, H.F.; Yasumoto, T. Isolation and structure of a new brevetoxin analog, brevetoxin B2, from greenshell mussels from New Zealand. *Tetrahedron* **1998**, *54*, 735–742. [CrossRef]
215. Morohashi, A.; Satake, M.; Naoki, H.; Kaspar, H.F.; Oshima, Y.; Yasumoto, T. Brevetoxin B4 isolated from greenshell mussels *Perna canaliculus*, the major toxin involved in neurotoxic shellfish poisoning in New Zealand. *Nat. Toxins* **1999**, *7*, 45–48. [CrossRef]
216. Plakas, S.M.; El Said, K.R.; Jester, E.L.; Granade, H.R.; Musser, S.M.; Dickey, R.W. Confirmation of brevetoxin metabolism in the Eastern oyster (*Crassostrea virginica*) by controlled exposures to pure toxins and to *Karenia brevis* cultures. *Toxicon* **2002**, *40*, 721–729. [CrossRef]
217. Ishida, H.; Nozawa, A.; Hamano, H.; Naoki, H.; Fujita, T.; Kaspar, H.F.; Tsuji, K. Brevetoxin B5, a new brevetoxin analog isolated from cockle *Austrovenus stutchburyi* in New Zealand, the marker for monitoring shellfish neurotoxicity. *Tetrahedron Lett.* **2004**, *45*, 29–33. [CrossRef]
218. Abraham, A.; El Said, K.R.; Wang, Y.; Jester, E.L.E.; Plakas, S.M.; Flewelling, L.J.; Henry, M.S.; Pierce, R.H. Biomarkers of brevetoxin exposure and composite toxin levels in hard clam (*Mercenaria* sp.) exposed to *Karenia brevis* blooms. *Toxicon* **2015**, *96*, 82–88. [CrossRef]
219. Plakas, S.M.; Wang, Z.; El Said, K.R.; Jester, E.L.E.; Granade, H.R.; Flewelling, L.; Scott, P.; Dickey, R.W. Brevetoxin metabolism and elimination in the Eastern oyster (*Crassostrea virginica*) after controlled exposures to *Karenia brevis*. *Toxicon* **2004**, *44*, 677–685. [CrossRef] [PubMed]
220. Steidinger, K.A.; Carlson, P.; Baden, D.; Rodriguez, C.; Seagle, J. Neurotoxic shellfish poisoning due to toxin retention in the clam *Chione cancellata*. *Harmful Algae* **1998**, *6*, 457–458.
221. Baden, D.G.; Mende, T.J. Toxicity of two toxins from the Florida red tide marine dinoflagellate, *Ptychodiscus brevis*. *Toxicon* **1982**, *20*, 457–461. [CrossRef]
222. Costas, C.; Louzao, M.C.; Raposo-García, S.; Vale, C.; Graña, A.; Carrera, C.; Cifuentes, J.M.; Vilariño, N.; Vieytes, M.R.; Botana, L.M. Acute toxicology report of the emerging marine biotoxin Brevetoxin 3 in mice: Food safety implications. *Food Chem. Toxicol.* **2023**, *182*, 114178. [CrossRef]
223. Selwood, A.I.; Ginkel, R.V.; Wilkins, A.L.; Munday, R.; Ramsdell, J.S.; Jensen, D.J.; Cooney, J.M.; Miles, C.O. Semisynthesis of S-desoxybrevetoxin-B2 and brevetoxin-B2, and assessment of their acute toxicities. *Chem. Res. Toxicol.* **2008**, *21*, 944–950. [CrossRef]
224. Ishida, H.; Nozawa, A.; Nukaya, H.; Tsuji, K. Comparative concentrations of brevetoxins PbTx-2, PbTx-3, BTX-B1 and BTX-B5 in cockle, *Austrovenus stutchburyi*, greenshell mussel, *Perna canaliculus*, and Pacific oyster, *Crassostrea gigas*, involved neurotoxic shellfish poisoning in New Zealand. *Toxicon* **2004**, *43*, 779–789. [CrossRef]
225. Steidinger, K.A.; Burklew, M.; Ingle, R. The effects of *Gymnodinium breve* toxin on estuarine animals. In *Marine Pharmacognosy*; Academic Press: Cambridge, MA, USA, 1973; pp. 179–202.
226. Baden, D.G. Brevetoxins: Unique polyether dinoflagellate toxins. *FASEB J.* **1989**, *3*, 1807–1817. [CrossRef]
227. Landsberg, J.H.; Flewelling, L.J.; Naar, J. *Karenia brevis* red tides, brevetoxins in the food web, and impacts on natural resources: Decadal advancements. *Harmful Algae* **2009**, *8*, 598–607. [CrossRef]
228. Naar, J.P.; Flewelling, L.J.; Lenzi, A.; Abbott, J.P.; Granholm, A.; Jacocks, H.M.; Gannon, D.; Henry, M.; Pierce, R.; Baden, D.G. Brevetoxins, like ciguatoxins, are potent ichthyotoxic neurotoxins that accumulate in fish. *Toxicon* **2007**, *50*, 707–723. [CrossRef]
229. Griffin, L.P.; Friess, C.; Bakenhaster, M.D.; Bassos-Hull, K.; Burnsed, S.W.; Brownscombe, J.W.; Cooke, S.J.; Ellis, R.D.; Gardiner, J.M.; Locascio, J. Assessing the potential for red tide (*Karenia brevis*) algal bloom impacts on Atlantic tarpon (*Megalops atlanticus*) along the southwestern coast of Florida. *Environ. Biol. Fish.* **2023**, *106*, 255–273. [CrossRef]
230. Walters, S.; Lowerre-Barbieri, S.; Bickford, J.; Tustison, J.; Landsberg, J.H. Effects of *Karenia brevis* red tide on the spatial distribution of spawning aggregations of sand seatrout *Cynoscion arenarius* in Tampa Bay, Florida. *Mar. Ecol. Prog. Ser.* **2013**, *479*, 191–202. [CrossRef]
231. Mardones, J.I.; Dorantes-Aranda, J.J.; Nichols, P.D.; Hallegraeff, G.M. Fish gill damage by the dinoflagellate *Alexandrium catenella* from Chilean fjords: Synergistic action of ROS and PUFA. *Harmful Algae* **2015**, *49*, 40–49. [CrossRef]

232. Mackieh, R.; Abou-Nader, R.; Wehbe, R.; Mattei, C.; Legros, C.; Fajloun, Z.; Sabatier, J.M. Voltage-gated sodium channels: A prominent target of marine toxins. *Mar. Drugs* **2021**, *19*, 562. [CrossRef] [PubMed]
233. Farabegoli, F.; Blanco, L.; Rodriguez, L.P.; Vieites, J.M.; Cabado, A.G. Phycotoxins in marine shellfish: Origin, occurrence and effects on humans. *Mar. Drugs* **2018**, *16*, 188. [CrossRef]
234. Poli, M.A.; Musser, S.M.; Dickey, R.W.; Eilers, P.P.; Hall, S. Neurotoxic shellfish poisoning and brevetoxin metabolites: A case study from Florida. *Toxicon* **2000**, *38*, 981–993. [CrossRef]
235. Wu, C.H.; Narahashi, T. Mechanism of action of novel marine neurotoxins on ion channels. *Annu. Rev. Pharmacol. Toxicol.* **1988**, *28*, 141–161. [CrossRef]
236. Morris, P.D.; Campbell, D.S.; Taylor, T.J.; Freeman, J.I. Clinical and epidemiological features of neurotoxic shellfish poisoning in North Carolina. *Am. J. Public Health* **1991**, *81*, 471–474. [CrossRef]
237. Berman, F.W.; Murray, T.F. Brevetoxins cause acute excitotoxicity in primary cultures of rat cerebellar granule neurons. *J. Pharmacol. Exp. Ther.* **1999**, *290*, 439–444.
238. Rossini, G.P.; Hess, P. Phycotoxins: Chemistry, mechanisms of action and shellfish poisoning. In *Molecular, Clinical and Environmental Toxicology*; Springer: Basel, Switzerland, 2010; Volume 100, pp. 65–122.
239. Soliño, L.; Costa, P.R. Differential toxin profiles of ciguatoxins in marine organisms: Chemistry, fate and global distribution. *Toxicon* **2018**, *150*, 124–143. [CrossRef] [PubMed]
240. Litaker, R.W.; Vandersea, M.W.; Faust, M.A.; Kibler, S.R.; Chinain, M.; Holmes, M.J.; Holland, W.C.; Tester, P.A. Taxonomy of *Gambierdiscus* including four new species, *Gambierdiscus caribaeus*, *Gambierdiscus carolinianus*, *Gambierdiscus carpenteri* and *Gambierdiscus ruetzleri* (Gonyaulacales, Dinophyceae). *Phycologia* **2019**, *48*, 344–390. [CrossRef]
241. Vernoux, J.P.; Lewis, R.J. Isolation and characterisation of Caribbean ciguatoxins from the horse-eye jack (*Caranx latus*). *Toxicon* **1997**, *35*, 889–900. [CrossRef]
242. Satake, M.; Murata, M.; Yasumoto, T. The structure of CTX3C, a ciguatoxin congener isolated from cultured *Gambierdiscus toxicus*. *Tetrahedron Lett.* **1993**, *34*, 1975–1978. [CrossRef]
243. Lewis, R.J.; Vernoux, J.P.; Brereton, I.M. Structure of Caribbean ciguatoxin isolated from *Caranx latus*. *J. Am. Chem. Soc.* **1998**, *120*, 5914–5920. [CrossRef]
244. Lewis, R.J.; Sellin, M.; Poli, M.A.; Norton, R.S.; MacLeod, J.K.; Sheil, M.M. Purification and characterization of ciguatoxins from moray eel (*Lycodontis javanicus*, Muraenidae). *Toxicon* **1991**, *29*, 1115–1127. [CrossRef] [PubMed]
245. Raposo-García, S.; Castro, D.; Lence, E.; Estévez, P.; Leão, J.M.; González-Bello, C.; Gago-Martínez, A.; Louzao, M.C.; Vale, C.; Botana, L.M. In silico simulations and functional cell studies evidence similar potency and distinct binding of Pacific and Caribbean ciguatoxins. *Expo. Health* **2023**, *15*, 641–660. [CrossRef]
246. Pearn, J. Neurology of ciguatera. *J. Neurol. Neurosurg. Psychiatry* **2001**, *70*, 4–8. [CrossRef]
247. Friedman, M.A. Ciguatera fish poisoning: Treatment, prevention and management. *Mar. Drugs* **2008**, *6*, 456–479. [CrossRef]
248. Lehane, L.; Lewis, R.J. Ciguatera: Recent advances but the risk remains. *Int. J. Food Microbiol.* **2000**, *61*, 91–125. [CrossRef]
249. Mak, Y.L.; Wai, T.C.; Murphy, M.B.; Chan, W.H.; Wu, J.J.; Lam, J.C.W.; Chan, L.L.; Lam, P.K.S. Pacific ciguatoxins in food web components of coral reef systems in the republic of kiribati. *Environ. Sci. Technol.* **2013**, *47*, 14070–14079. [CrossRef] [PubMed]
250. Jiang, X.W.; Li, X.; Lam, P.K.S.; Cheng, S.H.; Schlenk, D.; Mitcheson, Y.S.d.; Li, Y.; Gu, J.D.; Chan, L.L. Proteomic analysis of hepatic tissue of ciguatoxin (CTX) contaminated coral reef fish *Cephalopholis argus* and moray eel *Gymnothorax undulatus*. *Harmful Algae* **2012**, *13*, 65–71. [CrossRef]
251. Mak, Y.L.; Li, J.; Liu, C.N.; Cheng, S.H.; Lam, P.K.S.; Cheng, J.; Chan, L.L. Physiological and behavioural impacts of Pacific ciguatoxin-1 (P-CTX-1) on marine medaka (*Oryzias melastigma*). *J. Hazard. Mater.* **2017**, *321*, 782–790. [CrossRef] [PubMed]
252. Yan, M.; Mak, M.Y.L.; Cheng, J.; Li, J.; Gu, J.R.; Leung, P.T.Y.; Lam, P.K.S. Effects of dietary exposure to ciguatoxin P-CTX-1 on the reproductive performance in marine medaka (*Oryzias melastigma*). *Mar. Pollut. Bull.* **2020**, *152*, 110837. [CrossRef] [PubMed]
253. Yan, M.; Leung, P.T.Y.; Ip, J.C.H.; Cheng, J.P.; Wu, J.J.; Gu, J.R.; Lam, P.K.S. Developmental toxicity and molecular responses of marine medaka (*Oryzias melastigma*) embryos to ciguatoxin P-CTX-1 exposure. *Aquat. Toxicol.* **2017**, *185*, 149–159. [CrossRef]
254. Li, L.; Chen, S.; Xu, S.Y.; Li, D.W.; Li, H.Y.; Yang, W.D. Toxicity and underlying mechanism of the toxic dinoflagellate *Gambierdiscus caribaeus* to the fish *Oryzias melastigma*. *Ecotoxicol. Environ. Saf.* **2022**, *247*, 114223. [CrossRef]
255. Xu, Y.; He, X.; Lee, W.H.; Chan, L.L.; Lu, D.; Wang, P.; Tao, X.; Li, H.; Yu, K. Ciguatoxin-producing dinoflagellate *Gambierdiscus* in the beibu gulf: First report of toxic *Gambierdiscus* in Chinese waters. *Toxins* **2021**, *13*, 643. [CrossRef]
256. Clausing, R.J.; Losen, B.; Oberhaensli, F.R.; Darius, H.T.; Sibat, M.; Hess, P.; Swarzenski, P.W.; Chinain, M.; Dechraoui Bottein, M.Y. Experimental evidence of dietary ciguatoxin accumulation in an herbivorous coral reef fish. *Aquat. Toxicol.* **2018**, *200*, 257–265. [CrossRef]
257. Litaker, R.W.; Vandersea, M.W.; Faust, M.A.; Kibler, S.R.; Nau, A.W.; Holland, W.C.; Chinain, M.; Holmes, M.J.; Tester, P.A. Global distribution of ciguatera causing dinoflagellates in the genus *Gambierdiscus*. *Toxicon* **2010**, *56*, 711–730. [CrossRef]
258. Pisapia, F.; Holland, W.C.; Hardison, D.R.; Litaker, R.W.; Fraga, S.; Nishimura, T.; Adachi, M.; Nguyen-Ngoc, L.; Séchet, V.; Amzil, Z. Toxicity screening of 13 *Gambierdiscus* strains using neuro-2a and erythrocyte lysis bioassays. *Harmful Algae* **2017**, *63*, 173–183. [CrossRef]
259. Pérez-Arellano, J.-L.; Luzardo, O.P.; Brito, A.P.; Cabrera, M.H.; Zumbado, M.; Carranza, C.; Angel-Moreno, A.; Dickey, R.W.; Boada, L.D. Ciguatera fish poisoning, Canary Islands. *Emerg. Infect. Dis.* **2005**, *11*, 1981. [CrossRef] [PubMed]

260. Otero, P.; Pérez, S.; Alfonso, A.; Vale, C.; Rodríguez, P.; Gouveia, N.N.; Gouveia, N.; Delgado, J.; Vale, P.; Hirama, M. First toxin profile of ciguateric fish in Madeira Arquipelago (Europe). *Anal. Chem.* **2010**, *82*, 6032–6039. [CrossRef] [PubMed]
261. Boada, L.D.; Zumbado, M.; Luzardo, O.R.; Almeida-Gonzalez, M.; Plakas, S.M.; Granade, H.R.; Abraham, A.; Jester, E.L.E.; Dickey, R.W. Ciguatera fish poisoning on the West Africa Coast: An emerging risk in the Canary Islands (Spain). *Toxicon* **2010**, *56*, 1516–1519. [CrossRef] [PubMed]
262. Estevez, P.; Castro, D.; Leao, J.M.; Yasumoto, T.; Dickey, R.; Gago-Martinez, A. Implementation of liquid chromatography tandem mass spectrometry for the analysis of ciguatera fish poisoning in contaminated fish samples from Atlantic coasts. *Food Chem.* **2019**, *280*, 8–14. [CrossRef]
263. Costa, P.R.; Estevez, P.; Soliño, L.; Castro, D.; Rodrigues, S.M.; Timoteo, V.; Leao-Martins, J.M.; Santos, C.; Gouveia, N.; Diogène, J. An update on ciguatoxins and CTX-like toxicity in fish from different trophic levels of the Selvagens Islands (NE Atlantic, Madeira, Portugal). *Toxins* **2021**, *13*, 580. [CrossRef]
264. Silva, M.; Rodriguez, I.; Barreiro, A.; Kaufmann, M.; Neto, A.I.; Hassouani, M.; Sabour, B.; Alfonso, A.; Botana, L.M.; Vasconcelos, V. First Report of Ciguatoxins in Two Starfish Species: *Ophidiaster ophidianus* and *Marthasterias glacialis*. *Toxins* **2015**, *7*, 3740–3757. [CrossRef]
265. Chinain, M.; Gatti, C.; Darius, H.; Quod, J.P.; Tester, P. Ciguatera poisonings: A global review of occurrences and trends. *Harmful Algae* **2021**, *102*, 101873. [CrossRef]
266. Roué, M.; Darius, H.T.; Ung, A.; Viallon, J.; Sibat, M.; Hess, P.; Amzil, Z.; Chinain, M. Tissue distribution and elimination of ciguatoxins in *Tridacna maxima* (*Tridacnidae*, Bivalvia) fed *Gambierdiscus polynesiensis*. *Toxins* **2018**, *10*, 189. [CrossRef]
267. Caillaud, A.; De la Iglesia, P.; Darius, H.T.; Pauillac, S.; Aligizaki, K.; Fraga, S.; Chinain, M.; Diogène, J. Update on methodologies available for ciguatoxin determination: Perspectives to confront the onset of ciguatera fish poisoning in Europe. *Mar. Drugs* **2010**, *8*, 1838–1907. [CrossRef]
268. Rhodes, L.L.; Smith, K.F.; Murray, J.S.; Nishimura, T.; Finch, S.C. Ciguatera fish poisoning: The risk from an Aotearoa/New Zealand perspective. *Toxins* **2020**, *12*, 50. [CrossRef]
269. Lewis, R.J. The changing face of ciguatera. *Toxicon* **2001**, *39*, 97–106. [CrossRef] [PubMed]
270. Friedman, M.; Fernandez, M.; Backer, L.; Dickey, R.; Bernstein, J.; Schrank, K.; Kibler, S.; Stephan, W.; Gribble, M.; Bienfang, P. An updated review of ciguatera fish poisoning: Clinical, epidemiological, environmental, and public health management. *Mar. Drugs* **2017**, *15*, 72. [CrossRef]
271. Diogène, J.; Reverté, L.; Rambla-Alegre, M.; del Río, V.; de la Iglesia, P.; Campàs, M.; Palacios, O.; Flores, C.; Caixach, J.; Ralijaona, C. Identification of ciguatoxins in a shark involved in a fatal food poisoning in the Indian Ocean. *Sci. Rep.* **2017**, *7*, 8240. [CrossRef] [PubMed]
272. Dickey, R.W.; Plakas, S.M. Ciguatera: A public health perspective. *Toxicon* **2010**, *56*, 123–136. [CrossRef]
273. Butera, R.; Prockop, L.D.; Buonocore, M.; Locatelli, C.; Gandini, C.; Manzo, L. Mild ciguatera poisoning: Case reports with neurophysiological evaluations. *Muscle Nerve* **2000**, *23*, 1598–1603. [CrossRef]
274. Arena, P.; Levin, B.; Fleming, L.E.; Friedman, M.A.; Blythe, D. A pilot study of the cognitive and psychological correlates of chronic ciguatera poisoning. *Harmful Algae* **2004**, *3*, 51–60. [CrossRef]
275. Friedman, M.; Arena, P.; Levin, B.; Fleming, L.; Fernandez, M.; Weisman, R.; Bernstein, J.; Schrank, K.; Blythe, D.; Backer, L. Neuropsychological study of ciguatera fish poisoning: A longitudinal case-control study. *Arch. Clin. Neuropsychol.* **2007**, *22*, 545–553. [CrossRef]
276. Nicholson, G.M.; Lewis, R.J. Ciguatoxins: Cyclic polyether modulators of voltage-gated lion channel function. *Mar. Drugs* **2006**, *4*, 82–118. [CrossRef]
277. Pasinszki, T.; Lako, J.; Dennis, T.E. Advances in detecting ciguatoxins in Fish. *Toxins* **2020**, *12*, 494. [CrossRef] [PubMed]
278. Duan, G.F.; Liu, Y.; Zhang, L.N.; Li, H.Y.; Liu, J.S.; Yang, W.D. Cinnamaldehyde could reduce the accumulation of diarrhetic shellfish toxins in the digestive gland of the mussel *Perna viridis* under laboratory conditions. *Mar. Drugs* **2021**, *19*, 63. [CrossRef]
279. Yuan, K.K.; Duan, G.F.; Liu, Q.Y.; Li, H.Y.; Yang, W.D. Inhibition of diarrheal shellfish toxins accumulation in the mussel *Perna viridis* by curcumin and underlying mechanisms. *Toxins* **2021**, *13*, 578. [CrossRef] [PubMed]

MDPI AG
Grosspeteranlage 5
4052 Basel
Switzerland
Tel.: +41 61 683 77 34

Marine Drugs Editorial Office
E-mail: marinedrugs@mdpi.com
www.mdpi.com/journal/marinedrugs

www.ingramcontent.com/pod-product-compliance
Lightning Source LLC
Chambersburg PA
CBHW061834210926
51850CB00006B/889
* 9 7 8 3 7 2 5 8 6 9 8 8 6 *